NanoScience and Technology

Springer
Berlin
Heidelberg
New York
Hong Kong
London
Milan
Paris
Tokyo

NanoScience and Technology

Series Editors: P. Avoris K. von Klitzing H. Sakaki R. Wiesendanger

The series NanoScience and Technology is focused on the fascinating nano-world, mesoscopic physics, analysis with atomic resolution, nano and quantum-effect devices, nanomechanics and atomic-scale processes. All the basic aspects and technology-oriented developments in this emerging discipline are covered by comprehensive and timely books. The series constitutes a survey of the relevant special topics, which are presented by leading experts in the field. These books will appeal to researchers, engineers, and advanced students.

Mesoscopic Physics and Electronics
Editor: T. Ando, Y. Arakawa, K. Furuya, S. Komiyama, H. Nakashima
1997

Scanning Probe Microscopy
Analytical Methods
Editor: R. Wiesendanger
1998

Sliding Friction
Physical Principles and Applications
By B.N.J. Persson
2nd ed., 2000

Biological Micro- and Nanotribology
Nature's Solutions
By M. Scherge, S. Gorb
2001

Semiconductor Spintronics and Quantum Computation
Editors: D.D. Awschalom, D. Loss, N. Samarth
2002

Semiconductor Quantum Dots
Physics, Spectroscopy and Applications
Editors: Y. Masumoto, T. Takagahara
2002

Noncontact Atomic Force Microscopy
Editors: S. Morita, R. Wiesendanger, E. Meyer
2002

Nano-Optoelectronics
Concepts, Physics and Devices
Editor: M. Grundmann
2002

Single Organic Nanoparticles
Editors: H. Masuhara, H. Nakanishi, K. Sasaki
2003

Nanoelectrodynamics
Electrons and Electromagnetic Fields in Nanometer-Scale Structures
Editor: H. Nejo
2003

Epitaxy of Nanostructures
By V.A. Shchukin, N.N. Ledentsov, D. Bimberg
2004

new publication

Nanoscale Characterisation of Ferroelectric Materials
Scanning Probe Microscopy Approach
Editors: M. Alexe, A. Gruverman
2004

Applied Scanning Probe Methods
Editors: B. Bhushan, H. Fuchs, S. Hosaka
2004

Nanostructures Theory and Modelling
Editors: Ch. Delerue, M. Lannoo
2004

Bharat Bhushan
Harald Fuchs
Sumio Hosaka

Applied Scanning Probe Methods

With 338 Figures
Including 6 Color Figures

Springer

Professor Bharat Bhushan
The Ohio State University, Department of Mechanical Engineering,
206 W. 18th Avenue, Columbus, Ohio 43210-1107
e-mail: bhushan.2@osu.edu

Professor Dr. Harald Fuchs
Universität Münster, Wilhelm-Klemm-Str. 10, 48149 Münster, Germany
e-mail: fuchsh@uni-muenster.de

Professor Dr. Sumio Hosaka
Gunma University, Dept. Electronic Engineering, 1-5-1 Tenjin-cho, 376-8515 Kiryu-City
e-mail: hosaka@el.gunma-u.ac.jp

ISSN 1434-4904
ISBN 978-3-642-05602-4

Library of Congress Cataloging-in-Publication Data
Applied scanning probe methods / Bharat Bhushan, Harald Fuchs, Sumio Hosaka [eds.].
p. cm. – (Nanoscience and technology)
Includes bibliographical references and index.
ISBN 3.540-00527-7 (alk. paper)
1. Materials–Microscopy. 2. Scanning probe microscopy. I. Bhushan, Bharat, 1949- II. Fuchs, H. (Harald) III. Hosaka, Sumio. IV. Series.

TA417.23.A655 2003
620.1′1299–dc22

2003059049

Springer-Verlag is a part of Springer Science+Business Media
springeronline.com

Printed in Germany

Coverdesign: design & production, Heidelberg

Printed on acid-free paper 2/3020xv – 5 4 3 2 1 0

Preface

Since introduction of the Scanning Tunnelling Microscope (STM) by inventors G. Binnig and H. Rohrer in 1981 at the IBM Zurich Research Laboratory, Rüschlikon, Switzerland., many variations of tip probe based microscopies, referred to as Scanning Probe Microscopies (SPM) have been developed. Nanoscale Science and Technology are strongly driven by SPMs which allow investigation and manipulation of surfaces down to the atomic scale. While the pure imaging capabilities of SPM techniques dominated the application of these methods at their early development stages, the physics of probe-sample interactions and the quantitative analyses of elastic/plastic, tribological, electronic, magnetic, biological, and chemical surfaces have now become of increasing interest. With growing understanding of the underlying interaction mechanisms, SPMs have found applications in many fields outside basic research fields. In addition, a variety of derivatives of all these methods has been developed for special applications, some of them targeting far beyond microscopy.

For industrial applications, the large SPM-subfamily of Scanning Force Microscopy (SFM) exhibiting atomic resolution capability is of special interest. Originally denoted as Atomic Force Microscopy (AFM), the basic instrument was invented by G. Binnig, C. Quate and Ch. Gerber in 1986. The introduction of this technique was a direct consequence of the observation of sizable interaction forces between a near field probe and a sample surface which were already seen in earlier experiments done with Scanning Tunnelling Microscopy (STM). STM experiments performed, for example, on HOPG-graphite surfaces, indicated a strong influence of elastic surface properties on the measured surface corrugation which could not be traced back to electronic surface properties. Based on these types of results it appeared quite natural to introduce a method that is solely based on tip-sample force interactions rather than transport phenomena such as electron tunnelling as used in the STM. The AFM was the first purely force sensitive microscope. Its applicability on electrically insulating surfaces opened a wide range of applications. In addition, this method can also be applied under various environmental conditions, such as vacuum, air and liquids. Thus, a wide variety of applications including industrial quality control and the investigation of living biological materials became feasible with high resolution, and without any conductive coatings. Since for many practical applications atomic resolution is not the ultimate and only goal, the family of commercial instruments which measure forces is more generally referred to as Scanning Force Microscopes (SFMs). While in the beginning of SFMs, high resolution measurement of surface topography was the major application, it became clear very soon, that by applying more complex strategies, i.e. dynamic and lateral force techniques the quantitative measurement of local material properties also became possible. Beyond that, the local measurement of the distribution of locally varying electrical surface potentials and magnetic stray fields became accessible – altogether properties of great practical interest. Therefore, is appears quite natural that SFM related techniques play a major role in industrial applications today including topographic and dynamical surface studies such as of semiconductors, polymers, paper, ceramics, magnetic and biological materials.

Commercial production of SPMs started with STM in 1988 and has grown to several hundred million dollars industry. The number of these instruments is equally divided among the U.S., Japan and Europe with the following industry, university and government laboratories split: 50/50, 70/30, and 30/70, respectively.

This book selects some representative recent applications of SPM with a focus on practical applications. While not necessarily limited to SFM, the general applicability and variability of this method in industrial research and quality control led to a natural focus as reflected in the large number of chapters dealing with SFM related techniques. Selected articles on the application of near field optical and capacitance techniques for novel storage applications put a special focus on industrial R&D based on these techniques. The organization of the book is straightforward. The book is divided into three parts. Part I contains chapters dealing with the theoretical background of static and dynamic force microscopies including sensor technology and methods for tip characterization. Part II contains chapters on applications of scanning force microscopy such as micro- and nanotribology, polymer surfaces, stray field measurements and roughness investigations. The last Part III contains a selection of articles mainly written by industrial researchers. Examples for special application of SFM such as atomic manipulation and surface modification are presented. Furthermore, novel concepts for high resolution optical storage techniques as well as for the production of single electron devices based on SFM are presented.

The purpose of this book is to give both, a systematic physical and technical background for the academic researcher on recent progress of applied scanning probe microscopy and an overview of the current state of SPM applications for the industrial researcher and engineer. We hope that readers find this book stimulating and productive.

We would like to thank the authors for their excellent contributions in a timely manner. We would also like to thank Dr. Marion Hertel, Senior Editor Chemistry, and Mrs. Beate Siek of Springer-Verlag for their continued support and patience during the preparation of this book.

December 2002

Prof. Dr. Bharat Bhushan, USA
Prof. Dr. Harald Fuchs, Germany
Prof. Dr. Sumio Hosaka, Japan

Contents

List of Contributors

Boris Anczykowski
nanoAnalytics GmbH, Gievenbecker Weg 11, 48149 Münster, Germany
e-mail: anczykowski@nanoanalytics.com

Walter Arnold
Fraunhofer Institute for Nondestructive Testing (IZFP), University, Building 37, 66123 Saarbruecken, Germany
e-mail: arnold@izfp.fhg.de

Bharat Bhushan
The Ohio State University, Department of Mechanical Engineering, 206 W. 18th Avenue, Columbus, Ohio 43210-1107, USA
e-mail: bhushan.2@osu.edu

Harald Fuchs
Universität Münster, Wilhelm-Klemm-Str. 10, 48149 Münster, Germany
e-mail: fuchsh@uni-muenster.de

Redhouane Henda
School of Engineering, Laurentian University, Ramsey Lake Road, Sudbury, Ontario P3E 2C6, Canada
e-mail: rhenda@nickel.laurentian.ca

Sumio Hosaka
Gunma University, Graduate School of Engineering Dept. Nano Material Systems, 1-5-1 Tenjin-cho, 376-8515 Kiryu-City, Japan
e-mail: hosaka@el.gunma-u.ac.jp

Jack E. Houston
Surface & Interface Science Dept.-1114, Sandia National Laboratories, PO Box 5800, Albuquerque, NM 87185-1415, USA
e-mail: jehoust@sandia.gov

Rainer Kassing
Center for Interdisciplinary Nanostructure, Science and Technology,
Institut für Technische Physik, Universität Kassel, Heinrich-Plett-Str. 40,
34132 Kassel, Germany
e-mail: kassing@physik.uni-kassel.de

Juergen Keller
Fraunhofer Institute for Reliability and Microintegration (IZM),
Dept. Mechanical Reliability and Micro Materials,
Gustav-Meyer-Allee 25, 13355 Berlin, Germany
e-mail: juergen.keller@izm.fraunhofer.de

Egbert Oesterschulze
Center for Interdisciplinary Nanostructure, Science and Technology,
Institut für Technische Physik, Universität Kassel, Heinrich-Plett-Str. 40,
34132 Kassel, Germany
e-mail: oester@physik.uni-kassel.de

Frédéric Leroy
Physics Department, L'OREAL Recherche, 1,
Avenue Eugène Schueller, 93601 Aulnay sous Bois, France
e-mail: fleroy@recherche.loreal.com

Gustavo Luengo
Nano-physics Laboratory, L'OREAL Recherche, 1, Avenue Eugène Schueller,
93601 Aulnay sous Bois, France
e-mail: gluengo@recherche.loreal.com

Sergei Magonov
Digital Instruments/Veeco Metrology Group, 112 Robin Hill Rd.,
Santa Barbara, CA 93117, USA
e-mail: sergei.magonov@veeco.com

Arun Majumdar
Department of Mechanical Engineering, University of California at Berkeley,
Berkeley, CA 94720, USA
e-mail: majumdar@me.berkeley.edu

Kazuhiko Matsumoto
Osaka University, Institute of Scientific & Industrial Research, 8-1, Mihogaoka,
Ibaraki, Osaka, 567-0047, Japan
e-mail: k-matsumoto@sanken.osaka-u.ac.jp

Bernd Michel
Fraunhofer Institute for Reliability and Microintegration (IZM),
Dept. Mechanical Reliability and Micro Materials,
Gustav-Meyer-Allee 25, 13355 Berlin, Germany
e-mail: bernd.michel@izm.fraunhofer.de

Ndubuisi G. Orji
NIST, 100 Bureau Drive, Stop 8212, Gaithersburg, MD 20899-8212, USA
e-mail: ngorji@nist.gov

Jay Raja
Dept. of Mechanical Engineering and Engineering Science,
University of North Carolina Charlotte, Charlotte, NC 28223, USA
e-mail: jraja@uncc.edu

Michael Reinstädtler
Fraunhofer Insititute for Nondestructive Testing IZFP, University, Bldg. 37
66123 Saarbruecken, Germany
e-mail: reinstaedtler@izfp.fhg.de

Martha I. Sanchez
IBM Almaden Research Center , 650 Harry Rd. K17–D1, San Jose,
CA 95120-6099, USA
e-mail: msanchez@almaden.ibm.com

Volker Scherer
Federal Mogul, Technology Rings and Liners, Bürgermeister-Schmidt Str. 17,
51399 Burscheid, Germany
e-mail: volker_scherer@eu.fmo.com

André Schirmeisen
University of Münster, Wilhelm-Klemm-Str. 10, 48149 Münster, Germany
e-mail: schirmeisen@uni-muenster.de

Andreas Schubert
Fraunhofer Institute for Reliability and Microintegration (IZM),
Dept. Mechanical Reliability and Micro Materials, Gustav-Meyer-Allee 25,
13355 Berlin, Germany
e-mail: andreas.schubert@izm.fraunhofer.de

Li Shi
Department of Mechanical Engineering, ETC 7.136, Mail Code C2200,
University of Texas at Austin, Austin, TX 78712, USA
e-mail: lishi@mail.utexas.edu

Junji Tominaga
Centre for Applied Near-Field Optics Research, National Institute of Advanced, Industrial Science and Technology, Tsukuba Central 4 site,
1-1-1 Higashi Tsukuba-shi, Ibaraki 305-8562, Japan
e-mail: j-tominaga@aist.go.jp

John Villarrubia
Precision Engineering Division, National Inst. of Standards & Technology,
100 Bureau Dr. Stop 8212, Gaithersburg, MD 20899-8212, USA
e-mail: john.villarrubia@nist.gov

Dietmar Vogel
Fraunhofer Institute for Reliability and Microintegration (IZM),
Dept. Mechanical Reliability and Micro Materials, Gustav-Meyer-Allee 25,
13355 Berlin, Germany
e-mail: dietmar.vogel@izm.fraunhofer.de

Theodore V. Vorburger
NIST, 100 Bureau Drive, Stop 8212, Gaithersburg, MD 20899-8212, USA
e-mail: tvtv@nist.gov

Ryoichi Yamamoto
Gunma Industrial Technology Center, 884-1, Kamesato-cho, Maebashi-City,
Gunma, 379-2147, Japan
e-mail: yamamoto@tec-lab.pref.gunma.jp

Part I
Scanning Probe Microscopy

1 Dynamic Force Microscopy

André Schirmeisen · Boris Anczykowski · Harald Fuchs

1.1 Introduction

This chapter is an introduction to the concept of the dynamic mode of the atomic force microscope (AFM). While the first part is dedicated towards a systematic discussion of the different operational modes in dynamic AFM, some practical issues for the experimentalist are pointed out. Special care is taken to explain the quantitative relation of the experimental parameters with the physical magnitudes, like force and dissipation.

1.2 Motivation: Measurement of a Single Atomic Bond

The direct measurement of the force interaction between two distinct molecules has been the motivation for scientists for many years now. The fundamental forces responsible for the solid state of matter can be directly investigated, ultimately between defined single molecules. But it has not been until very recently that the chemical forces could be quantitatively measured for a single atomic bond [1]. How can we reliably measure forces, which may be as small as one billionth of one Newton? How can we identify one single pair of atoms as the source of the force interaction?

It is the same mechanical principle that is used to measure the gravitational force exerted by your body weight (e.g., with the scale in your bathroom), which can be employed to measure the forces between single atoms. A spring with a defined elasticity is compressed by an arbitrary force (e.g., your weight). The compression Δz of the spring (with spring constant k) is a direct measure of the force F exerted, which in the regime of elastic deformation obeys Hooke's law:

$$F = k \cdot \Delta z \tag{1.1}$$

The only difference to your bathroom scale is the sensitivity of the measurement. Typically springs with a stiffness of 0.1–10 N m^{-1} are used, which will be deflected by 0.1–10 nm upon application of an interatomic force of some nN. Experimentally, a laser deflection technique is used to measure the movement of the spring. The spring is a bendable cantilever, microfabricated from silicon wafers. If a sufficiently sharp tip, usually directly attached to the cantilever, is approached towards a surface within some nanometers, we can measure the interacting forces through changes in the deflected laser beam. This is a static measurement; hence it is called *static AFM*. Alternatively, the cantilever can be excited to vibrate at its resonant frequency. Under the influence of tip-sample forces the resonant frequency (and consequently also amplitude and phase) of the cantilever will change and serve as measurement parameters. This is called the *dynamic AFM*. Due to the multitude of possible operational modes, expressions like non-contact mode, intermittent-contact mode, tapping-mode, FM-mode, AM-mode, self-excitation, constant-excitation, or constant-amplitude mode AFM are found in the literature, which will be systematically categorized in the following paragraphs.

In fact, the first AFMs were operated in the dynamic mode. In 1986 Binnig, Quate, and Gerber [2] presented the concept of the atomic force microscope. The deflection of the cantilever with the tip was measured with sub-Angstrom precision by an additional scanning tunneling microscope (STM). While the cantilever was externally oscillated close to its resonant frequency the amplitude and phase of the oscillation were measured. If the tip is approached towards the surface, the oscillation parameters amplitude and phase are influenced by the tip-surface interaction, and can therefore be used as feedback channels. A certain set point, for example, the amplitude is given, and the feedback loop will adjust the tip-sample distance such that the amplitude remains constant. The controller parameter is recorded as a function of the lateral position of the tip with respect to the sample and the scanned image essentially represents the surface topography.

What is then the difference between the static and dynamic mode of operation for the AFM? The static deflection AFM directly gives the interaction force between tip and sample by using Eq. 1.1. In the dynamic mode we find that the resonant frequency, amplitude, and phase of the oscillation change as a consequence of the interaction forces (and also dissipative processes, as discussed in the last section).

To get a basic understanding of the underlying physics it is instructive to consider a very simplified case. Assume that the vibration amplitude is small compared to the range of force interaction. Since van der Waals forces range over

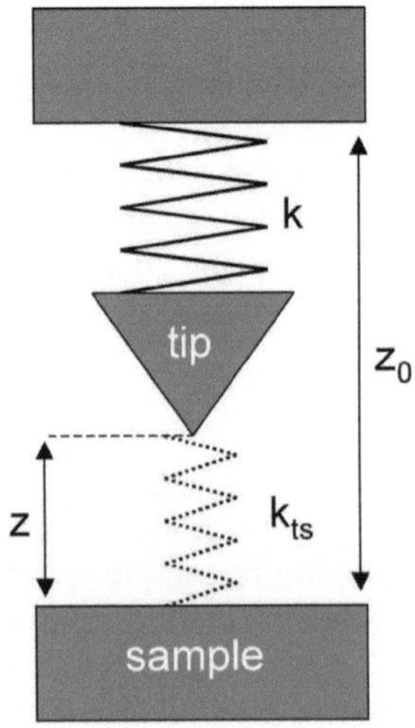

Fig. 1.1. Model of the AFM tip while experiencing tip-sample forces. The tip is attached to a cantilever with spring constant k, and the force interaction is modeled by a spring with a stiffness k_{ts} equal to the force gradient. Note that the force interaction spring is not constant, but depends on the tip-sample distance z

typical distances of 10 nm, the vibration amplitude should be less than 1 nm. We furthermore require that the force gradient $\partial F_{ts}/\partial z$ does not vary significantly over one oscillation cycle.

We can view the AFM set-up as a coupling of two springs (see Fig. 1.1). Whereas the cantilever is represented by a spring with spring constant k, the force interaction between tip and surface can be modeled by a second spring. The derivative of the force with respect to tip-sample distance is the force gradient, and represents the spring constant k_{ts} of the interaction spring. This spring constant k_{ts} is constant only with respect to one oscillation cycle, but varies with the average tip-sample distance as the probe is approached to the sample. The two springs are coupled in series; therefore we can write for the total spring constant of the AFM system:

$$k_{total} = k + k_{ts} = k - \frac{\partial F_{ts}}{\partial z} \tag{1.2}$$

From the simple harmonic oscillator (neglecting any damping effects) we find that the resonant frequency ω of the system is shifted by $\Delta\omega$ from the free resonant frequency ω_0 due to the force interaction:

$$\omega^2 = (\omega_0 + \Delta\omega)^2 = k_{total}/m^* = \left(k + \frac{\partial F_{ts}}{\partial z}\right)/m^* \tag{1.3}$$

Here m^* represents the effective mass of the cantilever. A detailed analysis of how m^* is related to the geometry and total mass of the cantilever can be found in the literature [3]. In the approximation that $\Delta\omega$ is much smaller than ω_0 we can write:

$$\frac{\Delta\omega}{\omega_0} \cong -\frac{1}{2k} \cdot \frac{\partial F_{ts}}{\partial z} \tag{1.4}$$

Therefore, we find that the frequency shift of the cantilever resonance is proportional to the force gradient of the tip-sample interaction.

Although the above consideration is based on a very simplified model, it shows qualitatively that in dynamic force microscopy we will find that the oscillation frequency depends on the force gradient, while static force microscopy measures the force itself. In principle, we can calculate the force curve from the force gradient and vice versa (neglecting a constant offset). It seems, therefore, that the two methods are equivalent and our choice will depend on whether we can measure the beam deflection or the frequency shift with better precision at the cost of technical effort.

However, we have neglected one important issue for the operation of the AFM so far: the mechanical stability of the measurement. In static AFM the tip is slowly approached towards the surface. The force between the tip and the surface will always be counteracted by the restoring force of the cantilever. In Fig. 1.2 you can see a typical force-distance curve. Upon approach of the tip towards the sample, the negative attractive forces, representing, for example, van der Waals or chemical interaction forces, increase until a maximum is reached. This turnaround point is due to the onset of repulsive forces, caused by coulomb repulsion, which will start to dominate upon further approach. The spring constant of the cantilever is represented by the slope of the straight line. The position of the z-transducer (typically a piezo element) that moves the probe, is at the intersection of the line with the horizontal axis. The position of the tip, shifted from the probe's base due to the lever bending, can be found at the intersection of the cantilever line with the force curve. Hence, the total force is zero, that is, the cantilever is in its equilibrium position (note that the spring constant line here shows attractive forces, although in reality the forces are repulsive, i.e., pulling back the tip from the surface). As soon as the position A in Fig. 1.2 is reached, we find two possible intersection points, upon further approach there are even three force equilibrium points. However, between point A and B the tip is at a local energy minimum, and therefore will still follow the force curve. But at point B, when the adhesion force upon further approach would become larger than the spring restoring force, the tip will suddenly jump to point C. We can then probe the predominantly repulsive force interaction by further reducing the tip-sample distance. Then, while retracting the tip, we will pass point C, because the tip is still in a local energy minimum. At position D the tip will jump suddenly to point A again, since the restoring force now exceeds the adhesion. From Fig. 1.2 we can see, that the sudden instability will happen at exactly the point at which the slope of the adhesion force exceeds the slope of the spring constant. Therefore, if the negative force gradient of the tip-sample interaction will, at any point, exceed the spring constant, a mechanical instability occurs. Mathematically speaking we demand for a stable measurement that:

$$\left.\frac{\partial F_{ts}}{\partial z}\right|_z > k \text{ for all points } z \tag{1.5}$$

The phenomenon of mechanical instability is often referred to as the 'jump-to-contact'.

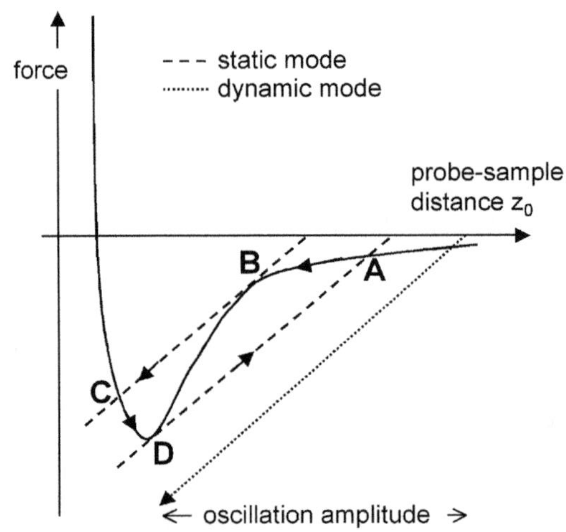

Fig. 1.2. Force-distance curve of a typical tip-sample interaction. In static-mode AFM the tip would follow the force curve until *point B* is reached. If the slope of the force curve becomes larger than the spring constant of the cantilever (*dashed line*) the tip will suddenly jump to *position C*. Upon retraction a different path will be followed along *D* and *A* again. In dynamic AFM the cantilever oscillates with amplitude *A*. Although the equilibrium position of the oscillation is far from the surface, the tip will experience the maximum attractive force at *point D* during some parts of the oscillation cycle. However, the total force is always pointing away from the surface, therefore avoiding an instability

Looking at Fig. 1.2, we realize that large parts of the force curve cannot be measured, if the jump-to-contact phenomenon occurs. We will not be able to measure the point at which the attractive forces reach their maximum, representing the temporary chemical bonding of tip and surface atoms. Secondly, the sudden instability, the jump-to-contact, will often cause the tip to change the very last tip or surface atoms. A smooth careful approach needed to measure the full force curve does not seem feasible. Our goal to measure the chemical interaction forces of two single molecules may become impossible.

There are several solutions to the jump-to-contact problem: on the one hand, we can simply choose a sufficiently stiff spring, so that Eq. 1.5 is fulfilled at all points of the force curve. On the other hand, we can resort to a trick to enhance the counteracting force of the cantilever: we will oscillate the cantilever with large amplitude, therefore making it virtually stiffer at the point of strong force interaction.

Consider the first solution, which seems simpler at first glance. Chemical bonding forces extend over a distance range of about 0.1 nm. Typical binding energies of a couple of eV will lead to adhesion forces of the order of some nN. Force gradients will therefore reach values of some 10 N m^{-1}. A spring for stable force measurements will have to be as stiff as 100 N m^{-1} to ensure that no instability occurs (a safety factor of 10 seems to be a minimum requirement, since usually one cannot be sure a priori that only one atom will dominate the interaction). In order to measure the nN interaction force, a static cantilever deflection of 0.01 nm has to be detected. With standard beam deflection AFM set-ups this becomes a challenging task.

This problem was solved [4, 5] by using an in situ optical interferometer measuring the beam deflection at liquid nitrogen temperature in a UHV environment. To ensure that the force gradients are smaller then the lever spring constant (50 N m^{-1}) the tips were fabricated to terminate in only three atoms, therefore minimizing the total force interaction. The only tool known to be able to engineer SPM tips down to atomic dimensions is the field ion microscope (FIM). This technique not only allows imaging of the tip apex with atomic precision but can also be used to manipulate the tip atoms by field evaporation [6], as shown in Fig. 1.3. Atomic interaction forces were measured with sub-nN precision, revealing force curves of only a few atoms interacting without mechanical hysteresis. However, the technical effort to achieve this type of measurement is considerable and most researchers today have resorted to the second solution.

The alternative solution can be visualized in Fig. 1.2. The straight dashed line now represents the force values of the oscillating cantilever with amplitude A assuming Hooke's law is valid. This is the tensile force of the cantilever spring, pulling the tip away from the sample. The restoring force of the cantilever is at all points stronger than the adhesion force. For example, the total force at point D is still pointing away from the sample, although the spring has the same stiffness as before. Mathematically speaking, the measurement is stable as long as the cantilever spring force ($F_{cb} = kA$) is larger than the attractive tip-sample force F_{ts} [7]. In the static mode we would already experience instability at that point; however, in the dynamic mode the spring is pre-loaded with a force stronger than the attractive tip-sample force. The equilibrium point of the oscillation is still far away

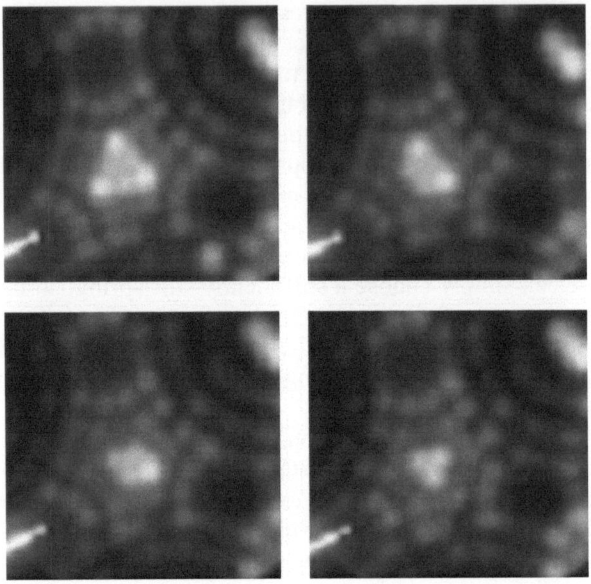

Fig. 1.3. Manipulation of the apex atoms of an AFM tip using field ion microscopy (FIM). Images are acquired at a tip bias of 4.5 kV. The last six atoms of the tip can be inspected in this example. Field evaporation to remove single atoms is performed by increasing the bias voltage for a short time to 5.2 kV. Each of the outer three atoms can be consecutively removed, eventually leaving a trimer tip apex

from the point of closest contact of the tip and surface atoms. The total force curve can be probed by varying the equilibrium point of the oscillation (i.e., by adjusting the z-piezo).

The diagram also shows that the oscillation amplitude has to be quite large, if fairly soft cantilevers are to be used. With lever spring constants of 10 N m^{-1}, the amplitude must be at least 1 nm to ensure forces of 1 nN can be reliably measured. In practical applications amplitudes of 10–100 nm are used to be on the safe side. This means that the oscillation amplitude is much larger than the force interaction range. The above simplification that the force gradient remains constant within one oscillation cycle does not hold anymore. Measurement stability is gained at the cost of a simple quantitative analysis of the experiments. In fact, dynamic AFM was first used to obtain atomic resolution images of clean surfaces [8], and it took another six years [1] for quantitative measurements of single bond forces to be obtained.

The technical realization of dynamic mode AFMs is based on the same key components as a DC-AFM set-up. The most common principle is the method of laser deflection sensing (see e.g. Fig. 1.4). A laser beam is focused on the backside of a microfabricated cantilever. The reflected laser spot is detected with a positional-sensitive diode (PSD). This photodiode is sectioned into two parts, which are read out separately (usually even a four-quadrant diode is used to detect torsional movements of the cantilever for lateral friction measurements). With the cantilever at equilibrium the spot is adjusted such that the two sections show the same intensity. If the cantilever bends up or down, the spot moves, and the difference signal between upper and lower section is a measure of the bending.

To enhance sensitivity, several groups have adopted an interferometer system to measure the cantilever deflection. A thorough comparison of different measurement methods with analysis of sensitivity and noise level is given by Marti [3].

Fig. 1.4. Representation of an AFM set-up with the laser beam deflection method. Cantilever and tip are microfabricated from silicon wafers. A laser beam is deflected from the backside of the cantilever and again focused on a photosensitive diode. The diode is segmented into four quadrants, which allows measurement of vertical and torsional bending of the cantilever (artwork from J. Heimel rendered with POV-Ray 3.0)

The cantilever is mounted on a device which allows the beam to oscillate. Often a piezo-element serves this purpose. The reflected laser beam is analyzed for oscillation amplitude, frequency, and phase difference between excitation and vibration. Depending on the mode of operation, a feedback mechanism will adjust oscillation parameters and/or tip-sample distance during the scanning. The set-up can be operated in air, UHV, and even in fluids. This allows the measurement of a wide range of surface properties from atomic resolution imaging [8] up to studying biological processes in liquid [9, 10].

1.3 Harmonic Oscillator: A Model System for Dynamic AFM

The oscillating cantilever has three degrees of freedom: the amplitude, the frequency, and the phase difference between excitation and oscillation. Let us consider the damped driven harmonic oscillator. The cantilever is mounted on a piezoelectric element that is oscillating with amplitude A_d at the frequency ω:

$$z_d(t) = A_d \cos(\omega t) \tag{1.6}$$

We assume that the cantilever spring obeys Hooke's law. Secondly, we introduce a friction force that is proportional to the speed of the cantilever motion, whereas α denotes the damping coefficient (Amonton's law). With Newton's first law we find for the oscillating system the following equation of motion for the position $z(t)$ of the cantilever tip (see also Fig. 1.1):

$$m\ddot{z}(t) = -\alpha\dot{z}(t) - kz(t) - kz_d(t) \tag{1.7}$$

We define $\omega_0^2 = k/m^*$, which turns out to be the resonant frequency of the free (undamped, i.e., $\alpha = 0$) oscillating beam. We further define the dimensionless quality factor $Q = m^*\omega_0/\alpha$, antiproportional to the damping coefficient. The quality factor describes the number of oscillation cycles, after which the damped oscillation amplitude decays to $1/e$ of the initial amplitude with no external excitation ($A_d = 0$). After some basic math this results in the following differential equation:

$$\ddot{z}(t) + \frac{\omega_0}{Q}\dot{z}(t) + \omega_0^2 z(t) = A_d \omega_0^2 \cos(\omega t) \tag{1.8}$$

The solution is a linear combination of two regimes [11]. Starting from rest, and switching on the piezo-excitation at $t = 0$, the amplitude will increase from zero to the final magnitude and reach a steady state, where amplitude, phase, and frequency of the oscillation stay constant over time. The steady-state solution $z_1(t)$ is reached after $2Q$ oscillation cycles and follows the external excitation with amplitude A_0 and phase difference φ:

$$z_1(t) = A_0 \cos(\omega t + \varphi) \tag{1.9}$$

The oscillation amplitude in the transient regime during the first $2Q$ cycles follows:

$$z_2(t) = A_t \cdot e^{-\omega_0 t/2Q} \cdot \sin(\omega_0 t + \varphi_t) \tag{1.10}$$

We emphasize the important fact that the exponential term causes $z_2(t)$ to diminish exponentially with time constant τ:

$$\tau = 2Q/\omega_0 \tag{1.11}$$

Under vacuum conditions, only the internal dissipation due to bending of the cantilever is present, and Q reaches values of 10,000 at typical resonant frequencies of 100,000 Hz. This results in a relatively long transient regime of $\tau \cong 30$ ms, which limits the possible operational modes for dynamic AFM (for a detailed analysis see ref. [11]). Changes in the measured amplitude, which reflect a change of atomic forces, will have a time lag of 30 ms, which is very slow considering one wants to scan a 200 × 200-point image within a few minutes. In air, however, viscous damping due to air friction dominates and Q goes down to less than 1,000, resulting in a time constant below the millisecond level. This response time is fast enough to use the amplitude as a measurement parameter.

If we evaluate the steady state solution $z_1(t)$ in the differential equation, we find the following well-known solution for amplitude and phase of the oscillation as a function of the excitation frequency ω:

$$A_0 = \frac{A_d \cdot Q \cdot \omega_0^2}{\sqrt{\omega^2\omega_0^2 + Q^2\left(\omega_0^2 - \omega^2\right)^2}} \tag{1.12}$$

$$\varphi = \arctan\left(\frac{\omega \cdot \omega_0}{Q \cdot \left(\omega_0^2 - \omega^2\right)}\right) \tag{1.13}$$

Amplitude and phase diagrams are depicted in Fig. 1.5. As can be seen from Eq. 1.12 the amplitude will reach its maximum at a frequency different from ω_0, if Q has a

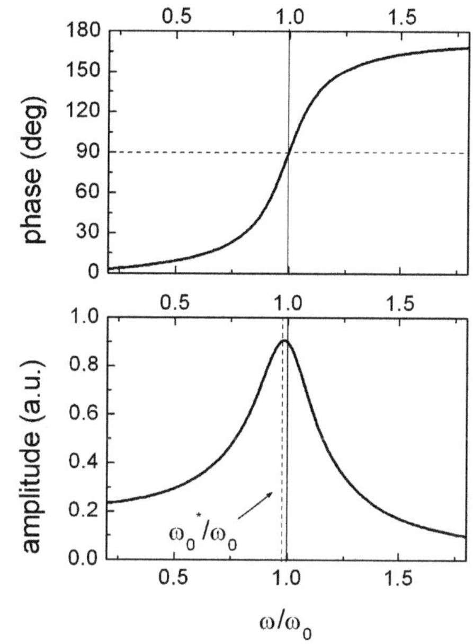

Fig. 1.5. Amplitude and phase versus excitation frequency curves for the damped harmonic oscillator, with a quality factor of $Q = 4$

finite value. The damping term of the harmonic oscillator causes the resonant frequency to shift from ω_0 to ω_0^*:

$$\omega_0^* = \omega_0 \sqrt{1 - \frac{1}{2Q^2}} \tag{1.14}$$

The shift is negligible for Q-factors of 100 and above, which is the case for most applications in vacuum or air. However, for measurements in liquids, Q can be smaller than 10 and ω_0 differs significantly from ω_0^*. As we will discuss later, it is also possible to enhance Q by using a special excitation method.

In the case that the excitation frequency is equal to the resonant frequency of the undamped cantilever $\omega = \omega_0$, we find the useful relationship:

$$A_0 = Q \cdot A_d \text{ for } \omega = \omega_0 \tag{1.15}$$

Since $\omega_0^* \approx \omega_0$ for most cases, we find that Eq. 1.15 holds true for exciting the cantilever at its resonance. From a similar argument, the phase becomes approximately 90° for the resonance case. We also see that in order to reach vibration amplitudes of some 10 nm, the excitation only has to be as small as 1 pm, for typical cantilevers operated in vacuum.

So far we have not considered an additional force term, describing the interaction between the probing tip and the sample. For typical large vibration amplitudes of 10–100 nm no general solution for this analytical problem has been found yet. The cantilever tip will experience a whole range of force interactions during one single oscillation cycle, rather than one defined tip-sample force. Only in the special case of a self-excited cantilever oscillation has the problem been solved semi-analytically, as we will see later.

1.4 Dynamic AFM Operational Modes

While the quantitative interpretation of force curves in contact AFM is straightforward using Eq. 1.1, we explained in the previous paragraphs that its application to assess short-range attractive interatomic forces is rather limited. The dynamic mode of operation seems to open a viable direction towards achieving this task; however, interpretation of the measurements generally appears to be more difficult. Different operational modes are employed in dynamic AFM and the following paragraphs are intended to distinguish these modes and categorize them in a systematic way.

The oscillation trajectory of a dynamically driven cantilever is determined by three parameters: the amplitude, the phase, and the frequency. Tip-sample interactions can influence all three parameters, in the following termed the internal parameters. The oscillation is driven externally, with excitation amplitude A_d and excitation frequency ω. These variables will be referred to as the external parameters. The external parameters are set by the experimentalist, whereas the internal parameters are measured and contain the crucial information about the force interaction. In scanning probe applications, it is common to control the probe-surface distance z_0 in order to keep an internal parameter constant (i.e., the tunneling current in STM or the beam deflection in contact AFM), which represents a certain tip-sample interaction. In z-spectroscopy mode the distance is varied in a certain range, and the changes of the internal parameters are measured as a fingerprint of the tip-sample interactions.

In dynamic AFM the situation is rather complex. Any of the internal parameters can be used for feedback of the tip-sample distance z_0. However, we already realized that in general the tip-sample forces could only be fully assessed by measuring all three parameters. This makes it difficult to obtain images where the distance z_0 is a representative of the surface at one force set-point. A solution to this problem is to establish additional feedback loops, which keep the internal parameters constant by adjusting the external variables.

In the simplest set-up, the excitation frequency is set to a predefined value, and the excitation amplitude remains constant by a feedback loop. This is called the AM mode (amplitude modulation) or also tapping mode. As stated before, in principle any of the internal parameters can be used for feedback to the tip-sample distance; in the AM mode the amplitude signal is used. A certain amplitude (smaller than the free oscillation amplitude) at a frequency close to the resonance of the cantilever is chosen, the tip is approached towards the surface under investigation, and the approach is stopped as soon as the set-point amplitude is reached. The oscillation-phase is usually recorded during the scan; however, the shift of the resonant frequency of the cantilever cannot be directly accessed, since this degree of freedom is blocked by the external excitation at a fixed frequency. It turns out that this mode is simple to operate from a technical perspective, but quantitative information about the tip-sample interaction forces has so far not been reliably extracted from AM-mode AFM. Still, it is so far one of the most commonly used modes in dynamic AFM operated in air and even in liquid. The strength of this mode is the qualitative imaging of a large variety of surfaces.

It is interesting to discuss the AM mode in the extreme situation that the external excitation frequency is much lower than the resonant frequency [12, 13]. This results in a quasi-static measurement, although a dynamic oscillation force is applied and therefore this mode can be viewed as a hybrid between static and dynamic AFM. Unfortunately, it has the drawbacks of the static mode, namely that stiff spring constants must be used and therefore the sensitivity of the deflection measurement must be very good, typically employing a high-resolution interferometer. Still it has the advantage of the static measurement in quantitative interpretation, since in the regime of small amplitudes (<0.1 nm) a direct interpretation of the experiments is possible. In particular, the force gradient at tip-sample distance z_0 is given by the change of the amplitude A and the phase angle φ:

$$\left.\frac{\partial F_{ts}}{\partial z}\right|_{z_0} = k\left(1 - \frac{A_0}{A}\cos\varphi\right) \tag{1.16}$$

In effect, the modulated AFM in contrast to the purely static AFM can enhance sensitivity due to the use of lock-in techniques, which allows one to measure amplitude and phase of the oscillation signal with high precision.

As stated before, the internal parameters can be fed back to the external excitation variables. One of the most useful applications in this direction is the self-excitation system. Here the resonant frequency of the cantilever is detected and selected again as the excitation frequency. In a typical set-up this is done with a phase shift of 90° by feeding back the detector signal to the excitation piezo (i.e., the cantilever is always excited in resonance). Influences of the tip-sample interaction forces on the resonant frequency do not change the two other parameters of the oscillation (i.e., amplitude and phase); only the oscillation frequency is shifted. Therefore it is sufficient to measure the frequency shift between the free oscillation and oscillation with tip-sample interaction. Since the phase remains at a fixed value the oscillating system is much better defined than before, and the degrees of freedom for the oscillation are reduced. To even reduce the last degree of freedom, the oscillation amplitude, another feedback loop can be established to keep the oscillation amplitude A constant by varying the excitation amplitude A_d. Now all internal parameters have a fixed relationship to the external excitation variables, the system is well defined and all parameters can be assessed during the measurement. As it turns out, this mode is the only dynamic mode for which a quantitative relationship between tip-sample forces and the change of the resonant frequency can be established.

In the following section we want to discuss the two most popular operational modes, tapping-mode and self-excitation mode, in more detail.

1.4.1 Amplitude-modulation/tapping-mode AFM

In tapping-mode or AM-AFM the cantilever is excited externally at a constant frequency close to its resonance. Oscillation amplitude and phase during approach of tip and sample serve as the experimental observation channels. Figure 1.6 shows

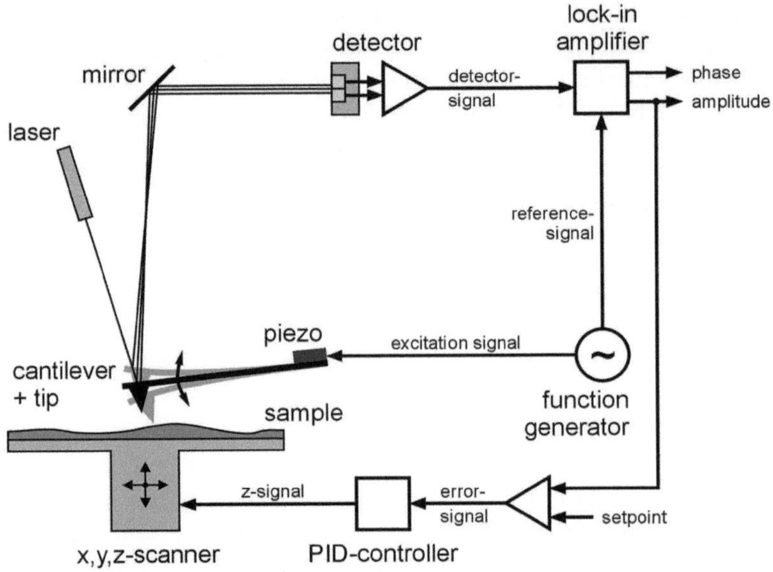

Fig. 1.6. Set-up of a dynamic force microscope operated in the AM or tapping mode. A laser beam is deflected by the backside of the cantilever and the deflection is detected by a split photodiode. The excitation frequency is chosen externally with a modulation unit, which drives the excitation piezo. A lock-in amplifier analyses phase and amplitude of the cantilever oscillation. The amplitude is used as the feedback signal for the probe-sample distance control

a diagram of a typical tapping-mode AFM set-up. The oscillation amplitude and the phase (not shown in diagram), detected with the photodiode, are analyzed with a lock-in amplifier. The amplitude is compared to the set-point, and the difference or error signal is used to adjust the z-piezo (i.e., the probe-sample distance). The external modulation unit supplies the signal for the excitation piezo, and at the same time the oscillation signal serves as the reference for the lock-in amplifier.

During one oscillation cycle with amplitudes of 10–100 nm the tip-sample interaction will range over a wide distribution of forces, including attractive and repulsive forces. We will therefore measure a convolution of the force-distance curve with the oscillation trajectory. This complicates the interpretation of AM-AFM measurements appreciably.

At the same time, the resonant frequency of the cantilever will change due to the appearing force gradients, as could already be seen in the simplified model from Eq. 1.4. If the cantilever was excited exactly at its resonant frequency before, it will be excited off-resonance after interaction forces are encountered. This in return changes amplitude and phase (see Eqs. 1.12, 1.13), which serve as the measurement signals. Consequently, a different amplitude will cause a change of the encountered effective force. We can see already from this simple gedanken-experiment that the interpretation of force curves will be highly complicated. In fact, there is no quantitative theory for AM-AFM available, which allows the experimentalist to unambiguously convert the experimental data to a force-distance relationship.

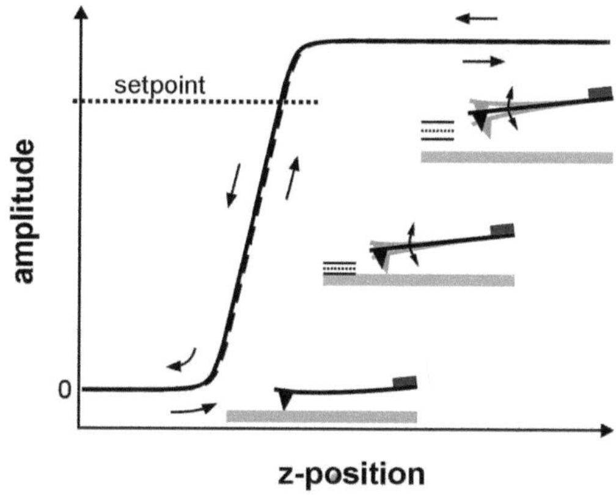

Fig. 1.7. Simplified model showing the oscillation amplitude in tapping mode AFM for various probe-sample distances

The qualitative behavior for amplitude versus z_0-position curves is depicted in Fig. 1.7. At large distances, where the forces between tip and sample are negligible, the cantilever oscillates with its free oscillation amplitude. Upon approach of the probe towards the surface the interaction forces cause the amplitude to change, resulting typically in an amplitude getting smaller with continuously decreasing tip-sample distance. This is expected, since the force-distance curve will eventually reach the repulsive part and the tip is hindered from indenting further into the sample, resulting in smaller oscillation amplitudes.

However, in order to gain some qualitative insight into the complex relationship between forces and oscillation parameters, we resort to numerical simulations. Anczykowski et al. [14, 15] have calculated the oscillation trajectory of the cantilever under the influence of a given force model. Van der Waals interactions were considered the only effective, attractive forces, and the total interaction resembled a Lennard-Jones-type potential. Mechanical relaxations of tip and sample surface were treated in the limits of continuum theory with the numerical MYD/BHW approach [16, 17], which allows the simulations to be compared to corresponding experiments. Figure 1.8 shows the force-distance curves for different tip radii underlying the dynamic AFM simulations.

The cantilever trajectory was analyzed by solving the differential Eq. 1.7 extended by the force-distance relationships from Fig. 1.8 using the numerical Verlet algorithm [18]. The results of the simulation for the amplitude and phase of the tip oscillation as a function of z-position of the probe are presented in Fig. 1.9. One has to keep in mind that the z-position of the probe is not equivalent with the real tip-sample distance at equilibrium position, since the cantilever might bend statically due to the interaction forces. The behavior of the cantilever can be subdivided into three different regimes. We distinguish three cases, where the beam is oscillated below its resonant frequency ω_0, exactly at ω_0 and above ω_0.[1]

Clearly, Fig. 1.9 exhibits more features than were anticipated from the initial simple arguments. Amplitude and phase seem to change rather abruptly at certain

[1] In the following we will refer to ω_0 as the resonant frequency, although the correct resonant frequency is ω_0^*, if the finite Q-value is taken into account.

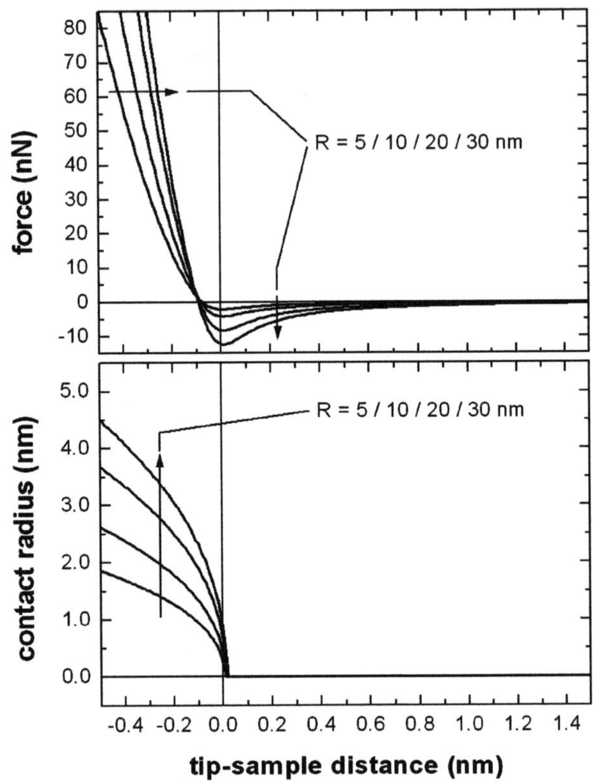

Fig. 1.8. Force curves and corresponding contact radius calculated with the MYD/BHW-model as a function of tip radius for a Si–Si contact. These force curves are used for the tapping-mode AFM simulations

points when the z_0-position is decreased. Besides, the amplitude or phase-distance curves do not resemble the force-distance curves from Fig. 1.8 in a simple direct manner. Additionally, we find a hysteresis between approach and retraction.

As an example, let us start the by discussing the discontinuous features in the AFM spectroscopy curves of the first case, where the excitation frequency is smaller than ω_0. Consider the oscillation amplitude as a function of excitation frequency in Fig. 1.5 in conjunction with a typical force curve as depicted in Fig. 1.8. Upon approach of probe and sample, attractive forces will lower the effective resonant frequency of the oscillator. Therefore, the excitation frequency will now be closer to the resonant frequency, causing the vibration amplitude to increase. This in return reduces the tip-sample distance, which again gives rise to a stronger attractive force. The system becomes unstable until the point $z_0 = d_{app}$ is reached at which repulsive forces stop the self-enhancing instability. This can be clearly observed in Fig. 1.9. Large parts of the force-distance curve cannot be measured due to this instability.

In the second case, where the excitation equals the free resonant frequency, only a small discontinuity is observed upon reduction of the z-position. Here a shift of the resonant frequency towards smaller values, induced by the attractive force interaction, will reduce the oscillation amplitude. The distance between tip and sample is therefore reduced as well, and the self-amplifying effect with the sudden in-

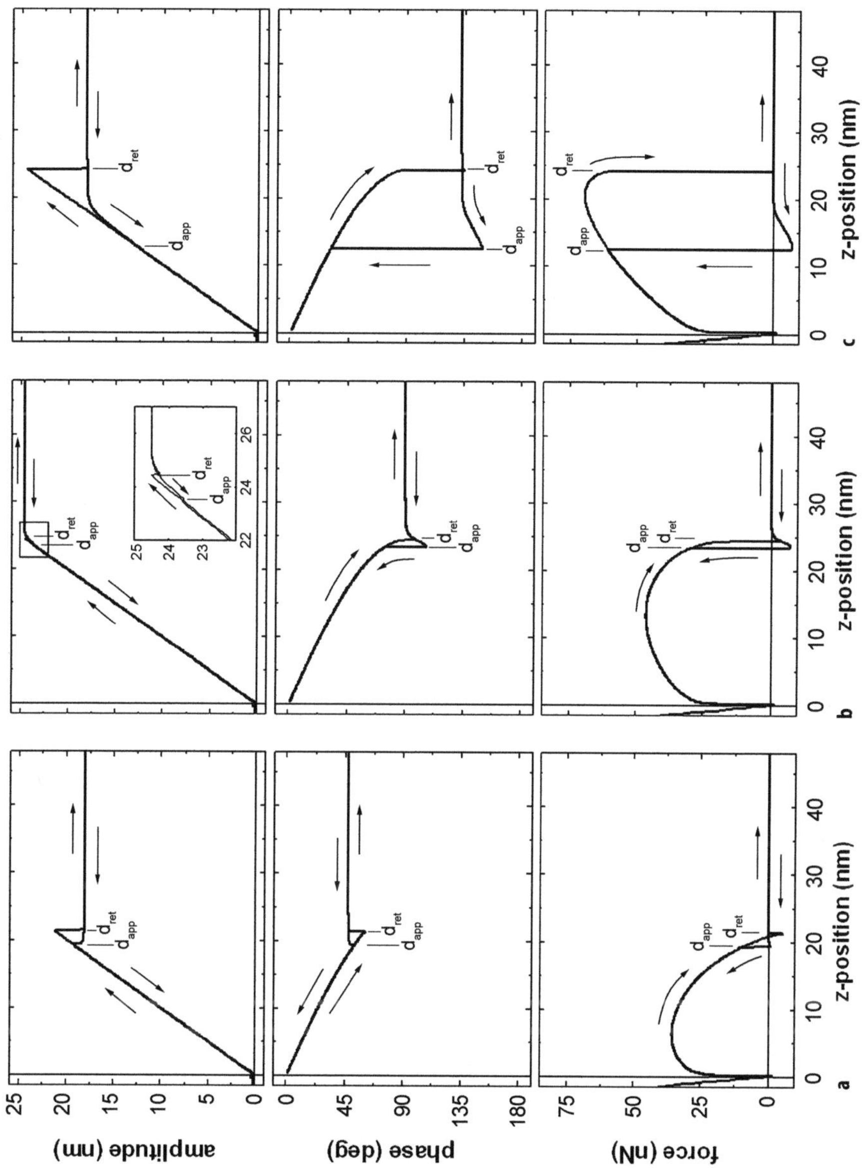

Fig. 1.9a–c. Amplitude and phase diagrams with excitation frequency **a** below, **b** exactly at, and **c** above the resonant frequency for tapping-mode AFM from the numerical simulations. Additionally, the *bottom diagrams* show the interaction forces at the point of closest tip-sample distance, i.e. the lower turn-around point of the oscillation

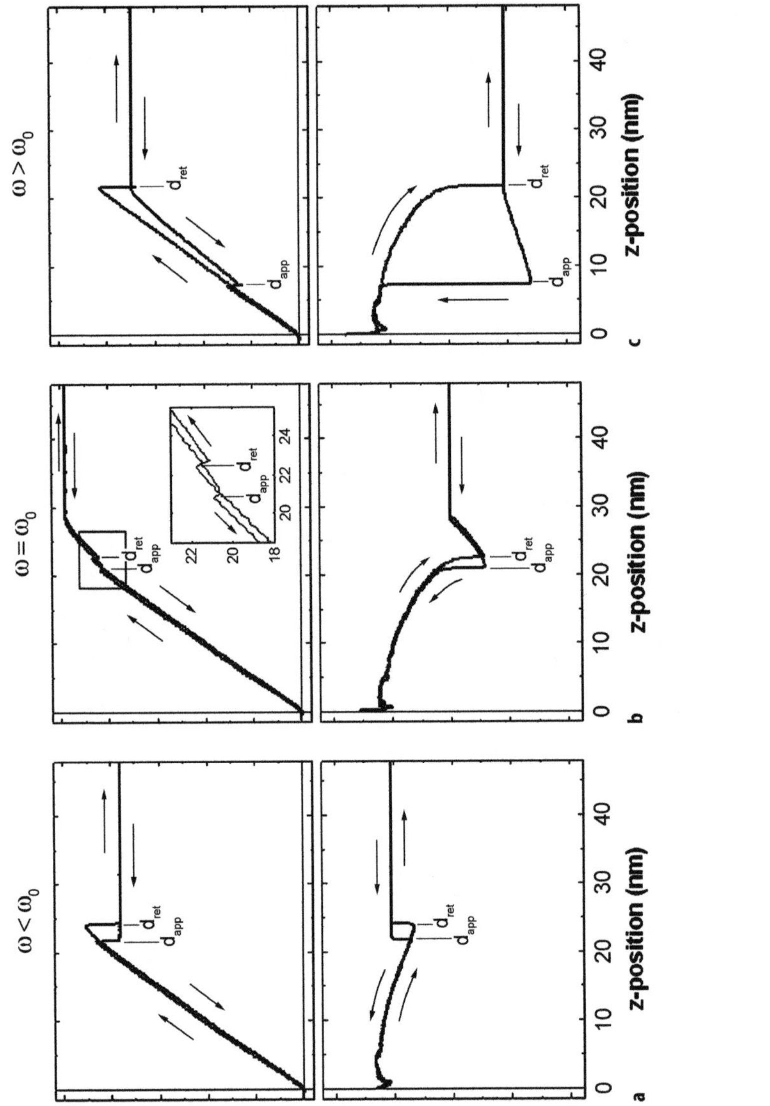

Fig. 1.10a–c. Amplitude and phase diagrams with excitation frequency **a** below, **b** exactly at, and **c** above the resonant frequency for tapping-mode AFM from experiments with a Si-cantilever on a Si-wafer in air

stability does not occur, as long as repulsive forces are not encountered. However, at closer tip-sample distances, repulsive forces will cause the resonant frequency to shift again towards higher values, increasing the amplitude with decreasing tip-sample distance. Therefore a self-enhancing instability will also occur in this case, but here at the crossover from purely attractive forces to the regime where repulsive forces occur. Correspondingly, a small kink in the amplitude curve can be observed in Fig. 1.9. An even clearer indication of this effect is manifested by the sudden change in the phase signal at d_{app}.

In the last case, with $\omega > \omega_0$, the effect of amplitude reduction due to the resonant frequency shift is even larger. Again, we find no instability in the amplitude signal during approach in the attractive force regime. Although, as soon as the repulsive force regime is reached, the instability occurs due to the induced positive frequency shift. Consequently, a large jump in the phase curve from values smaller than 90° towards values larger than 90° is observed. The small change in the amplitude curve is not resolved in the simulated curves in Fig. 1.9; however, this can be clearly seen in the experimental curves in Fig. 1.10.

In Fig. 1.10 corresponding experimental amplitude and phase curves are depicted. The measurements were performed in air with a Si-cantilever approaching a Si-wafer, with a cantilever resonant frequency of 299.95 kHz. Qualitatively, all prominent features of the simulated curves can also be found in the experimental data sets. Hence, the above model seems to capture the important factors necessary for an appropriate description of the experimental situation.

But what is the reason for this unexpected behavior? We have to turn to the numerical simulations again, where we have access to all physical parameters, in order to understand the underlying processes. The lower part of Fig. 1.9 also shows the interaction force between the tip and the sample at the point of closest approach (i.e., the sample-sided turn-around point of the oscillation). We see that exactly at the points of the discontinuities, the total interaction force changes from the net-attractive regime[2] to the attractive-repulsive regime, also termed the intermittent contact regime. As soon as a minimum distance is reached, the tip also starts to experience repulsive forces, which completely changes the oscillation behavior. In other words, the dynamic system switches between two oscillatory states.

Directly related to this fact is the second phenomenon, the hysteresis effect. We find separate curves for the approach of the probe towards the surface and the retraction. This seems to be somewhat counterintuitive, since the tip is constantly approaching towards and retracting from the surface, and the average values of amplitude and phase should be independent of the direction of the average tip-sample distance movement. A hysteresis between approach and retraction within one oscillation, due to dissipative processes, should directly influence amplitude and phase. However, no dissipation models were included in the simulation. In this case, the hysteresis in Fig. 1.9 is due to the fact that the oscillation jumps into different modes and the system exhibits bistability. This effect is often observed in oscillators under the influence of non-linear forces [20].

[2] We use the term net-attractive to emphasize that the total force is attractive, despite the fact that some minor contributions might still originate from repulsive forces.

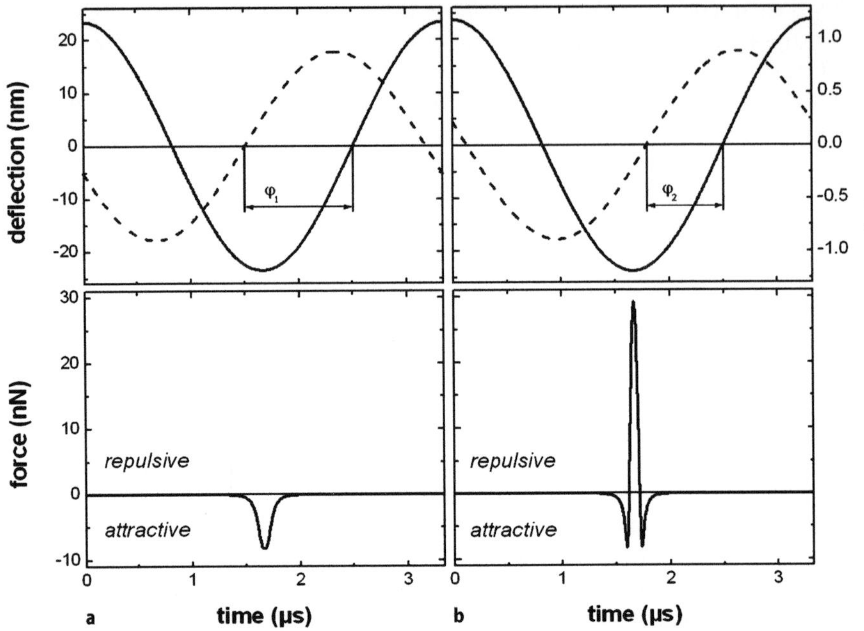

Fig. 1.11a,b. Simulation of the tapping-mode cantilever oscillation in the **a** net-attractive and **b** the intermittent contact regime. The *dashed line* represents the excitation amplitude and the *solid line* is the oscillation amplitude

For the interpretation of these effects it is helpful to look at Fig. 1.11, which shows the behavior of the simulated tip vibration and force during one oscillation cycle over time. The data is recorded at the same z-position at a point where hysteresis is observed, while Fig. 1.11a was taken during the approach and Fig. 1.11b during the retraction. Excitation is in resonance, representing case 2, where the amplitude shows a small hysteresis. Also note, that the amplitude is almost exactly the same in Fig. 1.11a and b. We see that the oscillation at the same z-position exhibits two different modes: while in Fig.1.11a the experienced force is net-attractive, in Fig. 1.11b the tip is exposed to attractive and repulsive interactions. Experimental and simulated data shows that the change between the net-attractive and intermittent contact mode takes place at different z-positions (d_{app} and d_{ret}) for approach and retraction. Between d_{app} and d_{ret} the system is in a bistable mode. Depending on the history of the measurement, for example, whether the position d_{app} during the approach (or d_{ret} during retraction) has been reached, the system flips to the other oscillation mode. While the amplitude might not be influenced strongly (case 2), the phase is a clear indicator of the mode switch. On the other hand, if point d_{app} is never reached during the approach, the system will stay in the net-attractive regime and no hysteresis is observed (i.e., the system remains stable).

In conclusion we find that, although a qualitative interpretation of the interaction forces is possible, the AM- or tapping-mode AFM is not suitable to gain direct quantitative knowledge of tip-sample force interactions. However, it is a very useful tool for imaging nanometer-sized structures in a wide variety of set-ups, in air or even in liquid. We find that there exist two distinct modes for the externally excited oscillation: the net-attractive and the intermittent contact mode, describing which kind of forces the tip-sample interaction is governed by. The phase can be used as an indicator, in which mode the system is running.

In particular it can be easily seen that if the free resonant frequency of the cantilever is higher than the excitation frequency, the system cannot stay in the net-attractive regime, due to a self-enhancing instability. Since in many applications involving soft and delicate biological samples strong repulsive forces should be avoided, the tapping-mode AFM should be operated at frequencies equal or above the free resonant frequency [21]. Even then statistical changes of tip-sample forces during the scan might induce a sudden jump into the intermittent contact mode, and the above-explained hysteresis will tend to keep the system in this mode. It is therefore of high importance to tune the oscillation parameters in such a way that the AFM stays in the net-attractive regime [22]. A concept which achieves this task is the Q-control system, which will be discussed in some detail in the forthcoming paragraphs.

A last word concerning the overlap of simulation and experimental data: while the qualitative agreement down to the detailed shape of hysteresis and instabilities is rather striking, we still find quite some quantitative discrepancies between the positions of the instabilities d_{app} and d_{ret}. This is probably due to the simplified force model, which only takes into account van der Waals and repulsive forces. Especially under ambient conditions, an omnipresent water meniscus between tip and sample will give rise to much stronger attractive and also dissipative forces than considered in the model. A very interesting feature is that the simulated phase curves in the intermittent contact regime tend to have a steeper slope in the simulation than in the experiments (see also Fig. 1.12). We will show later that this effect is a fingerprint of an effect that had not been included in the above simulation at all: dissipative processes during the oscillation, giving rise to an additional loss of oscillation energy.

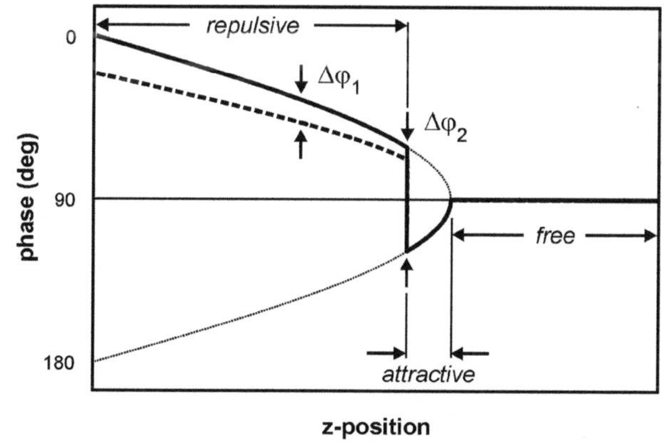

Fig. 1.12. Phase shift in tapping mode as a function of tip-sample distance

1.4.2
Self-Excitation Modes

Despite the wide range of technical applications of the AM- or tapping-mode of dynamic AFM, historically it was found that it is unsuitable for measurements in an environment extremely useful for scientific research: in vacuum or ultra-high vacuum (UHV) with pressures reaching 1×10^{-10} mbar. The STM has already shown how much insight can be gained from some highly defined experiments under those conditions. Consider Eq. 1.11 from the above section. The time constant τ for the amplitude to adjust to a different tip-sample force scales with $1/Q$. In vacuum applications Q of the cantilever is of the order of 10,000, which means that τ is in the range of some 10 ms. This is clearly too long for a scan of at least (100×100) data points. The temperature-induced drift of the sample will render useful interpretation of images an impossible task. However, the resonant frequency of the system will react instantaneously to tip-sample forces. This has led Albrecht et al. [11] to use a modified excitation scheme.

The system is always oscillated at its resonant frequency. This is achieved by feeding back the oscillation signal from the cantilever into the excitation piezo element. Figure 1.13 pictures the method in a block diagram. The signal from the PSD is phase shifted by 90° (and therefore always exciting in resonance) and used as the excitation signal of the cantilever. An additional feedback loop adjusts the excitation amplitude in such a way that the oscillation amplitude remains constant.

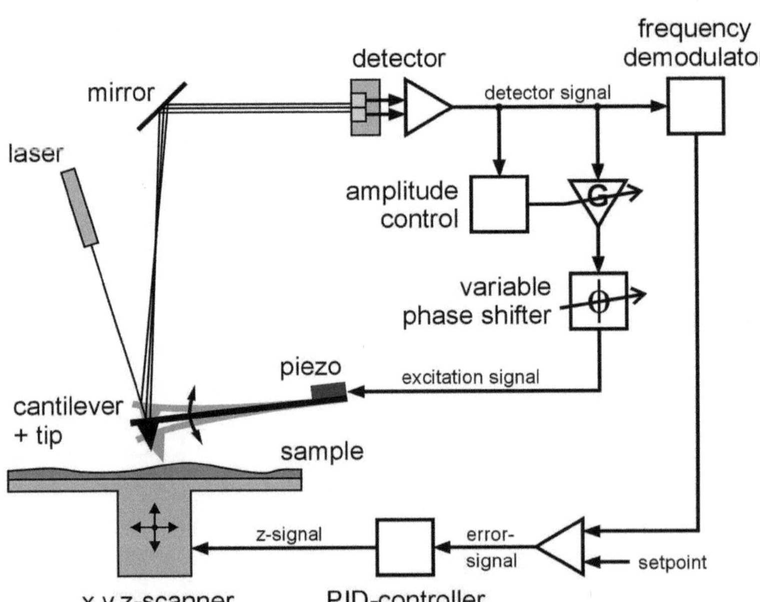

Fig. 1.13. Dynamic AFM operated in the self-excitation mode, where the oscillation signal is directly fed back to the excitation piezo. The detector signal is amplified with the variable gain G and phase shifted by phase ϕ. The frequency demodulator detects the frequency shift due to tip-sample interactions, which serves as the control signal for the probe-sample distance

This ensures that the tip-sample distance is not influenced by changes of the oscillation amplitude. The only degree of freedom of the oscillation system that can react to the tip-sample forces is the change of the resonant frequency. This shift of the frequency is detected and used as the set-point signal for surface scans. Therefore this mode is also called the FM (frequency-modulated) mode.

Let us take a look at the sensitivity of the dynamic AFM. If electronic noise, laser noise, and thermal drift can be neglected, the main noise contribution will come from thermal excitation of the cantilever. A detailed analysis of a dynamic system yields the following relationship for the minimum detectable force gradient [11]:

$$\left.\frac{\partial F}{\partial z}\right|_{\mathrm{MIN}} = \sqrt{\frac{4k \cdot k_B T \cdot B}{\omega_0 Q \left\langle z_{osc}^2 \right\rangle}} \tag{1.17}$$

Here B is the bandwidth of the measurement, T is the temperature, and $\langle z_{osc}^2 \rangle$ is the mean-square amplitude of the oscillation. Please note that this sensitivity limit was deliberately calculated for the FM mode. A similar analysis of the AM mode, however, yields virtually the same result [23]. We find that the minimum detectable force gradient (i.e., the measurement sensitivity), is inversely proportional to the Q-factor of the cantilever. This means, that it should be possible to achieve very high resolution imaging under vacuum conditions. In contrast, AM- or tapping-mode AFM cannot be usefully pursued with large Q. Only the FM mode makes it possible to take practical advantage of Eq. 1.17.

A breakthrough in high-resolution AFM application was the atomic resolution imaging of the Si(111)-(7 × 7) surface reconstruction by Giessibl [8] under UHV conditions. Today, atomic resolution imaging has become a standard feature guaranteed by industrially produced dynamic AFM systems. While STM has already proven to be an indispensable tool to gain detailed insight into surface structures of conductors with atomic resolution, the dynamic AFM has opened up the avenue into investigating non-conductive surfaces with equal precision (for example on aluminium oxide by Barth et al. [24]).

However, we were concerned with measuring atomic force potentials of a single pair of molecules. Clearly, FM-mode AFM will allow us to identify single atoms, and with sufficient care we will be able to ensure that effectively only one atom from the tip contributes to the total force interaction. Can we, therefore, fill in the last bit of information and find a quantitative relationship between the oscillation parameters and the force?

Gotsmann et al. [25] investigated this relationship by employing a numerical approach. During each oscillation cycle the tip experiences a whole range of forces. For each step during the approach the differential equation for the whole oscillation loop (including also the feedback system) was evaluated and the relationship between force and frequency shift was revealed. It was possible to determine the quantitative interaction forces of a metallic contact of nanometer dimensions.

However, it was shown that there also exists an analytical relationship, if some approximations are accepted [7, 26, 27]. Here, we will follow the route as indicated by Dürig [26], although different alternative ways have also been proven successful.

Consider the tip oscillation trajectory reaching over a large part of the force gradient curve in Fig. 1.2. We model the tip-sample interaction as a spring constant of stiffness $k_{ts}(z) = \partial F/\partial z\,|_{z_0}$ as in Fig. 1.1. For small oscillation amplitudes we already found that the frequency shift is proportional to the force gradient in Eq. 1.4. For large amplitudes, we can calculate an effective force gradient k_{eff} as a convolution of the force and the fraction of time that the tip spends between the positions x and $x + dx$:

$$k_{eff}(z) = \frac{2}{\pi A^2} \int_{z}^{z+2A} F(x) \cdot g\left(\frac{x-z}{A} - 1\right) dx \text{ with } g(u) = -\frac{u}{\sqrt{1-u^2}} \tag{1.18}$$

In the approximation that the vibration amplitude is much larger than the range of the tip sample forces the above equation can be simplified to:

$$k_{eff}(z) = \frac{\sqrt{2}}{\pi} A^{3/2} \cdot \int_{z}^{\infty} \frac{F(x)}{\sqrt{x-z}}\, dx \tag{1.19}$$

This effective force gradient can now be used in Eq. 1.4, the relationship between frequency shift and force gradient. We find:

$$\Delta f = \frac{f_0}{\sqrt{2}\pi \cdot k A^{3/2}} \cdot \int_{z}^{\infty} \frac{F(x)}{\sqrt{x-z}}\, dx \tag{1.20}$$

If we separate the integral from other parameters we can define:

$$\Delta f = \frac{f_0}{k A^{3/2}} \cdot \gamma(z) \text{ with } \gamma(z) = \frac{1}{\sqrt{2}\pi} \cdot \int_{z}^{\infty} \frac{F(x)}{\sqrt{x-z}}\, dx \tag{1.21}$$

This means we can define $\gamma(z)$ which is only dependent on the shape of the force curve $F(z)$, but independent of the external parameters of the oscillation. The function $\gamma(z)$ is also referred to as the "normalized frequency shift" [7], a very useful parameter, which allows one to compare measurements independent of resonant frequency, amplitude, and spring constant of the cantilever.

The dependence of the frequency shift on the vibration amplitude is an especially useful relation, since this parameter can be easily varied during one experiment. A nice example is depicted in Fig. 1.14, where frequency shift curves for different amplitudes were found to coincide very well in the $\gamma(z)$ diagrams [28].

This relationship has been nicely exploited for the calibration of the vibration amplitude by Guggisberg [29], which is a problem often encountered in dynamic AFM operation and worthwhile discussing. One approaches tip and sample and takes frequency shift curves versus distance, which show a reproducible shape. Then the z-feedback is permanently turned off and several curves with different amplitudes are taken. The amplitudes are typically chosen by adjusting the amplitude setpoint in volts. One has to take care that drift in z-direction is negligible. An analysis of the corresponding $\gamma(z)$ curves will show the same curves (as in Fig. 1.14), but these will be shifted in the horizontal axis. These shifts correspond to the change in amplitude, allowing one to correlate the voltage values with the z-distances.

For the often-encountered force contributions from electrostatic, van der Waals, and chemical binding forces the frequency shift has been calculated from the force

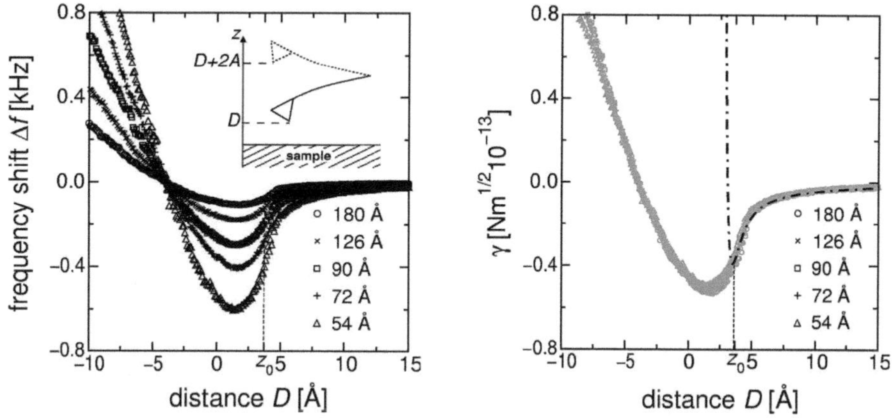

Fig. 1.14. **a** Frequency shift curves for different oscillation amplitudes for a silicon tip on a graphite surface in UHV, **b** γ curves calculated from the Δf curves in *a* (reprinted from Hölscher et al. [28] with permission, Copyright (2000) by The American Physical Society)

laws. In the approximation that the tip radius R is larger than the tip-sample distance z, an electrostatic potential V will yield a normalized frequency shift of (adapted from Guggisberg et al. [29]):

$$\gamma(z) = \frac{\pi\varepsilon_0 R \cdot V^2}{\sqrt{2}} \cdot z^{-1/2} \tag{1.22}$$

For van der Waals forces with Hamaker constant H and also with R larger than z we find accordingly:

$$\gamma(z) = \frac{H \cdot R}{12\sqrt{2}} \cdot z^{-3/2} \tag{1.23}$$

Finally, short-range chemical forces represented by the well-known Morse potential (with the parameters binding energy U_0, decay length λ, and equilibrium distance z_{equ}) yield:

$$\gamma(z) = \frac{U_0\sqrt{2}}{\sqrt{\pi\lambda}} \cdot \exp\left(-\frac{(z - z_{equ})}{\lambda}\right) \tag{1.24}$$

These equations allow the experimentalist to interpret directly the spectroscopic measurements. For example, the contributions of the electrostatic and van der Waals forces can be easily distinguished by their slope in a log-log plot (for an example see Guggisberg et al. [29]).

Alternatively, if the force law is not known beforehand, the experimentalist wants to analyze the experimental frequency shift data curves and extract the force or energy potential curves. We therefore have to invert the integral in Eq. 1.21 to find the tip-sample interaction potential V_{ts} from the $\gamma(z)$ curves [26]:

$$V_{ts}(z) = \sqrt{2} \cdot \int_z^{\infty} \frac{\gamma(x)}{\sqrt{x - z}}\, dx \tag{1.25}$$

By using this method, quantitative force curves were extracted from Δf-spectroscopy measurements on different, atomically resolved sites of the Si(111)-(7 × 7) reconstruction [1]. Comparison to theoretical MD simulations showed good quantitative agreement with theory and confirmed the assumption that force interactions were governed by one single atom at the tip apex. Our initially formulated goal seems to be achieved: with FM-AFM we have found a powerful method, which allows us to measure the chemical bond formation of single molecules. The last uncertainty, the exact shape and identity of the tip apex atom, can possibly be resolved by employing the FIM technique to characterize the tip surface in combination with FM-AFM.

All the above equations are only strictly valid in the approximation that the oscillation amplitudes are much larger than the distance range of the encountered forces. For amplitudes of 10 nm and long-ranged forces like electrostatic interactions this approximation might not be valid anymore. An iterative approach has been presented by Dürig [30] to overcome this problem. The interaction force is calculated from the frequency shift curves (gradient of Eq. 1.25). This force curve again is used to calculate the frequency shift with the exact relationship from Eq. 1.18. The difference between the experimental Δf curve and the reconstructed Δf curve can now be used to calculate a first-order correction term to the interaction force. This procedure can be iteratively followed until sufficient agreement is found.

In this context it is worthwhile to point at a slightly different dynamic AFM method. While in the typical FM-AFM set-up the oscillation amplitude is controlled to stay constant by a dedicated feedback circuit, one could simply keep the excitation amplitude constant (this has been termed *CA = constant amplitude* as opposed to the *CE = constant excitation* mode). It is expected that this mode is more gentle to the surface, because any dissipative interaction will reduce the amplitude and therefore increase the tip-sample distance. The tip is prevented from indenting deeply into the surface. This mode has been employed to image soft biological molecules like DNA or thiols in UHV [19]. However, quantitative interpretation of the obtained frequency spectra is more complicated, since the amplitude and tip-sample distance are altered during the measurement. Until now, we have always associated the self-excitation scheme with vacuum applications. Although it is difficult to operate the FM-AFM in constant-amplitude mode in air, since large dissipative effects make it difficult to ensure a constant amplitude, it is indeed possible to use the constant-excitation FM-AFM in air or even in liquid. Still, only few applications of FM-AFM under ambient or liquid conditions have been reported so far. In fact, a low budget construction-set (employing a tuning fork force sensor) for a CE-mode dynamic AFM set-up has been published on the internet [31].

1.5 Q-Control

We have already discussed the virtues of a high Q value for high sensitivity measurements: the minimum detectable force gradient was inversely proportional to the square root of Q. In vacuum Q mainly represents the internal dissipation of the cantilever during oscillation, an internal damping factor. Little damping is obtained by using high-quality cantilevers, which are cut (or etched) from defect-free, single-crystal silicon wafers. Under ambient or liquid conditions, the quality factor is

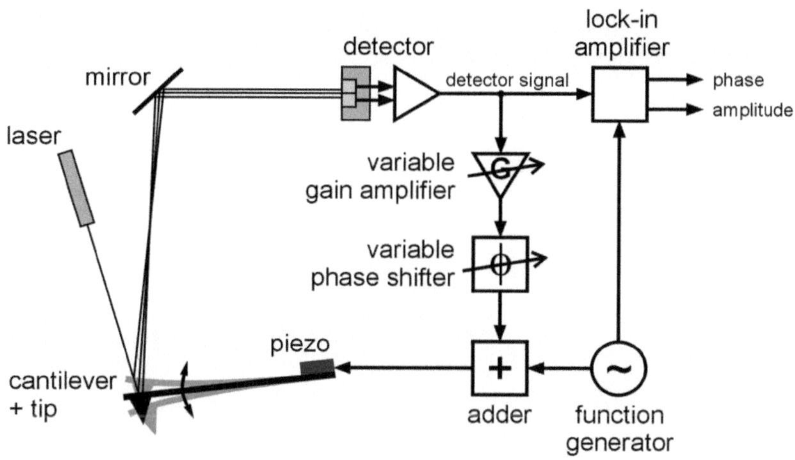

Fig. 1.15. Schematic diagram of operating a Q-control feedback circuit with an externally driven dynamic AFM. The tapping-mode set-up is in effect extended by an additional feedback loop

dominated by dissipative interactions between the cantilever and the surrounding medium, and Q values can be as low as 100 for air or even 5 in liquid. Still, we are asking, is it somehow possible to compensate for the damping effect by exciting the cantilever in a sophisticated way?

It turns out that the shape of the resonance curves in Fig. 1.5 can be influenced towards higher (or lower) Q values by an amplitude feedback loop. In principle, there a several mechanisms to couple the amplitude signal back to the cantilever, by the photothermal effect [32] or capacitive forces [33]. Figure 1.15 shows a method in which the amplitude feedback is mediated directly by the excitation piezo [34]. This has the advantage that no additional mechanical set-ups are necessary.

The working principle of the feedback loop can be understood by analyzing the equation of motion of the modified dynamic system:

$$m^*\ddot{z}(t) + \alpha\dot{z}(t) + kz(t) - F_{ts}(z_0 + z(t)) = F_{ext}\cos(\omega t) + Ge^{i\phi}z(t) \quad (1.26)$$

This *ansatz* takes into account the feedback of the detector signal through a phase shifter, amplifier, and adder as an additional force, which is linked to the cantilever deflection $z(t)$ through the gain G and the phase shift $e^{i\phi}$. To a good approximation we assume that the oscillation can be described by a harmonic trajectory. With a phase shift of $\phi = \pm\,\pi/2$ we find:

$$e^{\pm i\pi/2}z(t) = \pm\frac{1}{\omega}\,\dot{z}(t) \quad (1.27)$$

This means, that the additional feedback force signal $Ge^{i\phi}z(t)$ is proportional to the velocity of the cantilever, just like the damping term in the equation of motion. We can define an effective damping constant α_{eff}, which combines the two terms:

$$m^*\ddot{z}(t) + \alpha_{eff}\dot{z}(t) + kz(t) - F_{ts}(z_0 + z(t)) = F_{ext}\cos(\omega t)$$
$$\text{with}\quad \alpha_{eff} = \alpha \mp \frac{1}{\omega}G \quad \text{for}\quad \phi = \pm\frac{\pi}{2} \quad (1.28)$$

Fig. 1.16. Amplitude and phase diagrams measured in air with a Si-cantilever far away from the sample. The quality factor can be increased from 450 to 20,000 by using the Q-control feedback method

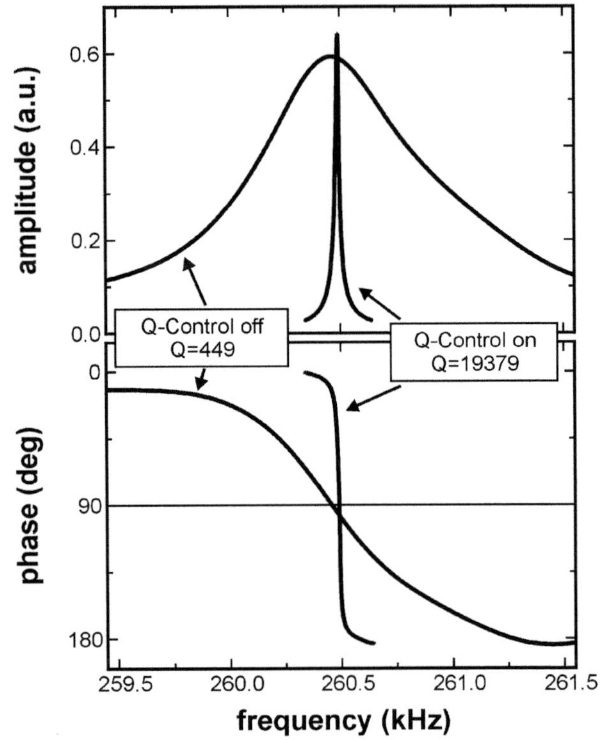

Equation 1.28 shows that the dampening of the oscillator can be enhanced or weakened by the choice of G, $\phi = +\pi/2$, or $\phi = -\pi/2$, respectively. The feedback loop therefore allows us to vary the effective quality factor $Q_{eff} = m\omega_0/\alpha_{eff}$ of the complete dynamic system. Hence, we term this system Q-control. Figure 1.16 shows experimental data on the effect of Q-control on the amplitude and phase as a function of the external excitation frequency [34]. While $Q = 449$ without Q-control the quality factor of the system operated in air is enhanced by a factor of more than 40 by applying the feedback loop.

The effect of improved image contrast is demonstrated in Fig. 1.17. Here a computer hard disk was analyzed with a magnetic tip in tapping mode, where the magnetic contrast is observed in the phase image. The upper part shows the recorded magnetic data structures in standard mode, whereas in the lower part of the image Q-control feedback was activated, giving rise to an improved signal, that is, magnetic contrast. A more detailed analysis of measurements on a magnetic tape shows that the signal amplitude (upper diagrams in Fig. 1.18), that is, the image contrast, was increased by a factor of 12.4 by the Q-control feedback. The lower image shows a noise analysis of the signal, indicating an increase of the signal-to-noise ratio by a factor of 2.3.

Note that the diagrams represent measurements in air with an AFM operated in AM mode. Only then we can make a distinction between excitation and vibration frequency, since in the FM mode these two frequencies are equal by definition. Al-

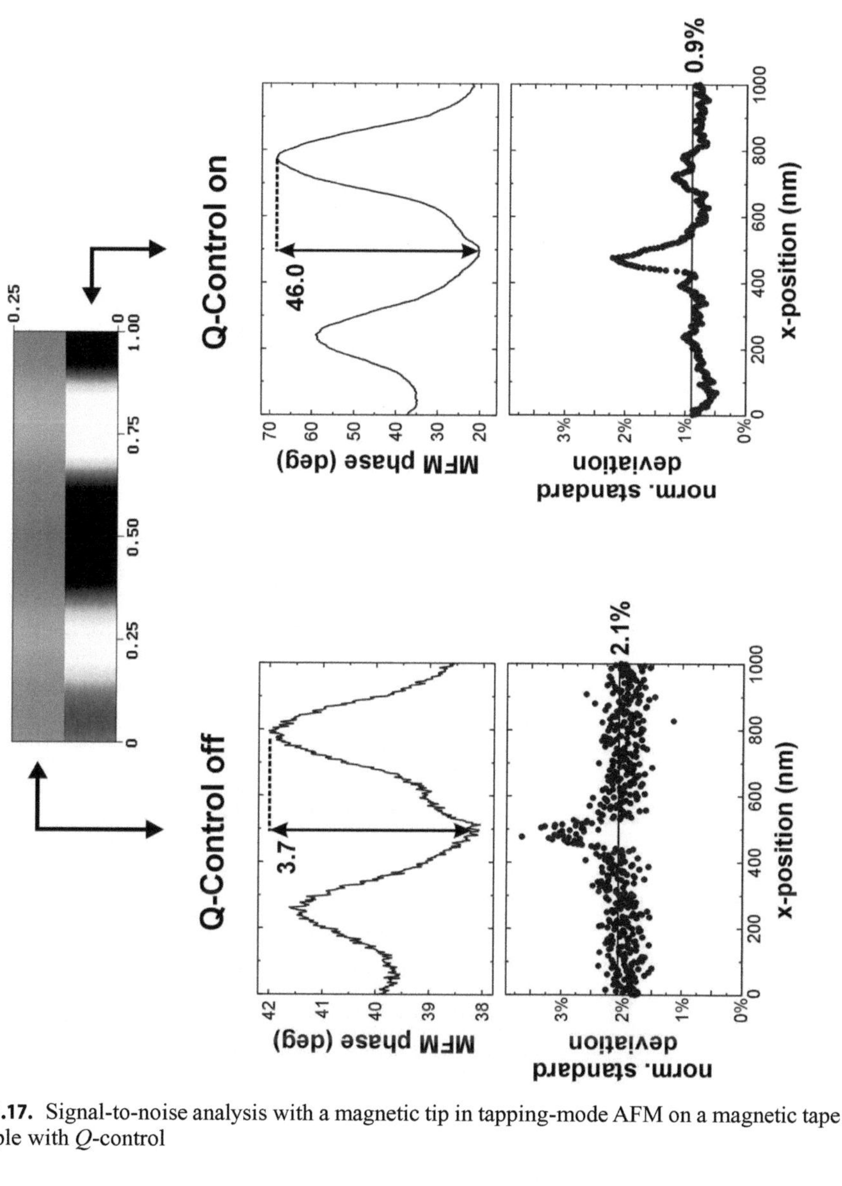

Fig. 1.17. Signal-to-noise analysis with a magnetic tip in tapping-mode AFM on a magnetic tape sample with Q-control

though the relationship between sensitivity and Q-factor in Eq. 1.17 is the same for AM and FM mode, it must be critically investigated if the enhanced quality factor by Q-control can be inserted in the equation for FM-mode AFM. In vacuum applications, Q is already very high, which makes it oblique to operate an additional Q-control module.

As stated before we can also use Q-control to enhance the dampening in the oscillating system. This would decrease the sensitivity of the system, but on the other hand the response time of the amplitude change is decreased as well. For tapping-

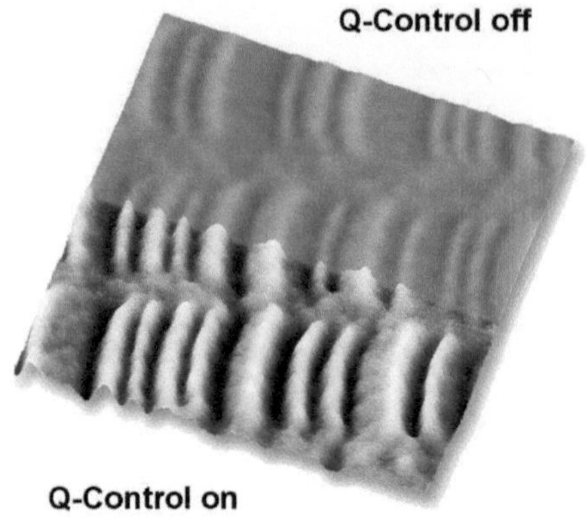

Fig. 1.18. Enhancement of the contrast in the phase channel due to Q-control on a magnetic hard disk measured with a magnetic tip in tapping-mode AFM in air. Scan size 5×5 µm, phase range 10° (www.nanoanalytics.com)

mode applications, where high-speed scanning is the goal, Q-control will be able to reduce the relaxation time [35].

A large quality factor Q does not only have the virtue of increasing the force sensitivity of the instrument. It has also the advantage of increasing the parameter space of stable AFM operation in tapping- or AM-mode AFM. Consider the resonance curve of Fig. 1.5. When approaching the tip towards the surface there are two competing mechanisms: on the one hand we bring the tip closer to the sample, which results in increasing attractive forces (see Fig. 1.2). On the other hand, for the case $\omega > \omega_0$, the resonant frequency of the cantilever is shifted towards smaller values due to the attractive forces, which causes the amplitude to become smaller preventing a tip-sample contact. This is the desirable regime, where stable operation of the AFM is possible in the net-attractive regime. But as explained before, below a certain tip-sample separation d_{app}, the system switches suddenly into the intermittent contact mode, where surface modifications are likely due to the onset of strong repulsive forces. The steeper the amplitude curve, the larger the regime of stable, net-attractive AFM operation. Looking at Fig. 1.5 we find that the slope of the amplitude curve is governed by the quality factor Q. A high Q therefore facilitates stable operation of the AM-AFM in the net-attractive regime.

An example can be found in Fig. 1.19. Here a surface scan of an ultrathin organic film is acquired in tapping mode under ambient conditions. First, the inner square is scanned without the Q enhancement, and then a wider surface area was scanned with applied Q-control. The high quality factor provides a larger parameter space for operating the AFM in the net-attractive regime, allowing good resolution of the delicate organic surface structure. Without the Q-control the surface structures are deformed and even destroyed due to the strong repulsive tip-sample interactions [36, 37, 38]. This also allowed imaging of DNA structures without predominantly depressing the soft material during imaging. It was then possible to observe a DNA diameter close to the theoretical value with the Q-control feedback [39].

In conclusion we have shown that by applying an additional feedback circuit to the dynamic AFM system it is possible to influence the quality-factor Q of the oscillator system. High-resolution, high-speed, or low-force scanning is then possible.

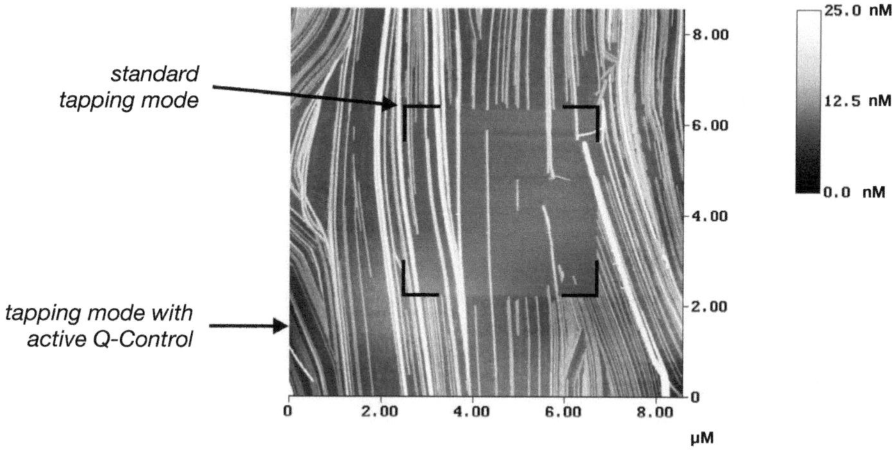

Fig. 1.19. Imaging of a delicate organic surface, a Langmuir-Blodgett film (ethyl-2,3-dihydroxy-octadecanoate) on a mica substrate, with Q-control. The topographical image clearly shows that the highly sensitive sample surface can only be imaged non-destructively with active with Q-control, whereas the periodic repulsive contact with the probe in standard operation without Q-control leads to a significant modification or destruction of the surface structure (data courtesy Dr Lifeng Chi and co-workers, Physikalisches Institut, Westfälische Wilhelms-Universität Münster, Germany)

1.6 Dissipation Processes Measured with Dynamic AFM

Dynamic AFM methods have proven their great potential for imaging surface structure at the nanoscale and we have also discussed methods which allow one to assess forces between distinct single molecules. However, there is another physical mechanism, which can be analyzed with the dynamic mode and has been mentioned in some paragraphs before: energy dissipation. In Fig. 1.12 we have already shown an example, where in tapping mode the phase signal cannot be explained by conservative forces alone, dissipative processes must also play a role. In constant-amplitude FM mode, where the quantitative interpretation of experiments has proven to be less difficult, there is an intuitive distinction between conservative and dissipative tip-sample interaction possible. While we have shown the correlation between forces and frequency shifts of the oscillating system, we have neglected one experimental input channel. The excitation amplitude, which is necessary to keep the oscillation amplitude constant, is a direct indication of the energy dissipated during one oscillation cycle. Dürig [40] has shown that in self-excitation mode (with an excitation-oscillation phase difference of 90°), conservative and dissipative interactions can be strictly separated. Partially this energy is dissipated in the cantilever itself, another part is due to external viscous forces in the surrounding medium. But more interestingly, some energy is dissipated at the tip-sample junction. This is the focus of the following paragraphs.

In contrast to conservative forces acting at the tip-sample junction, which at least in vacuum can be understood in terms of van der Waals, electrostatic, and chemical interactions, the dissipative processes are comparatively poorly understood. Stowe et al. [41] have shown that if a voltage potential is applied between tip and sample, charges are induced in the sample surface, which will follow the tip motion (here the oscillation is parallel to the surface). Due to the finite resistance of the sample material, energy will be dissipated during the charge movement. This effect has been exploited to image the doping level of semiconductors. Energy dissipation has also been observed in imaging magnetic materials. Liu et al. [42] found that energy dissipation due to magnetic interactions was enhanced at the boundaries of magnetic domains, which was accounted to domain wall oscillations. But also in the absence of external electromagnetic fields, energy dissipation was observed in close proximity of tip and sample within one nanometer. Clearly, mechanical surface relaxations must give rise to energy losses. One could model the AFM tip as a small hammer, hitting the surface at high frequency, possibly resulting in phonon excitations. From a continuum mechanics point of view, we assume that the mechanical relaxation of the surface is not just governed by elastic responses. Viscoelastic effects of soft surfaces will also render a significant contribution to energy dissipation. The whole area of phase imaging in tapping mode is concerned with those effects [43–46].

In the atomistic view the last tip atom can be envisaged to change position while yielding to the tip-sample force field. A strictly reversible change of position would not result in a loss of energy. Still, it has been pointed out by Sasaki et al. [47] that a change in atom position would result in a change in the force interaction itself. Therefore it is possible that the tip atom changes position at different tip-surface distances during approach and retraction, effectively causing an atomic-scale hysteresis to develop. In fact, Hoffmann et al. [12] have measured short-range energy dissipation for a tungsten tip on silicon in UHV. A theoretical model, explaining this effect on the basis of a two-energy-state system was developed. However, a clear understanding of the underlying physical mechanism is still lacking.

Nonetheless, the dissipation channel has been used to image surfaces with atomic resolution [48]. Instead of feedbacking the distance on the frequency shift, the excitation amplitude in FM mode has been used as the control signal. The Si(111)-(7 × 7) reconstruction was successfully imaged in this mode. The step edges of monoatomic NaCl islands on single-crystalline copper have also rendered atomic resolution contrast in the dissipation channel [49]. The dissipation processes discussed so far are mostly in the configuration where the tip is oscillated perpendicular to the surface. Friction is usually referred to as the energy loss due to lateral movement of solid bodies in contact. It is interesting to note in this context that Israelachvili [50] has pointed out a quantitative relationship between lateral and vertical (with respect to the surface) dissipation. He states that the hysteresis in vertical force-distance curves should equal the energy loss in lateral friction. An experimental confirmation of this conjecture at the molecular level is still missing.

Physical interpretation of energy dissipation processes at the atomic scale seems to be a daunting task at this point. Notwithstanding, we can find a quantitative relationship between the energy loss per oscillation cycle and the experimental parameters in dynamic AFM, as will be shown in the following section.

In static AFM it was found that permanent changes of the sample surface by indentations can cause a hysteresis between approach and retraction (e.g., see ref. [6]).

The area between the approach and retraction curves in a force-distance diagram represents the lost or dissipated energy, caused by the irreversible change of the surface structure. In dynamic-mode AFM the oscillation parameters like amplitude, frequency, and phase must contain the information about the dissipated energy per cycle. So far, we have resorted to a treatment of the equation of motion of the cantilever vibration, in order to find a quantitative correlation between forces and the experimental parameters. For the dissipation it is useful to treat the system from the energy conservation point of view.

By assuming a dynamic system is in equilibrium, the average energy input must equal the average energy output or dissipation. By applying this rule to an AFM running in a dynamic mode means that the average power fed into the cantilever oscillation by an external driver, denoted as $\bar{P}_{in}$, must equal the average power dissipated by the motion of the cantilever beam $\bar{P}_0$ and by tip-sample interaction $\bar{P}_{tip}$:

$$\bar{P}_{in} = \bar{P}_0 + \bar{P}_{tip} \tag{1.29}$$

The term $\bar{P}_{tip}$ is what we are interested in, since it gives us a direct physical quantity to characterize the tip-sample interaction. Therefore we have first to calculate and then measure the two other terms in Eq. 1.29, in order to determine the power dissipated when the tip periodically probes the sample surface. This requires an appropriate rheological model to describe the dynamic system. Although there are investigations in which the complete flexural motion of the cantilever beam has been considered [51], a simplified model, comprising a spring and two dashpots (Fig. 1.20), represents a good approximation in this case [52].

The spring, characterized by the constant k according to Hooke's law, represents the only channel through which power P_{in} can be delivered to the oscillating tip $z(t)$ by the external driver $z_d(t)$. Therefore the instantaneous power fed into the dynamic system is equal to the force exerted by the driver times the velocity of the driver (the force that is necessary to move the base side of the dashpot can be neglected, since this power is directly dissipated and therefore does not contribute to the power delivered to the oscillating tip):

$$P_{in}(t) = F_d(t)\, \dot{z}_d(t) = k\,[z(t) - z_d(t)]\, \dot{z}_d(t) \tag{1.30}$$

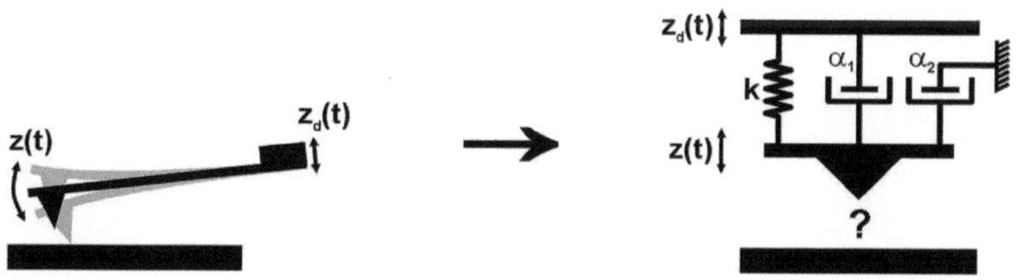

Fig. 1.20. Rheological model applied to describe the dynamic AFM system, comprising the oscillating cantilever and tip interacting with the sample surface. The movement of the cantilever base and the tip is denoted as $z_d(t)$ and $z(t)$, respectively. The cantilever is characterized by the spring constant k and the damping constant α. In a first approach damping is broken into two pieces α_1 and α_2: first, intrinsic damping caused by the movement of the cantilever's tip relative to its base; and second, damping related to the movement of the cantilever body in a surrounding medium, e.g. air damping

Assuming a sinusoidal steady-state response and that the base of the cantilever is driven sinusoidally (see Eq. 1.6) with amplitude A_d and frequency ω, the deflection from equilibrium of the end of the cantilever follows Eq. 1.9, where A and $0 \le \varphi \le \pi$ are the oscillation amplitude and phase shift, respectively. This allows us to calculate the average power input per oscillation cycle by integrating Eq. 1.30 over one period $T = 2\pi/\omega$:

$$\bar{P}_{in} = \frac{1}{T}\int_0^T P_{in}(t)\, dt = \frac{1}{2} k\omega A_d A \sin\varphi \tag{1.31}$$

This contains the familiar result that the maximum power is delivered to an oscillator when the response is 90° out of phase with the drive.

The simplified rheological model as it is depicted in Fig. 1.20 exhibits two major contributions to the damping term $\bar{P}_0$. Both are related to the motion of the cantilever body and assumed to be well modeled by viscous damping with coefficients α_1 and α_2. The dominant damping mechanism in UHV conditions is intrinsic damping, caused by the deflection of the cantilever beam (i.e., the motion of the tip relative to the cantilever base). Therefore the instantaneous power dissipated by such a mechanism is given by

$$P_{01}(t) = |F_{01}(t)\, \dot{z}(t)| = |\, \alpha_1 [\dot{z}(t) - \dot{z}_d(t)]\, \dot{z}(t)| \tag{1.32}$$

Note that the absolute value has to be calculated, since all dissipated power is "lost" and therefore cannot be returned to the dynamic system.

However, when running an AFM in ambient conditions an additional damping mechanism has to be considered. Damping due to the motion of the cantilever body in the surrounding medium (e.g., air damping), is in most cases the dominant effect. The corresponding instantaneous power dissipation is given by

$$P_{02}(t) = |F_{02}(t)\, \dot{z}(t)| = \alpha_2\, \dot{z}^2(t) \tag{1.33}$$

In order to calculate the average power dissipation, Eq. 1.32 and Eq. 1.33 have to be integrated over one complete oscillation cycle. This yields

$$\begin{aligned} \bar{P}_{01} &= \frac{1}{T}\int_0^T P_{01}(t)\, dt \\ &= \frac{1}{\pi}\alpha_1\omega^2 A \left[(A - A_d\cos\varphi) \arcsin\left(\frac{A - A_d\cos\varphi}{\sqrt{A^2 + A_d^2 - 2AA_d\cos\varphi}} \right) + A_d \sin\varphi \right] \end{aligned} \tag{1.34}$$

and

$$\bar{P}_{02} = \frac{1}{T}\int_0^T P_{02}(t)\, dt = \frac{1}{2}\, \alpha_2\omega^2 A^2 \tag{1.35}$$

Considering the fact that commonly used cantilevers exhibit a quality factor of at least several hundreds (in UHV several ten thousands), we can assume that the oscillation amplitude is significantly larger than the drive amplitude when the dynamic system is driven at or near its resonant frequency: $A \gg A_d$. Therefore Eq. 1.34 can be simplified in first-order approximation to an expression similar to Eq. 1.35. Combining the two equations yields the total average power dissipated by the oscillating cantilever

$$\bar{P}_0 = \frac{1}{2}\alpha\omega^2 A^2 \quad \text{with} \quad \alpha = \alpha_1 + \alpha_2 \tag{1.36}$$

where α denotes the overall effective damping constant.

We can now solve Eq. 1.29 for the power dissipation localized to the small interaction volume of the probing tip with the sample surface, represented by the question mark in Fig. 1.20. Furthermore by expressing the damping constant α in terms of experimentally accessible quantities such as the spring constant k, the quality factor Q, and the natural resonant frequency ω_0 of the free oscillating cantilever, $\alpha = \dfrac{k}{Q \cdot \omega_0}$, we obtain:

$$\bar{P}_{tip} = \bar{P}_{in} - \bar{P}_0 = \frac{1}{2}\frac{k\omega}{Q}\left[Q_{cant} A_d A \sin\varphi - A^2 \frac{\omega}{\omega_0}\right] \tag{1.37}$$

Note that so far no assumptions have been made on how the AFM is operated, except that the motion of the oscillating cantilever has to remain sinusoidal to a good approximation. Therefore Eq. 1.37 is applicable to a variety of different dynamic AFM modes.

For example, in FM-mode AFM the oscillation frequency ω changes due to tip-sample interaction, while at the same time the oscillation amplitude A is kept constant by adjusting the drive amplitude A_d. By measuring these quantities one can apply Eq. 1.37 to determine the average power dissipation related to tip-sample interaction. In spectroscopy applications usually $A_d\,(d)$ is not measured directly, but a signal $G(d)$ proportional to $A_d\,(d)$ is acquired representing the gain factor applied to the excitation piezo. With the help of Eq. 1.15 we can write:

$$A_d(d) = \frac{A_0 \cdot G(d)}{Q \cdot G_0} \tag{1.38}$$

while A_0 and G_0 are the amplitude and gain at large tip-sample distances where the tip-sample interactions are negligible.

Now let us consider the tapping-mode or AM-AFM. In this case the cantilever is driven at a fixed frequency and with constant drive amplitude, while the oscillation amplitude and phase shift may change when the probing tip interacts with the sample surface. Assuming that the oscillation frequency is chosen to be ω_0, Eq. 1.37 can be further simplified again by employing Eq. 1.15 for the free oscillation amplitude A_0. This yields

$$\bar{P}_{tip} = \frac{1}{2}\frac{k\omega_0}{Q_{cant}}\left[A_0 A \sin\varphi - A^2\right] \tag{1.39}$$

Equation 1.39 implies that if the oscillation amplitude A is kept constant by a feedback loop, like it is commonly done in tapping mode, simultaneously acquired phase data can be interpreted in terms of energy dissipation [43, 44, 53, 54]. When

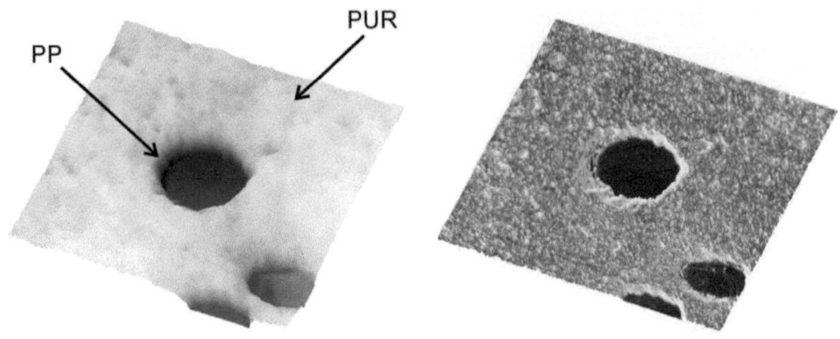

Fig. 1.21. Topography and phase-image in tapping-mode AFM of a polymer blend composed of polypropylene (PP) particles embedded in a polyurethane (PUR)-matrix. The dissipation image shows a strong contrast between the harder PP (little dissipation, dark) to the softer PUR (large dissipation, bright) surface

analyzing such phase images [50, 55, 56] one has also to consider the fact that the phase may also change due to the transition from net-attractive ($\varphi > 90°$) to intermittent contact ($\varphi < 90°$) interaction between the tip and the sample [15, 34, 57, 58]. For example consider Fig. 1.12, in which the phase shift in tapping mode is given as a function of z-position. If phase measurements are performed close to the point at which the oscillation switches from the net-attractive to the intermittent contact regime, a large contrast in the phase channel is observed. However, this contrast is not due to dissipative processes. Only a variation of the phase signal within the intermittent contact regime will give information of the tip-sample dissipative processes.

An example of a dissipation measurement is depicted in Fig. 1.21. The surface of a polymer blend was imaged in air, simultaneously acquiring the topography and dissipation. The dissipation on the softer polyurethane matrix is significantly larger than on the embedded, mechanically stiffer, polypropylene particles.

1.7 Conclusion

Dynamic force microscopy is a powerful tool, which is capable of imaging surfaces with atomic precision. It also allows us to look at surface dynamics and it can operate in vacuum, air, or even in liquid. However, the oscillating cantilever system introduces a level of complexity, which disallows a straightforward interpretation of acquired images. An exception is the self-excitation mode, where tip sample forces can be successfully extracted from spectroscopic experiments. However not just conservative forces can be investigated with dynamic AFM: energy dissipation also influences the cantilever oscillation and can therefore serve as a new information channel.

Open questions are still concerned with exact geometric and chemical identity of the probing tip, which significantly influences the imaging and spectroscopic results. Use of predefined tips like single-walled nanotubes or use of atomic resolution techniques like field ion microscopy to image the tip itself are possible approaches addressing this issue. Furthermore, only little is known about the dissipative processes at the scale where only few atoms between tip and sample interact. It is also desirable to learn more about the interpretation of images acquired in tapping mode.

References

1. Lantz MA, Hug HJ, Hoffmann R, van Schendel PJA, Kappenberger P, Martin S, Baratoff A, Güntherodt H-J (2001) Quantitative measurement of short-range chemical bonding forces. Science 291:2580–2583
2. Binnig G, Quate CF, Gerber CH (1986) Atomic force microscope. Phys Rev Lett 56:930–933
3. Marti O (1999) AFM instrumentation and tips. In: Bushan B (ed) Handbook of micro/nanotribology, 2nd edn, CRC Press, pp 81–144
4. Cross G, Schirmeisen A, Stalder A, Grütter P, Tschudy M, Dürig U (1998) Adhesion interaction between atoically defined tip and sample. Phys Rev Lett 80:4685–4688
5. Schirmeisen A, Cross G, Stalder A, Grütter P, Dürig U (2000) Metallic adhesion and tunneling at the atomic scale. New J Phys 2:29.1–29.10
6. Schirmeisen A (1999) Metallic adhesion and tunneling at the atomic scale. PhD thesis, McGill University, Montréal, Canada, pp 29–38
7. Giessibl FJ (1997) Forces and frequency shifts in atomic-resolution dynamic-force microscopy. Phys Rev B 56:16010–16015
8. Giessibl FJ (1995) Atomic resolution of the silicon (111)-(7 × 7) surface by atomic force microscopy. Science 267:68–71
9. Bezanilaa M, Drake B, Nudler E, Kashlev M, Hansma PK, Hansma HG (1994) Motion and enzymatic degradation of dna in the atomic force microscope. Biophys J 67:2454–2459
10. Jiao Y, Cherny DI, Heim G, Jovin TM, Schäffer TE (2001) Dynamic interactions of p53 with DNA in solution by time-lapse atomic force microscopy. J Mol Biol 314:233–243
11. Albrecht TR, Grütter P, Horne D, Rugar D (1991) Frequency modulation detection using high-Q cantilevers for enhanced force microscopy sensitivity. J Appl Phys 69:668–673
12. Hoffmann PM, Jeffery S, Pethica JB, Özer HÖ, Oral A (2001) Energy dissipation in atomic force microscopy and atomic loss processes. Phys Rev Lett 87:265502–265505
13. Jarvis SP, Lantz MA, Dürig U, Tokumoto H (1999) Off resonance AC mode force spectroscopy and imaging with an atomic force microscope. Appl Surf Sci 140:309–313
14. Anczykowski B, Krüger D, Fuchs H (1996) Cantilever dynamics in quasinoncontact force microscopy: spectroscopic aspects. Phys Rev B 53:15485–15488
15. Anczykowski B, Krüger D, Babcock KL, Fuchs H (1996) Basic properties of dynamic force spectroscopy with the scanning force microscope in experiment and simulation. Ultramicroscopy 66:251–259
16. Hughes BD, White LR (1979) 'Soft' contact problems in linear elasticity. Quart J Mech Appl Math 32:445–471
17. Muller VM, Yushchenko VS, Derjaguin BV (1980) On the influence of molecular forces on the deformation of an elastic sphere and its sticking to a rigid plane. J Coll Interf Sci 77:91–101
18. Verlet L (1967) Computer "experiments" on classical fluids I: thermodynamical properties of Lennard-Jones molecules. Phys Rev 159:98–103; Verlet L (1968) Computer "experiments" on classical fluids II: equilibrium correlation functions. Phys Rev 165:201–214
19. Uchihasi T, Ishida T, Komiyama M, Ashino M, Sugawara Y, Mizutani W, Yokoyama K, Morita S, Tokumoto H, Ishikawa M (2000) High-resolution imaging of organic monolayers using noncontact AFM. Appl Surf Sci 157:244–250
20. Gleyzes P, Kuo PK, Boccara AC (1991) Bistable behavior of a vibrating tip near a solid surface. Appl Phys Lett 58:2989–2991

21. San Paulo A, García R (2000) High-resolution imaging of antibodies by tapping-mode atomic force microscopy: attractive and repulsive tip-sample interaction regimes. Biophys J 78: 1599–1605
22. Krüger D, Anczykowski B, Fuchs H (1997) Physical properties of dynamic force microscopies in contact and noncontact operation. Ann Phys 6:341–363
23. Martin Y, Williams CC, Wickramasinghe HK (1987) Atomic force microscope force mapping and profiling on a sub 100-A scale. J Appl Phys 61:4723–4729
24. Barth C, Reichling M (2001) Imaging the atomic arrangement on the high-temperature reconstructed α-$Al_2O_3(0001)$ surface. Nature 414:54–57
25. Gotsmann B, Fuchs H (2001) Dynamic force spectroscopy of conservative and dissipative forces in an Al-Au(111) tip-sample system. Phys Rev Lett 86:2597–2600
26. Dürig U (1999) Relations between interaction force and frequency shift in large-amplitude dynamic force microscopy. Appl Phys Lett 75:433–435
27. Hölscher H, Allers W, Schwarz UD, Schwarz A, Wiesendanger R (1999) Determination of tip-sample interaction potentials by dynamic force spectroscopy. Phys Rev Lett 83:4780–4783
28. Hölscher H, Schwarz A, Allers W, Schwarz UD, Wiesendanger R (2000) Quantitative analysis of dynamic-force-spectroscopy data on graphite(0001) in the contact and noncontact regime. Phys Rev B 61:12678–12681
29. Guggisberg M, Bammerlin M, Meyer E, Güntherodt H-J (2000) Separation of interactions by noncontact force microscopy. Phys Rev B 61:11151–11155
30. Dürig U (2000) Extracting interaction forces and complementary observables in dynamic probe microscopy. Appl Phys Lett 76:1203–1205
31. http://sxm4.uni-muenster.de
32. Mertz J, Marti O, Mlynek J (1993) Regulation of a microcantilever response by force feedback. Appl Phys Lett 62:2344–2346
33. Rugar D, Grütter P (1991) Mechanical parametric amplification and thermomechanical noise squeezing. Phys Rev Lett 67:699–702
34. Anczykowski B, Cleveland JP, Krüger D, Elings VB, Fuchs H (1998) Analysis of the interaction mechanisms in dynamic mode SFM by means of experimental data and computer simulation. Appl Phys A 66:885–889
35. Sulchek T, Yaralioglu GG, Quate CF, Minne SC (2002) Characterization and optimisation of scan speed for tapping-mode atmoic force microscopy. Rev Sci Instr 73:2928–2936
36. Chi LF, Jacobi S, Anczykowski B, Overs M, Schäfer H-J, Fuchs H (2000) Supermolecular periodic structures in monolayers. Adv Mater 12:25–30
37. Gao S, Chi LF, Lenhert S, Anczykowski B, Niemeyer C, Adler M, Fuchs H (2001) High-quality mapping of DNA-protein complexes by dynamic scanning force microscopy. Chem Phys Chem 6:384–388
38. Zou B, Wang M, Qiu D, Zhang X, Chi LF, Fuchs H (2002) Confined supramolecular nanostructures of mesogen-bearing amphiphiles. Chem Commun 9:1008–1009
39. Pignataro B, Chi LF, Gao S, Anczykowski B, Niemeyer C, Adler M, Fuchs H (2002) Dynamic scanning force microscopy study of self-assembled DNA-protein nanostructures. Appl Phys A 74:447–452
40. Dürig U (2000) Interaction sensing in dynamic force microscopy. New J Phys 2:5.1–5.12
41. Stowe TD, Kenny TW, Thomson DJ, Rugar D (1999) Silicon dopant imaging by dissipation force microscopy. Appl Phys Lett 75:2785–2787
42. Liu Y, Grütter P (1998) Magnetic dissipation force microscopy studies of magnetic materials. J Appl Phys 83:7333–7338
43. Anczykowski B, Gotsmann B, Fuchs H, Cleveland JP, Elings VB (1999) How to measure energy dissipation in dynamic mode atomic force microscopy. Appl Surf Sci 140:376–382
44. Cleveland JP, Anczykowski B, Schmid AE, Elings VB (1998) Energy dissipation in tapping-mode atomic force microscopy. Appl Phys Lett 72:2613–2615
45. García R, Tamayo J, Calleja M, García F (1998) Phase contrast in tapping-mode scanning force microscopy. Appl Phys A 66:S309–S312
46. Tamayo J, García R (1997) Effects of elastic and inelastic interactions on phase contrast images in tapping-mode scanning force microscopy. Appl Phys Lett 71:2394–2396

47. Sasaki N, Tsukada M (2000) Effect of microscopic nonconservative process on noncontact atomic force microscopy. Jpn J Appl Phys 39:L1334–L1337
48. Lüthi R, Meyer E, Bammerlin M, Baratoff A, Howald L, Gerber C, Güntherodt H-J (1997) Ultrahigh vacuum atomic force microscopy: true atomic resolution. Surf Rev Lett 4:1025–1029
49. Bennewitz R, Foster AS, Kantorovich LN, Bammerlin M, Loppacher CH, Schär S, Guggisberg M, Meyer E, Shluger AL (2000) Atomically resolved edges and kinks of NaCl islands on Cu(111): experiment and theory. Phys Rev B 62:2074–2084
50. Israelachvili J (1992) Intermolecular and surface forces. Academic Press
51. Rabe U, Turner J, Arnold W (1998) Analysis of the high-frequency response of atomic force microscope cantilevers. Appl Phys A 66:S277–S282
52. Rodríguez TR, García R (2002) Tip motion in amplitude modulation (tapping-mode) atomic-force microscopy: Comparison between continuous and point-mass models. Appl Phys Lett 80:1646–1648
53. García R, Tamayo J, San Paulo A (1999) Phase contrast and surface energy hysteresis in tapping mode scanning force microsopy. Surf Interface Anal 27:312–316
54. Tamayo J, García R (1998) Relationship between phase shift and energy dissipation in tapping-mode scanning force microscopy. Appl Phys Lett 73:2926–2928
55. Chen X, McGurk SL, Davies MC, Roberts CJ, Shakesheff KM, Tendler SJB, Williams PM, Davies J, Dwakes AC, Domb A (1998) Chemical and morphological analysis of surface enrichment in a biodegradable polymer blend by phase-detection imaging atomic force microscopy. Macromolecules 31:2278–2283
56. Pickering JP, Vancso GJ (1998) Apparent contrast reversal in tapping mode atomic force microscope images on films of polystyrene-b-polyisoprene-b-polystyrene. Polymer Bulletin 40:549–554
57. Kühle A, Sørensen AH, Bohr J (1997) Role of attractive forces in tapping tip force microscopy. J Appl Phys 81:6562–6569
58. Kühle A, Sørensen AH, Zandbergen JB, Bohr J (1998) Contrast artifacts in tapping tip atomic force microscopy. Appl Phys A 66:S329–S332

2 Interfacial Force Microscopy: Selected Applications

Jack E. Houston

2.1 Introduction

The interfacial force microscope (IFM) was developed in the early 1990s out of a research program aimed at fundamental studies of interfacial adhesion. The desire was to evaluate the adhesive bond and its failure involving controlled interfacial surfaces. One of the most important aspects of this interfacial control involves morphological defects, which are most easily handled by utilizing a very small surface for at least one of the interfaces, for example, by using a small probe. The atomic force microscope (AFM) had recently burst onto the scene [1, 2] and, thus, all the necessary control features were already in place. The idea then was to measure the interfacial force as a function of the relative separation of the two surfaces. The behavior of this force upon approach would give a "fingerprint" of the bonding (e.g., van der Waals, electrostatic, covalent, etc.) and the withdrawal would give information on the bond failure, providing a detailed measure of the interfacial bonding process.

Unfortunately, all AFM instruments measured (and still measure) the interfacial force with a displacement force sensor, that is, by measuring the deflection of an elastic element with the force given by the spring constant times the degree of deflection. This widely used sensor scheme is extremely simple and sensitive but has several serious shortcomings. First, the act of force measurement changes the interfacial separation and does not give a direct measure of the force versus separation. Second, if the change in separation is large enough the sensor will become unstable and the two surfaces will uncontrollably crash together. And finally, the measurement of force stores elastic energy in the sensor so that when some relaxation process occurs (e.g., dislocation nucleation) this energy is dumped into the relaxation making it difficult to study the fundamental aspects of the process. The second of these shortcomings causes a large portion of the approach and withdrawal behavior to be lost to the sensor instability, especially for a high-sensitivity sensor. Thus, in order to proceed with the adhesion program, it was necessary to develop a stable force sensor. Again, the instability problem arises because the sensor has to deflect in order to sense the force and, if the force increases as a result of the motion, it can produce an uncontrollable situation. The solution involves eliminating the deflection. This can be done with a self-balancing, force-feedback arrangement and it is this type of sensor that is at the heart of the interfacial force microscope (IFM) [3]. In the present chapter, the details of the sensor design and operation will be briefly discussed followed by the presentation of several examples of recent applications. These examples are chosen not only to illustrate the kind of unique information that can be obtained, but also to demonstrate the type and breadth of materials problems that can be approached. In addition, they are specifically aimed at showing the close relationship between adhesive, nanomechanical, and tribological properties.

The majority of the examples presented will involve published results from the IFM group at Sandia. The space available in the present chapter, unfortunately, does not lend itself to covering the efforts from several other groups, which involve a very broad set of IFM applications to an unusual and sometimes very practical group of materials and materials problems. These will be briefly outlined and referenced in the closing remarks. Where appropriate, additional data and interpretation will be blended utilizing later unpublished results and these will be indicated as such.

2.2 The IFM Sensor

There are any number of ways to implement a self-balancing force sensor. The two ingredients common to all such devices are a deflection sensing technique to detect the presence of a force and some scheme for opposing the force in order to null the deflection. The arrangement that was chosen for the initial IFM sensor is illustrated in Fig. 2.1 [3]. It consists of a differential capacitor with a common plate suspended by a set of torsional springs. The common plate is suspended above two conductive pads defining the capacitors C_1 and C_2. A tip is attached to one side of the common plate such that, when an upward vertical force is applied to it, the common plate will rotate counter clockwise increasing the value of C_2 and decreasing C_1. This motion is detected by an RF-bridge whose ac output is proportional to the degree of imbalance. The deflection could also be detected by a laser lever or in-

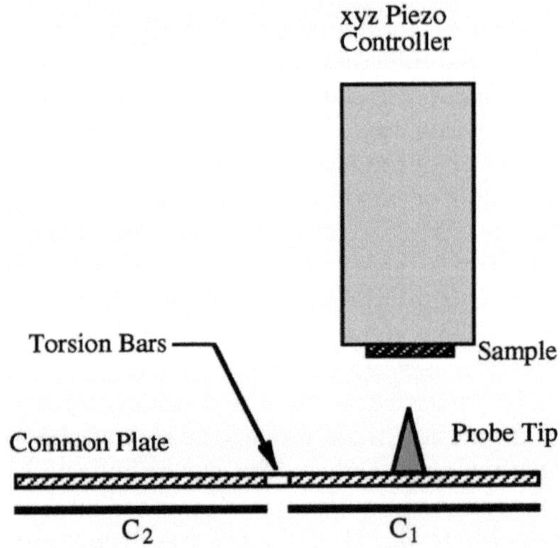

Fig. 2.1. A schematic diagram of the IFM sensor. The common plate of a differential capacitor is suspended above two capacitor pads C_1 and C_2 by torsion bars. A sharp probe is placed on one side of the common plate, which can interact with a sample mounted on a xyz piezo controller. Such interactions are detected by changes in the differential capacitance and balanced by appropriate voltages applied to the capacitor pads

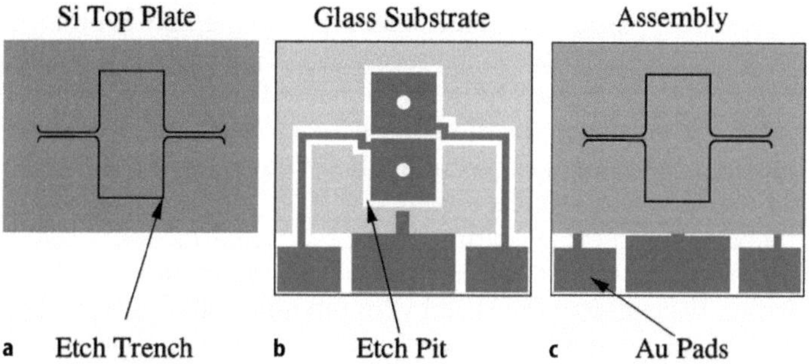

Fig. 2.2a–c. A schematic representation of the sensor assembly. **a** A 25-μm trench etched completely through a 100-μm-thick Si(001) rectangle defines the capacitor common plate and its torsion bars; **b** the Pyrex substrate showing the deposited Au capacitor pads, electrical leads and bonding pads, which are located in an approximately 5-μm pit etched into the glass; **c** the assembled sensor consisting of the top plate and substrate anodically bonded together. The common plate makes electrical contact with the center pad during the bonding

terferometer technique, but the RF-bridge scheme is simple, self-contained and quite sensitive. The ac bridge-imbalance signal is synchronously demodulated and sent to a full PID (proportional-integral-derivative) controller, which balances the bridge by applying the proper voltages to the capacitor pads. The electrostatic forces on the pads create torques that balance those resulting from the applied tip forces, thus balancing the sensor. Within the frequency response of the controller, and the mechanical properties of this classic second-order system (mass on a spring with damping), the applied force is automatically balanced by the feedback force and the sensor does not move. Thus, the mechanical instability is eliminated.

In its present form, the sensor is constructed in two parts. The common plate and torsion bars, illustrated in Fig. 2.2a, are defined by a 25-μm-wide trench etched through a 100-μm Si(001) wafer by a Bosch process [4]. The substrate is made of Pyrex glass and consists of Au capacitor pads, leads, and contacts deposited in an approximately 5-μm pit that defines the capacitor gap (Fig. 2.2b). The two pieces are then assembled by anodic bonding (Fig. 2.2c) [5].

Of course, the operation is not quite as simple as indicated above. One of the problems with this particular sensor design is that voltages applied to the capacitors result in only attractive electrostatic forces, which arises from the fact that the force varies with the square of the applied voltage. If an attractive force is applied between tip and sample in Fig. 2.1, it can be balanced by the proper voltage applied across C_1, whereas, a repulsive tip/sample force requires that the voltage be applied to C_2. However, balancing the repulsive force in this manner results in the bending of the common plate, as well as its downward displacement. The torsion bars are designed to have a small torsional spring constant and stiff springs in both the bending and up/down modes. However, the sensor still moves as a result of the applied forces.

The situation can be partially corrected, and the force/voltage relationship linearized, by applying a constant voltage V_0 to both pads and having the PID control voltage applied with opposite sign to the pads, for example, $V_0 + V$ on C_1 and $V_0 - V$ on C_2 (this is referred to as double-sided feedback). Under these circumstances, the sensor force applied to the tip will be proportional to V_0V and, since the total torque remains constant during the application of the tip/sample force, the bending of the common plate is eliminated. However, the total of the forces on the sensor are not balanced and the up/down motion remains. That is, the two electrostatic forces placed on the pads are in opposite directions, balancing the total torque, but the common plate has a net vertical force equal to the applied force. Thus, the common plate will still displace vertically by an amount determined by the vertical spring constant of the torsion bars. While small, this motion could result in instability under certain circumstances.

The situation can be completely corrected by applying the PID control voltage *only* to the capacitor pad directly beneath the tip (i.e., C_1 in Fig. 2.1; this mode is referred to as single-sided feedback). Now, both the total torque and the net vertical force remain balanced during the application of a tip/sample force. Under these circumstances, the sensor effectively has "zero compliance". When a tip/sample force is applied, the sensor balances it but does not move. Thus, the sensor instability is eliminated. The price that is paid for this mode of operation is that the relationship between the PID controller voltage and applied force is no longer linear, but is proportional to $V_0V - 0.5V^2$. Since in the present controller the PID output voltage V ranges ± 10 V, V_0 must be equal to 10 V in order to keep the system stable when V is near its maximum range.

In single-sided feedback, the force applied to the probe is given by $F_\mathrm{p} = \mathrm{c}G_\mathrm{f}(V_0V - 0.5V^2)$, where G_f is the sensor's force constant expressed in force per volt squared and c is a constant depending on the location of the probe (i.e., the lever arm for the torque that the applied force produces). G_f is dependent *only* on the capacitance values C_1 and C_2. In its present form, the sensor has capacitances of approximately 15 pf resulting in G_f values of about 1.5 μN V^{-2}. The transient response can be tuned to have a rise time of less than about 1 ms and the noise floor is approximately 2 nN with a maximum force capability ranging to near 0.5 mN. Because the force sensitivity depends only on the sensor geometry, it is a constant for a particular sensor and is easily calibrated with a laboratory balance in order to produce quantitative results.

To illustrate the effect of the force feedback in eliminating the sensor instability, Fig. 2.3 shows a direct comparison of the force profile (or loading curve) taken on a Si(001) surface with an approximately 300-nm radius tip.

Such data is taken by hovering on the surface at a small normal force, retracting by a predetermined amount (≈30 nm in Fig. 2.3), approaching the surface at a constant rate (2.0 nm s^{-1}), followed by withdrawal at the same rate after a preset normal force is attained (1.0 μN in Fig. 2.3). The Si surface has been chemically cleaned and the data were taken in air. The results shown in Fig. 2.3a, probe force in μN plotted against relative tip displacement in nm, illustrate the fingerprint of a contaminated surface commonly observed under a normal laboratory environment taken under single-sided feedback. It consists of a smooth force variation, extending to a range of about 8 nm, with considerable adhesive hysteresis upon withdrawal. In contrast, Fig. 2.3b illustrates the behavior in the absence of feedback. This mode of operation represents the results for a simple displacement sensor

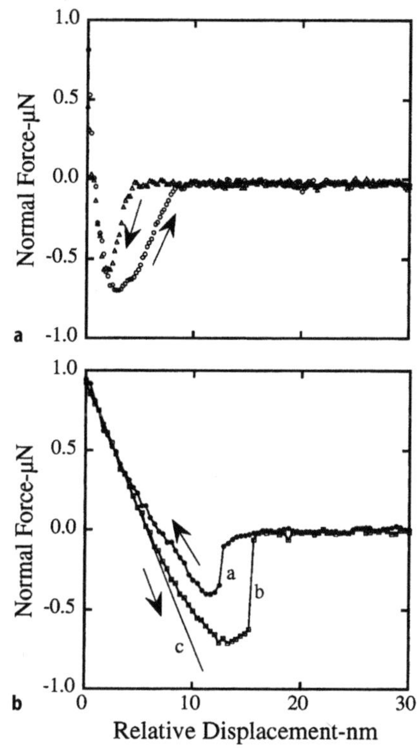

Fig. 2.3a,b. A comparison of the force profiles for a chemically cleaned Si sample taken in air with **a** single-sided feedback and **b** with no feedback (deflection mode). In the absence of feedback, snap in/snap out instabilities are clearly evident (regions *a* and *b*) as well as the influence of the torsion-bar compliance (lower slope in region *c*)

and is similar to that commonly reported in, for example, AFM adhesion studies, although with a stiffer spring [6]. It can readily be seen that the profile is dominated by both jump-to-contact and pull-off instabilities, (points a and b, respectively) as well as the rotational spring constant of the torsion bars (identified by the nearly linear slope in the region c). Clearly, the instabilities eliminate large portions of the details of the bonding interaction and its failure, as well as the true compliance behavior of the tip/sample interaction. In addition, the quantitative relationship between relative displacement and the true deformation of the tip, sample, and any intervening material depends on an accurate knowledge of the spring constant of the torsion bars, as well as all the subsequent electronics, which are very tedious to calibrate.

The results of Fig. 2.3 demonstrate the primary advantage of the IFM sensor: its elimination of sensor instability. Some additional advantages include 1) the quantitative nature of the sensor, resulting from its dependence only on geometry, and ease of calibration using a laboratory balance, 2) the fact that there is no sensor motion as a result of a force measurement, that is, it is effectively a "zero-compliance" sensor with no energy being stored in the sensor during for measurements, and 3) since it balances the total torque applied to the tip, it is sensitive to both normal and lateral forces. However, the measurement of lateral and normal forces is not independent. In order to decompose the total measured torque into normal and lateral tip-applied forces, two schemes have been developed. The first involves a "friction-loop" method, which relies on the fact that the lateral force depends on the direction of tip motion, and changes sign as the motion is reversed, while the normal force does not. The second uses a small lateral modulation of the tip position and separates the lateral and normal forces in the frequency domain.

In the remainder of this chapter, these advantages will be illustrated with the results of applications ranging from interfacial adhesion through nanomechanical properties to friction for self-assembled molecular-monolayer lubricants. These illustrations were chosen to cover the range from research-oriented applications to problems which impact more practical materials problems.

2.3 Selected Applications

2.3.1 Adhesion

There is hardly an aspect of everyday life that is not impacted in some way by interfacial adhesion, its formation and failure, and there are countless examples of studies in this area (for a general introduction see, for example, ref. [7]). In this section, several studies of the subject are outlined using the unique capabilities of the IFM. The separation of the subject of adhesion from that of mechanical properties is somewhat misleading, since there are few examples where the mechanics of the support system for the interfacial surfaces is not a major contributor to the adhesive strength, and this relationship will be illustrated in the adhesion examples. In contrast, the nanomechanics section will concentrate on examples where at least one of the surfaces has been passivated to minimize the adhesion allowing the study of these properties virtually in the absence of adhesive bonding. The first example involves "fingerprint" force profiles delineating a particular type of interfacial interaction (van der Waals) characterized by the force versus separation behavior prior to interfacial contract.

Early in the IFM development, the use of self-assembled molecular monolayer (SAM) films was introduced as a means of producing controlled interfacial surfaces [8–10]. Au is a reasonably inert metal and the use of alkanethiol SAMs produces well-ordered and closely packed layers because the reasonably strong thiolate bond to the Au tends to displace the weakly bound hydrocarbon materials usually found as contaminants on the surface after normal cleaning.

Figure 2.4 shows an example of a 5.1-μm parabolic W probe interacting with a octadecanethiol self-assembled molecular adsorbed on a Au surface. The resulting long-range attractive force has been fit using the expression for the van der Waals force between a sphere and a plane normalized by the tip radius of curvature (i.e., ref. [11]),

$$\frac{F}{R} = -\frac{A}{6\,(d - d_0)^2}. \tag{2.1}$$

The comparison is demonstrated by the solid line in Fig. 2.4 and can be seen to be quite close, within the limits imposed by the signal-to-noise ratio.

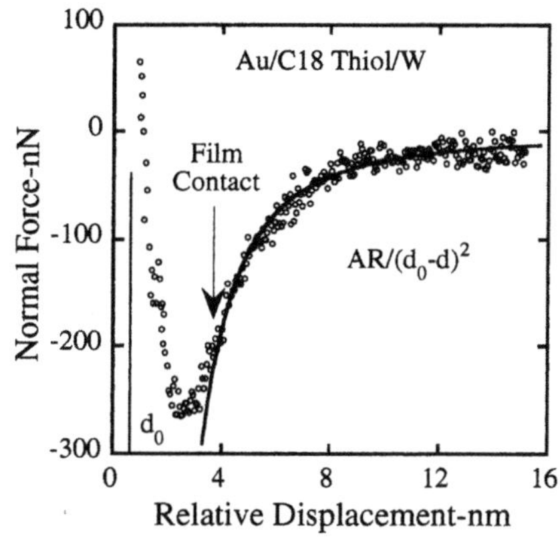

Fig. 2.4. The approach portion of the force profile for an 5.1-μm-radius W tip interacting with a Au surface functionalized with a octadecanethiol ($SH(CH_2)_{17}CH_3$) self-assembled monolayer film. The *solid line* represents a simple van der Waals fit to the data for a Hamaker constant of 1.9×10^{-19} J and a relative-displacement offset parameter d_0 of 1.6 nm

The fit results in a value for the Hamaker constant and offset parameter of $A = 1.9 \times 10^{-19}$ J and $d_0 = 1.6$ nm, respectively. The A value is a little high for hydrocarbon interactions [11, 12], but the value of d_0 suggests that, at small separations, the principal contribution comes from the van der Waals interaction between the Au substrate and the W probe. Supporting this is the fact that the distance from d_0 to the point where the curve first deviates from van der Waals behavior is approximately 3.7 nm, as shown by the vertical line in Fig. 2.4, which is somewhat larger than the nominal thickness of the thiol film of about 2.5 nm. A better comparison would require detailed calculations of this multilayer combination [11], which includes the mechanical properties of the monolayer, and is certainly beyond the scope of the present discussion. Deviation from the fit at smaller distances is, of course, due to the physical contact between the probe and the SAM surface, where the interfacial force is still attractive but just beginning its repulsive rise.

In earlier work, force profiles, of the type illustrated in Fig. 2.4, were used to obtain quantitative values for the bonding strength between self-assembled monolayer films functionalized with various combinations of chemically distinct end groups [13]. In these experiments, Au probes and substrates are both coated with SAM films and the chemical functionality of the outermost groups, or end groups, was cycled through various pairs of three groups (i.e., methyl ($-CH_3$), amine ($-NH_2$) and carboxylic acid (–COOH)). The maximum adhesive force values, normalized by the tip radius, were used along with contact mechanics models to obtain values for the strength of the interfacial interactions. If the equilibrium area of the contact is known, then it is possible to integrate force curves of the kind shown in Fig. 2.4 to obtain the *actual* work of adhesion per unit of surface area. However, because of the small tip size used, this quantity is not known for the scanning force microscopies, so it is common to use simple contact-mechanics models to relate the peak attractive force to the work of adhesion. Two reasonably simple and intuitive models have often been used to estimate the work of adhesion from the peak value of the attractive force (the so-called pull-off force). The Johnson-Kendall-Roberts (JKR) [14] model considers only contact forces and takes into account the assumed elastic properties of the substrate under the adhesive interaction using the Hertz model [15]. In contrast, the Derjaguin-Muller-Toporov

(DMT) model [16] includes long-range forces but effectively has them act on the tip as a whole, again using the Hertz model for the substrate compliance. This latter model has often been considered to give better quantitative values for the experimental work of adhesion and was used in the earlier work to obtain such values from the force-profile data.

In the DMT model, the work of adhesion is given by the relationship [16]

$$W_{12} = \frac{F}{2\pi R}, \tag{2.2}$$

where F/R refers to the peak attractive force divided by the tip radius of curvature and W_{12} is the work of adhesion. By using this approach, the work of adhesion for the various pairs of end groups is calculated from the force-profile data and the results are presented, along with other parameters, in Table 2.1.

The F/R values in Table 2.1 represent the peak attractive force in the withdrawal portion of the force profiles normalized by the probe tip radius, W_{12} is the work of adhesion calculated from these values through Eq. (2.2), γ is the surface energy calculated as half the work of adhesion for the symmetric combinations. For the asymmetric NH_2/COOH case, the interfacial energy γ_{12} is obtained from the Dupre equation [11], that is,

$$W_{12} = \gamma_1 + \gamma_2 - \gamma_{12}. \tag{2.3}$$

The intermolecular bond energies are calculated from the work of adhesion per unit of surface area and the known unit cell area of the thiol molecules ($\approx 0.21\ \text{nm}^2$) [17] after accounting for the van der Waals contribution using the Fowkes relationship [18], that is,

$$W = 2\,\gamma_{\text{vdw}} + 2\,\gamma_{\text{H-bond}}\,. \tag{2.4}$$

The value of γ_{vdw} is taken to be approximated by that for the CH_3/CH_3 interaction. As an example, for the NH_2 versus NH_2 pair, $\gamma_{\text{H-bond}}$ turns out to be about 40 mJ m^{-2} (i.e., $100 - 2 \times 30 = 40$ mJ m^{-2} from Table 2.1). Multiplying this by the unit cell area and converting to bond energy per mole yields the tabulated figure of approximately 1.3 kcal mol^{-1}. The other two bond energies are calculated in a similar

Table 2.1. The normalized peak attractive force (pull-off force) F/R was used, along with the DMT relationship of Eq. (2.2), to estimate the work-of-adhesion values W_{12}. For the symmetric combinations (W_{11} and W_{22}), the surface energy values γ are half the work of adhesion. For the asymmetric NH_2/COOH case, Eq. (2.3) was used to calculate the interfacial energy γ_{12}. The bond energies per mole and similar values, obtained from more classical chemical procedures, are calculated by procedures described in the text. Since the van der Waals interaction is not a "two-center" bond, its bond energy is not applicable

Interaction combination	F/R (N m^{-}1)	W_{12} (mJ m^{-2})	γ (mJ m^{-2})	Bond energy (kcal mol^{-1})	Comparison (kcal mol^{-1})
CH_3/CH_3	0.4	60	30	NA	NA
NH_2/NH_2	0.7	100	50	1.3	3.4
COOH/COOH	1.4	228	114	5.0	14.0
NH_2/COOH	4.3	680	–516 (γ_{12})	16.0	23.0

manner. The "comparison" values shown in the last column of Table 2.1 are taken from gas-phase dimerization measurements involving methylamine (CH_3NH_2) for NH_2–NH_2 and formic acid (HCOOH) for COOH–COOH interactions [19], while the NH_2–COOH value comes from calorimetry work [20].

For the methyl–methyl interaction, where only van der Waals forces are involved, the calculated work of adhesion yields twice the surface energy, since individual bonds are not formed. The value of 30 mJ m^{-2} agrees well with the figure of 28 mJ m^{-2} obtained for crystalline hydrocarbon films by the surface forces apparatus [21] and values ranging from 23–31 mJ m^{-2} obtained from bulk contact-angle measurements [22].

For the other symmetric combinations, the higher surface energies reflect the contribution, beyond the van der Waals level, of intermolecular hydrogen bonding. Bonding of this kind is often discussed in terms of acid/base or donor/acceptor interactions and involves the formation of a bond by the sharing of electrons from a Lewis base (donor) and a Lewis acid (acceptor) [23]. In the cases discussed here, the base (or donor) sites are the lone-pair electrons at the oxygen or nitrogen atoms forming the amine (NH_2) or carboxylic acid (COOH) end groups, while the acid (or acceptor) sites are at the partially ionized H sites. Thus, surfaces consisting of molecules terminated by these groups can exhibit both acid and base character. If the surfaces are pure, or "monopolar" acid or base groups, then the adhesion between symmetric combinations of them would essentially involve only van der Waals interactions. From the data for these two symmetric combinations in Table 2.1, it can be concluded that the NH_2–NH_2 interaction (–NH_2 is known to be principally basic in nature) involves only a weak acid/base bond. In contrast, the symmetric carboxylic acid (COOH) interaction shows essentially four times the bond strength and indicates a considerably increased acid/base contribution.

In the case of the asymmetric amine/carboxylic acid combination, which now involves a base directly interacting with an acid, Table 2.1 indicates a considerably stronger interfacial bond and a large negative value for the interfacial energy γ_{12} of Eq. (2.3). This negative value represents the energy gained due to the strong acid/base interaction and is the difference between the observed work of adhesion and the combined surface energies of the symmetric combinations of the amine and carboxylic acid surfaces. The negative value is an unusual occurrence, since such systems are normally unstable to the formation of a bimolecular solution making the formation of an extended interface impossible. However in this case, the stability is provided by the strong thiolate bond to the substrate, which maintains the self-assembled structure.

The agreement of the results of Table 2.1 with similar values obtained from more classical chemical means is surprisingly close, although the experimental values are always smaller. This discrepancy is undoubtedly due to a combination of steric inhibition in the monolayer-film contact, that is, the fact that the end groups might not meet across the interface in an orientation suitable for efficient bond formation, and to the fact that not all bonds have time to form during the force-profile measurement. The first of these has recently been dramatically demonstrated in friction and bonding measurements for COOH-terminated SAM films with odd versus even numbers of methylene units (CH_2) in the alkyl backbones of the self-assembled molecules, as will be discussed later [24].

Another application along a similar vein involves an analysis of the electrostatic force between a Au probe and Au substrate in the presence of liquid hexadecane [25]. It has been shown by several authors that liquid hexadecane forms a structured "lying-down" phase near, for example, the surface of an Au substrate due to the strong metal/hydrocarbon van der Waals interaction [26–28]. The purpose of the study was to investigate the behavior of this "interphase" region near the Au surfaces with respect to temperature and the imposition of an electric field. Figure 2.5a shows the force profile, during both approach and withdrawal, for an approximately 1-μm-radius Au tip interacting with a single-crystal Au substrate, both of which are immersed in liquid hexadecane, with no voltage between the sample and tip.

The presence of the hexadecane liquid virtually eliminates both the strong metal/metal as well as the much weaker van der Waals interaction between the two Au surfaces (i.e., no negative force values) and a small hysteresis is noted between the approach and withdrawal curves. The hysteresis indicates some sort of energy loss has taken place as a result of compressing the two surfaces and the solid-like intervening interphase. If a 2-V potential is placed between tip and sample, the behavior shown in Fig. 2.5b results. A reasonably large attractive interaction, and a greater degree of hysteresis, is clearly evident. Also shown in Fig. 2.5b are the electrostatic fits (solid lines) from the relationship expected for the electrostatic interaction from the Derjaguin approximation [16],

$$F = 2\pi R \frac{K\varepsilon_0}{(d - d_0)} V^2 = \frac{\mathrm{B}}{(d - d_0)} V^2, \tag{2.5}$$

where R is the tip radius, K the dielectric constant of the liquid, ε_0 is the free-space permittivity, V the applied voltage, and d_0 is the offset parameter. The constant B identifies the strength of the interaction with units of nN nm V^{-2}.

The fits result in two parameters, the strength coefficient for the interaction and the offset parameter, which identifies the point on the relative separation axis at which the two metal surfaces would make contact in the absence of the interphase film. The offset parameter for the approach curve, along with the interphase contact point (i.e., the relative separation at which the force begins to turn repulsive; given by the arrow in Fig. 2.5b) indicates an interphase thickness of about 3.6 nm, or about 7 molecular layers [29]. The hysteresis seen in Fig. 2.5b between the approach and withdrawal curves indicates that the interphase film is disturbed by the approaching tip, so that withdrawal takes place with a thinner film and a smaller offset parameter. However, even after some plastic deformation of the initial film, there still remains a solid-like film of approximately 2.3-nm total thickness, or about three molecular diameters. It is interesting to note from Fig. 2.5a that this interphase film passivates the normally strong interaction between the tip and sample even at applied stresses capable of damaging the Au surfaces.

The values for the strength coefficient $\mathrm{B} = 2\pi RK\varepsilon_0$ turn out to be –460 and –380 nN m V^{-2} for the approach and withdrawal, respectively. With a measured tip radius of approximately 1 μm, the only parameter in the coefficient is the value of K, which is calculated to be 2 and 1.7 for the approach and withdrawal, respectively (the handbook value is two). The difference between the two values would indicate that upon withdrawal the hexadecane in the vicinity of the tip is not as well orga-

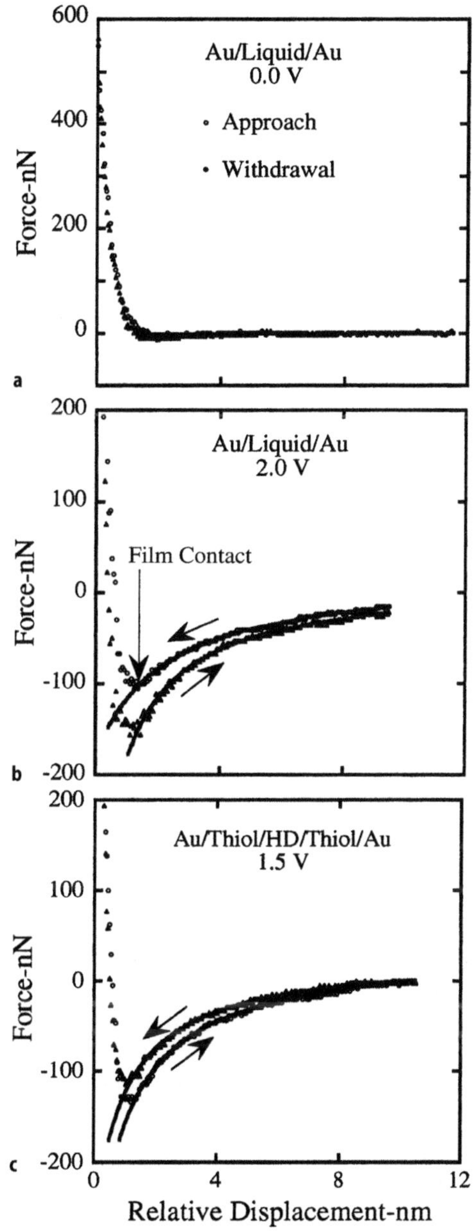

Fig. 2.5a–c. The approach and withdrawal force profiles for an Au tip and single-crystal Au surface immersed in liquid hexadecane. **a** result with zero bias on the sample, which indicates essentially no attractive interaction and only a small hysteresis between approach and withdrawal; **b** shows the same combination with a 2-V bias on the substrate. The electrostatic fits yield strength coefficients of 460 and 380 nN nm V^{-2} and corresponding relative displacement offset parameters of 11.8 and 10.4 nm for the approach and withdrawal curves, respectively. The *solid line* indicates the approximate position of the tip/interphase-film contact; **c** presents similar results with the addition of a hexadecanethiol SAM adsorbed onto the substrate surface. Here the electrostatic fits yield strength parameters of 317 and 400 nN nm V^{-2} for the approach and withdrawal curves, respectively, and the same 11.5-nm value for the offset parameter

nized as upon approach and has a somewhat smaller density and, thus, smaller dielectric constant.

The results shown in Fig. 2.5c are for the same Au tip/substrate combination with a 1.5-V bias on the substrate, but with the substrate surface covered with a hexadecanethiol SAM film. The interposed SAM film is expected to dramatically reduce the strong van der Waals interaction between the Au substrate and the liquid hexadecane. The parameters from the electrostatic fits confirm this expectation. The offset parameters turn out to be the same for both approach and withdrawal predicting a film thickness of approximately 2.5 nm, which is only slightly larger than the approximate 2.1-nm thickness of the SAM film, or an interphase thickness of only approximately one layer. In addition, the strength parameters turn out to be 317 and 400 nN nm V^{-2}, indicating a slight increase in the dielectric constant for the withdrawal curve – as if the field was slightly increasing the local density. Thus, the conclusion is that the hexadecanethiol SAM has virtually eliminated the liquid interphase for the interaction with only a single Au surface. These two examples clearly demonstrate the kind of detailed information that is available in the interfacial interaction profiles when probed by a quantitative and stable force sensor.

The value of being able to quantitatively study the behavior of the adhesive interaction involving reasonably well-characterized interfacial surfaces over the entire range of interfacial separation is hard to overemphasize. There is an enormous amount of information available in this seemingly simple measurement. Of course, as is always the case, the results are strongly dependent on the careful control of the experimental conditions, for example, substrate roughness, the care given to the tip/substrate preparation, the tip shape and size, and the quality of the environment (humidity, solvent impurities, etc.). Characterizing the force behavior as a function of interfacial separation in terms of specific surface-force models (e.g., van der Waals, electrostatic, hydrophobic, etc.) is difficult, since in any real system the force is made up of combinations of sources. In addition, as mentioned earlier, the use of scanning probes to characterize an interfacial interaction in terms of an energy-of-adhesion value (energy per unit area of interface) is difficult because of a lack of an accurate measure of the equilibrium contact area. However, an accurate value of the energy of adhesion can be obtained by integrating the withdrawal profile from the equilibrium position (the point at which the normal force is zero) to complete separation, and similar measurements done with the same tip for differing interfacial conditions can be of enormous value.

2.3.2 Nanomechanical Properties

Along with the use of SAM films to study adhesion, as discussed earlier, it was discovered that these films offered a high degree of chemical passivation of the Au surface [10]. In the absence of passivation, the interaction of a clean W tip and Au surface is violent due to high surface energies and negative interfacial energies. At the monolayer level, Au forms a strained layer lattice on W and would rather be on the W surface than its own [30]. The interaction is illustrated in Fig. 2.6a for data taken with a single-crystal Au(111) surface and a 400-nm-radius W tip in a UHV

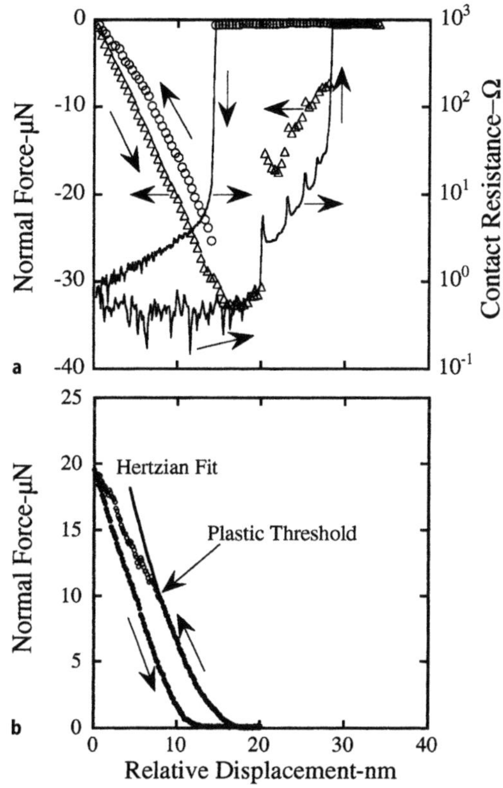

Fig. 2.6a,b. Force profiles for the interaction of a Au(111) surface and 400-nm-radius W tip, both cleaned in an ultra-high vacuum environment: **a** shows the behavior after initial contact between the clean surfaces for the force, indicated by *symbols*, and the contact resistance, shown as the *solid line*. The contact resistance was measured with a 1 µA, constant-current source; **b** illustrates the behavior of a similar contact involving a single-crystal Au(111) surface covered by a hexadecanethiol self-assembled monolayer film. The adhesive bond found in **a** is virtually eliminated by the passivating SAM even after an approximately 3.0-nm indentation of the Au surface. The *solid line* represents the Hertzian fit to the approach portion of the profile

IFM system with sputter cleaning facilities for both the substrate and tip [31]. Both force and contact resistance are displayed for a 1-V maximum, constant-current source maintaining a 1-µA current. Approach and withdrawal are indicated by arrows.

Here it can be seen that only small van der Waals and electrostatic forces (hard to identify on the force scale presented in the figure) are present up until the surfaces are very close to contact, at which point the surfaces suddenly jump into contact up to an attractive force of approximately –26 µN, while the contact resistance falls to approximately 20 Ω.

The initial jump-to-contact results from the compliance of the tip and sample (not the sensor) and is typical of a JKR-type contact [14]. Following the initial jump, the Au surface is somewhat plastic as indicated by the further decrease in contact resistance down to about 6 Ω – a behavior not described by the JKR model, which includes only elastic interactions. The total plastic deformation is clearly seen in the hysteresis loop upon withdrawal and amounts to approximately 4 nm. The force profile then reaches a maximum attractive force of about 32 µN, during which time the contact resistance remains essentially constant. Past the maximum, the plastic behavior is typical of the formation of a "nanowire" and its subsequent deformation and eventual fracture [32]. This plasticity involves several relaxation events with a sudden decrease in the force and increase in the contact resistance. In addition, immediately after each event there is a small increase in attractive force and decrease in contact resistance indicating a brief healing of the nanowire, probably by Au surface diffusion.

In the presence of an alkanethiol SAM assembled on the Au substrate surface, however, the strong bonding is eliminated and only very small van der Waals forces are observed, as illustrated in Fig. 2.6b. In addition, the passivating properties persisted even under stresses large enough to seriously damage the Au surface, which is illustrated by the hysteresis seen in Fig. 2.6b. Here, the initial elastic deformation is closely fit by the classic Hertzian relationship for a parabolic probe interacting with a planar surface, as illustrated by the solid line in Fig. 2.6b [15]. The elastic behavior is followed by a plastic deformation to a depth of about 3 nm. Note also that after the indentation there is no hint of a bonding interaction. The SAM molecules have become trapped between the tip and substrate by the nested methylene groups and continue to passivate the tip/substrate bonding even under these severe stress conditions.

This latter observation presented an opportunity to use the IFM in an indenter mode in order to study the mechanical properties of the Au lattice at the nanometer level. The reason that this is particularly interesting stems from the fact that the mechanical properties of materials at small scales are often found to be quite different than at larger scales [33, 34]. One reason for this scaling behavior is that crystalline defects strongly influence mechanical properties and, as the sample volume decreases, the likelihood of encountering defects also decreases [35]. This effect has already been extensively exploited in an earlier body of work with whiskers, but sample preparation and handling presented a major drawback [36]. This fact has allowed the measurement of the theoretical lattice strength for Au in several IFM nanoindentation studies [37–42], as well as a characterization by embedded-atom-type theoretical modeling [43].

To gain a better understanding of how these defect-free nanoscopic measurements change when more macroscopic conditions exist, singular, well-defined surface defects were introduced into the indentation stress fields to identify the role of individual defects in reducing the yield stresses [44]. By measuring the effect of one such defect (in the form of a single surface step), it is possible to begin bridging the gap between the shear strengths of ideal crystals (about 1 GPa) and macroscopic, well-annealed single crystals (about 1 MPa [45]). Whereas some nanoindentation work on this very effect preceded these efforts [46], the IFM measurements were performed systematically on these simple defects as a function of probe/defect separation.

Again, using hexadecanethiol-coated single crystals and flame-annealed microballs [47, 48], targeted regions with wide terraces, such as that shown in Fig. 2.7a, can be identified.

This figure represents a constant repulsive-force image of the surface covering an area 750 × 750 nm. Two steps can be seen: a single step having a height of 0.25 nm and a double step with a 0.5-nm height. Intrastep separations within multistep defects are smaller than the lateral resolution of the imaging technique (in this case, about 10 nm), such that step bunches appeared as single step edges. Once suitable steps are identified, indentations are made at specified probe-step separations. As

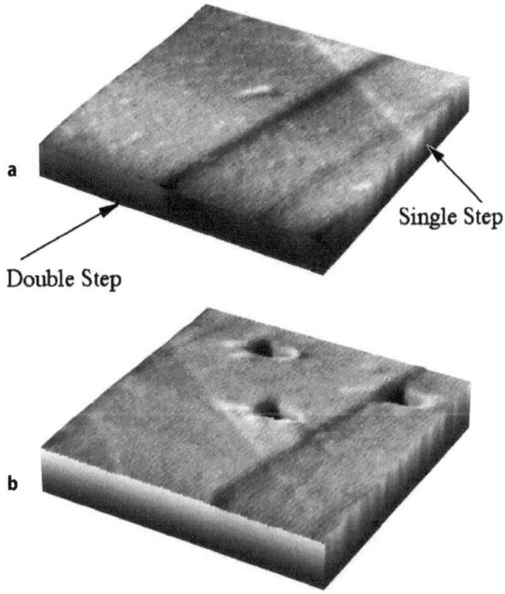

Fig. 2.7a,b. Constant repulsive-force images of a single-crystal Au(111) surface covered by a hexadecane-thiol SAM: **a** illustrates the surface before indentation and shows several steps running across the surface; **b** shows the result of three plastic indentations of the surface at various positions with respect to the surface steps

an example, Fig. 2.7b shows three indentations made in close proximity both to each other and to several steps.

Measurements of this kind indicated that the step lowers the yield point considerably and the length over which this effect occurs was larger than expected. As an example of the effect, Fig. 2.8 shows force profiles contrasting the mechanical behavior far from and close to a single surface step.

Figure 2.8a illustrates the result for an indentation made approximately 200 nm away from the nearest step. Here it can be seen that the force builds up according to the expected Hertzian relationship up to a level just above 15 μN, at which point it suddenly relaxes to zero, followed by a decreasing series of minor relaxation events. In contrast, the behavior of Fig. 2.8b, taken near the step, shows a slightly decreased maximum force at initial yield and the degree of relaxation is considerably smaller. Note that the total indentation depth (width of the hysteresis loop) is 25–30 nm in these two cases.

Sudden relaxations of the sort shown if Fig. 2.8 are commonly seen in nanoindentation. However, these are normally observed as so-called pop in events, in which the force remains constant while the displacement suddenly jumps deeper into the material [49]. The difference in behavior here results from the fact that essentially all indentation instrumentation is operated in a force mode, where a spring-supported probe is loaded at a constant displacement rate while measuring the resulting force (or load) by the degree of spring deflection. When the material suddenly relaxes in a plastic event, the spring dumps its stored energy into the plastic process and the tip moves into the material, while the force remains virtually constant. In contrast, the IFM is a displacement-controlled device, where no energy is stored in the act of force measurement. Thus, when plastic relaxation occurs, the force suddenly falls without tip motion.

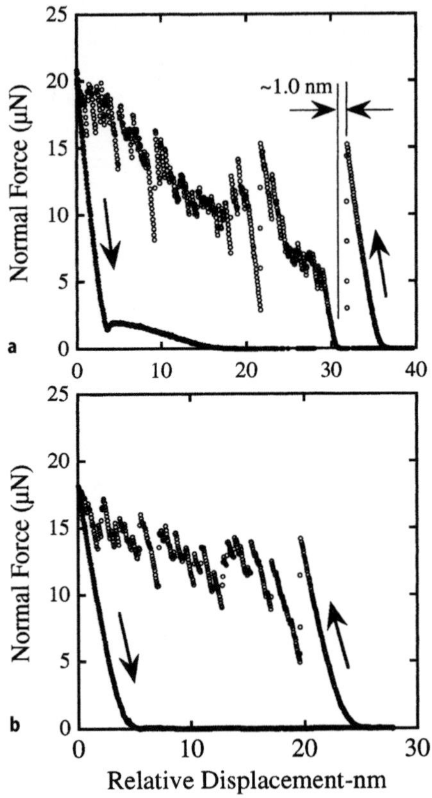

Fig. 2.8a,b. Force profiles for a W tip indenting a surface similar to that shown in Fig. 2.7: **a** illustrates the plastic behavior of the surface with the tip located about 200 nm from the nearest step. Initially, the force rises according to the Hertzian relation until reaching a force just over 15 μN, at which point the force completely relaxes, loosing contact with the indented surface by approximately 1.0 nm, as illustrated by the *vertical solid lines*. Further indentation involves minor force relaxations and the total depth of the indent after withdrawal is about 30 nm; **b** illustrates the behavior for an indentation very near a step. The neighboring step reduces the peak force and results in a considerably less dramatic relaxation event. The total indentation depth here is about 20 nm

The advantage of the displacement mode is that individual relaxation events can be detected because no energy is available to cause the relaxation process to proceed beyond the initial event. This fact was used in earlier work to show that for tip radii below about 400 nm, the stress necessary for the initial nucleation of dislocations unexpectedly varies as the reciprocal of tip radius [39]. One of the remarkable aspects of the result of Fig. 2.8a, taken far from any steps, is that the first event involves a complete relaxation of the tip-applied force. In fact, after this event, the tip has to move about an additional 1.0 nm in order to catch up with the relaxed surface. To the author's knowledge, this is the first time such an effect has been seen and it is a tribute to a true nanoindentation experiment done with a displacement device using a non-compliant sensor.

The degree of stress relaxation depends critically on the ratio of the force required to nucleate the dislocation-relaxation process and that necessary to continue the dislocation-motion once nucleation has occurred. The first of these involves a barrier characterized by the unstable stacking-fault energy [50] while the second results from the internal friction of the dislocation motion [51, 52]. If the relaxation process is viewed as analogous to a block on a plane pulled by a force acting through a spring, the initial spring extension to produce sliding would be that necessary to overcome the static friction given by the static friction coefficient times

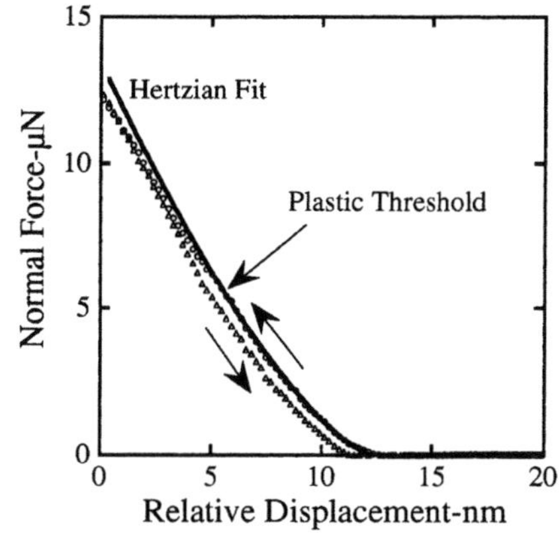

Fig. 2.9. A force profile illustrating a minor relaxation event. In this case, the relaxation does not represent a sudden fall in normal force and the eventual percentage of the force decrease is less than 0.5%

the weight of the block divided by the spring constant. How far the block travels after sliding begins, under the influence of the stretched spring in this "stick-slip" motion, depends on the ratio of the static to sliding friction coefficients. If this ratio is greater than two, the block will slide a distance larger than the spring elongation and over-run the driven end. Apparently, this is the situation for the indentation illustrated in Fig. 2.8a. The effect of a neighboring step in Fig. 2.8b is to reduce the "static" friction coefficient and, thus, produce a less than complete relaxation, as was found in the calculations.

At this point, it is not clear what gives rise to the overshoot of the relaxing material shown in Fig. 2.8a. However, as was pointed out in earlier work, minor events occasionally appear in IFM experiments of the type shown in Fig. 2.9, where the relaxation is not sudden and its extent is small [38].

The relative frequency of these events was found to decrease as the tip radius increases, virtually disappearing at tip radii of about 200 nm. In contrast, the major events of the kind shown in Fig. 2.8a produce considerable pileup of the kind normally associated with indentation. In addition, the general shape of the pileup was found to reflect the surface symmetry of the particular Au lattice orientation probed [37].

This behavior is in stark contrast to that found in an embedded-atom modeling study, which involved a tip modeled as a parabolic, repulsive surface, having a radius of a few tens of nm and a repulsive tip/substrate interaction [43]. The analysis of the results of this study utilized a centrosymmetry scheme to increase the contrast for those atoms not occupying normal-lattice sites. This allowed a direct visualization of both partial and full dislocations, along with "snapshots" illustrating the position of their initial origins. The results showed that the minor events involved a pair of dislocations both nucleated at the points of maximum resolved shear stress below the tip. The dislocations glide along (111) slip planes, intersect the surface at the contact periphery and eventually collide and become locked below the tip along its axis. For the initial nucleation event, the force relaxation was observed to be only about 20%, in contrast with the results of Fig. 2.8. Thus, in this case, the extent of the relaxation may ultimately be controlled by the tip size as a result of this dislocation-locking mechanism. Since the experimental results of

Fig. 2.8 involved tips an order of magnitude larger, it may be that this "locking" limit no longer applies and that there is enough force, and energy stored in the deformed lattice, to drive the relaxation to the point where cross slip takes over and the dislocations end up at the surface outside the contact periphery, creating pileup. Considerably more measurements and larger-scale modeling will have to be done before a definitive answer to this question is obtained.

Figure 2.10 summarizes the results of the step study for many individual indentations done with various tip radii, step heights, and at various distances (*d*) from neighboring steps. The average normal stress at the yield point is plotted as a function of distance from the step to the center axis of the probe. The mean stress is calculated from the relationship,

$$\bar{\sigma}_y = \frac{F}{\pi R \, \delta_y}, \tag{2.6}$$

where $\bar{\sigma}_y$ is the calculated mean stress, F the applied force, R the tip radius and δ_y is the surface deformation at yield. The two broken lines identify the range of yield stress found on defect-free Au(111) crystals [37], which agree favorably with this study's measurement when $d > 100$ nm. However, it can be seen from Fig. 2.10a that whenever d is less than about 100 nm the yield stresses are found to be lowered. The typical contact diameter ($2a_y$) for one data point is depicted as a horizontal error bar, which is intended to illustrate that, even when the contact area did not overlap the step, yield stresses were lowered. Ultimately, when the two overlap, $\bar{\sigma}_y$ is decreased by about 50% from its defect-free value. Note also that there does not appear to be significant asymmetry (within the limits set by the noise) of this effect on the two sides of the step edge. What, then, controls the range of influence of the step? It appears that there is a volume under and around the indenter in which yield occurs and if the step is within that volume, the yield stress is affected. This is shown by measuring indentation diameters (including pileup if there is any) on defect-free surfaces. It is found that the average value is about $3a_y$, which is consistent with other measurements of indentation pileup [53]. When the step-probe separation was normalized by the contact radius at yield, as shown in Fig. 2.10b, it was also found that $3a_y$ was the approximate distance within which the lower yield stresses were observed. The similarity of these two distances suggested that there is some volume within which plasticity occurs and, if there is a defect within this volume, the activation barrier for yield is lowered. In the present case the radius of the volume appears to be about $3a_y$.

Recently embedded-atom calculations, including the presence of surface steps, have been made of a similar experiment [55]. The force-profile results are illustrated in Fig. 2.11 comparing the case of the 4.0-nm-radius tip located 2.0 nm from the step on the down-step side and with the tip centered on the step. Two features are evident: first, the on-step profile is slightly less stiff than the separated case and two, the plastic relaxation event occurs at a lower force level (as was observed in the experiments). However, the calculations also showed, in contradiction to the experimental conclusions, that the average applied stress at initial relaxation occurred when the tip actually contacted the step edge at yield. The authors concluded that the contact-mechanics calculations underestimated the contact radius leading to the conclusion that the step affected the threshold stress at separations up to three times

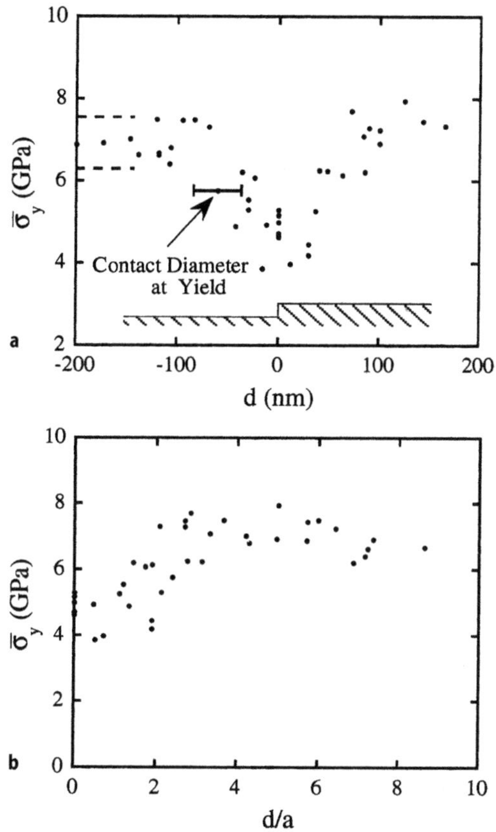

Fig. 2.10. **a** Mean applied stress as a function of the distance from the nearest step for a variety of step heights and tip radii. The two *horizontal dashed lines* indicate the scatter in mean-stress values for many indentations taken far from neighboring step edges; **b** the results of **a** normalized by each individual contact radius at yield

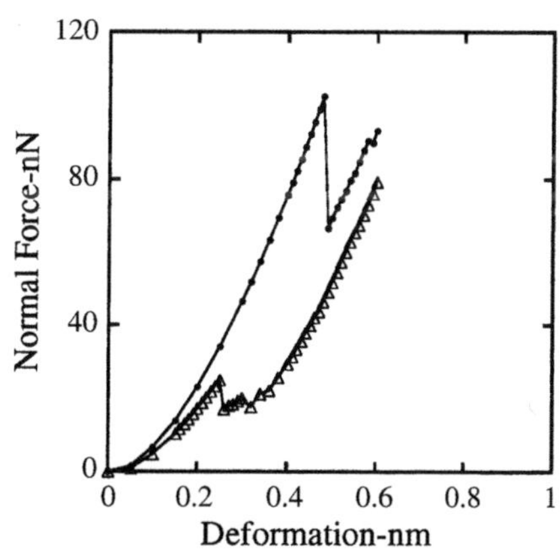

Fig. 2.11. Embedded-atom modeling results for a 4.0-nm-radius repulsive-wall probe indenting a Au(111) surface 2.0 nm (*solid circles*) and 0.0 nm (*open triangles*) from a single atomic step

the contact radius. Whether a tip-size effect, similar to that discussed before, is responsible for this discrepancy is not presently known. Additional measurements with various tip sizes will be required to obtain a definitive answer.

The material presented in this section is, again, aimed at emphasizing the unique information available to a stable and quantitative force-probe sensor in areas dealing with the nanomechanical properties of solids. It is, in addition, directed toward exploiting the advantages in the absence of adhesion and with carefully controlled and shaped tips. Although all of this work involved reasonably hard and easily shaped W probes, which work very well for Au surface and soft materials such as polymers, they are not capable of dealing with more technically relevant metals and alloys. For these cases, diamond tips are required, which presents a problem with respect to shape control. One viable alternative is the use of specifically shaped diamond which has been accomplished by focused-ion beam etching [56–58]. These have been used effectively in several surface and thin-film studies [57].

2.3.3 Molecular-Level Friction Studies

Considerable interest has recently been generated in the potential use of self-assembled monolayer (SAM) films as passivating agents and lubricants for such applications as micro-electromechanical systems (MEMS) [59]. In addition, broad discussions have appeared in the literature addressing the details of the mechanisms of friction at the molecular level, especially with regard to determining the relationship between adhesion and friction [60]. Extensive experimental work has already been done in these areas using the surface forces apparatus [61–64] and the atomic force microscope [65, 66] and recent reviews of the general area have appeared [67]. Self-assembled monolayer films offer a particularly useful way of answering the question as to the role of adhesion and mechanical properties in determining tribological behavior. The mechanical aspects can be probed by changing, for example, the chain length and overall packing density by using end groups of various sizes [68]. The adhesive interactions can be readily varied by using two sets of molecules (one on the tip and the other on the substrate) with identical backbone structure but with end groups having a variety of chemically distinct functionalities. Because of the unique capabilities of the IFM sensor, that is, its ability to make quantitative measurements of both the normal and lateral friction forces throughout the full range of interfacial separation, such a detailed study is readily doable.

In the initial experiments, the surfaces of Au tips and single-crystal Au substrates were covered by SAM films having various end-group combinations of CH_3 and COOH [24]. The chemically inert CH_3 combination represents friction with a totally mechanical origin, while interactions of COOH groups have the capability of producing significant adhesion through inter-group hydrogen bonding (i.e., bonding between groups on the two different surfaces). The Au tips are electrochemically etched to a parabolic end form (radius and shape determined by SEM). It is well known that surface roughness has a dramatic effect on friction [60]. Atomically flat Au(111) surfaces were prepared by flame annealing a small ball on the end of a 200-μm diameter, high-purity, polycrystalline-Au wire [69], which exposed several very flat (111) facets with dimensions of approximately 100 μm. The

Au substrates are atomically flat and the tips are smooth, but undoubtedly have atomic-level roughness. However, the measurements are all relative using the same substrate and tip. Therefore, the overall conclusions should be relatively unaffected by the slight remaining roughness. The friction forces were obtained by the lateral modulation method. Here, a 2.5-nm sine-wave dither was imposed on the lateral position of the tip at 100 Hz and the resulting modulation in force was synchronously detected. The relationship between detected voltage and friction force required a careful measurement of the length of the tip and its position relative to the torsion-bar axis.

The COOH-terminated films are generated from 1 mM ethanolic solutions (all solutions use the same ethanolic concentration) of either 16-mercaptohexadecanoic acid ($HS(CH_2)_{15}COOH$) or *n*-hexadecanethiol ($HS(CH_2)_{15}CH_3$), referred to here according to the number of methylene (CH_2) units in the alkyl backbone, as COOH(15) and $CH_3(15)$, respectively. The same 500-nm-radius Au tip was used for all combinations.

Data from this study are presented in two related forms, that is, force/friction profiles, where both the normal and friction forces are plotted against relative displacement, and direct plots of the lateral friction force as a function of the normal force. An example of these is given in Fig. 2.12.

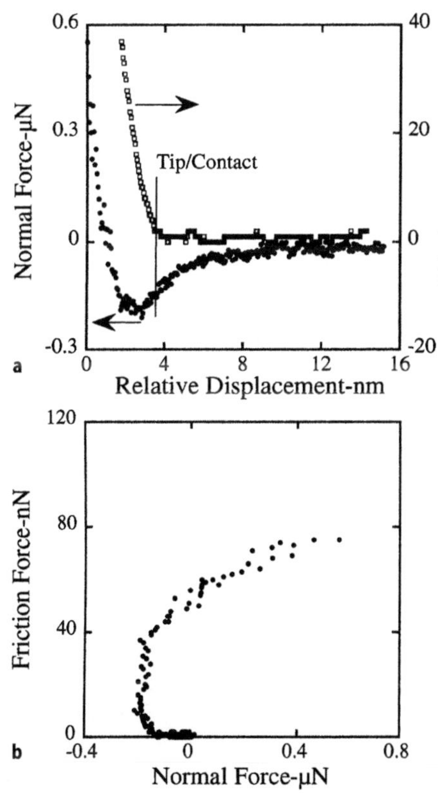

Fig. 2.12. **a** Friction and normal forces plotted against relative displacement showing that the friction force begins to rise at the point where the normal force just deviate from the van der Waals behavior illustrated in Fig. 2.4; **b** the data of **a** plotted as friction force versus normal force

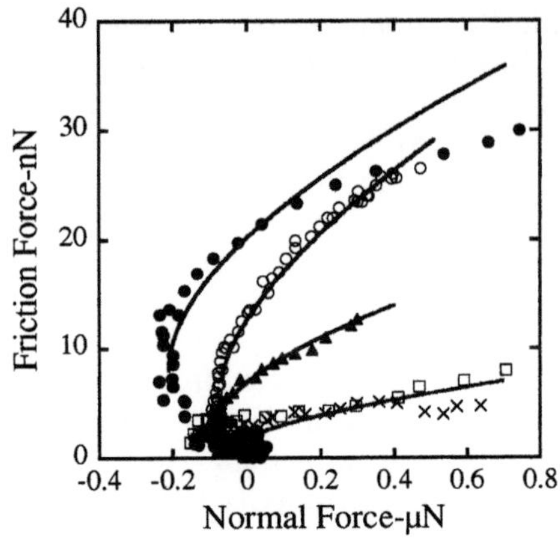

Fig. 2.13. Friction force versus normal force for the interaction of the combinations: *filled circles* 2 × COOH(10), *open circles* 2 × COOH(15), *filled triangles* CH_3(15)/COOH(15) and *open squares* CH_3(15)/COOH(10). The *solid curves* represent the JKR fits to the respective combinations with the resulting parameters tabulated in Table 2.2. *X* represent the 2 × CH_3(15) results and are not fit, since they are very similar to those for CH_3(15)/COOH(10)

As indicated in the description of Fig. 2.5, the point at which the attractive force deviates from its rapid rise, due to the attractive van der Waals interaction, indicates initial tip/sample contact. From Fig. 2.12a, it is clear that the friction force begins to rise near this point. The behavior of the friction force as a function of normal force (Fig. 2.12b) is reminiscent of results from the work of Carpick, et al. [70], where it was shown to be due to the fact that the friction force varies directly with contact area. This is in sharp contrast to classic results involving macroscopic surfaces where the friction varies directly with the applied force, independent of contact area.

The results of the study outlined above are summarized in the friction versus normal-force data shown in Fig. 2.13.

In this figure, the molecular combinations are identified by a shorthand notation, that is, 2 × COOH(10) indicates that both the tip and substrate are functionalized by alkanethiol molecules with 10 methylene (CH_2) units and carboxylic acid (COOH) end groups, whereas CH_3(15)/COOH(15) indicates a methyl (CH_3)-terminated, 15 methylene-unit SAM on the tip and a 15 methylene unit, COOH-terminated SAM on the substrate. In comparing the behavior of the CH_3(15)/COOH(15) (open squares) and 2 × COOH(10) (filled circles) combinations, it is immediately clear that the adhesive bonding in the latter case has a dramatic effect. There are only van der Waals adhesive forces present in the former and very small friction. This combination represents friction having a purely mechanical origin. In contrast, the latter combination shows significant hydrogen bonding and greatly enhanced friction. The friction in this case is due to the breaking of the inter-film bonds (i.e., between the two films), as a result of the lateral motion of the tip. Nominally, the difference in the level of friction between the two combinations can be taken as the purely adhesive component of the friction. However, this ignores the mechanical effect of films in contact with differing molecular lengths. In order to get a better measure of the separate contributions of these two friction mechanisms, it is helpful to quantitatively characterize the adhesive and friction strengths through the simple contact-mechanics models mentioned above.

Following Carpick et al. [70–73], these data were analyzed using the Johnson-Kendall-Roberts (JKR) contact-mechanics model [14] relating the area of contact to the normal force. The resulting fits to the data are shown as the solid lines in Fig. 2.13 [74]. The approximate characterization in terms of the contact areas allows the friction to be given in terms of a friction shear strength (τ) and this parameter is a variable in the JKR fitting procedure. The friction, or shear force, is then given by the product of τ and the contact area A. In this procedure, the equation describing the friction force versus the normal force is given as [73]

$$F_{\mu} = \left(1 + \sqrt{1+L}\right)^{4/3} \tag{2.7}$$

where the friction and normal forces F_{μ} and L have been normalized by their critical values defined by the relationships

$$F_{\mu} = \frac{F}{F_c},\ L = \frac{f}{|f_c|}. \tag{2.8}$$

The critical values F_c and f_c represent the friction and normal-force values at the maximum in the attractive force (commonly called the "pull-off force") and are, in turn, defined by the relationships

$$F_c = \tau A_c = \pi\tau\left(\frac{3\pi\gamma_{12} R^2}{2E^*}\right)^{2/3} \text{ and } f_c = -\frac{3}{2}\pi R\gamma_{12}, \tag{2.9}$$

where A_c is the critical contact area (the area at the maximum attractive force), R the tip radius and γ_{12} is the energy of adhesion (corresponding to the work per unit area required to separate the surfaces). The subscript indicates a general form where the two interacting surface are different. In the case where they are the same, $\gamma_{12} = 2\gamma$, where γ is the surface energy of the two interacting surfaces. E^* is the reduced modulus given by the expression,

$$\frac{1}{E^*} = \frac{1-\upsilon_t^2}{E_t} + \frac{1-\upsilon_s^2}{E_s}, \tag{2.10}$$

where E_t, E_s, v_t, and v_s are the modulus and Poisson ratio values for the tip and substrate materials, respectively.

The analysis of the data to obtain a value for the friction shear strength τ requires a knowledge of three unknowns (Eq. 2.9), namely, the tip radius R (500 nm in this case), the energy of adhesion γ_{12}, and the composite modulus E^*. Both γ_{12} and E^* values are obtained from a JKR analysis of the normal force versus relative displacement data, an example of which is shown as the solid curve in Fig. 2.14 for the 2 × COOH(10) interaction.

The JKR analysis further requires a knowledge of the contact radius a and the film deformation δ. The contact radius is given by the equation

$$a = \left(\frac{R|f_c|}{E^*}\right)^{\frac{1}{3}} \left(1 + \sqrt{1+L}\right)^{\frac{2}{3}}, \tag{2.11}$$

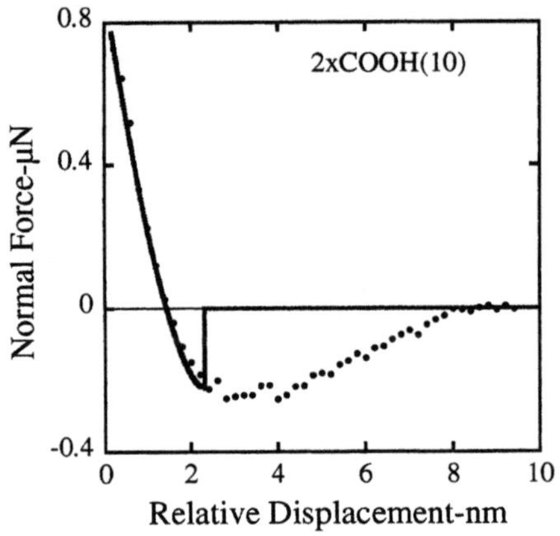

Fig. 2.14. Normal force versus relative-tip displacement for the 2 × COOH(10) interaction (*closed circles*). The *solid curve* indicates the JKR fit to the force profile. Similar levels of fitting are obtained for all the combinations to obtain values for E^* and γ_{12}

and the relationship between δ and a takes the form

$$\delta = \frac{a^2}{R}\left[1 - \frac{2}{3}\left(\frac{a_c}{a}\right)^{\frac{3}{2}}\right], \tag{2.12}$$

where a_c is the value of a at the point of maximum adhesive force.

The advantage of obtaining the value of E^* in this way is that the modulus and Poisson's ratio values necessary to calculate E^* from Eq. (2.10) are not known for these very thin films. In addition, the E^* value is a combination of the parameters for the two films as well as the substrate and tip. All of these factors are taken into account by obtaining the E^* value from experiment. (Note that, because only contact forces are considered in this model, the calculated force is zero until contact is made, after which it suddenly jumps to its maximum attractive value. The experiment, of course, is not limited by this assumption and shows long-range attractive forces.) The resulting parameters are tabulated, along with those obtained from data similar to that of Fig. 2.14 for the other end-group combinations, in Table 2.2. With the γ_{12} and E^* values determined through this procedure, it is possible to use Eqs. (2.5)–(2.7) to fit the friction versus normal force data and obtain the value for the friction shear strength τ. The results of this analysis are shown as the solid lines in Fig. 2.13 (except for the 2 × CH_3(15) case, which is very similar to that obtained for CH_3(15)/COOH(10)) and compiled in Table 2.2 under the same fitting parameters.

Table 2.2. A tabulation of the results of a JKR analysis of the data of Fig. 2.13 obtained by methods describedin the text

End groups	E^* (GPa)	γ_{12} (mJ m^{-2})	τ (MPa)	$(CH_2)_{n1+n2}$
2×COOH(10)	16	93	6.5	20
2×COOH(15)	4	42	3.2	30
CH_3(15)/COOH(15)	4	42	1.6	30
CH_3(15)/COOH(10)	10	42	1.5	25
2×CH_3(15)	5	37	0.8	30

The shortest molecular combination, that is, 2 × COOH(10) with a total of 20 CH_2 units (the value of total length is given as the last column in Table 2.2), shows the highest E^* value followed by CH_3(15)/COOH(10) with 25 units. The other combinations have 29 or 30 units and the lowest E^* values. The inverse relationship between E^* and total molecular length is the result of the increased strain for the shorter films, which would ordinarily be divided out in actual modulus measurements. In fact, the compliance behavior of the molecules is highly nonlinear and the fitting only works over a small range of repulsive force. However, lacking detailed molecular-level calculations, this simple characterization is the only reasonable path available to quantify the relative values of the frictional parameters.

As expected, Table 2.2 shows that the 2×COOH(10) has the highest energy of adhesion, greater by more than a factor of two than all the other combinations. The fact that the 2 × COOH(15) and CH_3(15)/COOH(15) combinations have the same low value of γ_{12} indicates that the COOH(15) surface is virtually hydrophobic. The surface energy γ for the symmetric combinations is just half the γ_{12} values from Table 2.2. For 2 × COOH(10), this value would be about 46 mJ m^{-2}, while the 2×COOH(15) and 2 × CH_3(15) have values of approximately 20 mJ m^{-2}. The first is about 65% of the 73 mJ m^{-2} measured for water and the last two are somewhat less than the approximate 27 mJ m^{-2} found for *n*-hexadecane liquid. Here again, this latter value emphasizes the surprising level of hydrophobicity of the COOH(15) surface, which is only slightly greater than that obtained from the very hydrophobic 2 × CH_3(15) interaction.

The values for the friction shear strength τ vary widely and are representative of a combination of both mechanical and adhesive components. The fact that both the 2 × COOH(15) and CH_3(15)/COOH(15) combinations have elevated levels of shear strength, but only van der Waals levels of adhesion, indicates that there is significant *intra-film* bonding due to the changed orientation of the COOH group for the odd number of methylene units. This is the so-called odd-even effect, which has been studied for these films by infrared spectroscopy [29, 76]. As before, the lateral motion of the tip disturbs these bonds and the resulting energy dissipation gives rise to friction. However, in this case the "adhesive" component is due to bonding *within* the plane of the film.

The τ values from Table 2.2 suggest that the level of *inter-film* bonding involved in the 2 × COOH(10) interaction is virtually double that of the *intra-film* bonding, that is, bonding between groups within a single surface film, for the 2 × COOH(15) interaction. However, a similar comparison between the τ values for 2 × CH_3(15) and CH_3(15)/COOH(10) indicates that this assumption is not justified. The higher τ value for the shorter combination indicates that the mechanical disturbance of the shorter-length films is greater, which several authors have found this to be the case [77, 78].

To better view the systematic behavior of the separate mechanical and adhesive contributions of the friction, it can be simply assumed that the mechanical and adhesive contributions to the friction for the various interaction combinations are additive. Under this assumption, the 2 × CH_3(15) results are taken to represent a purely mechanical interaction, which permits the mechanical contribution for a single layer of long molecules to be taken as just half the τ value of Table 2.2 or approximately 0.4. Carrying this process further, the long/short combination

[CH_3(15)/COOH(10)] (purely mechanical according to the inter-film γ_{12} value) would have a mechanical contribution for a single layer of short molecules of (1.5–0.4) = 1.1. By subtracting these mechanical contributions from the total τ values for the other combinations, approximate values for the mechanical and bonding contributions are obtained, which are summarized in Table 2.3. Thus, the 2 × CH_3(15), 2 × COOH(15) and CH_3(15)/COOH(15) combinations have the same levels of mechanical friction. They involve the same length molecules and have no inter-film adhesion beyond the van Waals level. The CH_3(15)COOH(10) and 2 × COOH(10) combinations, involving two short molecules, have increased mechanical-component values of 1.1 and 2.2, respectively, reflecting the effect of total molecular length on the frictional loss [77, 78]. Carrying this process throughout Table 2.2 results in the mechanical and bonding components of the frictional shear stresses given in Table 2.3.

The largest value for the τ-bonding component occurs for the 2 × COOH(10) combination, which has the highest level of inter-film bonding, indicating that the friction results from the breaking of these strong bonds by the lateral motion of the tip. Interestingly, the two combinations involving COOH(15) both show significant values of bonding friction in spite of a lack of an *inter-film* interaction, indicated a considerable level of *intra-film* bonding for the odd-length COOH-terminated films, which are also broken by the tip motion. This, along with the van der Waals-level γ_{12} value for the *inter-film* work of adhesion, further supports the hydrophobic nature of the COOH(15) surface.

The large difference between the τ-bonding values for 2 × COOH(10) and 2 × COOH(15) would imply that the bonding strength per unit area is greater for the inter-film or "adhesive" interaction of the former than for the intra-film or "cohesive" interaction of the latter. This is plausible considering the steric flexibility of the molecules meeting head on, where the end groups can maximize their inter-film bonding, compared with interactions between the end groups of neighboring molecules leaning about 30° from the normal in the SAM packing arrangement. However, the factor of three in the τ-bonding values between 2 × COOH(15) and CH_3(15)/COOH(15) is less obvious. Since, nominally, there would be twice as many *intra-film* bonds for two interacting COOH(15) relative to a single film, a factor of two between the τ-bonding values would be expected. However, the differing τ-bonding values for 2 × COOH(15) and 2 × COOH(10), indicate that unsatisfied bonds remain on the 2 × COOH(15) film and these may be available for *inter-film*

Table 2.3. A tabulation of approximate values for the bonding and adhesive components of the friction shear strength obtained by simply assuming that these components were additive in the results given in Table 2.2

End groups	τ (MPa) total	τ (MPa) mechanical	τ (MPa) bonding
2×COOH(10)	6.5	2.2	4.3
2×COOH(15)	3.2	0.8	2.4
CH_3(15)/COOH(15)	1.6	0.8	0.8
CH_3(15)/COOH(10)	1.5	1.1	0.4
2×CH_3(15)	0.8	0.8	0.0

bonding, giving rise to increased friction. Along these same lines, the CH_3(15)/COOH(10) interaction turns out to have a non-zero bonding component of τ of 0.8 which, again, would imply that there is also a small intra-film bonding component for the COOH(10) film. The answer as to whether this is all true, or whether the discrepancy reflects the crude approximation used to calculate these values, will have to await further, more careful measurements which include a wider variety of combinations of both end-group functionality and molecular length.

These results involve friction at relative low film/film loading, where the variation of the contact area is a major factor. However, at higher loading, the character of the friction force changes with respect to the normal load, as was found in earlier work [79]. For the higher loads, a friction loop method for obtaining the friction force must be implemented in order to circumvent the small lateral compliance of the tip. In this method, the tip is scanned along the surface normal to the torsion-bar axis, under constant normal-force control – first in one direction and then in the other. Since the sign of the lateral friction force is opposite to the tip velocity, the "line scans" produce a friction loop. In one direction the total torque is increased by the friction torque, while in the opposite direction it is decreased. To keep the total torque constant, the constant-torque controller moves out for the former and in for the latter direction. Knowing the mechanical properties of the film (from force profiles such as that shown in Fig. 2.14) permits the conversion of the vertical motion into lateral force values. An example of the results of this kind of measurement for a 190-nm-radius W tip acting against a self-assembled monolayer film of hexadecanethiol [$SH(CH_2)_{15}$ CH_3] on a single-crystal Au(111) surface is shown in Fig. 2.15 [79].

At loads beyond the initial compression of the film, the friction becomes linear with respect to the normal force permitting the definition of a friction coefficient.

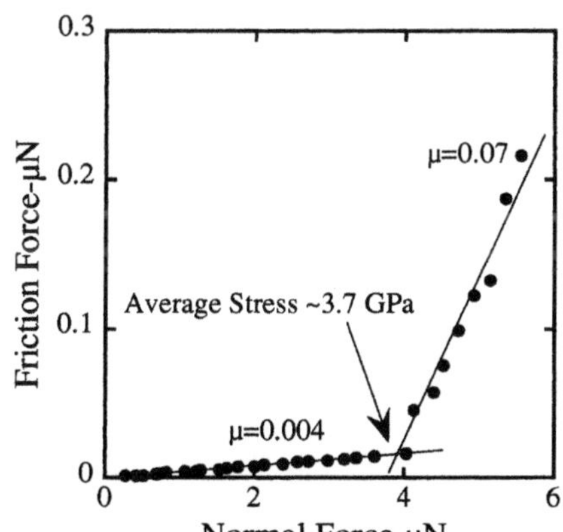

Fig. 2.15. Friction versus normal force taken at higher loads by the friction loop method for a 190-nm-radius W tip acting against a self-assembled monolayer film of hexadecanethiol ($SH(CH_2)_{15}CH_3$) on a single-crystal Au(111) surface. The friction force is directly proportional to the load permitting a friction coefficient value to be assigned. At calculated average applied stresses below about 3.7 GPa the coefficient is 0.004 while above this load the coefficient jumps to 0.07

Up to a calculated average applied stress of approximately 3.7 GPa, the coefficient of friction is calculated to be 0.004. This value is very low and illustrates the benefit of these SAMs as tribological coatings. However, above the 3.7-GPa stress level the coefficient suddenly jumps to 0.07 indicating a dramatic increase in the energy dissipated by the film. The linear behavior has been explained as being due to the fact that the energy is dissipated in a thin film and the volume of the disturbed molecules turns out to vary directly as the normal force [79]. Earlier calculations on similar systems indicated that the density of "gauche" defects increases, and significant movement of the head-group position occurs, at the higher loads [80–82]. Beyond this, it is not clear what gives rise to the change in dissipation. However, additional insight can be gained by looking at the actual film compliance as a function of the applied load. This can be done by using the known behavior of the tip and substrate to "back out" the mechanical character of the film itself.

Figure 2.16 shows a typical force profile taken to a maximum normal force of about 10 μN. Both approach and withdrawal portions of the profile are shown indicating a significant level of contact hysteresis. The solid line for the approach curve is the Hertzian fit [15] to the profile and demonstrates a close fit in the mid-force region, while significant deviations are seen in both the low and high-force regions. The ×10 expansion shows more clearly the lack of a fit in the low-force region.

The close mid-force fit implies that the compliance in this region is dominated by the Au substrate and W tip. In order to obtain the compliance of the film itself, the tip/substrate behavior is removed by assuming that the system can be viewed as two springs in series – one representing the tip/substrate system (as characterized by the Hertzian fitting parameters) and the other by the film itself. The film compliance can then be obtained by correcting the deformation axis according to the tip/substrate compliance. The equation for making this correction takes the form

$$d' = d + \left(\frac{F}{4/3\sqrt{R}E^*} \right)^{2/3} , \tag{2.13}$$

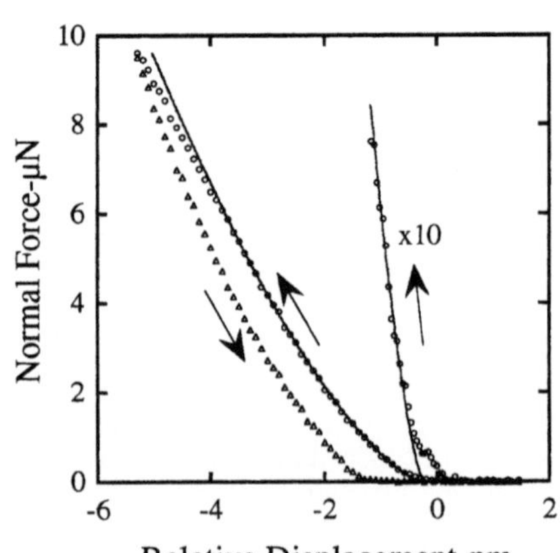

Fig. 2.16. The force profile for the system of Fig. 2.14 taken to a maximum normal force of about 10 μN. Both approach and withdrawal curves are shown indicating a significant degree of hysteresis. The *solid line* on the approach portion represents the Hertzian fit to the data and shows close agreement except at the low forces (illustrated by the ×10 plot) and at forces above about 6 μN

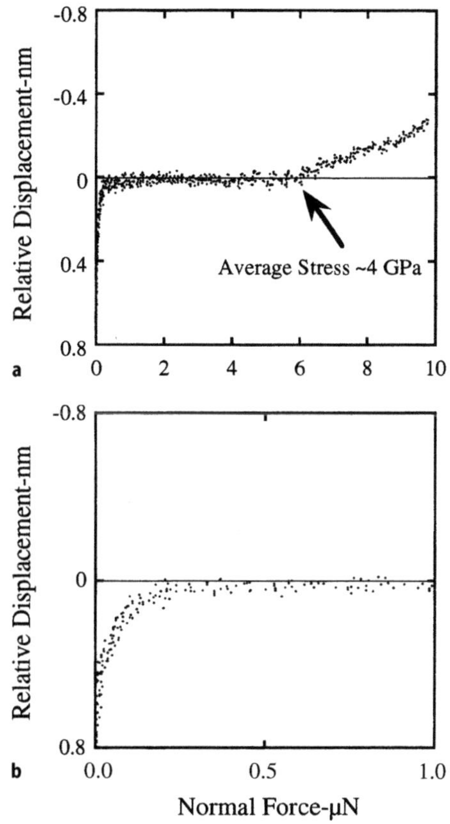

Fig. 2.17. **a** The film compliance of the hexadecane SAM of Fig. 2.16 obtained by correcting the relative-displacement axis for the deformation due to the Au substrate and W tip. This deformation is determined from the Hertzian fit to the force profile data in the mid range of normal forces; **b** an expanded view of the low normal-force region of the compliance shown in **a**

where d' is the corrected relative displacement, d is the displacement from the force profile, F is the normal force and the denominator is the Hertzian coefficient from the fit of Fig. 2.16. The result of applying this procedure to the approach portion of Fig. 2.16 is shown in Fig. 2.17, plotted as compliance (i.e., film deformation as a function of normal force). Figure 2.17a is the entire compliance range, while Fig. 2.17b shows an expanded view of the light-load region.

As suggested by the earlier calculations, the early stages of contact shows high compliance involving a nestling of the molecules to increase their average angle with the surface normal [80, 81]. Once the available space for further packing diminishes, however, the compliance decreases. A similar behavior is found under the leg of a chair on a plush carpet. At the higher stresses, however, the picture is not so clear. The creation of gauche defects gives an additional avenue for increased packing and could also facilitate intermolecular slippage. Under loading, there are enormous shear stresses trying to drive the molecules from under the tip. However, this motion is hindered by the interlocking of the methylene groups along the backbone of the molecules. If the methylene groups can rotate to form gauche defects, the molecules could slip laterally and decrease the average thickness of the film providing additional compliance. Note that the additional compliance shown in

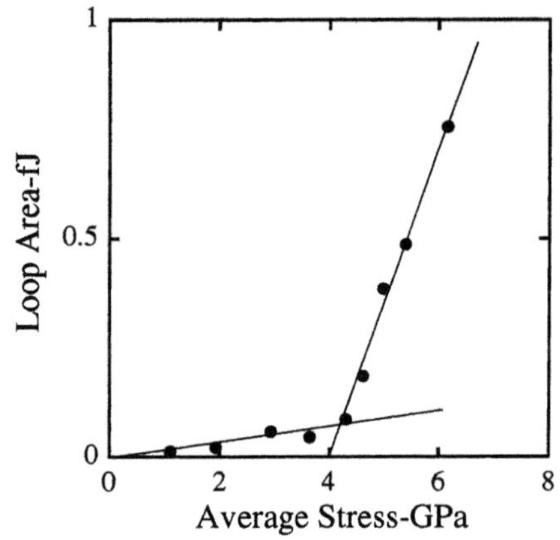

Fig. 2.18. Hysteresis-loop areas (given in fJ) for force profiles similar to that shown in Fig. 2.16, plotted against the calculated average applied stress corresponding to the maximum or turn around force. The *sharp break* in the loop area occurs at approximately the same stress level as that found in the direct friction measurements shown in Fig. 2.15

Fig. 2.17a is only about 0.3 nm in going from six to ten μN of applied force, a small effect indeed. At any rate, this added compliance would certainly result in a greatly enhance level of friction, as observed in Fig. 2.15. It should be emphasized that even at these high loads the hysteretic behavior is anelastic, that is, subsequent profiles at the same point will have the same characteristic loop. No permanent damage is done to either the film or the Au substrate.

A more sensitive way of looking at the friction-related film compliance is through the hysteresis loops, as shown in Fig. 2.16. It was demonstrated earlier that there is a direct relationship between the hysteresis-loop area and the friction force [79]. As a further demonstration of this relationship, Fig. 2.18 shows a plot of the hysteresis-loop area (in fJ) as a function of the calculated average stress at the maximum force in the force profile.

Although the scaling between loop area and friction force is not easily obtained, the close agreement between the overall behavior presented in Figs. 2.15 and 2.18, and the high-force break point in Fig. 2.17, clearly demonstrate the consistency of the kind of analysis present here.

2.4 Conclusions

In this brief review, an attempt has been made to present selected results from IFM work to illustrate the power of making nanoscale scanning-probe force measurements with a stable and quantitative force sensor. An additional aim was to illustrate from these examples the close tie that connects adhesion, mechanical, and tribological properties. Unfortunately, space restraints do not permit a discussion of the broad range of materials problems presently being explored by the technique. Examples of these include studies of: the viscous behavior of water near the surface of biologically relevant surfaces [83]; the properties of the solid-like interphase region for hexadecane liquid near Au surfaces [84]; the nanomechanical properties of interphase poly(aminopropyl)siloxane films on glass and SiO_2 surfaces [85]; the necessity of an applied stress to achieve conduction through alkanethiol SAMs on Au [86]; the tribological properties of alkyl SAM films grown on the SiO_x surface of Si by a new chemical route [87, 88]; the nanomechanical properties of Kelvar(R) [89]; the effect of an electric field on the morphology of Au surfaces [90, 91]; the nanomechanical properties of a variety of tribochemical, antiwear films [92]; the nanomechanical properties of the interphase region near glass fibers in an epoxy matrix [93]; and the nanomechanical properties of phase segregation at homopolymer surfaces [94].

In closing, it is important to point out a difficulty with the IFM technique that has limited its implementation. It is a typical research instrument and is not commercially available (i.e., it is "home built" and not a "turn key" instrument). As a result, the technique is very versatile and can easily be tailored for specific materials application. However this versatility comes with considerable overhead in terms of establishing the capability and learning to master its various idiosyncrasies. This is especially true of the sensors, which are presently hand assembled in small quantities and require reasonably sophisticated fabrication technologies. However, these are generally robust and have long lifetimes involving the changing of many probe tips of various materials. Many of these problems should be greatly diminished by the introduction in the near future of a fully independent two-dimensional IFM sensor using a two-axis laser-interferometer scheme for displacement detection. This sensor should be considerably easier to use, more stable, faster, and have greatly increased sensitivity.

Acknowledgements. The author would like to thank many graduate, post-doctoral students, and colleagues who have contributed to the work outlined in the present chapter. This work was supported by the US Department of Energy under Contract DE-AC04–94AL85000. Sandia is a multiprogram laboratory operated by Sandia Corporation, a Lockheed-Martin Company, for the US Department of Energy.

References

1. Binnig G, Quate CF, Gerber C (1986) Phys Rev Lett 56:930
2. Martin Y, Williams CC, Wickramasinghe HK (1987) J Appl Phys 61:4723
3. Joyce SA, Houston JE (1991) Rev Sci Instrum 62:710
4. Bosch R (1996) Method for anisotropically etching silicon. Patent No 5501893
5. Rogers T, Kowal J (1995) Sens Actuators A 46:113
6. Burnham NA, Colton RJ, Pollock HM (1993) Nanotechnology 4:64
7. Weis P (1962) Adhesion and cohesion. Academic Press, London
8. Nuzzo RG, Allara DL (1983) J Am Chem Soc 105:4481
9. Bain CD, Troughton EB, Tao Y-T, Evall J, Whitesides GM, Nuzzo RG (1989) J Am Chem Soc 111:321
10. Joyce SA, Thomas RC, Houston JE, Michalske TA, Crooks RM (1992) Phys Rev Lett 68: 2790
11. Israelachvili J (1992) Intermolecular & Surface Forces. Academic Press, New York
12. Colton RJ, Barger WR, Baselt DR, Corcoran SG, Koleske DD, Lee GU (1998) In: Van Ooij WL, Anderson HR Jr (eds) Mittal Festschrift, p 21
13. Thomas RC, Houston JE, Crooks RM, Kim T, Michalske TA (1995) J Am Chem Soc 117: 3830
14. Johnson KL, Kendall K, Roberts AD (1971) Proc Royal Soc London A324:301
15. Hertz H (1882) J Reine Angew Math 92:156
16. Derjaguin BV, Muller VM, Toporov YP (1975) J Colloid Interface Sci 53:314
17. Strong L, Whitesides GM (1988) Langmuir 4:546
18. Fowkes FM (1962) J Phys Chem 66:682

19. Pimentel GC, McClellan AL (1960) The hydrogen bond. Freeman, San Francisco
20. Ebeltoft H, Sjoblom J, Saeten JO, Olofsson B (1994) Langmuir 100:2262
21. Israelachvili JN (1992) J Vacuum Sci Technol B10:2961
22. Zisman WA, Fox HWJ (1952) J Colloid Sci 7:428
23. Good RJ (1993) J Adhesion Sci Technol 7:1015
24. Kim HI, Houston JE (2000) J Am Chem Soc 122:12045
25. Son K-A, Oliver AC, Houston JE, unpublished results
26. Yamada R, Uosaki K (2000) J Phys Chem B104:6021
27. Marchenko A, Xie ZX, Cousty J, Van LP (2000) Surf Interface Anal 30:167
28. Marchenko A, Lukyanets S, Cousty J (2002) Phys Rev B65:045414
29. Nuzzo RG, Dubois LH, Allara DL (1990) J Am Chem Soc 112:558
30. Houston JE, Peden CHF, Blair DS, Goodman DW (1986) Surf Sci 167:427
31. Cabibil H, Houston JE, unpublished results
32. Pascual JI, Mendez J, Gomez-Herrero J, Baro AM, Garcia N, Landman U, Luedtke WD, Bogachek EN, Cheng HP (1995) Science 267:1793
33. Gane N, Bowden FP (1968) J Appl Phys 39:1432
34. Gane N (1970) Proc Royal Soc London A317:367
35. Hirth JP, Lothe J (1982) Theory of dislocations. Wiley, New York
36. Brenner SS (1958) Growth and perfection of crystals. Wiley, New York
37. Kiely JD, Houston JE (1998) Phys Rev B57:12588
38. Kiely JD, Jarausch KF, Houston JE, Russell PE (1999) J Mater Res 14:2219
39. Michalske TA, Houston JE (1998) Acta Mater 46:391
40. Tangyunyong P, Thomas RC, Houston JE, Michalske TA, Crooks RM, Howard AJ (1993) Phys Rev Lett 71:3319
41. Tangyunyong P, Thomas RC, Houston JE, Michalske TA, Crooks RM, Howard AJ (1994) J Adhesion Sci Technol 8:897
42. Thomas RC, Houston JE, Michalske TA, Crooks RM (1993) Science 259:1883
43. Kelchner CL, Plimpton SJ, Hamilton JC (1998) Phys Rev B 58:11085
44. Kiely JD, Hwang RQ, Houston JE (1998) Phys Rev Lett 81:4424
45. Mitchell TE (1964) Prog Appl Mat Res 6:117
46. Corcoran SG, Colton RJ, Lilleodden ET, Gerberich WW (1997) Phys Rev B55:16057
47. Hsu T, Cowley JM (1983) Ultramicroscopy 11:239
48. Schneir J, Sonnenfeld R, Marti O, Hansma PK, Demuth JE, Hamers RJ (1988) J Appl Phys 63:717
49. Woirgard J, Tromas C, Audurier V, Girard JC (1998) Mat Res Soc Symp Proc 522:65
50. Rice JR (1992) J Mech Phys Solids 40:239
51. Hurtado JA, Kim KS (1999) Proc Royal Soc London Ser A Math Phys Eng Sci 455:3363
52. Hurtado JA, Kim KS (1999) Proc Royal Soc London Ser A Math Phys Eng Sci 455:3385
53. Bahr DF, Gerberich WW (1996) Metall Mater Trans A27A:3793
54. Hirth JP (1963) The relation between the structure and mechanical properties of metals. Her Majesty's Stationary Office, London
55. Zimmerman JA, Kelchner CL, Klein PA, Hamilton JC, Foiles SM (2001) Phys Rev Lett 87:165507
56. Russell PE, Stark TJ, Griffis DP, Phillips JR, Jarausch KF (1998) J Vacuum Sci Technol B16:2494
57. Jarausch KF (1999) Nanomechanics of thin films and surfaces. North Carolina State University
58. Phillips GN, Siekman M, Abelmann L, Lodder JC (2002) Appl Phys Lett 81:865
59. Maboudian R (1998) Surface Sci Reports 30:207
60. Persson BNJ (1998) Sliding friction. Springer, Berlin Heidelberg New York
61. Heuberger M, Luengo G, Israelachvili JN (1999) J Phys Chem B103:10127
62. Luengo G, Campbell SE, Srdanov VI, Wudl F, Israelachvili JN (1997) Chem Mater 9:1166
63. Israelachvili JN, Chen YL, Yoshizawa H (1994) J Adhesion Sci Technol 8:1231
64. Berman AD, Ducker WA, Israelachvili JN (1996) Langmuir 12:4559
65. Tsukruk VV, Bliznyuk VN (1998) Langmuir 14:446

66. Noy A, Vezenov DV, Lieber CM (1997) Ann Rev Mater Sci 27:381
67. Tsukruk VV (2001) Adv Mater 13:95
68. Graupe M, Koini T, Kim HI, Garg N, Miura YF, Takenaga M, Perry SS, Lee TR (1999) Mater Res Bull 34:447
69. Demir U, Shannon C (1994) Langmuir 10:2794
70. Carpick RW, Agrait N, Ogletree DF, Salmeron M (1996) J Vacuum Sci Technol B14:1289
71. Carpick RW, Agrait N, Ogletree DF, Salmeron M (1996) Langmuir 12:3334
72. Carpick RW, Ogletree DF, Salmeron M (1997) Appl Phys Lett 70:1548
73. Carpick RW, Ogletree DF, Salmeron M (1999) J Colloid Interface Sci 211:395
74. Houston JE, Kim HI (2002) Acc Chem Res 35:547
75. Hertzberg RW (1989) Deformation and fracture mechanics of engineering materials. Wiley, New York
76. Sun L, Kepley LJ, Crooks RM (1992) Langmuir 8:2101
77. Leo A, Charych DH, Salmeron M (1997) J Phys Chem B101:3800
78. Xiao X, Hu J, Charych DH, Salmeron M (1996) Langmuir 12:235
79. Kiely JD, Houston JE (1999) Langmuir 15: 4513
80. Tupper KJ, Brenner DW (1994) Langmuir 10:2335
81. Tupper KJ, Brenner DW (1994) Thin Solid Films 253:185
82. Tupper KJ, Colton RJ, Brenner DW (1994) Langmuir 10:2041
83. Kim HI, Kushmerick JG, Houston JE, Bunker BC (2003) Langmuir (in press)
84. Oliver AC, Houston JE, unpublished results
85. Cabibil H, Celio H, Lozano J, White JM, Winter R (2001) Langmuir 17:2160
86. Son KA, Kim HI, Houston JE (2001) Phys Rev Lett 86:5357
87. Kim HI, Boiadjiev V, Houston JE, Zhu XY, Kiely JD (2001) Tribol Lett 10:97
88. Zhu XY, Houston JE (1999) Tribol Lett 7:87
89. Graham JF, McCague C, Warren OL, Norton PR (2000) Polymer 41:4761
90. Cabibil H, Houston JE, Mayer TM, Franklin GF (1998) Mat Res Soc Symp Proc 522:463
91. Mayer TM, Houston JE, Franklin GE, Erchak AA, Michalske TA (1999) J Appl Phys 85:8170
92. Graham JF, McCague C, Norton PR (1999) Tribol Lett 6:149
93. Winter RM, Houston JE (1998) Proceedings of the SEM Spring Conference on Experimental and Applied Mechanics, p 355
94. Graham JF, Kovar M, Norton PR, Pappalardo P, VanLoon J, Warren OL (1998) J Mater Res 13:3565

3 Atomic Force Microscopy with Lateral Modulation

Volker Scherer · Michael Reinstädtler · Walter Arnold

'ει μεν ουν τι κινεται, ενδεχεται και αλλωσ εχειν'
'Now if something is moved it is capable of being otherwise than as it is'
(Aristotle, Metaphysics 12, Part 7, written 350 B.C.E)

3.1 Introduction

Friction is present in our every day lives. There are unwanted phenomena such as energy loss in the relative motion of contacting surfaces (e.g., gears, bearings or sealings). For example, the friction loss in a modern internal combustion engine is approximately 10%, in terms of indicated power at full load. In terms of fuel consumption, a savings of up to 26% has been calculated for the hypothetical case of a 'frictionless' engine. It is believed that potential savings in fuel consumption could be 7–9% in real engines [1]. Other parts such as sliding members would not move at all if friction were not taken care of. Estimates claim a loss of up to 1.6% of the gross national product in developed countries, that is 116 billion US-Dollars for the year 1995 in the US, due to inappropriate friction management [2]. On the other hand, friction is essential to an incredible number of processes provided by nature and civilization. If friction were not present, there would be no controlled blood flow; also, we would have difficulties in slowing down vehicles. Although the phenomenon friction does affect our being and doing at least as much as gravity or electricity does, we have not yet understood its origin. Friction has been intensely investigated in macroscopic length scales, at both, low and high velocities. Just recently, much of the tribology research has been devoted to the micrometer level and below [3, 4]. It was at the end of the 1960s when Bowden and Tabor at the Cavendish Laboratory of the University of Cambridge studied friction and tribology of two bodies contacting each other on an area as small as a few micrometers squared [5]. Their investigations employing the surface force apparatus (SFA) [6] and later research of Israelachvili and Tabor [7] have indicated that friction, though independent of apparent contact area, as Amontons had stated in the 17th century, is indeed proportional to the true area of contact. A less sophisticated but not least efficient way to do micro-tribology experiments was provided by the invention of the friction force microscope (FFM) or lateral force microscope (LFM) by Mate et al. at the IBM research center in Almaden [8]. Numerous developments of this technique were done in the following years, many of them aiming at improving the ease of use and converting the read-out signals from the bent and

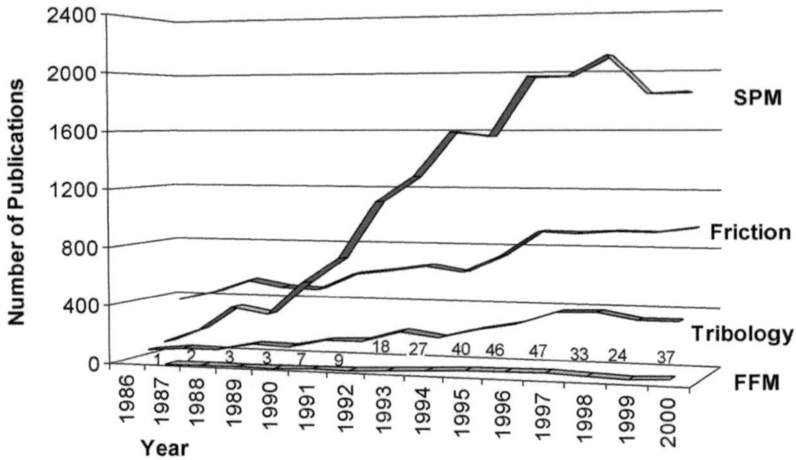

Fig. 3.1. Number of papers published per year in the field of scanning probe microscopy, tribology, and friction, as well as friction force microscopy. Data acquired from CAPlus and SCISEARCH database research

twisted cantilever beam into units of normal and friction force. There is an almost steady increase of published work employing the FFM from 1987 until today, accounting for about forty papers, annually. In Fig. 3.1, the number of articles concerning scanning probe microscopy (SPM) are shown, with friction and tribology and as a subgroup of SPM, FFM. It is interestingly to note that from the early 1990s on, the number of publications in the field of SPM exceeds the total number of papers published on both friction and tribology.

One direction of the development of FFM is to make the instrument more dynamic. FFM is a dynamic method per se; without lateral motion, we would not be able to measure wear nor friction. However, typical friction studies are performed at several Hz reciprocating motion or below, with a stroke length of several nm to some μm. Higher scanning rates could not be properly adjusted because of the limited speed of the feedback loop. This loop keeps the applied normal force constant as the sensor is scanned across the sample surface. Here, we will call the friction force microscopy method dynamic or laterally modulated if the sideways motion of either cantilever or sample is performed at a frequency > 1 kHz. A list of publications on laterally modulated AFM is given in Appendix 3.

3.2 Lateral Forces

To understand the physics of bodies in relative motion on the nanometer scale, we want to recall the classics that have started being developed perhaps more than 500 years ago. Leonardo da Vinci, to most of us known for his artwork, was probably as much a pioneer in engineering and science. Whereas his marvelous paintings and drawings were safely preserved for centuries, his engineering intelligence faded quickly with every original manuscript that disappeared in various private hands and later in museums before being printed, or even discovered. In 1699, more than 200 years later, French scientist Guillaume Amontons re-discovered Leonardo's findings which are today known as the first two laws of friction:

1. The force of friction is directly proportional to the applied load.
2. The force of friction is independent of the apparent area of contact.

We can formulate these two laws with F_N being the (normal) applied load, A_{app} the apparent area of contact, and F_F the friction force:

$$F_F = \mu F_N \tag{3.1}$$

$$F_F \neq f(A_{app}) \tag{3.2}$$

In Eq. (3.1), the proportionality constant μ is the (dynamic) friction coefficient. In addition to these rules, the French physicist Charles–Augustin de Coulomb added a third law: once motion has started, the force of friction does not depend on velocity of relative motion v.

$$F_F \neq f(v) \tag{3.3}$$

These three essential laws of friction should hold for most of the technical applications in macro-tribology. Many attempts have been made to explain the physics of friction on a fundamental basis. Surface roughness and molecular adhesion were believed to be the main constituents of the friction force. Coulomb formulated a concept that friction is composed of what he called cohesion, by normal load and plastic deformation.

$$F_F = C + \frac{F_N}{\mu_{def}} \tag{3.4}$$

In Eq. (3.4), the first term C is attributed to adhesive or cohesive forces and the second term accounts for plastic deformation or ploughing. For normal loads much greater than the effect of adhesive or cohesive forces, the inverse of μ_{def} is exactly what we have defined as the coefficient of friction μ in Eq. (3.1). In his studies, Coulomb also found that, under dry conditions, a maximum force must be exceeded before two bodies start sliding past each other at a lower constant force. Today, we call this maximum force the static (or sticking) friction force and the quotient of static friction force and normal load is the static coefficient of friction μ_{stat}. If we speak about Coulomb-type friction, we mean dry friction with static and sliding component. Coulomb suspected that static friction was caused by the interlocking of surface asperities, but this mechanism was ruled out as a potential cause in the middle of the 20th century. We know that, surprisingly, friction can indeed increase with the smoothness of the opposing surfaces, for example the faces of ceramic seals.

Bowden and Tabor found that molecular adhesion was a good concept to explain friction [5]. They also found that friction, though independent of the apparent area of contact, was in fact proportional to the true area of contact A_R.

$$F_F = \tau A_R \tag{3.5}$$

The proportionality constant τ is the average lateral force per unit area; it is called the critical shear stress or shear strength. The true area of contact is believed to be proportional to the adhesion force. Thus, this concept is often called the adhesion model of friction. Bowden and Tabor assumed that the dissipation energy is transferred into the plastic deformation of the surface asperities, even on the microscopic length-scale. The shear strength seems to be a constant specific to the material of the opposing bodies. We can use a generalized expression for the shear strength

$$\tau = \tau_0 + \alpha \underbrace{\frac{F_N}{A_R}}_{P} \tag{3.6}$$

Here, α is a constant, P is the normal load per unit area, and τ_0 is also a constant. Combining Eq. (3.5) and Eq. (3.6) leads to

$$F_F = \tau_0 A_R + \alpha F_N \tag{3.7}$$

For large normal pressures, such that $\alpha P >> \tau_0$, it follows that $F_F \approx \alpha F_N$ and then, α is nothing but Amontons' friction coefficient μ. Bowden and Tabor reported a shear strength of 140 MPa from macroscopic dry friction experiments with metal pairs [5].

3.3 Contact Mechanics

If two bodies are in contact with each other in a point under a specific load F_N, they are being compressed and hence, deformed. As a consequence, a contact of an area A is established. According to the Hertzian theory, the contact radius a of two spheres in contact is

$$a = \sqrt[3]{\frac{3F_N R}{4E^*}} \tag{3.8}$$

In the Hertzian formula, E^* is the reduced elastic modulus where

$$\frac{1}{E^*} = \frac{\left(1 - \nu_1^2\right)}{E_1} + \frac{\left(1 - \nu_2^2\right)}{E_2} \tag{3.9}$$

in which $E_{1,2}$ is the Elastic modulus and $\nu_{1,2}$ is the Poisson's ratio for the spheres with effective radius R given by

$$\frac{1}{R} = \left(\frac{1}{R_1} + \frac{1}{R_2}\right) \tag{3.10}$$

Here, R_1 and R_2 are the radii of the spheres. More advanced theories like Derjaguin–Muller–Toporov (DMT) [9] or Johnson Kendall Roberts (JKR) [10] also take the adhesive force F_{adh} into account that acts in the interface. The net normal force is composed of attractive and repulsive force. It is

$$F_{DMT} = F + F_{adh} = F_N \tag{3.11}$$

using the DMT model and

$$F_{JKR} = \left(\sqrt{F_N} + \sqrt{F_{adh}}\right)^2 \qquad (3.12)$$

using the JKR model. An analytical transition between both models has been developed by Maugis [11]. In a number of publications, these models were elucidated in view of applicability in AFM [12, 13]. A so-called force-versus-distance curve contains an attractive/adhesive part (negative force) and a repulsive/elastic part (positive force). Most AFM experiments are performed in ambient air. If so, adhesive forces caused by a water meniscus can be as high as 10^{-7} N and drop almost to zero when the water meniscus is broken during tip retraction. The externally applied normal load F is given by the product of cantilever spring constant and static cantilever deflection that is controlled by the operator.

All abovementioned contact mechanic models are valid for a sphere–sphere contact. By making one radius infinite we can approximate a sphere–flat-type contact. It has been shown that the Hertzian model is suitable for an AFM tip in contact with a sample for high normal loads [14]. But for low static loads used in most of the experiments, Schwarz et al. pointed out that commercially available cantilever tips are not spherical at the end and other contact models must be applied [15]. For a first approximation, the Hertzian model still holds even at normal loads of, let's say, 10 nN.

3.4 Conventional Friction Force Microscopy

In the conventional FFM, the sensing element, a fine tip integrated in a cantilevered spring, is scanned across the sample surface (see Fig. 3.2). The scanning motion can be implemented by means of piezoelectric tube scanners. Commercial instruments can be designed so that the scanner is at the same time the sample holder and the sample rests on the top of the tube scanner (e.g., the MultiMode design, Veeco Metrology Group, Santa Barbara, CA). Other designs have the cantilever mounted on the bottom side of the scanner (e.g., the Dimension series of Veeco Metrology).

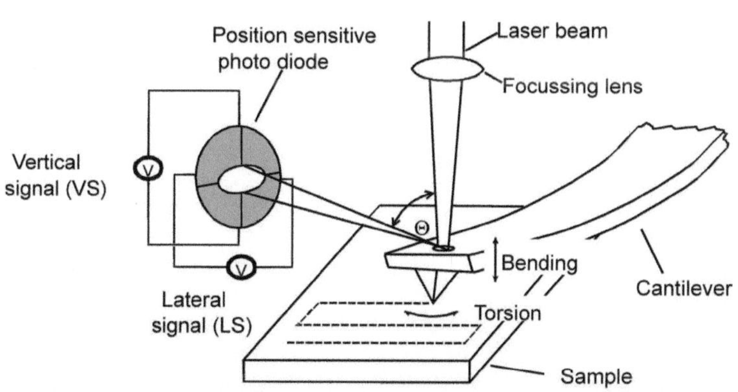

Fig. 3.2. Schematic drawing of the FFM. A microfabricated cantilever with an integrated tip is scanned across a sample surface. Deflection due to normal forces and torsion due to lateral forces are recorded by means of a four-quadrant photodiode

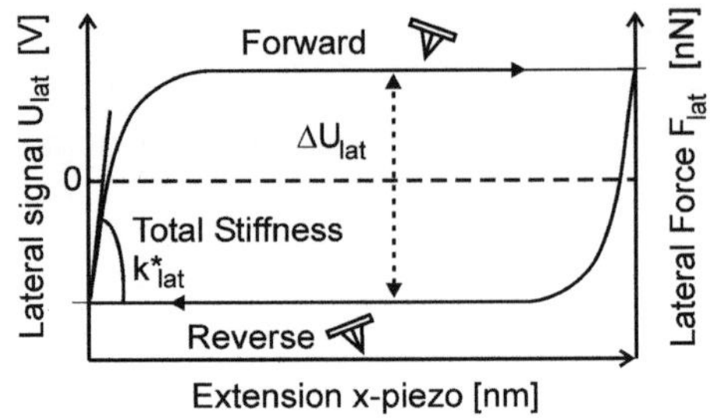

Fig. 3.3. Friction loop (schematic): amount of cantilever twist versus lateral position of the cantilever forward and backward stroke in one full cycle. *Linear part* is believed to be elastic contact (no slip). After critical torsion amplitude is reached, full sliding occurs

As the sensor is scanned across the sample surface, normal forces cause the cantilever to bend vertically. Most commercial instruments use the so-called optical lever method for measuring the cantilever deflection. A laser light is reflected from the backside of the cantilever. Since the lever is tilted, the light is reflected by an angle and led by a mirror into a position-sensitive photo-diode element. The voltage output from the photo diode can then be calibrated into units of normal deflection. As the cantilever is scanned sideways, lateral forces acting between sensor tip and sample surface lead to a twist of the cantilever. The reflected laser beam is laterally deflected on the photo diode. If we plot the lateral deflection of the laser spot versus the lateral position of the sample (or cantilever), we obtain the so-called friction loop as displayed in Fig. 3.3. The friction loop is the result of one full cycle sideways movement. At the beginning of a cycle, the cantilever twist increases in proportion to the lateral position of the sample. In this linear part, we believe, that there is no slip at the center of the tip–sample contact. As the lateral motion is further increased, we observe a deviation from linearity of the cantilever response until, at a maximum torsion, the tip is in full sliding motion and the twist remains constant. When the motion is reversed at the second half-cycle, we obtain the mirrored signal of the first half. The difference of maximum and minimum voltage at forward and reverse scan is the friction loop amplitude. If the position modulation is quantitatively known, we can convert the read-out voltage from the photo diode into units of cantilever twist angle θ, provided that elastic or plastic deformations are negligible. Problems of converting voltage output of the lateral deflection of the laser spot into twist angle have been discussed in detail [3]. If the cantilever is of width w and thickness t and its shear modulus is G, its torsional stiffness c_T is

$$c_T = \frac{wt^3}{3} G \tag{3.13}$$

The units of c_T are Nm^2. A cantilever that is twisted by an angle θ exerts an angular momentum M_θ at a distance L from the clamped end (see Fig. 3.4).

$$M_\theta \big|_{x=L} = \frac{c_T \theta}{L} \tag{3.14}$$

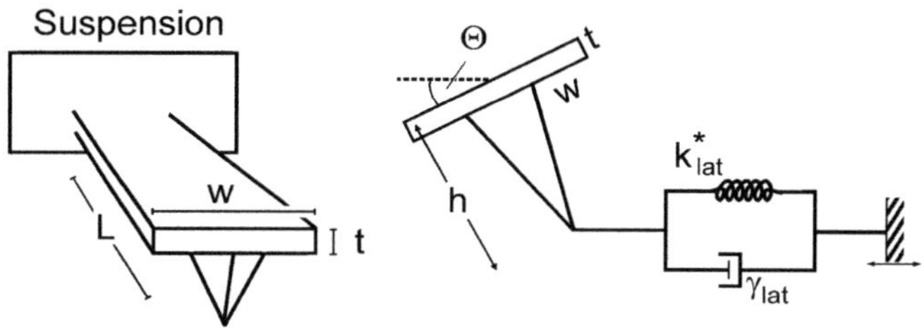

Fig. 3.4. Schematic drawing of the cantilever. Clamped-free state (*left*) and in contact with a laterally modulated sample modeled by damped elastic contact with a modulated lateral force (*right*)

A lateral force F_{lat} at the end of the sensor tip of height h, leads to a lateral displacement Δx where $\sin\theta = \Delta x/h$. If the tip height is much greater than Δx, then $\sin\theta \approx \theta = \Delta x/h$. The balance of angular momentum exerted by the lateral force and by the restoring spring force leads then to

$$F_{lat} = \underbrace{\frac{wt^3G}{3L}\frac{1}{h^2}}_{k_{c,lat}}\Delta x \tag{3.15}$$

The constant $k_{c,lat}$ is called the lateral restoring constant, has the units N m^{-1}, and relates the lateral force exerted at the tip end with its lateral deflection. This way we have an equivalent to the cantilever bending formula

$$F_N = k_{c,vert}\Delta y \tag{3.16}$$

which is used to calculate the normal force exerted by deflecting a cantilever in normal direction. The constant $k_{c,vert}$ is the cantilever bending stiffness and is usually referred as to the cantilever spring constant. It comprises Young's modulus E, width, thickness, and length of the cantilever.

$$k_{c,vert} = \frac{wt^3}{4L^3}E \tag{3.17}$$

As we noted above, the slope of the linear part in Fig. 3.3 can be converted into units of lateral force if no sliding occurs within the contact and elastic deformation of tip and sample material is negligibly small. However, in fact the slope is the stiffness of the whole system k^*_{tot}. If we assume that all deformation lies in the cantilever twist and in the sheared contact, the deformation can be modeled as two springs in sprips with spring constants $k_{c,lat}$ and k^*_{lat}.

$$\frac{dF_{lat}}{dx} = k^*_{tot} = \left(\frac{1}{k_{c,lat}} + \frac{1}{k^*_{lat}}\right)^{-1} \tag{3.18}$$

The total stiffness of the system is determined by the lateral restoring constant of the cantilever and the resistance to elastic deformation of the contact. The latter shall be referred to as the lateral or shear contact stiffness k^*_{lat}. It has an equivalent for normal loading, the vertical or compression contact stiffness k^*_{vert}. The compression contact stiffness at contacts commonly encountered in contact-mode force microscopy is typically in the range of 100–2,000 N m^{-1} [16]. It is therefore much greater than the bending stiffness of force microscopy cantilevers which are typically 0.01–10 N m^{-1}. In this case, deformation of the contact is almost negligible, if compared with the cantilever bending. The lateral restoring constant of commercial force microscopy cantilevers, however, is usually several hundred N m^{-1}. In this case, the deformation of the contact is more pronounced relative to the cantilever twist.

3.5 Lateral Modulation Below Resonance

Basically, there are two options of lateral modulation: (i) excitation of the sample and (ii) excitation of the cantilever. The most straightforward method is to use the piezoelectric tube scanner that is implemented for scanning sample or cantilever. Maivald et al. first reported lateral modulation experiments in 1991, employing a modified Nanoscope II (Veeco Metrology) [17]. The design of this instrument is schematically drawn in Fig. 3.5. The lateral position of the sample was oscillated by an amplitude of 25 nm at a frequency of 5 kHz, while the slow raster-scan was maintained to image a 100×100-nm^2 area of a gold sample. The lateral modulation did not interfere with the conventional imaging mode in the AFM. While scanning the sample, the high-frequency torsional amplitude can be acquired and analyzed by means of an rms-to-DC converter. This rather simple electronic circuit provides a DC voltage output with the DC voltage being proportional to the torsional amplitude. The DC voltage can be fed back in a separate data channel of the AFM, for example through the Signal Break Out Box (SBOB) from Veeco Metrology. As a result, Maivald et al. were able to display the torsional amplitude image simultaneously with the conventionally acquired topography image. In their lateral mod-

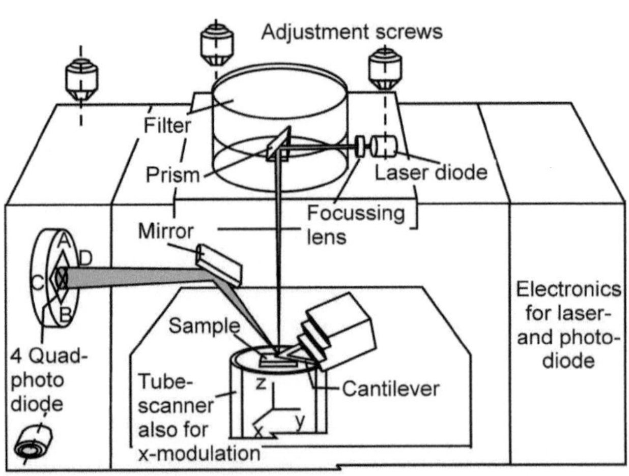

Fig. 5. Head of an atomic force microscope Nanoscope II (Veeco Metrology) with implemented FFM technique. The scanning tube can be used for lateral modulation experiments by adding a modulation voltage to the x-axis electrodes on the scanner. The modulation does not interfere with conventional imaging and data acquisition

ulation image, edges of monoatomic gold steps were highlighted, contrary to the topography image in which only a groove was visible. Several groups continued using similar approaches, either for measuring elastic properties, or for measuring friction on the nanoscale.

3.5.1 (Visco-)Elastic Properties by Lateral Modulation

Ge et al. investigated the glass transition temperature T_G of monodispersed polystyrene thin films [18]. In this method, which they refer as to shear modulation force microscopy (SFMM), a sinusoidal voltage drive signal with a frequency of 1.4 kHz is applied to the X-piezo element of the AFM. Lateral oscillatory motion is introduced to the sensor tip by lateral forces, and the tip–sample contact is assumed to be viscoelastic with no sliding. T_G is measured by qualitatively monitoring the lateral deflection amplitude Δx of the laser spot on the position-sensitive photo-diode and thus, the cantilever torsional amplitude. Ramping the temperature from 310 to 400 K, the torsional amplitude increased linearly with temperature. The slope of the Δx-versus-temperature curve did indeed change dramatically at the glass transition temperature (342–379 K for the samples under inspection). With this technique, it was possible to measure the glass transition temperature as a function of molecular weight of the polymer films and the influence of film thickness.

Wahl et al. studied viscoelastic effects in nanometer-scale contacts under shear [19]. By analyzing phase and amplitude of the cantilever twist, they obtained the relaxation time of poly(vinyl ethylene) samples. The influence of ageing on relaxation time was studied as well as the shear contact stiffness.

To reveal the shear contact stiffness on a nanoscale by lateral modulation, the excitation amplitude must be sufficiently small, cantilevers must be sufficiently compliant, and normal load sufficiently high, in order not to evoke sliding. This concept was realized in an AFM experiment by Carpick et al. [14]. The relative lateral position between the cantilever base and the sample was sinusoidally modulated by a few Angstroms; sliding-free contact was assumed. With a lock-in amplifier, the amplitude of the torsional vibration was recorded as a function of applied normal load. The ratio of the in-phase amplitude dF_{lat} and the lateral modulation amplitude dx provides the total lateral stiffness of the system, k^*_{tot}. If the restoring constant of the cantilever $k_{c,lat}$ is known, the shear contact stiffness k^*_{lat} can be derived. By using a lock-in amplifier, one can also analyze the out-of-phase response of the cantilever in torsional motion. Pure elastic response of the system can only contribute to in-phase motion; out-of-phase response clearly indicates the presence of slip. Slip on the other hand takes place if the contact (of area A) is sheared by a lateral force F_{lat} at a critical shear strength $\tau = F_F/A$. The critical force is the friction force F_F, and even if the contact is further laterally displaced, F_F does not increase any more. With the modulation technique, we have access to the shear contact stiffness. With conventional FFM, friction can be measured as a function of normal load. For a sphere–plane contact, the shear contact stiffness is proportional to the contact radius a [10].

$$k^*_{lat} = 8\,G^* a \tag{3.19}$$

Hence, the critical shear strength τ is a function of the reduced shear modulus G^*, the friction force, and the shear contact stiffness:

$$\tau = \frac{F_F}{\pi a^2} = \frac{64\, G^{*2} F_F}{\pi\, k_{lat}^{*\ 2}} \tag{3.20}$$

G^* is the reduced shear modulus

$$\frac{1}{G^*} = \frac{2-\nu_1}{G_1} + \frac{2-\nu_2}{G_2} \tag{3.21}$$

with the shear moduli G_1 and G_2 and Poisson number ν_1 and ν_2 of probe and sample material, respectively. Carpick et al. have determined this shear strength in an FFM experiment. By applying a normal load in the range of 30 nN adhesive to 40 nN repulsive normal force, they found constant shear strength between a Si_3N_4 probe tip and a cleaved mica sample of approximately 680 MPa. We can see from Eq. (3.20) that the tip radius does not need to be known a priori for calculating the shear strength. Carpick et al. determined the shear stiffness k_{lat}^* to be 0–70 N m^{-1}, for the given load range. The authors make it clear that the contact deformation can be equal or higher than the lateral lever deformation, at loads and conditions typically encountered in FFM experiments.

3.5.2 Friction Properties by Lateral Modulation

Colchero et al. propose a method to measure friction on the nanoscale by laterally oscillating the cantilever at a frequency of 1–20 kHz. They introduced lateral modulation by applying an AC voltage to the X-piezo [20]. The modulation amplitude was about 10 nm and the sample was cleaved mica. They made use of a lock-in amplifier that converts the high-frequency signal of the cantilever torsion into a DC voltage that is proportional to the high-frequency amplitude. A correction factor κ was calculated that compensates for the non-sinusoidal shape of input signal to the lock-in amplifier, due to sliding of the tip. As the X-piezo was laterally modulated, a force-versus-distance curve was taken, and at the same time the output of the lock-in amplifier was recorded. If the torsional vibration is plotted against the position of the Z-piezo in the AFM, a qualitative friction-versus-load curve was obtained. It was not explicitly mentioned, but the authors neglected elastic deformation of the contact and the cantilever torsion was only led by stick and slip processes. Finally, they used lateral modulation for imaging a sample of Si/SiO_2. The slow image-scanning motion of the cantilever was superimposed by a small amplitude (5 nm) high-frequency (20 kHz) lateral oscillation. Two regions could be identified having different friction coefficients as proven by dynamic friction-versus-load curves.

A lateral dithering of samples can also be used for studying lubricants. O'Shea et al. oscillated a mica sample lubricated with *n*-dodecanol (a surfactant) laterally with a frequency in the range 50–300 Hz and a dither amplitude lower than 5 nm [21]. By investigating this liquid–solid interface at room temperature, they found that the liquid is strongly layered. The lubrication properties of the adsorbed molecules could be shown at a high pressure around 100 MPa.

3.5.3
Lateral Modulation by Piezoelectric Transducers

Use of the piezoelectric tube scanner for lateral modulation has a drawback: the modulation frequency is typically limited to several kHz. Above this frequency, the tube may vibrate in resonant modes and hence, the vibration amplitude cannot be controlled. Thus, any result above this frequency will be questionable.

To overcome this drawback, one can use external sources for lateral modulation: either a piezoelectric shear plate, polarized to vibrate in-plane, or a piezoelectric shear wave transducer as they are commonly used for nondestructive testing.

Shear or transverse-wave transducers are nothing but shear plates with anti-wear coatings at their front side, a housing, and resistances and impedances for optimizing the frequency bandwidth for transmitting and receiving ultrasonic shear waves. Although they are at least one order of magnitude more expensive than shear plates, they are advantageous for two reasons: optimum mechanical damping on the backside and a mechanical design based on years of experience. They can be used as extremely broadband sources for in-plane vibrations. That means that the excitation amplitude in dynamic AFM experiments does not change abruptly by continuously varying the excitation frequency. We have good experience in using fingertip-type shear wave transducers, for example those provided by Panametrics, Waltham, MA (V156 and V157, 5-MHz center frequency). These transducers were originally designed for nondestructive testing of "macroscopic" components. There, usually, shear waves are superimposed by parasitic longitudinal waves with amplitudes generally 30 dB below the shear wave amplitudes. In the AFM case, samples are generally rather thin and tiny. If a thin sample is mounted on the shear wave transducer and covers only a part of its surface, the amplitude of parasitic out-of-plane vibrations can be of the same order of magnitude as the in-plane vibration. We used delay lines to overcome this disadvantage. The delay lines were aluminium circular plates with the same diameter as the transducer, and a thickness of several mm. The delay line was coupled onto the transducer surface and the sample was mounted on top of the delay line by a suited couplant. We noticed that by using the delay technique, parasitic out-of-plane vibrations were substantially reduced.

Already in 1994, Yamanaka et al. published some ideas on lateral ultrasonic force microscopy, called "lateral" UFM [22]. A piezoelectric transducer was bonded onto a sample holder (scanning tube) of an AFM, and the sample was glued on the transducer. A Si_3N_4-type cantilever with a spring constant of 0.01 N m^{-1} and a bending resonant frequency of 40 kHz was used. The transducer mounted beneath the sample excited a lateral modulation of 10 kHz frequency. Surface steps of an HOPG sample were enhanced in the lateral modulation image, similar to what was expected to be an edge artifact in FFM. In addition to the lateral oscillation, a bending vibration was excited in the cantilever via another piezoelectric element on the cantilever mount. The excitation frequency was 5.6 MHz, much greater than the lowest bending resonance in contact. Striking features different from the surface steps were resolved in this lateral-UFM image. The authors note that this asymmetric contrast could be due to subsurface edge dislocations with extra atomic planes on one side. The fact that these features appear much more distinguishable

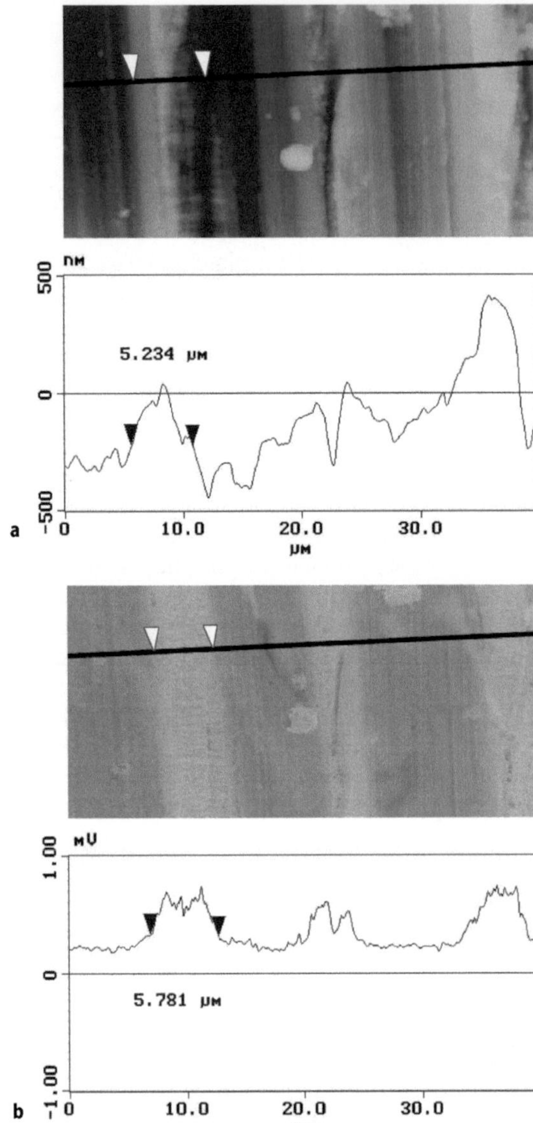

Fig. 3.6. **a** Topography of the biphase PE polymer with high-density (hard) phase indicated by *arrows* and low-density (softer) phase. *Below*: cross-sectional view along the *white line* in the scan. **b** Ultrasonic amplitude of cantilever torsion at resonance when scanned across the PE sample

as in the other imaging modes was explained by the higher subsurface sensitivity of the combined lateral modulated UFM.

The lateral modulation technique can also be used for imaging the lateral stiffness. We have examined a two-phase polyether blend consisting of alternating layers of low-density (LDPE) and a high-density (HDPE) phase. The thickness of the alternating layers was about 5–10 µm. The sample was prepared by cutting across the layers with a kryo-microtom and was then mounted on the ultrasonic transducer. The ultrasonic transducer was excited at 10 kHz. The amplitude was 10 V leading to a surface amplitude that was estimated to be 0.5 nm. Topography and torsional amplitude of the cantilever were recorded at the same time (see Figs. 3.6a and b). The amplitude image clearly resolves different contrast of LDPE and HDPE despite the much-distorted topography (due to the preparation). The higher amplitude on the HDPE can be ascribed to higher contact stiffness.

We have learned that lateral modulation can provide useful information about the elastic properties on the nanometer-length scale [23]. The purpose of conventional FFM, however, was to investigate friction. Several research groups also introduced lateral modulation for friction experiments on the nanometer scale.

One can combine FFM with lateral modulation and lock-in analyzing techniques. Krotil et al. proposed what they called dynamic scanning friction force microscopy (DSFFM) for friction force studies on polymer systems [24]. With the probe in contact, the sample is oscillated laterally by applying a sinusoidal modulation to a shear piezo mounted underneath the sample. The modulation amplitude was typically from a few nanometers up to several tens of nanometers. The lateral force signal detected by the four-quadrant photodiode was fed into a lock-in amplifier, in order to measure the cantilever torsional amplitude and phase at the modulation frequency. Just the onset of sliding, the friction loop amplitude r and phase ϕ of the detected signal are given by

$$r = cF_H \tag{3.22}$$

$$\phi = 0 \tag{3.23}$$

F_H is the sticking friction and c is a calibration factor for the photo-diode sensitivity. Friction force is at a maximum when the tip sticks to the surface (static friction force). As the modulation amplitude is increased above a critical value, the tip begins to slip (sliding friction force) and delay of the detected phase signal occurs. The value of the dynamic friction F_G can be determined from the amplitude value for high modulation amplitude

$$\lim_{A_m \to \infty} r = c\frac{4}{\pi}F_G \tag{3.24}$$

With this method, Krotil et al. were able to determine both static and sliding friction force values.

Whether using the scanner or an external PZT (lead zirconate titanate), one can laterally modulate the sample position with sufficiently high amplitude so that a slip occurs. Lateral dithering while the cantilever is scanned across the sample can be used for imaging. For example, Munz et al. characterized automotive brake pads in this way [25]. The lateral displacement modulation was induced by means of a piezoelectric out-of-plane transducer. Instead of being mounted beneath, the transducer was attached to the side of the cantilever suspension. Suspension and cantilever were sideways excited at a frequency of 59 kHz; the torsional amplitude of the cantilever was recorded by means of a lock-in amplifier. A brake pad material composed of polybutadiene (PBD), antimony sulphide (Sb_2S_3, solid lubricant), both dispersed in a polymeric matrix of phenolic resin was the object for investigation. Analysis of the cantilever torsional amplitude and phase, while raster-scanning the sample, gave contrast at the scanned areas of different friction.

3.6 Resonance Techniques

So far, we have learned about lateral modulation techniques with modulation frequencies typically up to several kHz. Lateral and normal forces obey the corresponding straightforward spring force law of the form $F = -Ds$ with the spring constant D and displacement s. Resonance phenomena did not need to be taken into account. However, cantilevers are elastic beams with characteristic bending and torsional vibration modes. By means of continuum elasticity theories, the resonance frequencies can be calculated. The resonance frequencies depend on the cantilever geometry, the forces acting at the beam and the boundary conditions. They can be calculated analytically for rectangular beams or numerically for more complex structures such as V-shaped cantilevers [16, 26]. In atomic force acoustic microscopy (AFAM), out-of-plane sample vibrations couple into bending resonances of the cantilever via the tip–sample interaction. When the probe tip is kept in contact with the vibrating sample, so-called contact resonances appear. These can be analyzed to exploit qualitative and quantitative information about the sample elasticity on the nanometer scale [27].

3.6.1 Free Resonances

Likewise, we can utilize torsional vibrations. The equation of motion for torsional vibrations in a rectangular beam is

$$c_T \frac{\partial^2 \Theta}{\partial x^2} - \rho J \frac{\partial^2 \Theta}{\partial t^2} = 0 \tag{3.25}$$

where Θ is the angle of twist, $c_T = wt^3G/3$ is the torsional stiffness, and $J = w^3t/12$ is the polar area moment of inertia. ρ is the mass density and G the elastic modulus of the beam[1], and w and t its width and thickness, respectively. A general solution for Eq. (3.25) can be written as

$$\Theta(x) = A \sin \eta x + B \cos \eta x \tag{3.26}$$

where constants A and B are given by the boundary conditions. Cantilevers are usually clamped at one end and the other end is either free or in contact with the sample. The boundary conditions for the cantilever vibration take the form

$$\theta|_{x=0} = 0 \tag{3.27}$$

at the clamped end and, due to equilibrium of moments,

$$c_T \left. \frac{\partial \Theta}{\partial x} \right|_{x=L} = F_{lat} h \tag{3.28}$$

[1] Note that the appropriate crystallographic direction (normal to the cross-sectional plane) of the cantilever shear modulus must be chosen for calculation. For most commercially available cantilevers, this is G <110>. Flexural vibrations require Young's modulus of orientation normal to the area $w \times L$, usually E<100>.

at the end with the tip contacting the sample surface, where F_{lat} is the lateral force acting on the tip, and h is the tip length.

For the clamped-free boundary condition, the lateral force and thus the moment at the cantilever free end is zero. The distribution of angular deflection is

$$\Theta(x) = A \sin \eta x \tag{3.29}$$

The torsional wave number η is obtained by the balance of the torsional moments [Eq. (3.28)] and leads to the eigenfrequencies for free torsional vibrations [28]

$$f_{n,tor} = \frac{2n-1}{2} \underbrace{\frac{1}{L}\frac{t}{w}\sqrt{\frac{G}{\rho}}}_{\Delta f} \tag{3.30}$$

It is noteworthy that, unlike the bending resonances of a cantilever, the torsional modes are equally spaced by Δf, because Eq. (3.30) exhibits no dispersion.

The corresponding governing equation for flexural or bending mode frequencies $f_{n,bend}$ is

$$f_{n,bend} = \frac{(k_n L)^2}{2\pi\, L^2}\sqrt{\frac{\rho A}{EI}} \tag{3.31}$$

The area moment I is $wt^3/12$, for a cantilever of rectangular cross-section. The product of wavenumber k_n and length L is a constant for the nth mode (n=1, 2...) and it is universal for bending vibrations of a rectangular beam.

Introducing clamped-free torsional resonances in the cantilever can be performed either through a modified cantilever holder or by using the vibration of a sample beneath the cantilever. For example, a shear wave transducer is excited by a sinusoidal drive signal and the frequency is swept (see Fig. 3.7). The transducer is oriented so that the shear polarization is perpendicular to the length of the can-

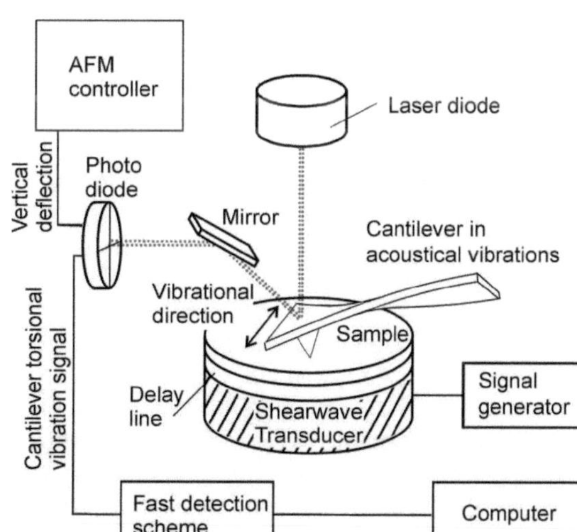

Fig. 3.7. Experimental set-up for exciting and detection of cantilever torsion, either free or in contact with the sample surface

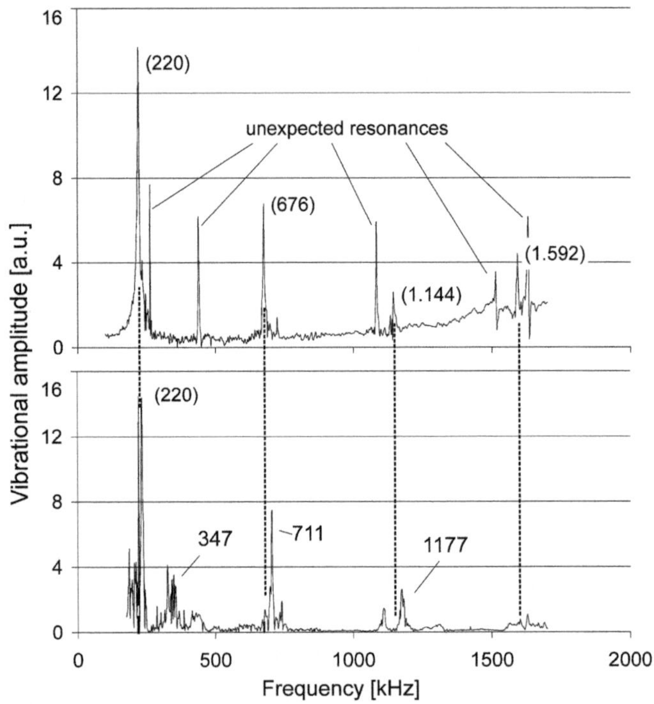

Fig. 3.8. Free (*top*) and contact (*bottom*) torsional resonance spectrum of a Si cantilever (455×43×1.9 μm^3), excited by a transducer mounted beneath the sample

tilever. The cantilever is positioned above the transducer with a vertical distance of some micrometers. Although there is no intimate contact, lateral forces couple into the tip-cantilever system and excite torsional vibrations in the cantilever (see Fig. 3.8, top graph). Several attempts have been made by our group to explain the mechanism of contactless excitation of torsional vibration by shear transducers. Though, we do not yet have a clear understanding of the interaction, being air-transmitted ultrasound, electromagnetic, or other.

Through (Eq. 3.30) and Eq. (3.31), elastic and/or geometrical properties of a cantilever can be evaluated by analyzing torsional and bending resonances of the clamped-free cantilever. Elastic properties for such thin beams are of particular interest, since they can differ from bulk values and it is difficult to measure them by other means. By using a home-built cantilever holder with an extra piezoelectric element, Nakano et al. excited clamped-free torsional resonances in a cantilever [29]. They modified the cantilever suspension in an AFM unit manufactured by Seiko Instruments Inc., Japan, implementing a vertically polarized and a shear-polarized PZT. The combined transducer is attached to the cantilever support by silicone grease couplant. Drive signals with identical phase were applied to both PZTs and thus, a directional vibration oblique to the plane of the cantilever was excited in the cantilever support. The deflection and torsional signals were measured by using the optical lever system. Young's modulus and Poisson's ratio can be estimated from experimental and calculated resonance frequencies. Elastic data of a Si_3N_4 beam were measured; the results deviated from literature elastic data for the bulk material.

3.6.2 Contact Resonances (Visco-)Elastic Contact

For a viscoelastic tip–sample interaction, the lateral force acting on the tip in Eq. (3.28) is

$$F_{lat} = -k^*_{lat}\left[h\,\theta(L,t) - a(t)\right] - d_K h\left[h\,\dot{\theta}(L,t) - \dot{a}(t)\right] \tag{3.32}$$

where k^*_{lat} is the shear contact stiffness, d_K the viscoelastic damping constant, and $a{=}a_0\sin(\omega t)$ is the lateral surface displacement of angular frequency ω. The solution of the beam equation (Eq. 3.25) with the clamped/surface-coupled boundary conditions (Eq. 3.27, Eq. 3.28, and Eq. 3.32) takes the form

$$\theta(\omega, L) = a_0 \frac{\sin\left(\frac{\omega}{4f_0}\right)\sqrt{{k^*_{lat}}^2 + (d_K\omega)^2}}{\sqrt{\left(\frac{c_T}{4f_0}\omega\cos\left(\frac{\omega}{4f_0}\right) + k^*_{lat}\sin\left(\frac{\omega}{4f_0}\right)\right)^2 + \left(d_K\omega\sin\left(\frac{\omega}{4f_0}\right)\right)^2}} \tag{3.33}$$

where f_0 is the frequency of the first torsional clamped-free eigenmode of the cantilever as determined from Eq. (3.30). For a purely elastic contact, Eq. (3.33) reduces to

$$\theta(\omega, L) = a_0 \frac{k^*_{lat}\sin\left(\frac{\omega}{4f_0}\right)}{\sqrt{\left(\frac{c_T}{4f_0}\,\omega\cos\left(\frac{\omega}{4f_0}\right) + k^*_{lat}\sin\left(\frac{\omega}{4f_0}\right)\right)^2}} \tag{3.34}$$

Note that damping-less contact as well as undamped cantilever vibration are hypothetical cases. Such cases would result in catastrophic resonance with infinite amplitude. By substituting Eq. (3.32) and Eq. (3.34) in Eq. (3.28), the shear contact stiffness can be expressed as a function of torsional contact resonance frequency f_{tor}

$$k^*_{lat} = -\frac{\pi f_{tor}\frac{G}{3}\frac{w^2t^2}{h^2}\sqrt{\frac{\rho}{G}}}{\tan\left(\pi f_{tor} L\frac{w}{t}\sqrt{\frac{\rho}{G}}\right)} \tag{3.35}$$

With Eq. (3.35), we have means to determine k^*_{lat}, at a given load, from the cantilever in elastic contact with the sample. As the tip is engaged at a constant normal load, the lateral surface vibration couples into the cantilever via the elastic tip–sample contact.

The method has been tested on different sample materials. A rectangular silicon cantilever (CONT-type, dimensions 450×50×2 μm³ as given by the manufacturer) was first vibrated with the free end to obtain the free resonances (see Fig. 3.8, top graph). Then, the tip was engaged with a monocrystalline silicon sample at an externally applied normal load of approximately 20–50 nN. The spectrum of the lever

Table 3.1. Calculated and measured resonance frequencies for rectangular Si cantilever 455×43×1.9 μm³. $E_{Si<100>}$=169 GPa, $G_{Si<110>}$=51 GPa, ρ_{Si}=2.33 kg cm⁻³

Mode number (*n*)	Bending resonance frequency (f_n) (kHz)		Torsional resonance frequency (f_n) (kHz)	
	Calculated	Measured	Calculated	Measured
1	13	13	227	220
2	79	79	681	660
3	223	222	1,135	1,144
4	434	437	1,589	1,592
5	718	724	2,043	–
6	1,072	1,084	2,496	–

Table 3.2. Shear contact stiffness k^*_{lat} calculated from measured contact resonances f_{tor} of a rectangular Si cantilever 455×43×1.9 μm³.Tip length was 15 μm

Mode number (*n*)	Silicon		Glass		HDPE		LDPE	
	f_{tor}	k^*_{lat}	f_{tor}	k^*_{lat}	f_{tor}	k^*_{lat}	f_{tor}	k^*_{lat}
1	347	129	380	229	339	112	280	36
2	711	51	790	252	696	25	684	5
3	1,177	120	1,198	190	1,152	47	1,140	14
4	1,604	58	1,596	28	1,600	43	1,592	13
			1,644	224				

in contact with the sample is shown in the bottom graph. Besides the resonances that were expected from Eq. (3.30), somc unexpected resonances can be observed in the upper graph.[2] The bending vibration spectrum (see also Table 3.1) exhibits bending resonances at 437 and 1,084 kHz that are likely to couple into the lateral signal channel. For further explanation of unexpected resonances see section *imaging torsional vibration* in this chapter. From the bottom graph of Fig. 3.8, we notice that the resonances are shifted to higher frequency if the tip is in contact with the sample. If the measured contact resonance f_{tor} is entered in Eq. (3.35), the shear contact stiffness can be calculated. This has been done for samples of monocrystalline <100> silicon, borosilicate glass, low-density polyether (LDPE), and high-density polyether (HDPE) material. The torsional resonances were measured at a net normal load of roughly 50 nN. The contact stiffness was calculated for a tip height of 15 μm (see Table 3.2). It is not as straight forward to deduce the sample elasticity from the shear contact stiffness. From Eq. (3.19), the reduced shear modulus G^* is the ratio of k^*_{lat} and the eightfold of the contact radius a. A solution can be found only with additional information about the elastic properties of the tip. The unknown tip radius and hence the contact radius may be obtained by using cal-

[2] Note: cantilever dimensions are not accurately determined by the manufacturer. Equation (3.30) and Eq. (3.31) can be used for precisely predicting the cantilever geometry, in this case 445×43×1.9 μm³.

ibration samples, as is similarly done in the AFAM technique where one exploits the vertical contact resonances to measure the local indentation modulus [30, 31]. On the other hand, by using appropriate contact models one can approximate the contact radius. By using the Hertz model [10, 32], one can write

$$G^* = \frac{k^*_{lat}}{8} \sqrt[3]{\frac{4\,E^*}{3\,R\,F_N}} \tag{3.36}$$

where R is the radius of the probe tip. A solution can be found if the elastic data for the tip E, G, and ν are known. A Poisson number for the sample material must be assumed a priori, and entered in the calculation.

It must be noted that the Hertz model covers the case of a sphere indenting another sphere or a flat. Putman and Igarashi showed that under ambient conditions, an Si_3N_4 tip in contact with a sample indeed performs a Hertz-type contact: friction force scaled with the 2/3-power of normal load [33]. For the same tip under gaseous conditions, the friction force increased linearly with load, indicating a multi-asperity contact as predicted by Greenwood and Williamson [34]. Putman and Igarashi further explained this phenomenon by contact contaminants that convert the quasi single-contact into a multiple-contact resulting in a linear friction–load relation. We do not intend to stress this out further but strongly recommend that one give attention to the models used for interpreting shear contact stiffness data. Yamanaka and Nakano proposed to combine vertical and lateral displacement techniques in order to independently derive the compression contact stiffness k^*_{vert} and shear contact stiffness k^*_{lat} [28]. With these data it is possible to deduce directly the Poisson number ν of the sample material under investigation [10]. Then, Young's modulus and shear modulus of the sample can also be calculated.

Another method to introduce torsional vibrations in the cantilever is depicted in Fig. 3.9a. One can simply excite the thickness-mode piezoelectric element in the cantilever holder. Predominantly, bending vibrations will be excited in the cantilever. However, due to slight deviations from symmetry of the cantilever shape or of the piezoelement-mount, mode coupling into torsional vibration can occur. The

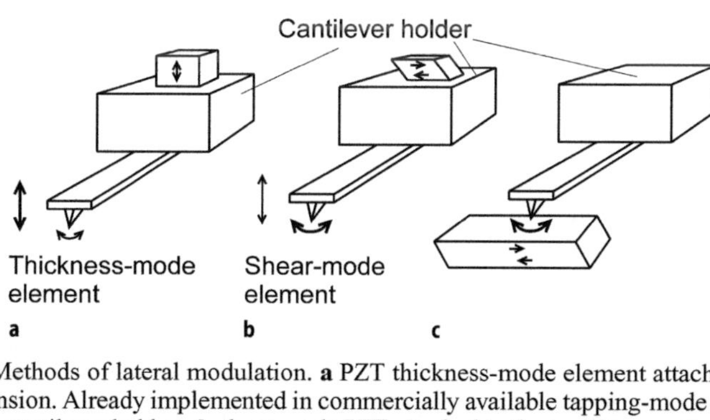

Fig. 3.9a–c. Methods of lateral modulation. **a** PZT thickness-mode element attached to the cantilever suspension. Already implemented in commercially available tapping-mode or force-modulation mode cantilever holders, **b** shear-mode PZT attached to the cantilever suspension, **c** shear-mode PZT or shear wave transducer mounted beneath the cantilever. Sample can be mounted onto the transducer surface

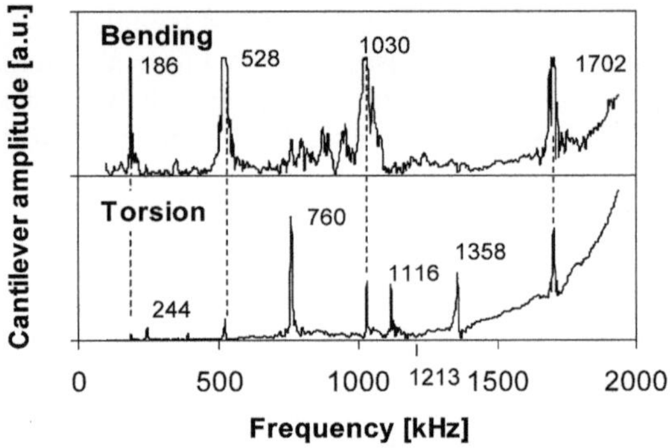

Fig. 3.10. Clamped-free torsional and bending resonance in a Si cantilever (FM-type, Nanosensors, 225×46×1.3 μm^3) acquired by analyzing signals of vertical and lateral deflection channel. Excitation via the tapping-mode cantilever holder, 0.1-V sinusoidal drive signal. Mode coupling indicated by *dotted lines*. Additional resonances from 0.5 to 1.0 MHz

cantilever bending and torsional amplitude are shown in Fig. 3.10 for a frequency sweep from 100 to 1,900 kHz. The experiment was performed in a Dimension 3000 AFM (Veeco Metrology) and the tapping-mode cantilever holder piezoelectric element was excited at 100 mV. A silicon cantilever 225×46×1.3 μm^3 (FM-type Nanosensors GmbH&Co KG, Germany) was utilized for this experiment. An electronic circuit shown in detail in Appendix 2 processed the signals of the vertical and lateral deflection of the laser spot on the position-sensitive photo-diode. This external signal processing was necessary, since the lateral signal output in commercially available instruments is generally limited in bandwidth (an equivalent circuit for a horizontally/vertically split photodiode is drawn in Appendix 1). A lock-in amplifier converted the output of the vertical and lateral channel; the DC output of the lock-in amplifier was recorded as a function of excitation frequency. Various resonances can be found in bending and torsional vibration. The steady increase in amplitude at increasing frequency above 1,300 kHz is an artifact due to inappropriate wire shielding. Resonance frequencies are in good agreement with those calculated for the given cantilever (see Table 3.3). The dotted

Table 3.3. Calculated and measured resonance frequencies forrectangular Si cantilever 225×46×1.3 μm^3. $E_{\mathrm{Si}<100>}$=169 GPa, $G_{\mathrm{Si}<110>}$=51 GPa, ρ_{Si}=2.33 kg cm^{-3}

Mode number (n)	Bending resonance frequency (f_n) (kHz)		Torsional resonance frequency (f_n) (kHz)	
	Calculated	Measured	Calculated	Measured
1	30	–	244	244
2	187	186	732	760
3	528	528	1,220	1116, 1,358
4	1,028	1,030	1,708	1,702
5	1,698	1,702	2,197	–
6	2,537	–	2,685	–

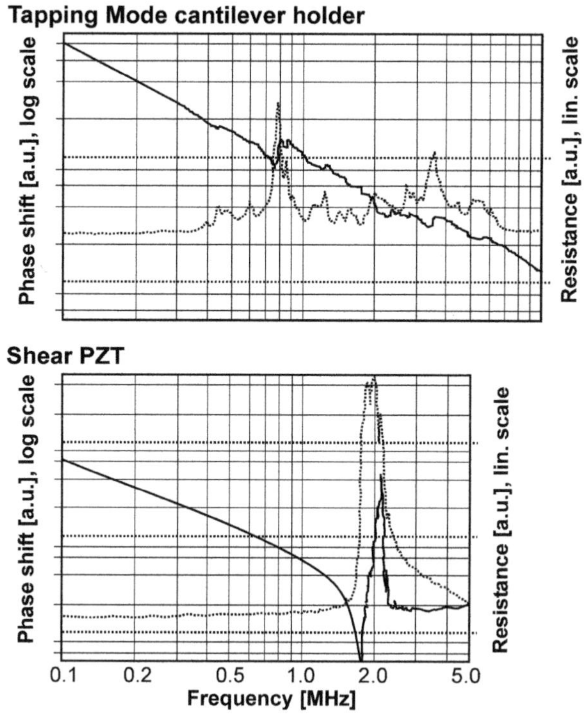

Fig. 3.11. Electrical resistance and phase spectrum of a PZT element in a tapping-mode suspension (*top*) and of a custom-made shear PZT (typical size 10×10×1 mm^3). Resonances in the tapping-mode PZT between 0.5 and 1.0 MHz that may cause unwanted cantilever vibration

lines mark those resonances where we observe strong mode coupling or cross talk between vertical and lateral deflection of the laser spot on the photodiode. That is, torsional resonances appear where they are not predicted from calculation (e.g., at 528 and at 1,030 kHz). At these frequencies, however, there are strong bending vibrations that couple into torsion as discussed later. Another peculiarity can be noticed in the torsional spectrum at 1,116 and 1,358 kHz: the calculation predicts a resonance at 1,213 kHz. We suppose there is mode splitting and it is interesting to note that the two satellite resonances are split by a ratio 2:3 in frequency. Such phenomena are discussed in further detail in the section *imaging torsional vibration*.

By applying the excitation of bending and torsional vibrations in the cantilever by either using the PZT in the cantilever holder or by extra PZTs attached to the cantilever support is disadvantageous for one reason: in Fig. 3.10, several bending resonances can be observed in the frequency range 700–1,000 kHz which are no cantilever resonances. The PZT in the holder or the extra PZT that is attached to the holder is small due to the limited space in the AFM equipment and thus, not optimized in bandwidth. We have verified that the PZT has eigenresonances in this frequency range by an impedance analyzer, and the spectrum is shown in Fig. 3.11. The electrical resistance and phase spectra of the PZT attached to the cantilever holder reveal several resonances above 300 kHz. For comparison, the spectra of a shear-polarized PZT plate of approximately 10×10×1-mm^3 size is also demonstrated in Fig. 3.11. The impedance does not change much between 100 and

1,800 kHz, and so does the surface displacement amplitude. Only at 2 MHz, does the shear plate exhibit a resonance. Thus, the shear plate is superior to the cantilever holder PZT in terms of frequency bandwidth.

3.6.3 Contact Resonances Sliding Contact

Studies on friction force with lateral excitation in a frequency range of 1 MHz have been reported by Scherer and Arnold [35]. Similar to the off-resonance, sliding is likely to occur in the resonance experiment, if the surface amplitude is sufficiently high, the lever stiff, and the damping low. By using a set-up as described above, a monocrystalline silicon sample was vibrated in-plane from a transducer below. Sweeping the excitation frequency from 230 to 250 kHz and analyzing the torsional vibrations by means of a lock-in amplifier leads to a curve as shown in Fig. 3.12. The resonance curves were subsequently acquired at various excitation amplitudes ranging from 0 to 10 V. We estimate a maximum displacement of approximately 0.5 nm at 10 V driving amplitude. At relatively low excitation amplitude, the resonance maximum increases linearly with the excitation voltage and the shape of the resonance curves is symmetric, as predicted by Eq. (3.34). Above a critical excitation amplitude (U=4 V), the shape of the resonance curve changes. The peak flattens out in the center and the shoulder points move away from the center frequency with further increase of excitation amplitude. A qualitative explanation is that at low excitation voltages, the AFM tip sticks to the sample surface and follows the surface motion in an elastic contact. The resonance curve is Lorentzian with a well-defined maximum. When the lateral surface displacement exceeds a critical value, at a certain frequency below and above resonance, the restored energy in the cantilever becomes high enough to initiate single or multiple slip events between the tip and the sample surface. This leads to a flattening of the resonance curve and causes energy dissipation by sliding friction. By analyzing the center frequency in more detail, we find that the resonance tends to shift to lower frequency when the excitation amplitude is increased. This can be interpreted as a reduction

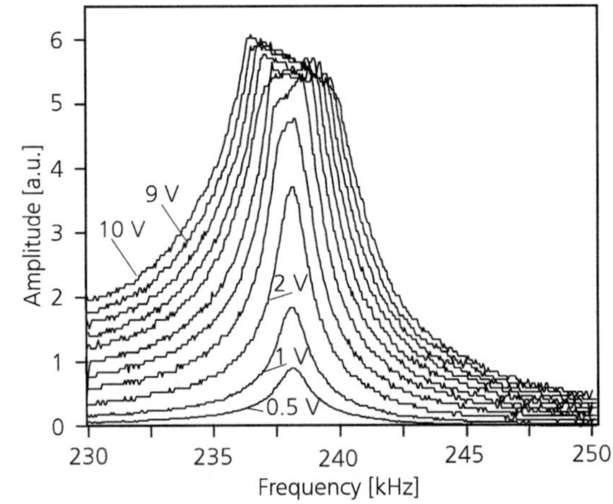

Fig. 3.12. Set of torsional resonance spectra acquired using a silicon cantilever of 0.1 N m^{-1} spring constant on a monocrystalline silicon sample. The *bottom* resonance curve was taken with 0.5-V excitation applied to the shear wave transducer below the sample. The next curve was obtained with 1 V excitation. Then the excitation voltage was increased in steps of 1 V until 10 V

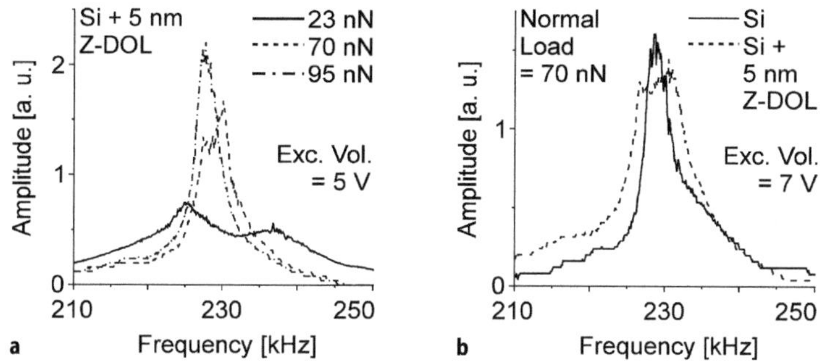

Fig. 3.13a–c. Spectra of torsional resonances for an AFM cantilever (CONT-type, 450×50×2 μm³) in contact with a monocrystalline silicon sample: **a** lubricated with 5-nm PFPE, different excitation amplitudes, **b** variation of applied load, and **c** comparison of bare and lubricated sample

in stiffness, so-called softening. We suggest that the apparent softening below the critical amplitude results from micro-slip in the tip–sample contact as theoretically postulated by the Mindlin theory [36]. Micro-slip caused by an oscillating tangential force entails the slipping of an annulus of the tip–sample contact-area before the center of the contact starts to be laterally displaced. In macroscopic contacts, the existence of microslip has been verified. We have calculated the contact stiffness change due to micro-slip. At the critical amplitude the contact was about 5% more compliant as compared with the static contact stiffness. This is in good agreement with the experiment.

Not only dry, but also lubricated contacts can be studied with such technique. Monocrystalline silicon samples were coated with chemically bonded Z-DOL (a perfluoropolyether with reactive end groups $HOCH_2CF_2O(CF_2CF_2O)_m(CF_2O)_n$-$CF_2CH_2OH$, $m/n \approx 2/3$, molecular weight ≈2,000) [37]. Z-DOL is widely used for lubricating magnetic disks and metal-evaporated magnetic tapes [38]. The lubricant thickness in the experiment was 5 nm as measured with an ellipsometer. The resonance shape of the probe in contact with the lubricated sample was recorded for different normal loads (see Fig. 3.13a). At high normal load (95 nN), the resonance is almost symmetrical. With decreasing static load, we observe a deviation from the Lorentzian resonance shape. Figure 3.13b shows the resonance curve taken at a normal load of 70 nN for the lubricated silicon and for an untreated silicon sample for reference. At the same excitation amplitude, the samples provided different resonance shapes. On uncoated silicon, the resonance curve was almost symmetrical, whereas on the lubricated surface, a resonance flattening was observed. We believe that this flattening is due to sliding friction.

Very recently, we have implemented friction into the equation of motion using a point-mass oscillator model. The external force acting on the tip end was set to have a limit indicating that above this lateral force, sliding occurs. In a numerical simulation experiment, we observed the resonance-flattening phenomenon as a consequence of sliding. Moreover, we have indications that the tilt angle of the flattened part is also a function of surface amplitude thus, sliding velocity. Hence, the resonance shape could be exploited to even investigate velocity dependence of friction on the nanoscale.

With the resonance technique, it will be possible to measure friction on a scale much smaller than conventional FFM. In the latter technique, stroke lengths of 100 nm minimum are required to achieve stable measurements under ambient conditions. With the torsional resonance method, a displacement modulation of the order of one times the contact radius can be applied. The equation of motion is a partial differential equation (the wave equation). The lateral forces compromising the elastic and frictional forces are contained in the boundary conditions. Since small changes in the boundary conditions, such as friction force, have a large impact on the shape of the resonance curve and thus, the resonance method is rather sensitive to measure friction.

3.6.4 "Buckling" Resonances

There is another possibility for measuring friction with ultrasonic in-plane vibrations. Spychalski–Merle et al. used a lateral sample modulation technique polarized parallel to the long axis of the cantilever [39]. Lateral forces acting between tip and sample lead to a buckling of the cantilever, detected by the vertical displacement sensor in the AFM. The resonance spectra show linear dependence of the resonance maximum from excitation amplitude, similar to those shown in the previous section. At high surface amplitudes, the resonance curve is much distorted and sideband maxima occur. The authors explain the existence of the sidebands by the nonlinearity introduced by stick-slip phenomena. Provided that the stiffness for bending and buckling are equal, the established calibration via force–distance curves can be used to extract friction force quantitatively.

3.6.5 Self Excitation

Drobek et al. reported a new method to extract elasticity data from resonant cantilever vibration without external excitation [26]. A V-shaped Si_3N_4 cantilever of 0.13 N m^{-1} spring constant with its first bending resonance at 38 kHz was used. As the tip was indented into the sample material with a given load, its thermal noise vibration spectrum was analyzed by means of a spectrum analyzer. By FEM simulation, they found an empirical relationship between torsional resonant frequency f_{tor} and shear contact stiffness k^*_{lat}. The noise spectrum of torsional vibration showed a resonance at the predicted frequency for the clamped-free cantilever. When the tip was in contact with the sample surface, the resonance in the noise spectra shifts with increasing normal load to higher frequencies. Experiments were performed on surfaces of silicon, aluminium, and cadmium telluride. All experiments showed that shear contact stiffness is proportional to one-third power of normal load. By using the Johnson–Kendall–Roberts (JKR) contact model [10], they were able to derive the reduced shear modulus G^* from the k^*_{lat}-versus-load curve. The measured data are in good agreement with the reduced shear modulus that they obtain from calculation employing literature data for the materials and assuming a certain tip radius.

Kawagishi et al. extended this resonance technique without external excitation for imaging purposes [40]. They used a "contact-mode" cantilever (typically

450×50×2 μm³, Nanosensors GmbH, Wetzlar, Germany) and a lock-in amplifier for screening the cantilever resonance. A torsional resonance at 450 kHz was found when the cantilever tip was in contact with a highly oriented pyrolithic graphite (HOPG) sample, leading to an amplitude of the tip end on the order of 10 pm. When the cantilever was then scanned across the sample surface and the vibration amplitude was recorded at a fixed frequency of 440 kHz, a change in contrast could be observed on different terraces of the HOPG sample. It was not denoted in the paper, but change in such contrast could be due to a change in actual resonant frequency (that would be filtered away by the scheme of detection) or change in vibrational amplitude.

3.6.6 Imaging Torsional Vibration

In the spectra of torsional cantilever vibration, we often noticed resonances in the experiment where they were not expected from calculation. Reinstädtler et al. demonstrated that mode coupling due to asymmetry of mass distribution in the cantilever accounts for a number of additional resonances [41, 42]. The coupled modes can be observed in many AFM cantilevers. Typical torsional spectra of an AFM cantilever (type CONT) are shown in Fig. 3.14. Besides torsional modes, flexural modes also appeared in the lateral channel. However, never were torsional modes observed in the vertical channel. By calculating the first torsional eigenfrequency using Eq. (3.30), the 212 kHz peak in the torsional spectrum can be identified as a torsional resonance. However, a second peak appeared at 265.5 kHz with no corresponding peak in the vertical spectrum at the same frequency. In order to verify the nature of the unexpected resonances, we measured the vertical deflection components of the clamped-free cantilever in torsional resonance. With an optical Michelson heterodyne interferometer set-up as shown in Fig. 3.15, the vertical displacement component was measured quantitatively. The suspension of the cantilever was attached to a shear ultrasonic transducer exciting torsional vibrations. By using a motor-driven linear translation stage, the focal spot of the interferometer was scanned over the entire area of the cantilever. The amplitude and phase of

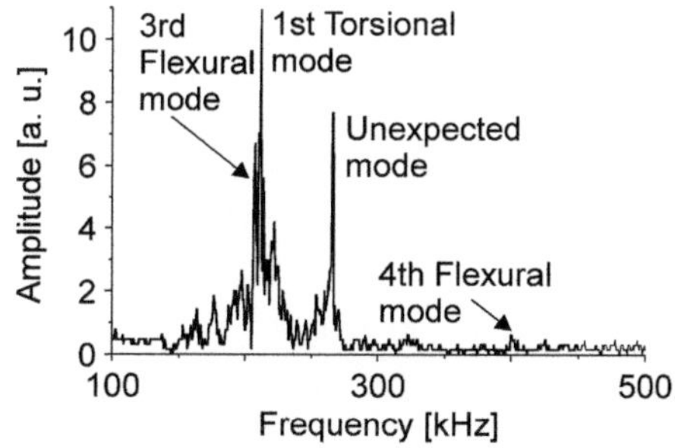

Fig. 3.14. Frequency spectrum of an AFM cantilever (CONT-type, 450-μm long) acquired from the lateral deflection signal excited by a shear wave transducer beneath the cantilever. Besides torsional modes and cross-coupled flexural modes, unexpected modes can be observed

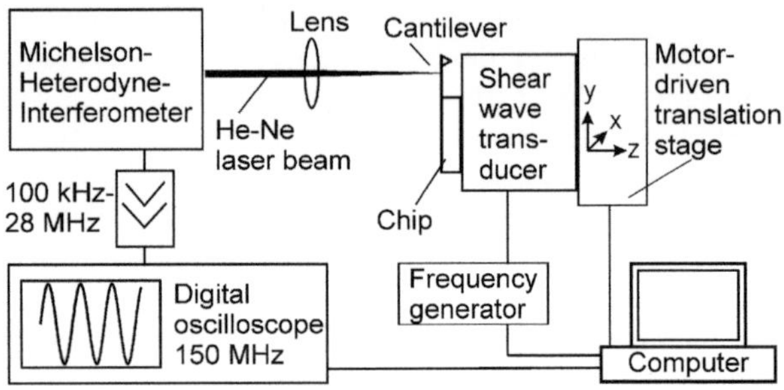

Fig. 3.15. Experimental setup for imaging bending and torsional resonances of an AFM-cantilever using a scanning optical interferometer

each time-signal were obtained by using digital Fourier transformation. The first three torsional modes of the 450-µm-long rectangular silicon cantilever are shown in Fig. 3.16. Amplitude and phase-lag distribution of the second and third torsional mode agree well with the calculated distribution (see Fig. 3.17, plotting Eq. (3.29) with clamped-free boundary condition, using Origin software). It is interesting to note that the first two modes with eigenfrequencies close to one another do not correspond with the modes expected. Moreover, the vibration is asymmetric; nodes do not coincide with the neutral line (longitudinal axis) of the cantilever. We suggest that the origin of the vibrational asymmetry is due to asymmetry in mass distribution. Most likely, the tip with its additional mass may be off-centered. Considering that the tip is off-centered, flexural, and torsional modes are coupled. The coupling is even more pronounced if the tip is in force-interaction with the sample. The resonance peak at 212 kHz comprises both third mode of flexural and first mode of torsional vibration, in accordance with the calculation. By forced vibration, they are being excited simultaneously. Another resonance was also observed at 265 kHz. The corresponding mode does not fit into an analysis based on mode coupling as explained above. Most likely, this mode originates from nonlinear coupling into flexural motion about an axis perpendicular to the plan view. Such a mode would

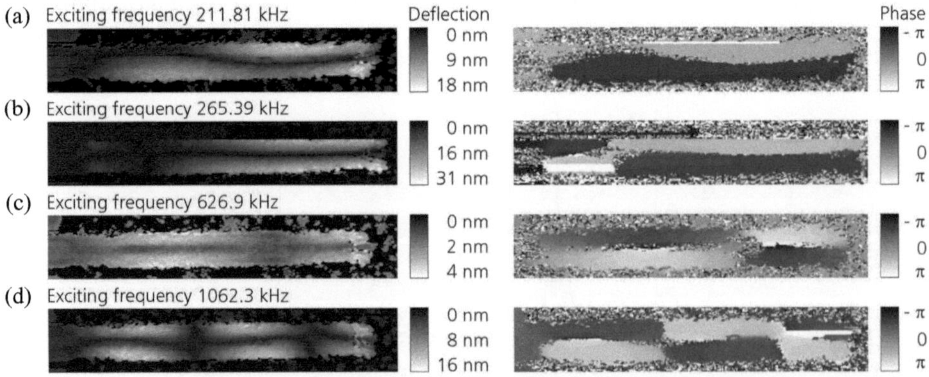

Fig. 3.16. Out-of-plane amplitude (*left*) and phase (*right*) distribution of the first (a), second (c), and third (d) torsional vibration mode of a silicon cantilever (CONT-type, Nanosensors, 450-µm long). Part (b) shows the unaspected mode. Scanned heterodyne interferometer, scan size approximately 550×100 µm^2

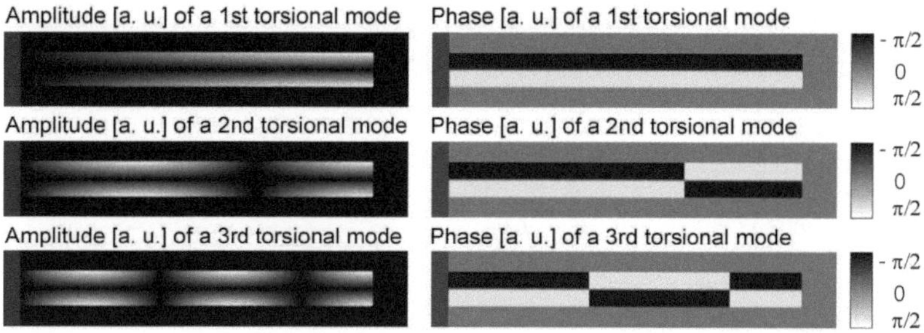

Fig. 3.17. Amplitude (*left*) and phase (*right*) distribution of the first (*top*), second (*middle*), and third (*bottom*) torsional vibration mode of the cantilever. Data calculated by solving Eq. (3.29) under clamped-free boundary conditions, using Origin software program

have a frequency approximately 22.5 times (being the ratio of cantilever width and thickness) the first bending resonance [43, 44]. This value matches with the observed frequency at 265 kHz.

3.7 Sonolubrication

The friction between two sliding surfaces can be substantially reduced by vibrating the surfaces at ultrasonic frequencies (>20 kHz) but the mechanism of ultrasonic lubrication, or referred to as sonolubrication, is not well understood. Even though several reports on this phenomenon have appeared in the past [45–48], it has not been extensively studied. It has not been used in industrial applications because the lubricity effect is rather weak at relatively high normal loads. With decreasing size of mechanical parts in micro-engineering and decreasing normal forces, ultrasound is being considered to actively control friction [49, 50].

An FFM can be used for microscale studies, classical pin-on-flat (POF) test apparatuses for macroscale studies [51]. With these two methods, we have investigated the influence of ultrasonic surface vibration on friction. Normal loads were of the order of 10 nN in the FFM and 1 mN in the POF test apparatus that is schematically drawn in Fig. 3.18. Samples are insonified by ultrasonic transducers in and out of plane. The experiments were conducted with a piece of silicon wafer

Fig. 3.18. Pin-on-flat tester with strain gauge detection, F_N=1–500 mN

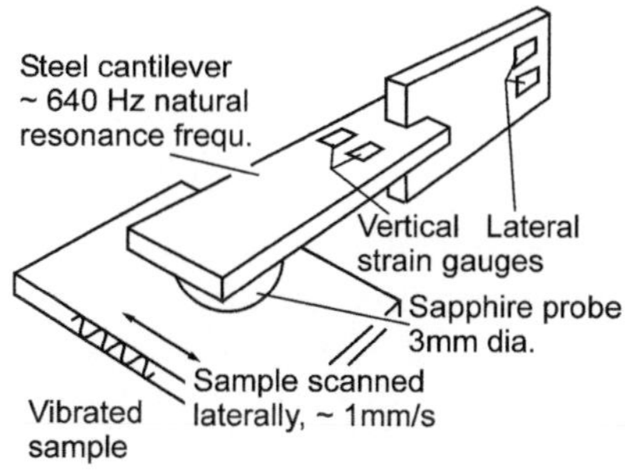

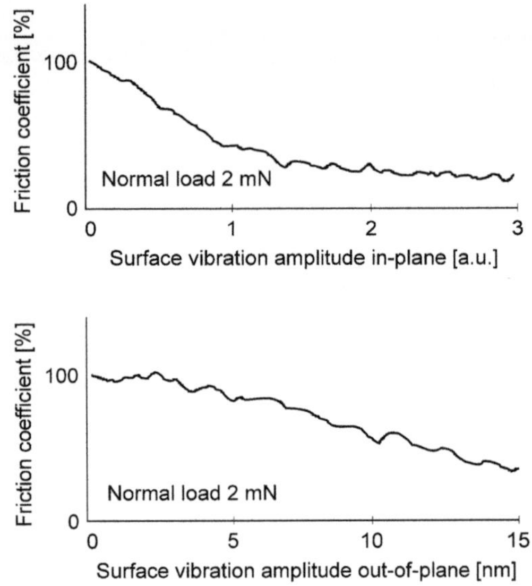

Fig. 3.19. Reduction of friction coefficient when the silicon sample surface is vibrated in-plane (*top*) and out-of-plane (*bottom*) at a frequency of 1 MHz, measured with the pin-on-flat test apparatus. Friction coefficient is recorded relative to the friction coefficient of the static surface, approximatly 0.2

mounted on top of the longitudinal and shear wave transducer. The voltage applied to the transducers was of the order of 1 V and the surface deflections for both out-of-plane and in-plane modes are in the nm regime. In the POF test apparatus, the amount of lateral bending of the cantilever is measured by using the strain gauge beams to obtain friction force. The coefficient of friction is calculated by dividing the friction force with the applied normal force. The coefficient of friction measured with the POF apparatus on a static (i.e., non-vibrating surface) was typically 0.2. For simplicity, the coefficient of friction in this study is given as a fraction of the value of the static surface (i.e., 100%). Figure 3.19 shows the reduction of the friction coefficient, measured with a POF when the surface is vibrated in-plane (top graph) and out-of-plane (bottom graph) at a frequency of about 1 MHz. It was reported that parasitic out-of-plane components are superimposed on the in-plane experiment. For both experiments a substantial reduction of friction was observed, up to a fraction of 20% in the case of in-plane and approximately 40% for out-of-plane vibrations. This reduction of friction does not only occur at 1-MHz frequency, but also at much higher frequencies, a reduction of friction was reported when the surface was exited to vibrate at 25 MHz [51].

A quasi single-asperity contact experiment can be carried out in FFM. The results for a vertically and laterally modulated sample surface are shown in Fig. 3.20. As with the POF, we observe a reduction of friction, obtained by measurement of the amount of cantilever torsion, when the surface is vibrated. Note that in the FFM experiment, the normal forces are now as low as 100 nN, about four orders of magnitude lower than with the POF test apparatus. Simultaneously to the friction data, the tip–sample separation, hereafter referred as to lift-off, was recorded with the split photodiode, as the surface amplitude was increased linearly. There is a nonlinear increase in lift-off, with both in-plane and out-of-plane surface vibrations. In the experiments, the force exerted by the cantilever was held constant at 100 nN.

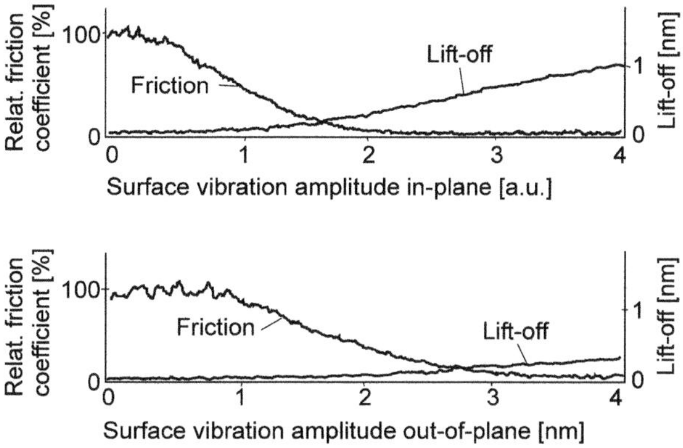

Fig. 3.20. Reduction of friction on silicon measured in the FFM with in-plane (*top*) and out-of-plane (*bottom*) vibration. Friction coefficient was qualitatively determined from the friction loop amplitude. Normal average deflection (lift-off) was recorded. Normal load was about 100 nN

We believe that there is no loss of contact at low excitation amplitude, that is, from 0 to 1 (arbitrary units) in the in-plane case and from 0 to 1.8 (a.u.) in the out-of-plane experiment. Above these amplitudes, intermittent contact can take place which is not of interest to us. Similar AFM studies on lubricated contacts were reported, but have to be studied in further detail and under controlled conditions (e.g., humidity) [52].

3.7.1 Theory of Out-Of-Plane Sonolubrication

Dinelli et al. explored also a reduction of friction in an FFM if the surface is vibrated out-of-plane [53]. The friction reduction begins already at low vibration amplitudes well before the break of contact between tip and sample surface. To explain this effect, they suggest that a liquid layer is present between the tip and the sample that exhibits viscoelastic behavior. The problem of two lubricated surfaces under oscillating normal load has been treated some time ago. The application is in machinery, since most bearings do encounter dynamic loading. Leading papers were published by Christensen [54] and Tabor and Bowden [5]. More recently, Hess and Soom predicted that friction could be reduced between two solid bodies if the normal load is modulated [55, 56]. If the contact is of Hertzian type, one can find the equation of motion for a system as shown in Fig. 3.21. A static force P_0 deflects a sphere of mass m from its equilibrium position y_0, the sphere is indented in the flat. The external static load is superimposed by a dynamic load $\alpha P_0 \cos \omega t$. The restoring force F of the elastic contact is not constant but it depends on the actual con-

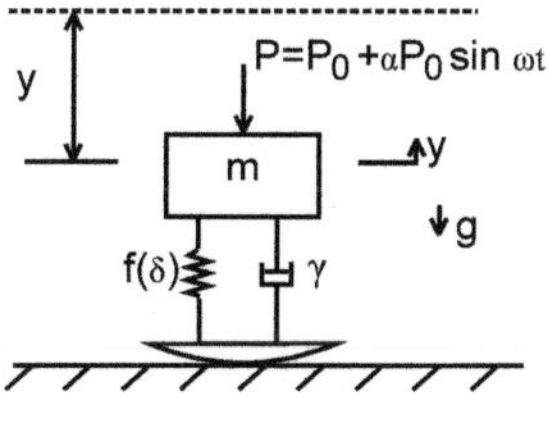

Fig. 3.21. Model for contact resonance of a point-mass oscillator with Hertzian contact and damping

tact area and thus, on the indentation depth δ. The Hertzian contact can be modeled as a nonlinear spring. Contact area A and normal force P are related by

$$A \propto P^{2/3} \tag{3.37}$$

At the onset of contact loss,

$$P = P_0 (1 + \cos \omega t) \tag{3.38}$$

Index 0 indicates the static case, with no vibration. The ratio between actual contact area and contact area at static load is

$$\frac{A}{A_0} = \frac{[P_0 (1 + \cos \omega t)]^{2/3}}{P_0^{2/3}} = (1 + \cos \omega t)^{2/3} \tag{3.39}$$

Then, the average contact area <A> at the onset of intermittent contact is

$$\frac{\langle A \rangle}{A_0} = \frac{\omega}{2\pi} \int_0^{2\pi/\omega} (1 + \cos \omega t)^{2/3} \, dt = 0.92 \tag{3.40}$$

Assuming the adhesive model of friction where friction is proportional to real area of contact, then the average friction <F> is

$$\frac{\langle F \rangle}{F_0} = \frac{\langle A \rangle}{A_0} = 0.92 \tag{3.41}$$

The friction between two solids can be reduced up to 8%, if a dynamic normal load is superimposed on the static normal load, even if there is steady, that is, non intermittent, contact. A similar effect exists in the case of an AFM contact. Here, both the nonlinearly with distance-varying Hertzian contact stiffness and to a stronger extent the adhesion forces contribute to the reduction of the contact area. This so-called mechanical diode effect was first observed by Kolosov and Yamanaka [57], calculated quantitatively by Hirsekorn et al. [58], and subsequently used by others (e.g., see Dinelli et al. [59]). However, if the surfaces are planar and rough, a normal vibration does not introduce a reduction in friction of unlubricated pairs [56].

3.7.2 Theory of In-Plane Sonolubrication

A hypothesis for the reduced friction on a laterally modulated surface was proposed by Scherer et al. [60]. Elastohydrodynamic lubrication (EHL) is believed to act between the two surfaces in relative motion where condensed water is the fluid separating the bodies. Indenting a silicon sphere, orientation <100> of 100-nm radius in a perfectly smooth silicon flat that is oriented <100>, leads to deformations on the order of 1 nm assuming Hertzian contact with a Young's modulus for sphere and flat of 170 GPa, Poisson ratio 0.181, and normal load 100 nN. If a fluid is present between any two contacting surfaces that slide against each other, a lubrication film is built up. Its thickness is a function of normal load, sliding velocity, geometry, and elasticity of the opposing surfaces and the physical properties of the fluid involved. If the two surfaces are sufficiently hard, as it is the case for metals, hard EHL is to be considered. The governing equation to calculate the film thickness for the case of a sphere sliding against a flat surface can be taken from the literature [61]:

$$h_{\min} = 1.79\ R^{0.47}\ \alpha^{0.49}\ \eta_0^{0.68}\ U^{0.68}\ E^{-0.12}\ W^{-0.07} \tag{3.42}$$

In this empirical power-law dependence, h_{min} is the minimum film thickness, E the reduced elastic modulus given by the Hertz contact theory, R the radius of the sphere, η_0 the viscosity of the fluid, and α is the pressure coefficient for the liquid. U is the sliding velocity and W the normal load (all in SI units). The viscosity of the fluid increases exponentially with pressure according to

$$\eta = \eta_0\ e^{(\alpha P)} \tag{3.43}$$

with the hydrostatic pressure P. In the case of absorbed water being the liquid, α can be quantified by 6.7×10^{-10} m^2 N^{-1} as given by Hamrock [62]. The tip radius was estimated to be 100 nm and the normal load was set to 10^{-7} N. Both sample and tip were made of single crystal Si oriented in the <100> direction with an elastic modulus and a Poisson's ratio of 1.3×10^{11} N m^{-2} and 0.18, respectively, resulting in a reduced elastic modulus of E=0.7×10^{11} N m^{-2}. When the sample is vibrated at a frequency of 1 MHz with maximum amplitude of 10 nm, the sliding velocity U can be as high as 10 mm s^{-1}. For a water viscosity at atmospheric normal pressure of η_0=10^{-3} Ns m^{-2} (=1 cP), the minimum film thickness is still below 1. It was found, however, that the viscosity of molecular thin films could be up to 10,000 times the bulk viscosity of the same liquid [63]. We realize that at a viscosity 1,000 times greater than bulk water, a lift-off of 1 nm can be obtained with sliding speeds of 10 mm s^{-1}. We think that the marginal lift-off at surface amplitudes below the threshold in Fig. 3.20 is due to the build-up of an elastohydrodynamic lubrication film. Above the threshold, the cantilever is intermittently bounced off the surface due to either parasitic out-of-plane surface motions or mode conversions in the cantilever, so that the average height is greater than zero.

Behme and Hesjedahl have picked up the open question with regards to parasitic out-of-plane vibrations in an in-plane experiment employing shear wave transducers [64]. They proposed surface acoustic wave (SAW) transducers as a source of well-defined surface oscillations. On specific layered piezoelectric substrate systems, one can excite Rayleigh waves and Love waves. The Rayleigh-type waves are polarized in the sagittal plane that is formed by the direction of wave propagation and the direction normal to the surface. The surface oscillates on elliptical trajectories with large vertical and smaller lateral oscillation component. Love waves are purely in-plane with their polarization perpendicular to the wave propagation. The authors performed their experiment with different interdigital transducers (IDT) at frequencies ranging from 200 MHz to 3 GHz. To measure the effect of ultrasonic lubrication, they recorded the friction loop in an FFM when the surface was vibrated by either mode. The results clearly showed that for Rayleigh waves, friction was reduced and that the amount of reduction scales with the SAW amplitude. Like with the longitudinal wave and shear wave transducers, the friction is reduced to zero at sufficiently high vibration amplitude. Purely in-plane polarized Love waves did not exhibit a lubricity effect. The authors conclude that in-plane vibrations do not alter friction, contrary to our findings. The proposed method using the SAW transducers, however, opens new possibilities for research. Experiments with controlled atmosphere and well-defined thin lubricant films can be done to study EHL on nanometer-size contacts. Furthermore, the rheology of lubricants at high frequencies can be studied.

3.8 Related Methods

Besides the FFM, there are similar techniques to investigate friction on the nanometer scale. In this section, a brief reference to these related methods will be made, as far as they employ lateral modulation.

3.8.1 Surface Force Apparatus

By the end of the 1950s, Spaarnay developed an instrument that could be used to measure forces between two parallel metallic plates that were separated by only a few nanometers distance [65]. Later, Tabor and Winterton, and Israelachvili and Tabor further improved this method and called it the surface force apparatus (SFA) [6, 7]. This instrument is today commercially available from SurForce Corporation, Santa Barbara, CA. Two atomically smooth cylindrical surfaces with well-defined radii of curvature are brought in contact under controlled atmosphere. A fluid film separates the bodies. By using optical interferometry techniques, the gap between the bodies can be determined with a resolution of at least 0.1 nm. The SFA was originally developed to measure normal forces between two bodies. Chan and Horn added the lateral displacement option to the SFA [66]. The lateral displacement was introduced by a leaf spring system. During a measurement, the vertical distance, true area of contact, elastic deformation, and lateral force could be registered. Typical sliding velocities were in the range of 0.1–20 μm s^{-1} with a stroke length of several hundred micrometers. Van Alsten and Granick implemented a piezoelectric transmitter and receiver, mounted on the lateral springs to excite and measure the lateral displacement [67]. They were able to measure the lateral force with a resolution of 10^{-6} N; the lateral displacement amplitude ranged between several nanometers up to 10 μm. They reported investigations on thin lubricant films of molecular thickness. It was observed that there is a transition from purely viscous to solid-like behavior with decreasing film thickness [68]. Georges et al. even improved the set-up that required transparent samples in order to measure contact area [69, 70]. With capacitive displacement sensors and piezoelectric transducers, they were able to measure sliding friction between a flat and a sphere in quasi-static and dynamic state.

3.8.2 Quartz Crystal Balance

Quartz crystals have been used for decades to measure weight with highest precision in the so-called quartz crystal microbalance (QCM) [71]. The resonant frequency of a quartz resonator (typically 1–30 MHz, Q-factor $\approx 10^5$) is changed proportionally to the mass that it is loaded with. If, for example, a monolayer film is deposited on to the resonator surface, not only the frequency, but also its Q-factor is changed [72]. The smaller the Q-factor (Q parameterizes the sharpness of the resonance, i.e., the ratio of the resonant frequency to the full-width at half maximum), the more energy is dissipated due to rearranging of atoms in the monolayer that is being shaken by the resonating quartz. In the QCM experiment, any change over the full (macroscopic) surface area of the resonator affects the result. In order to extract information on the nanometer length scale, Daly and Krim, and Sasaki et al. combined scanning tunneling microscopy with the QCM [73, 74]. A nanomechanical experiment employing QCM as lateral displacement actor and sensor, both at the same time, was demonstrated by Laschitsch et al. [75, 76]. They coated the surface of the QCM with perfluorpolyether (PFPE) lubricant of less than 1 nm thickness. The QCM resonant frequency and Q-factor (bandwidth of the resonance curve) were recorded using an impedance analyzer. With a manipulator, a pin with an Si_3N_4 sphere of 2.3 mm radius was engaged with the laterally vibrating QCM surface. For a lubricant thickness of up to 0.11 nm, they observe a maximum in bandwidth just at the point of engagement. Width and height of the maximum decrease with increasing lubricant thickness. This is attributed to dissipation occurring during intermittent contact between the substrate and the sphere. If lubricant film thickness is greater than 0.25 nm, the sphere discontinuously jumps into contact; the sphere follows the QCM surface vibration. Another technique, the nanoindenter, was combined with QCM by Borovsky et al. [77]. Probe tips, of etched tungsten (approximately 12 µm radius) and of fused silica sphere (0.5 mm radius) were indented in the laterally vibrating QCM silver electrode by a normal load up to 0.5 mN. Metal–metal and glass–metal contacts were found to introduce positive frequency shifts in the QCM, directly proportional to the contact area.

It would be interesting to compare lateral modulation experiments by AFM, SFA, nanoindenter, and QCM. The gap between nano/micro and macroscale tribology could be bridged if the techniques were further developed and employed to investigate friction, wear, and lubrication.

Appendix 1

Electronic circuit for converting the four signals from the split-photodiode if splitting is vertical and horizontal. Voltage outputs of the four elements are converted into voltages that correspond to vertical and lateral displacement of the laser spot. Two different amplifications A_1 and A_2 can be chosen in order to compensate for the reflectivity of the cantilever back side (silicon cantilevers have lower reflectivity than gold-coated Si_3N_4 cantilevers).

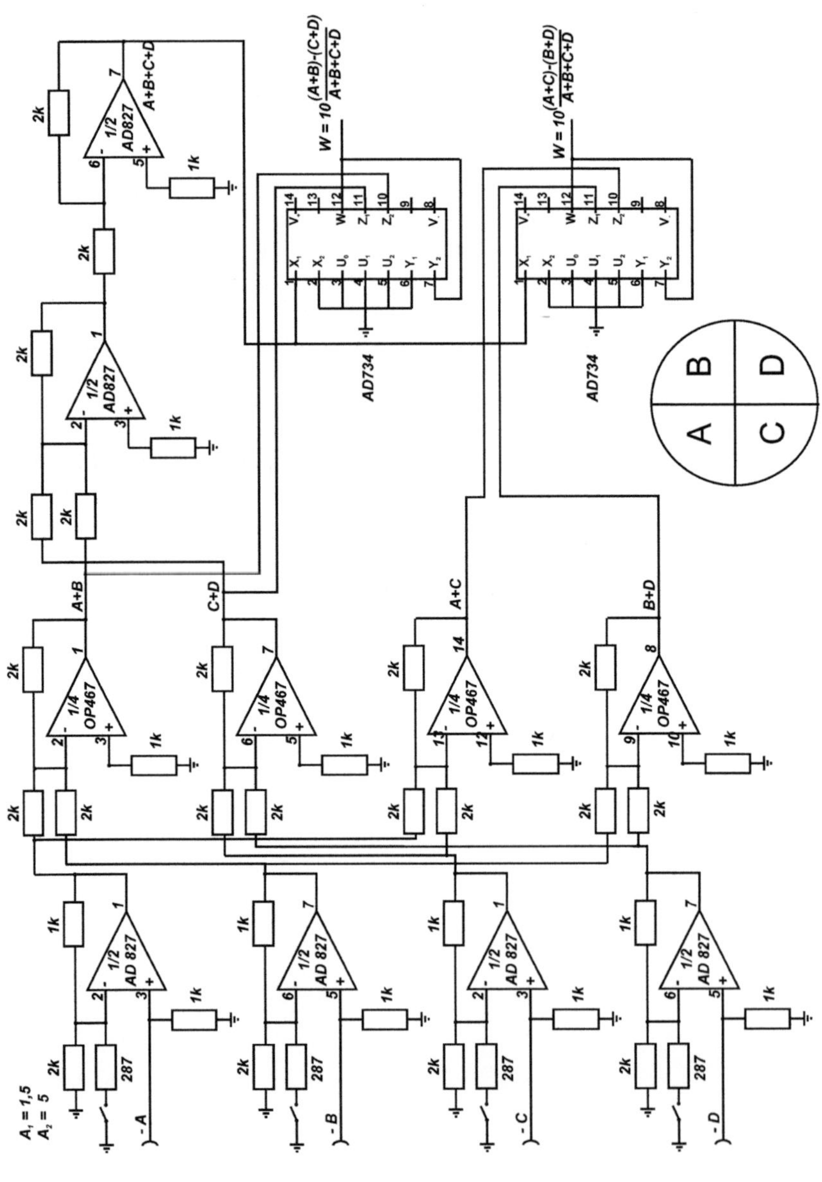

Appendix 2

Electronic circuit for diagonally split photodiodes.

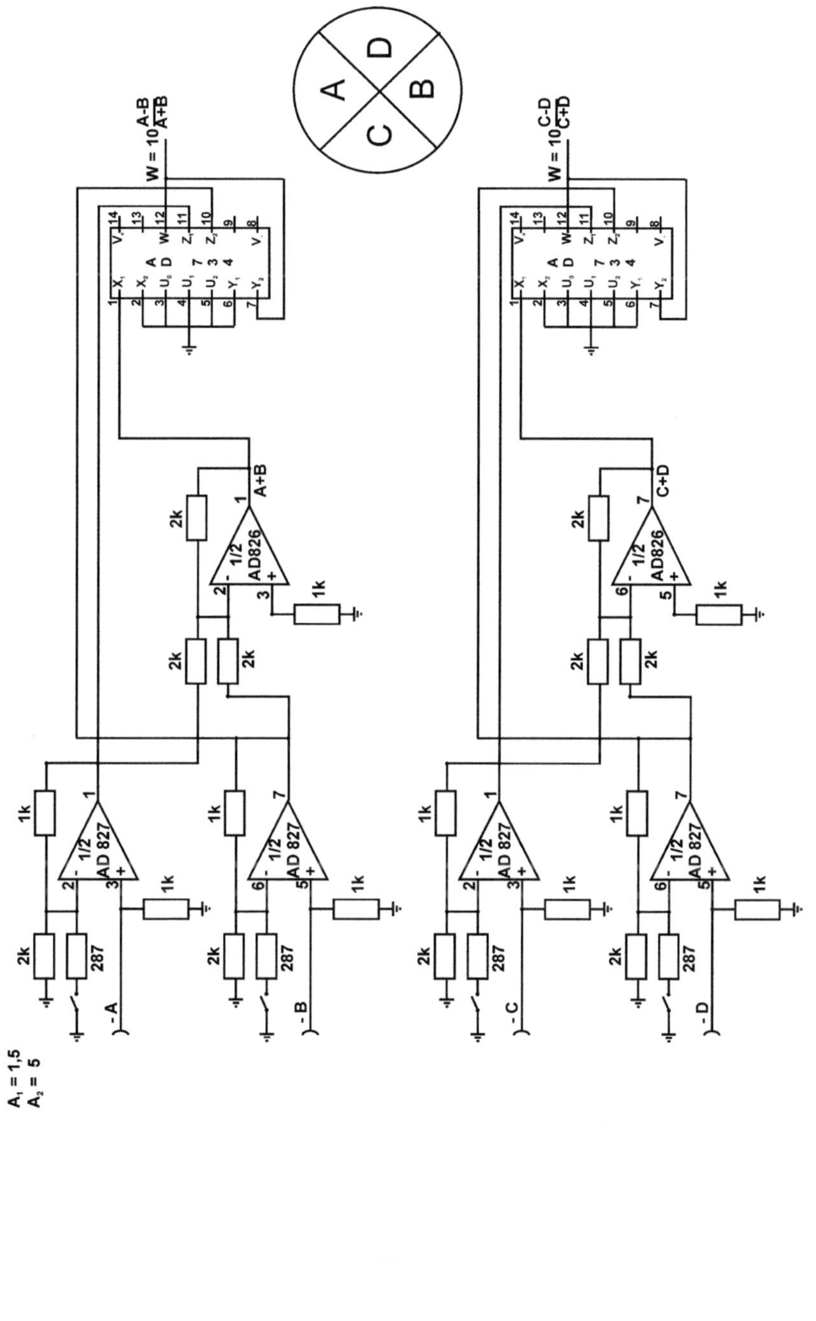

Appendix 3

	Excitation	Sample/ cantilever	Lateral excitation frequency [Hz]	Amplitude [nm]	Normal force [nN]	Cantilever	Data acquisition	Detected property	Name	Authors
1	Sample	Gold	5 k	25	20	Si_3N_4 (0.6 N/m) diamond tip	Modified Nanoscope II	Edges on gold planes	Lateral modulation mode	Maivald et al. [17]
2	Sample	n-dodecanol	50–300	< 5	0–125	Si_3N_4 (0.58 N/m)	Lock-in amplifier	Lubrication	–	O'Shea et al. [21]
3	Sample (buckling)	NdFeB film, HOPG, nickel compact disk stamper matrix	Below cantilever resonance	a few nm	–	Pyramidal Si_3N_4	Lock-in amplifier and laser interferometer	Friction	–	Gödden-henrich et al. [78]
4	Sample	HOPG	10 k (lateral)	–	–	Si_3N_4 (0.09 N/m, deflection resonance: 40 kHz)	–	Friction	Lateral force modulation mode	Yamanaka et al. [79]
			10 k (lateral) +5.6 M (vertical)	–	–			Subsurface edge dislocation	Lateral ultrasonic force microscopy	
5	Sample	Gold thin film	16 k	1	1,2	Si_3N_4 (0.09 N/m)	–	Static and kinetic friction	Lateral force modulation atomic force microscopy (LM-AFM)	Yamanaka and Tomita [80]
		Si wafer (100) in water	1 k	3	0–12	Si_3N_4 (0.09 N/m)		Kinetic friction by friction curve force		
		Magnetic tape	–	> 3	0–400	Si (3.2 N/m)		Friction and sample deformation		

6	Sample	Langmuir-blodgett fil in water or air	–	–	20	Si_3N_4 (0.09 N/m)	Lock-in amplifier	Friction by friction force curve	Lateral force modulation atomic force microscopy (LM-AFM)	Yamanaka et al. [81]
7	–	Cleaved mica	–	a few angstrom	–40–40	Si_3N_4 (0.58 N/m)	Lock-in amplifier	Lateral stiffness of a sample	–	Carpick et al. [14]
8	Sample	Styrene-butadiene thin films	0.02–100	0–2000	–	Si_3N_4 (–)	Envelope of oscillation	Loss and storage modulus	–	Fretigny et al. [82]
9	Sample	Carbon fibers in epoxy matrix	37 k	10	–	Si_3N_4 (0.6 N/m)	Lock-in amplifier	Youngs Modulus	X-Modulation	Mazeran and Loubet [83]
10	Cantilever	Si_3N_4 cantilever	0–300 k (resonance)	–	–	Si_3N_4 (0.09 N/m)	–	Elastic properties of a cantilever	–	Nakano et al. [29]
11	Sample	Z-Dol (lubricant)	370–380 k	–	145	Si_3N_4 (0.6 N/m)	Envelope of oscillation	Lubrication	Acoustic friction force microscopy (AFFM)	Scherer et al. [52]
12	Cantilever	NbS_2	0–1 M	–	–	Si(–)	Low-frequency friction loop	Frictional forces	Lateral force modulation	Kerssemakers and De Hosson [84]
13	Sample	Poly(vinyl-ethylene)	0–1 k	0–3	– 10–2	Si_3N_4 (0.06 N/m)	Lock-in amplifier	Response time and shear contact stiffness	Shear force modulation	Wahl et al. [19]
14	Sample	Polymer (polyolefine)-silicon	10 k	–	< several tens of nN	Si_3N_4 (0.05 N/m)	Lock-in amplifier	Static and dynamic friction	Dynamic scanning friction force microscopy (DSFFM)	Krotil et al. [24]

	Excitation	Sample/cantilever	Lateral excitation frequency [Hz]	Amplitude [nm]	Normal force [nN]	Cantilever	Data acquisition	Detected property	Name	Authors
15	Sample	Metal particle tape	1.59 M (resonance)	–	100	Si(–)	rms-to-DC converter	Viscoelasticity	Accoustic friction force microscopy (AFFM)	Scherer et al. [60]
16	Cantilever	Chemically functionalized Si cantilerver for AFM	11.6 k (resonance)	4	–	Optical fiber cantilever	Laser feedback interferometer + lock-in amplifier	Influence of interaction force on damping of SNOM tip-vibration amplitude	–	Bernal et al. [85]
17	Sample	Polymer	1.4 k	3	Typically 25	–	Lock-in amplifier	Glass transition temperatures (viscoelasticity)	Shear modulation force microscopy (SMFM)	Ge et al. [18]
18	Cantilever (out of plane) + sample (inplane)	ABC-triblock copolymer	kHz range (our of plane), 150 k (in-plane)	3.25–6.5	–	–	Lock-in amplifier	Adhesive, static and dynamic frictional forces	Combined dynamic X mode (CODY mode)	Krotil et al. [86]
19	Sample (buckling)	HOPG	100 k (resonance)	1–10	–	Si (–)	Lock-in amplifier	Frictional damping	Dynamic scanning friction microscopy (DSFM)	Spychalski-Merle et al. [39]
20	Cantilever	Dots of CH_3 or OH-terminated Au substrate in air or water	9 k (resonance)	0–4	–	Optical fiber cantilever	Interferometric feedback system	Influence on wettability on oscillation amplitude damping	–	Vaccaro et al. [87]

21	Sample	Piezoelectric substrate	200 M–3G	–	–	Si (–)	Low-frequency friction loop	Frictional forces	Lateral force modulation scanning acoustic force microscopy (LFM-SAFM)	Behme and Hesjedahl [64]
22	Thermomechanical	Si, poly-Al and CdTe	180 k–230 k	–	– 15–45	Si_3N_4 (0.1 N/m)	Spectral analyzer	Lateral stiffness	Passive overtone	Drobek et al. [26]
23	Cantilever/sample	SiO_2 mesa on Si substrate/SAM	400–500 k (resonance)	0.4–4	< 20	Si_3N_4 (0.06 N/m)	Lock-in amplifier/ABS amplifier	Lateral dissipation by resonance tracking	Lateral cantilever resonance mode (LCR mode)	Kawagishi et al. [40]
24	Sample	polybutadiene, Sb_2S_3, resin matrix				Si	Lock-in amplifier	Friction	Harmonically modulated LFM (HM-LFM)	Munz et al. [25]
25	Sample/Cantilever	Si and polyfluoropolyether (PFPE) thin films	210 k–260 k	–	0–225	Si (0.1 N/m)	Lock-in amplifier	Frictional force, damping	–	Reinstaedtler et al. [88]
26	Cantilever	Si Cantilever	100 k–3 M	–	0	Si (0.131 N/m)	Lock-in amplifier	Cantilever torsional resonance amplitude and frequency	–	Reinstaedtler et al. [42]
27	Sample/Cantilever	Si, borsilica glass, polyethylene (PE) composite	1 k–1.5 M		0–100	Si (0.1 N/m)	Lock-in amplifier			Scherer et al. [see this chapter]

References

1. Basshuysen R, Schaefer F (2002) (eds) Handbuch Verbrennungsmotor, 2nd edn. Vieweg, Braunschweig
2. Krim J (1996) Sci Am 275:74
3. Bhushan B (1999) Handbook of micro/nanotribology, 2nd edn. CRC Press, Boca Raton, Florida
4. Bhushan B (2001) Fundamentals of tribology and bridging the gap between the macro- and micro/nanoscales. Kluwer, Dordrecht, NL
5. Tabor D, Bowden FP (1986) The friction and lubrication of solids. Clarendon Press, London
6. Tabor D, Winterton RHS (1969) Proc R Soc London Ser A 331:19
7. Israelachvili JN, Tabor D (1973) Nature 241:148
8. Mate CM, McClelland GM, Erlandsson R, Chiang S (1987) Phys Rev Lett 59:1942
9. Derjaguin BV, Muller VM, Toporov YP (1975) J Coll Interface Sci 53:314
10. Johnson KL (1985) Contact mechanics. University Press, Cambridge
11. Maugis D (1992) J Coll Interface Sci 150:243
12. Burnham NA, Behrend OP, Oulevey F, Gremaud G, Gallo PJ, Gourdon D, Dupas E, Kulik AJ, Pollock HM, Briggs GAD (1997) Nanotechnology 8:67
13. Anczykowski B, Krüger D, Babcock KL, Fuchs H (1996) Ultramicroscopy 66:251
14. Carpick RW, Ogletree DF, Salmeron M (1997) Appl Phys Lett 70:1548
15. Schwarz UD, Zwörner O, Köster P, Wiesendanger R (1997) Phys Rev B 56:6987
16. Rabe U, Janser K, Arnold W (1996) Rev Sci Instrum 67:3281
17. Maivald P, Butt H-J, Gould SAC, Prater CB, Drake B, Gurley JA, Elings VB, Hansma PK (1991) Nanotechnology 2:103
18. Ge S, Pu Y, Zhang W, Rafailovich M, Sokolov J, Buenviaje C, Buckmaster R, Overney RM (2000) Phys Rev Lett 85:2340
19. Wahl KJ, Stepnowski SV, Unertl WN (1998) Tribology Lett 5:103
20. Colchero J, Luna M, Baro AM (1996) Appl Phys Lett 68:2896
21. O'Shea SJ, Welland ME, Rayment T (1992) Appl Phys Lett 61:2240
22. Yamanaka K, Ogiso H, Kolosov O (1994) Appl Phys Lett 64:178
23. Scherer V (2002) Ultraschall-Kraftmikroskopie mit lateraler Anregung. Dissertation, Universität des Saarlandes, Naturwissenschaftlich-Technische Fakultät III
24. Krotil HU, Weilandt E, Stifter Th, Marti O, Hild S (1999) Surf Interface Anal 27:341
25. Munz M, Schulz E, Sturm H (2002) Surf Interface Anal 33:100
26. Drobek T, Stark RW, Heckl WM (2001) Phys Rev B 64:045401
27. Rabe U, Amelio S, Kopycinska M, Hirsekorn S, Kempf M, Göken M, Arnold W (2002) Surf Interface Anal 33:65, and references contained therein
28. Yamanaka K, Nakano S (1998) Appl Phys A 66:313
29. Nakano S, Maeda R, Yamanaka K (1997) Jpn J Appl Phys 36:3265
30. Kester E, Rabe U, Presmanes L, Tailhades P, Arnold W (2000) J Phys Chem Sol 61:1275
31. Rabe U, Amelio S, Kopycinska M, Hirsekorn S, Kempf M, Gölen M, Arnold W (2002) Surf Interface Anal 33:65
32. Hertz H (1896) J Reine Angew Math 92:156
33. Putman CAJ, Igarashi M (1995) Appl Phys Lett 66:3221
34. Greenwood A, Williamson JBP (1966) Proc R Soc London A 295:300
35. Scherer V, Arnold W (1997) Friction force microscopy at ultrasonic frequencies In: Bhushan B (ed) Micro/nanotribology and its applications. Kluwer Academic
36. Goodman LE, Brown CB (1962) J Appl Mech 29:17
37. Ausimont SpA, Bollate, Italy
38. Bhushan B (1996) Tribology and mechanics of magnetic storage devices, 2nd edn. Springer, Berlin Heidelberg New York
39. Spychalski-Merle A, Krischker K, Göddenhenrich T, Heiden C (2000) Appl Phys Lett 77:501
40. Kawagishi T, Kato A, Hoshi Y, Kawakatsu H (2002) Ultramicroscopy 91:37
41. Reinstaedtler M (2002) Diploma thesis, Universität des Saarlandes
42. Reinstaedtler M, Scherer V, Rabe U, Turner JA, Arnold W (2003) Surf Sci (in press)

43. Turner JA (2002) private communication
44. Nayfeh AH (2000) Nonlinear Interactions: analytical, computational, and experimental methods. Wiley, New York
45. Fridman HD, Levesque P (1959) J Appl Phys 30:1572
46. Seireg A, Weiter EJ (1963) Wear 6:66
47. Pohlmann R, Lehfeldt E (1966) Ultrasonics 178
48. Lenkewicz W (1969) Wear 13:99
49. Frederick JR (1965) Ultrasonic engineering. Wiley, New York
50. Tam AC, Bhushan B (1987) J Appl Phys 61:1646
51. Scherer V (1997) Active friction control using ultrasonic vibration In: Bhushan B (ed) Tribology Issues and opportunities in MEMS. Kluwer
52. Scherer V, Bhushan B, Rabe U, Arnold W (1997) IEEE Trans Magn 33:4077
53. Dinelli F, Biswas SK, Briggs GAD, Kolosov OV (1997) Appl Phys Lett 71:1177
54. Christensen H (1961) Proc R Soc London Ser A 266:312
55. Hess DP, Soom A (1991) Trans ASME 80:80
56. Hess DP, Soom A (1991) Trans ASME 80:87
57. Kolosov O, Yamanaka K (1993) Jpn J Appl Phys Lett 32:L1095
58. Hirsekorn S, Rabe U, Arnold W (1997) Nanotechnology 8:57
59. Dinelli F, Castell MR, Ritchie DA, Mason NJ, Briggs GAD, Kolosov OV (2000) Phil Mag A 80:2299
60. Scherer V, Arnold W, Bhushan B (1999) Surf Interface Anal 27:578
61. Hutchings IM (1992) Tribology: friction and wear of engineering materials. E Arnold, London
62. Hamrock BJ (1994) Fundamentals of fluid film lubrication. McGraw-Hill, New York
63. Israelachvilli JN (1999) Surface forces and microrheology of molecularly thin liquid films. In: Bhushan B (ed) Handbook of micro/nanotribology, 2nd edn. CRC Press, Boca Raton, Florida
64. Behme G, Hesjedahl T (2001) J Appl Phys 89:4850
65. Spaarnay MJ (1958) Physica 24:751
66. Chan DYC, Horn RG (1985) J Chem Phys 83:5311
67. Van Alsten J, Granick S (1988) Phys Rev Lett 61:2570
68. Van Alsten J, Granick S (1990) Tribol Trans 33:436
69. Georges JM, Millot S, Loubet JL, Tonck A (1993) J Chem Phys 98:7345
70. Georges JM, Tonck A, Mazuyer D (1994) Wear 175:59
71. Lu C, Czanderna A (1984) Applications of piezoelectric quartz microbalances. Elsevier, Amsterdam
72. Krim J, Widom A (1988) Phys Rev B 38:12184
73. Daly C, Krim J (1995) Applications of a combined scanning tunneling microscope and quartz microbalance. In: Bray MT, Cohen SH, Lightbody ML (eds) Atomic force microscopy/scanning tunneling microscopy. Kluwer Academic
74. Sasaki A, Katsumata A, Iwata F, Aoyama H (1994) Appl Phys Lett 64:124
75. Laschitsch A, Johannsmann D (1999) J Appl Phys 85:3759
76. Laschitsch A, Bailey LE, Tyndall GW, Frank CW, Johannsmann D (2001) Appl Phys Lett 78:2601
77. Borovsky B, Krim J, Syed Asif SA, Wahl KJ (2001) J Appl Phys 90:6391

4 Sensor Technology for Scanning Probe Microscopy

Egbert Oesterschulze · Rainer Kassing

4.1 Introductory Remarks

The invention of the scanning tunneling microscopy (STM) by Binnig and Rohrer [1–3] and the atomic force microscopy (AFM) by Binnig et al. [4] has unleashed the development of a new class of analytical tools that revolutionized surface science during the last two decades. The capability to study various surface properties on an atomic scale [5] coupled with the capability to manipulate or arrange molecular structures [6, 7] has blossomed into what recently is denoted *nanoscience technology*. The heart of any scanning probe microscope tool is certainly the sensing element simply termed "probe". It consists in almost all cases of a sharpened tip with extremely small radius of curvature that determines the interaction volume with the local sample surface and thus the lateral resolution during the scanning process. We will concentrate in this report only on AFM-based probes, that is, cantilever probes consisting of a mechanical beam with an integrated sharp tip, because they have proven to be the most flexible arrangement for future developments in SPM. Details on the various AFM operation modes and their operation limits are not within the scope of this report and may be found elsewhere ([8–11] and references therein). The same holds for the various forces involved in that process (e.g., see refs. [12–15]).

Asking us to give an overview of recent probe developments resembles a Sisyphus work because scanning probe microscopy (SPM) has found widespread use not only in physics but also in biology, chemistry, mechanical, and electrical engineering etc. [5]. Therefore, we will focus in this report only on some aspects of future developments of SPM probe technology. Figure 4.1 shows a multifunctional

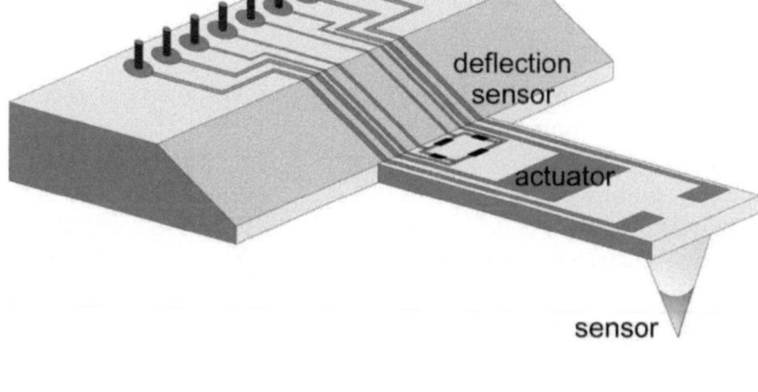

Fig. 4.1. Set-up of a multifunctional stand-alone SPM cantilever probe

probe with sensors integrated into the tip to sample different surface properties of the sample simultaneously. The cantilever is equipped with a stress-sensitive sensor integrated into the cantilever base as well as an actuator on the cantilever to establish a stand-alone operation of the cantilever in the AFM mode. Parallel operation of hundreds of these probes would guarantee an appropriate bandwidth to allow on-line imaging in TV quality.

4.2 Material Aspect of Probe Fabrication

The development of SPM probes adapted to certain applications has to be separated into at least two different steps. The first covers the selection of an appropriate sensor concept for the desired application. The second involves the choice of a proper material that allows one not only to obtain probes of proper sensitivity but also to meet the boundary conditions of the experimental environment (e.g., chemical inertness, mechanical hardness, low abrasion etc.). Before embarking on examples of probes for various applications in accordance with step 1, we emphasize the importance of the proper material choice. In the early stages of AFM cantilever probes (i.e. a mechanical beam with an integrated sharp tip at its very end) were manually fabricated [4]. However, it quickly became clear that reliable experiments de-

Table 4.1. Properties of some important seperate: important materials used for MEMS fabrication of SPM probes

Properties	(001) Si [16]	(001) GaAs [17]	Diamond [18]
Mechanical			
Mechanical density ρ_m (g cm^{-3})	2.329	5.317	3.515
Young's modulus $E_{[110]}$ (GPa)	168.4	121,5	1,163.6
Young's modulus $E_{[100]}$ (GPa)	129.2	85.5	1,050.3
$\sqrt{E_{[110]}/\rho}$ (m s^{-1})	8,503.3	4,780.3	18,194.5
Torsional modulus G (GPa)	80	60	–
Hardness (load (g))	1,150 (25)	680 (100)	18,000 (100)
Thermal			
Heat capacity c_p (J Kg^{-1}K^{-1})	700	500	516
Heat conductivity k (W mK^{-1})	150	45.5	2,000
Melting point T_m (K)	1,415	1,513	graphite formation
Thermal expansion coefficient (µV K^{-1})	2.6	6.86	0.8
Seebeck coefficient (µV K^{-1})	–1,600..1,500	–680..130	
Optical and electrical			
Refractive index n (λ = 633 nm)	3.4	3.878	2.41
Static dielectric constant ε	11.8	13.18	5.40
Gap energy E_G (eV)	1.12 (indirect)	1.42	5.45
Electron mobility μ_n (cm^2 V^{-1}s^{-1})	120–1,300	9,200 (SI)[a]	1,900
Hole mobility μ_p, (cm^2 V^{-1}s^{-1}))	70–500	400 (SI)	1,600
Electrical resistivity ρ (Ω cm)	10^5	10^9 (SI)	10^{16}

[a] *SI* semi-insulating material.

mand reproducible fabrication of probes with well-defined material properties. Some noteworthy physical properties of the most common materials (i.e., silicon, gallium arsenide, and diamond) used for the fabrication of **m**icro**e**lectro**m**echani-cal **s**ystems (MEMS) are summarized in Table 4.1.

4.2.1
Mechanical Properties of Cantilever Probes

To grasp the influence of the material and the geometrical properties of a cantilever beam in an AFM experiment we give a very brief theoretical introduction to the mechanics of cantilever beams. A more detailed description may be found in ref. [9] and references therein. The displacement $z(y)$ of a cantilever directed in the y-direction is governed by the following differential equation in frequency space:

$$\frac{d^4 z(y)}{dy^4} - \kappa_n^4 z(y) = 0 \quad \text{with } \kappa_\mathrm{n} \text{ defined as } \omega_n = \kappa_n^2 \sqrt{\frac{I}{\Lambda}} \sqrt{\frac{E}{\rho}} \tag{4.1}$$

The eigenfrequency ω_n of the n-th eigenmode thus depends on the geometrical cantilever properties given by the momentum of inertia I of the beam and its cross-sectional area A and on the mechanical properties, that is, Young's modulus E and the mechanical density ρ. In the simple case of a beam of length l and rectangular cross-section the solution of the associated eigenvalue problem results in the following eigenvalues for the first five flexural modes:

$$\kappa_\mathrm{n} l = 1,875;\ 4,694;\ 7,855;\ 10,996;\ 14,137 \tag{4.2}$$

Taking the compliance of the beam defined as:

$$k = 3E\,\frac{I}{l^3} \tag{4.3}$$

into account, yields the well-known relations of ω_1 and k of the first flexural mode:

$$\omega_1 = 1.0149\frac{d}{l^2}\sqrt{\frac{E}{\rho}} \quad \text{and} \quad k = \frac{1}{4}E\frac{bd^3}{l^3}\,. \tag{4.4}$$

A more detailed description of the mechanics of a beam made of isotropic and homogeneous material is given elsewhere [19]. This theoretical approach was also extended to study the mechanical behavior of cantilever probes fabricated from layered materials [20–22].

The influence of the mechanical material properties on the frequency of the first flexural mode of a beam in [Eq. (4.1)] is given in terms of the ratio $\sqrt{E/\rho}$. In accordance with Table 4.1 single-crystalline silicon offers obviously quite adequate properties. This explains its widespread use as a standard material for the fabrication of MEMS and in particular for SPM probes [23]. However, in fact all values of $\sqrt{E/\rho}$ mentioned in Table 4.1 differ only by a factor of 3.81. It is striking to note, that the same variation of ω can be easily achieved for instance by increasing the cantilever length l by a factor of 1.95 while keeping all other parameters constant. Only for cantilever beams with extreme geometrical dimensions, that is, where their technological fabrication processes constitute the limiting case, does the material choice of the cantilever beam become decisive. For the influence of the properties of crystalline materials see ref. [24].

4.2.1.1
Silicon

The enormous popularity of silicon or thin films of silicon nitride or silicon dioxide as a probe material is caused not only by their extraordinary properties but also by the huge amount of technological processes and experiences available for their processing [25, 26]. In the case of silicon, fabrication of probes relies in most cases on bulk micromachining. In particular sharp tips are obtained by underetching of masking pads during wet or plasma etching processes [27]. The radius of curvature of silicon tips of typically 10–20 nm can be further improved by repeated thermal oxidation at reduced temperatures and subsequent removal of the oxide layer [28, 29]. This sharpening process exploits the unique rheological behavior of thermally grown silicon dioxide layers as will be pointed out below for another purpose in some detail. Meanwhile, tips with radius of curvature of less than 5 nm are available (e.g., see refs. [24, 30]).

4.2.1.1.1
Homogeneity of Cantilever Probes

The total thickness variation of silicon wafers of typically 1–5 μm limits the reproducible fabrication of cantilevers with exactly the same mechanical properties as is obvious from Eq. (4.4). Etch stop layers were successfully introduced to obtain silicon membranes of homogeneous thickness and thus to reduce the scatter of mechanical cantilever properties substantially [32]. Thin films of etch-resistant material as well as p/n junctions obtained by doping have been proposed and used for this purpose [33, 34]. Recently, surface micromachining utilizing SOI (silicon on insulator) wafer became very popular because the thickness of cantilevers are well controlled by the thin cap silicon layer (e.g., see refs. [35, 36]). If serial fabrication is not imperative, focused ion beam (FIB) is also an adequate tool in particular if extremely small and precise geometries are of interest (i.e., probes with ultrasmall compliance) [37].

4.2.1.1.2
Probes Made of Thin Deposited Layers

Another option to solve the problem is the fabrication of cantilevers from thin deposited layers (e.g., silicon nitride [38]). In this case tip fabrication relies on molding from anisotropically etched pyramid-shaped etch grooves [38] or fabrication of Spindt-type tips made by evaporation through an orifice [39]. One advantage with respect to bulk micromachining is the possibility to control the thickness of the cantilever precisely by the deposition or oxidation process. However, the fragility of the thin cantilever membrane demands a mechanically stable holder to be mounted in a parallel process to the cantilever. Various elaborated and complex processes were considered for this purpose, for example, anodic bonding [38], silicon to silicon bonding [40], soldering [41], gluing [42] etc.

4.2.1.1.3
Stand-Alone Cantilever Probes and Nanometric Oscillators

Silicon as a substrate material features additional benefits from the huge amount of technological processes established in VSLI fabrication of electronic devices in the past [23]. Various approaches for the integration of electrical sensors in cantilever probes were assumed. In this section we embark on sensors and actuators that are intended for the stand-alone operation of silicon-based AFM probes. Piezoresistive transducers consisting of a single resistor or of four resistors forming a Wheatstone bridge were integrated into the base of silicon cantilever probes to detect both its bending and twisting [43–49]. The compact probe design makes it perfect for experiments under UHV conditions or with strongly restricted geometry. It avoids any kind of alignment that is necessary in the case of conventional optical [50–54] or electrical deflection methods [55–57]. A typical example of an array of silicon cantilevers with integrated piezoresistive readout is shown in Fig. 4.2. It constitutes also an important prerequisite for AFM imaging with increased bandwidth utilizing an array of cantilever probes. Quate et al. have demonstrated in their pioneering work the individual control of at least 50 cantilevers, integrating both a piezoresistive readout and a microactuator based on thin ZnO films [44]. High-speed imaging [47, 58], parallel lithography [59–61], and parallel data storage with the highest data densities [62] have been presented meanwhile. In this context the top-notch *millipede* project of IBM has to be mentioned as one of the first routinely working ultrahigh SPM data storage devices [63, 64]. A further increase of temporal bandwidth is achieved by utilizing nanometric oscillators with resonance frequencies of typically some GHz [65–69]. A parallel array of millions of nanometric oscillators was proposed and presented by Kawakatsu et al. [70, 71].

a 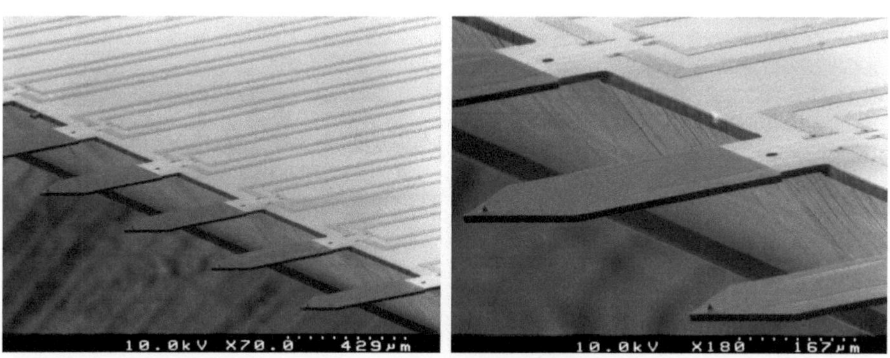 b

Fig. 4.2. **a** Array of silicon cantilever probes with integrated piezoresistive Wheatstone bridge. Each cantilever is perforated at the center of the Wheatstone bridge to improve the sensitivity by increasing the mechanical strain in the remaining material [31] (with courtesy of NaScaTec GmbH, Kassel, Germany)

4.2.1.1.4
Sensors Based on a Cantilever Set-Up

The unintentional temperature sensitivity of piezoresistors integrated in cantilevers was exploited for the detection of heat emerging from the chemical reaction of very small amounts of chemicals on the cantilever beam and led to the development of so-called chemical noses. Meanwhile various approaches have been discussed in the literature [72–78]. The first introduced exploited the bimetal effect of layered cantilevers [79, 80].

4.2.1.1.5
Sensors Integrated into the Probes Tip

So far our discussion has only concentrated on the cantilever as a mechanical tool that transforms forces affecting the tip into bending of the beam. However, in particular the tip itself offers a realm to implement ultrasmall sensors. This entails the difficulty to integrate sensors in a three-dimensional structure of only some micrometer size which is on the other hand a real challenge and stimulus for this kind of work. But it also includes the fascinating prospect to offer the capability to investigate different surface properties simultaneously. Owing to the rich variety of sensors integrated in the tip (e.g., see refs. [5, 24]) presented meanwhile we can only envisage a few examples in the second part of this paper.

4.2.1.2
Carbon

Carbon was used in various chemical modifications for probe fabrication. Next to electron beam deposited (EBD) carbon tips, sp^2- and sp^3-bonded carbon, that is, nanotubes and polycrystalline diamond, have attracted enormous attention as tip materials owing to their extraordinary properties.

4.2.1.2.1
Electron Beam Deposited Tips

Electron beam assisted gas-phase deposition of carbon containing material was developed in 1990 by Akama et al. [81] (see also ref. [82]) to fabricate carbon tips of high aspect ratio, very small radius of curvature, and appropriate hardness for most common applications [83, 84]. However, a few years later Dai et al. provided evidence that carbon nanotubes are a much better choice because of there excellent physical properties [85].

4.2.1.2.2
Carbon nanotube tips

The synthesis of carbon nanotubes was reported by Iijima for the first time in 1991 [86]. Carbon nanotubes consist of single or multiple seamless hollow cylinders made from folded graphene that is composed of sp^2-bonded carbon hexagons [87].

The diameter of single- or multi-walled nanotubes is in the range of some nanometers, although typical lengths of some micrometers have been reported [88]. This enormous aspect ratio together with their very small and reproducible tip radius of curvature make them perfect AFM tips for high-resolution imaging [89–91] as well as imaging in deep structures [85]. As can be conjectured from the similar structure of graphite in its basal plane, nanotubes exhibit a tremendous value of Young's modulus of about 0.4–1.3 TPa for typical rope diameters of 20–2 nm [92–94]. More fascinating, nanotubes are both stiff and gentle because buckling occurs if the load exceeds a critical force, the so-called Euler force [85]. The enormous strain resilience of carbon nanotubes avoids any damage of both the nanotube and the sample during buckling in contrast to almost any conventional silicon or silicon nitride tip [95–98]. Furthermore, Wong et al. [99] and Terrones et al. [100] successfully demonstrated the potential of nanotubes for chemical-sensitive imaging by attaching molecules to the very carbon nanotube tip that serve as molecular sensors.

Although carbon nanotubes attracted exceptional attention in the SPM community, no convincing batch fabrication process could be established right now. Almost all fabrication schemes presented so far rely on the manual attachment to conventional AFM cantilever tips [85, 101–103].

4.2.1.2.3
Diamond Probes

In applications that require probes of high compliance, tip wear becomes an important issue. Table 4.1 indicates that sp^3-bonded carbon in form of diamond with its unique mechanical properties is the most prominent candidate to fulfill the requirements of mechanical hardness and low abrasion. But it has also other extraordinary properties, for example, a very high thermal conductivity and chemical inertness. In particular, the large electrical band gap of 5.45 eV and the possibility to control the carrier density by boron and nitrogen doping over more than 10 decades excels diamond as the material for high-temperature semiconducting devices [104].

In the early stages grains of single-crystalline diamond were utilized to manually fabricate probe tips by fracture [108, 109] and grinding [110, 111]. With the introduction of CVD diamond growth the first probes made of sharpened metal wires with statistically distributed CVD-grown diamond grains were introduced [112, 113]. But the latter suffer among other things from their insufficient reproducibility. First approaches based on the molding of diamond tips were introduced by two groups [114–117]. Some typical examples are shown in Fig. 4.3. Their usefulness was demonstrated utilizing them for the electrical characterization of electronic devices with high-resolution nanopotentiometry, spreading resistance microscopy, and capacitance microscopy [118]. Improved processes, namely the tip-on-tip approach [107] and the *double molding process* [119] were later introduced to improve the quality of molded diamond probes.

However, the polycrystalline texture of the CVD-deposited diamond films adds tremendous difficulties to batch-processing of AFM probes, for example, incomplete filling of the pyramidal molds owing to a small nucleation density, insufficient control of the selective growth of CVD diamond on the silicon substrate, low qual-

Fig. 4.3a–d. SFM images of AFM cantilever probes made of polycrystalline diamond utilizing the conventional molding technique [105, 106]. **d** An example of a tip-on-tip diamond probe [107]

ity of the diamond film at the tip, an intricate process for the attachment of a holder to the diamond membrane etc. [120, 121]. The application of molded diamond probes in an AFM set-up was hindered because of the strong optical scattering of the laser beam necessary for the beam deflection measurement from the very rough growth surface of the polycrystalline diamond surface constituted of (111)-oriented grains. Coating of commercial silicon tips was suggested to overcome these problems but at the expense of an increased radius of curvature [116, 118, 122].

Recently, the *projection mask technique* was introduced to overcome most of the above mentioned problems [120]. This method utilizes an already structured substrate to define the vertical cantilever geometry as shown in Fig. 4.4. The lateral geometry of the cantilever is defined consecutively as a two-dimensional pattern on a conventional optical projection mask. This pattern is transferred via a proximity lithography and etching process on the already structured substrate. A mechanical holder necessary for the case of handling is finally fabricated from the substrate to accomplish the cantilever probe. The projection mask technique is capable of probe fabrication from the substrate as well as from thin films deposited onto the substrate.

The feasibility of this method was demonstrated in the case of all-diamond cantilever probes. SEM images of typical diamond probes are given in Fig. 4.5. These probes were successfully applied to the nanofabrication of ultrasmall single-electron in-plane gate transistors fabricated by mechanical scribing with the diamond tip while monitoring the depth of scribing electrically [121].

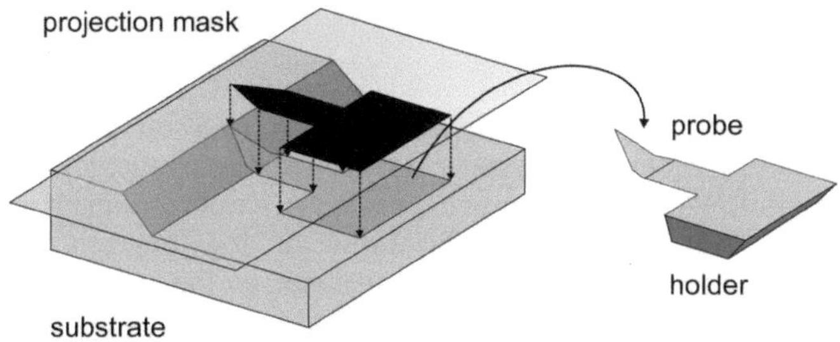

Fig. 4.4. Scheme of the projection mask technique. The outline of the cantilever is defined via a proximity lithography and etching process while the vertical geometry is given by the already structured substrate. The probe can be fabricated from both the substrate or a thin deposited film [120]

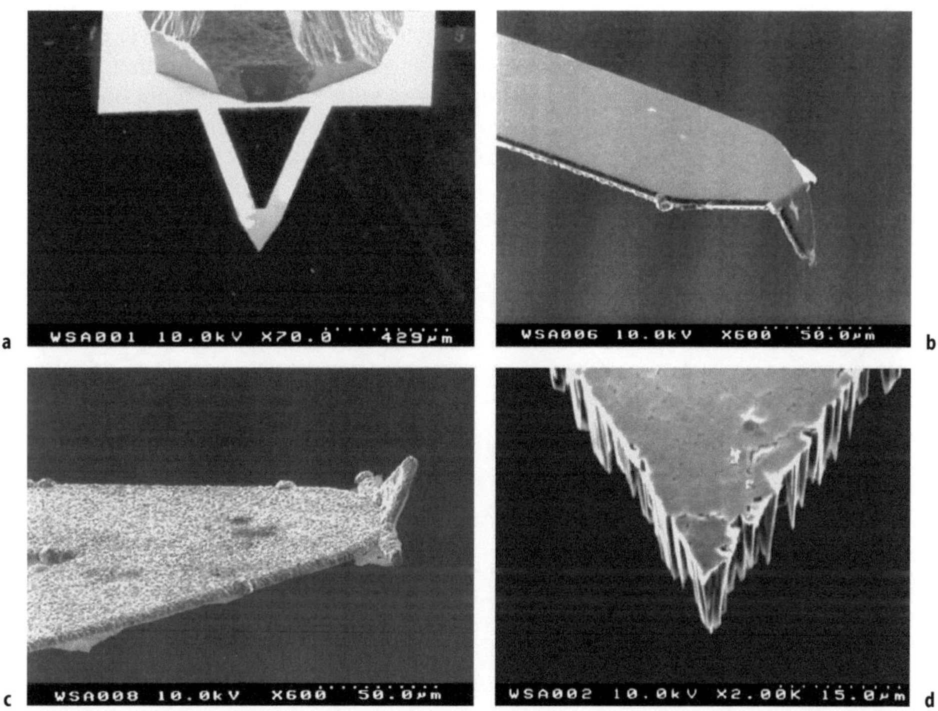

Fig. 4.5a–d. SEM images of all diamond AFM cantilever probes made by the projection mask technique: **a** diamond membrane with a v-shaped cantilever probe mounted on a silicon holder, **b** single cantilever with a pointed tip, **c** bottom view of the cantilever in **a** revealing the rough growth surface of HFCVD diamond, **d** sharpening of the diamond grains by plasma treatment [120]

4.2.1.3
Gallium Arsenide

Gallium arsenide (GaAs) gained much attraction due to its unique electronic and optical properties [123]. In this context it is considered substitute for the rich class of compounds of III/V- and II/VI-semiconductors. In particular the direct band gap predestines GaAs for a variety of opto-electronic and low-dimensional electronic devices (Fig. 4.6). Below we will focus for instance on the application of a very compact surface emitting quantum well laser for an active optical near-field probe. But also for the investigation of ultrafast phenomena GaAs is one of the predominant substrate materials for the establishment of photoconductive switches as will be discussed later. Another noteworthy application of GaAs is its use as an efficient photon-excited source of spin-polarized electrons [124].

In first attempts GaAs sensors were made by cleaving single-crystalline GaAs wafers twice to obtain very sharp tips at the intersection of the two edges [126]. Meanwhile, it became common practice to batch-fabricate GaAs cantilever probes with integrated tips and the process works in general analogous to that based on silicon [127]. However, the compound GaAs with its covalent/ionic bonding between the Ga amid As atoms leads to an intricate etching behavior [123]. On one hand this exacerbates the fabrication of MEMS structures by wet chemical etching but on the other hand it can also be exploited to obtain extreme tip shapes as shown in Fig. 4.7 that are not feasible in case of silicon or diamond material [128].

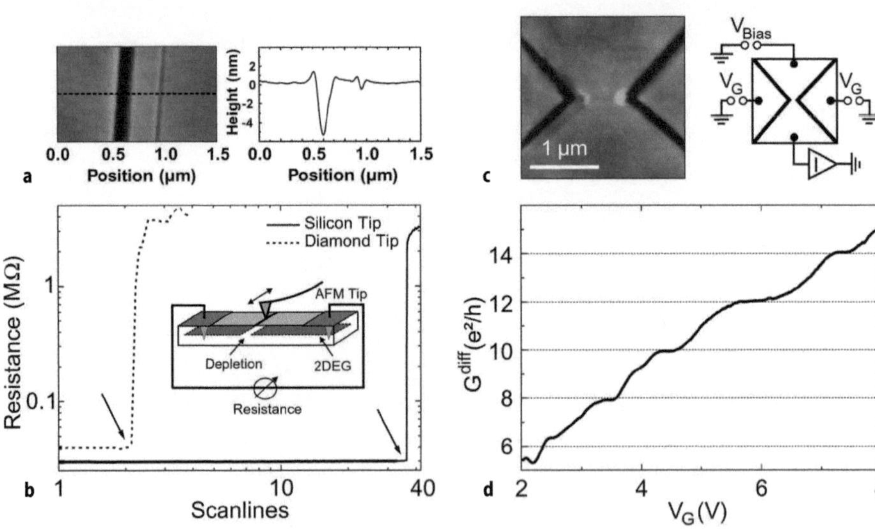

Fig. 4.6. **a** SFM image and corresponding profile of two grooves scribed with a silicon and a diamond tip into a stack of thin films on top of an insulating GaAs substrate, **b** number of scanlines necessary to separate conducting layers carrying a two-dimensional electron gas, **c** SFM image of a scribed in-plane gate transistor with its circuit diagram, and **d** its characteristics obtained at a temperature of 20 mK [121] (with courtesy of R. J. Haug and U. Keyser, Universität Hannover, Experimentalphysik, Hannover, Germany)

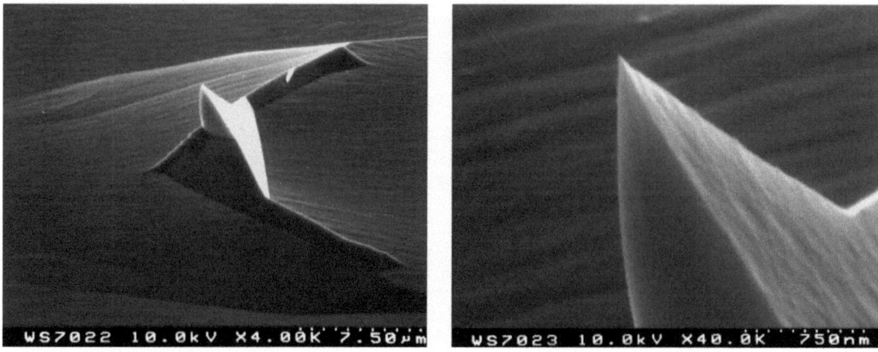

Fig. 4.7. SEM images of GaAs tips obtained by etching with aqueous solutions of $H_2SO_4 : H_2O_2 : H_2O$ [125]

In the following discussion probes for different applications will be discussed in more detail.

4.3 Scanning Near-Field Optical Microscopy

4.3.1 Principle of Near-Field Optics

The resolving power of optics is restricted by two spatial low pass filter processes. The first is inherent to the nature of light because the wavenumber $\boldsymbol{k} = 2\pi/\lambda$ of monochromatic light of a given wavelength λ limits the amount of the object spatial frequency component $k_{object} = \sqrt{k_x^2 + k_y^2}$ in the object plane. However, propagation in free space requires a positive and real component k_z in the propagation direction:

$$k_z = \sqrt{k_0^2 - k_{object}^2} = \sqrt{k_0^2 - \left(k_x^2 + k_y^2\right)}. \tag{4.5}$$

Hence, higher object spatial frequencies than k give rise to imaginary k_z and thus to evanescent waves, that is, waves that are exponentially damped in free space and are not accessible by far-field optics.

The second low pass filter stems from the geometry of the lens itself owing to the diffraction at the lens barrel. In terms of Fourier optics it is apparent from the simplifying scheme in Fig. 4.8 that only Fourier components with $k_x \leq k_x^{\max}$ are capable of passing the lens of focal length f and diameter D.

Assuming that:

$$\frac{k_x^{\max}}{k_0} = \sin\left(\upsilon^{\max}\right) \approx \tan\left(\upsilon^{\max}\right) = \frac{D}{2b} = NA, \tag{a}$$

where NA is the numerical aperture of the lens. It limits the largest spatial frequency $k_x^{\max}$ and thus the lowest spatial wavelength length $\lambda_x^{\max}$ of the object, respectively, that is passing the lens:

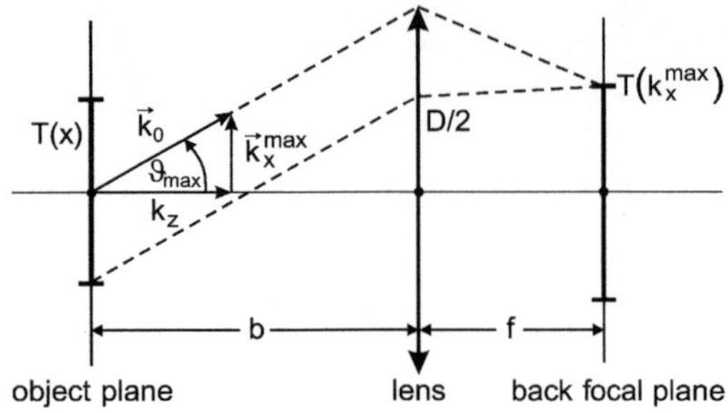

Fig. 4.8. Imaging of an object with transmission function $T(x)$ with a lens of focal length f and diameter D results in the Fourier transform $F\{T(x)\} = T(k_x)$ in its back focal plane [129]

$$k_x^{\max} = k_0 NA \quad \text{and} \quad \lambda_x^{\max} = \frac{\lambda_0}{NA}. \tag{b}$$

The latter expression is in accordance with the arbitrarily defined Rayleigh's criterion for the resolving power of a lens [130].

To overcome the limitations of far-field optics requires obviously the detection of the evanescent, that is, the high spatial frequency near-field components. This is achieved with an optical link that picks up part of the near-field component in the vicinity of the sample surface amid converts it into propagating waves. Optical links can be catalogued as aperture carrying wave-guiding probes, scattering (apertureless) probes, and immersion lens probes [131]. In the following we restrict the discussion to the first type of probe. For a more extensive discussion see, for example, refs. [24, 131] and references therein.

4.3.2 Probes for Scanning Near-Field Optical Microscopy (SNOM)

4.3.2.1 *Aperture-Based Near-Field Optical Probes*

Aperture probes consists in principle of a sub-wavelength-sized aperture in an optically opaque screen. If the latter is brought in close vicinity to the sample surface to pick up near-field optical information, diffraction is in good approximation negligible and the aperture size rather than the wavelength determines the lateral resolution. For the detection of near-field components the aperture is in almost every case integrated into the very tip of a pointed tapered metal-coated waveguide. Owing to the cut-off effect the latter shows unfavorable waveguide properties, that us, tapering the waveguide below half the wavelength gives rise to strong damping because of the lack of any propagating waveguide mode [132–135].

From this brief discussion we can draw the conclusion that an appropriate aperture probe should offer at least the following: a pointed tip with a sub-wavelength aperture, a very small taper region to reduce transmission losses, a proper mechanical compliance to keep the aperture mechanically in track with the sample surface during scanning, that is, image formation, and most importantly reproducible optical and mechanical probe properties for reliable imaging.

Most of the requirements discussed can easily be fulfilled if near-field optics is combined with scanning force microscopy integrating an aperture tip into a con-

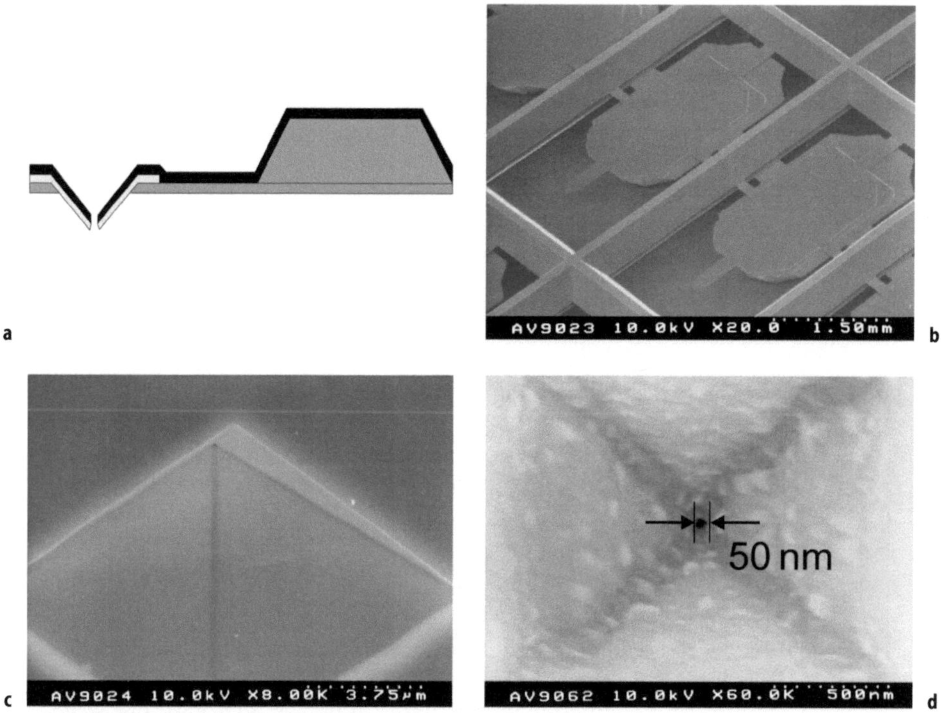

Fig. 4.9. **a** Cross-section of a SNOM aperture probe carrying a hollow tip made of a thin silicon dioxide layer. The latter is pierced at its apex and coated with an opaque aluminium film to obtain aperture SNOM tips. **b** The complete SFM/SNOM cantilever probe, **c–d** SEM images of the pyramidal hollow tip and the aperture of 50 nm size, respectively [136]

ventional SFM cantilever probe fabricated by MEMS technology [40, 137–139]. Hence we introduced an adapted technology to integrate a hollow silicon dioxide aperture tip into a conventional silicon cantilever probe as can be seen from the schematic cross-section of this tip in Fig. 4.9a [136]. The finished aperture SNOM/SFM probe is shown in Fig. 4.9b whereas c and d show zoom images of the aluminum coated tip forming a 50-nm sized aperture in the opaque metal layer at the tips apex, respectively. Aperture fabrication exploits the pronounced oxide retardation at concave (and convex) silicon structures. Owing to the self-adjusting nature of this process for *top-to-bottom* nanostructure fabrication we will now discuss this process in some detail.

In the case of planar silicon substrates oxidation follows in good approximation the simplifying Deal/Grove model that predicts a diffusion-limited (reaction rate-limited) growth of the oxide layer if the grown silicon dioxide layer exceeds (is less than) about 20–40 nm [140]. However, the Deal/Grove model does not account for the intrinsic mechanical stress originating from the volume expansion by a factor of 2.25 oxidizing a single Si atom to obtain an SiO_2 molecule [141–143]. Thus, Kao et al. emphasized the influence of the intrinsic stress effecting both the temperature-dependant diffusion coefficient and the reaction rate that give rise to remarkable oxide growth retardation at concave and convex surface structures in comparison to planar ones [144, 145]. Furthermore, they showed that viscous flow in the oxide layer at temperatures above the glass temperature of SiO_2 (965 °C) counteracts slightly the amount of oxide retardation via stress relaxation assuming a nonlocal viscosity [144]. Hence low oxidation temperatures are assumed to be best for strong oxide retardation.

However, Senez et al. evidenced that silicon dioxide has to be treated as viscoelastic material with a nonlinear stress-dependent viscosity [147]. The rheological behavior can be described in terms of a Maxwell liquid:

$$\dot{\varepsilon} - \left(\frac{\dot{\sigma}}{G} + \frac{\sigma}{\eta}\right) = 0 \tag{4.6}$$

where $\dot{\varepsilon}$, $\dot{\sigma}$ and σ denote the corresponding strain rate, stress rate, and stress. G and η are the modulus of rigidity and the viscosity of silicon dioxide, whereas the ratio η/G is the stress relaxation time τ. The most persuasive description of the nonlinear behavior of the viscosity of thermally grown silicon dioxide is given by the Eyring model [148]:

$$\eta\left(\sigma_s\right) = \eta_0 \frac{\sigma_s/\sigma_c}{\sinh\left(\sigma_s/\sigma_c\right)} \tag{4.7}$$

as proposed by several authors [147, 149].

Figure 4.10a,b show the calculated stress distribution and the outline of a silicon dioxide layer grown at oxidation temperatures of 900 °C in a V-groove structure assuming a rigid silicon substrate. Obviously, oxide retardation occurs at the concave part of the V-groove in comparison to the sidewalls of the groove which is also proven experimentally (see Fig. 4.10c). However, in contrast to the prediction of Kao et al. oxide growth retardation is more pronounced in case of high oxidation temperatures. This is due to the fact that the relaxation time is proportional to the viscosity which scales nonlinear with the local mechanical stress. Thus, at low temperatures the high intrinsic stress generates in accordance with the Eyring model a low viscosity and thus a shorter relaxation time as assumed by Kao et al. Thus the longer oxidation time necessary at low oxidation to obtain a similar silicon dioxide thickness allows a more pronounced relaxation of stress and this in turn results in a less pronounced oxide retardation.

To accomplish apertures in the silicon dioxide layer the structures shown in Fig. 4.11c are opened from the bottom by removing first the silicon substrate and subsequently part of the oxide layer. Due to the retarded oxide thickness at the apex, it opens during the etching process at the apex reproducibly to form an aperture. In the case of SNOM aperture tips of course pyramidal-shaped structures are used instead of grooves.

Of course aperture tip formation is not restricted to cantilever structures. Figure 4.11 shows other structures like a circular pad or b–d array arrangements of apertures tips. The first are used to fabricate a hybrid SNOM probe build up from a cleaved fiber probe mounted onto the MEMS-based aperture chip. Aperture arrays may be interesting for parallel SNOM imaging and data storage [150–152] but also other applications (e.g., mass filters, calibration samples etc.).

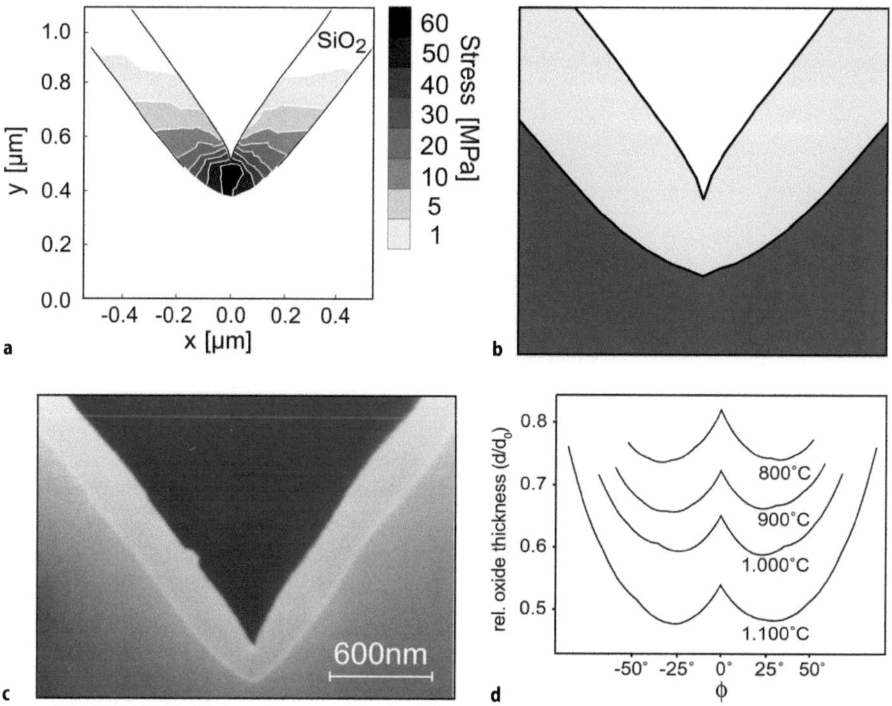

Fig. 4.10. **a** Calculated stress distribution in the oxide layer of a V-groove silicon structure that was wet oxidized at a temperature of 900 °C. **b** Corresponding contour plot of the oxide layer revealing the reduced oxide thickness at the apex of the V-groove. **c** The theoretical predictions are experimentally confirmed by SEM imaging of cross-sections of V-groove structures. **d** Additional simulations suggest that increasing oxidation temperature leads to a more distinct retardation of oxide growth [136]. Simulations were performed with the numerical code FLOOPS from M.E. Law and A. Tasch [146]

It should be emphasized that due to the large opening angle of 70.5° the taper of our aperture tip is rather short and thus the transmission rather high in comparison to tapered fiber probes. Nevertheless, it is not appropriate for some interesting applications like Raman spectroscopy or single molecule detection. For this purpose coaxial arrangements [154–156] as well as antenna structures like the bow-tie antenna were proposed [157, 158]. The latter consists as shown in Fig. 4.12a of two triangular metal electrodes that face each other and are separated by a tiny gap. Simulations underline that the bow-tie antenna has the capability of focusing electromagnetic waves even at optical wavelengths into the gap of the bow-tie antenna (see Fig. 4.12b) with almost unit efficiency [159]. However, the realization of gaps of less than 50-nm size on a three-dimensional structure is currently a real challenge.

Figure 4.13 shows first results of a microfabricated bow-tie antenna tip based on our silicon dioxide aperture probe [158]. The two arms of the antenna were accomplished by focused ion beam removal of triangular parts of the metal layer that was first deposited onto the whole aperture tip. Figure 4.13b proves that gaps of about 50 nm dimension can be realized in this serial process.

Fig. 4.11. **a** SEM image of circular silicon pads with integrated hollow aperture tip. Pads are to be mounted on cleaved fibers to obtain hybrid sensors for SNOM applications. **b** SEM image of an array of aperture tips in a mechanically stable silicon membrane. They are proposed to be used as calibration samples for SNOM. **c, d** Zoom images of single pits (in collaboration with WiTec GmbH (Ulm) Germany)

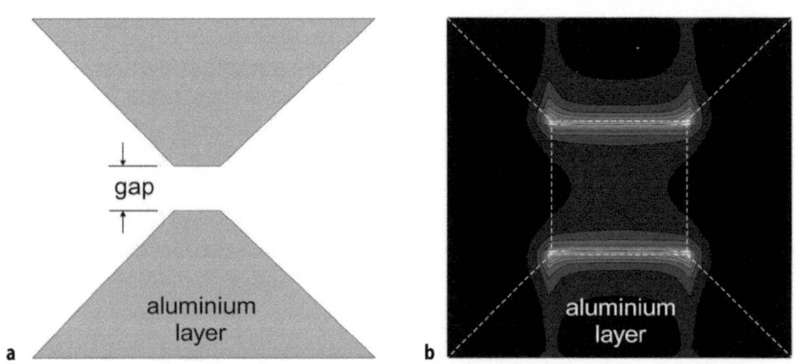

Fig. 4.12. **a** Scheme of a bow-tie antenna probe consisting of two triangular electrodes facing each other and leaving a tiny gap between them [153]. **b** Electric field energy distribution in the plane of the bow-tie antenna showing its focusing capability (*white color* indicates high density) [24]

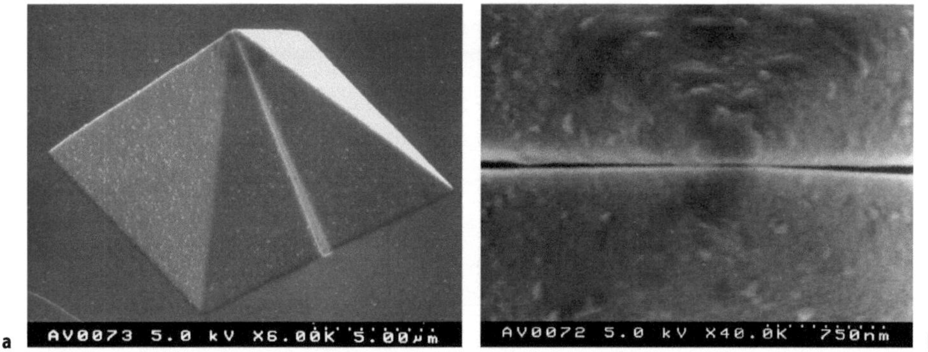

Fig. 4.13. SEM images of a hollow silicon dioxide tip coated with aluminium. The latter was partly removed by focused ion beam milling to establish a pointed dipolar bow-tie antenna (in collaboration with H.U. Danzebrink and Ch. Lehrer) [158]

4.3.2.2
Modification of Aperture Probes

In our discussion aperture tips were only assumed to be useful as passive wave guiding structure. However, the hollow geometry of the silicon dioxide tip suggests that it may also be exploited for active SNOM devices. Figure 4.14a shows the scheme of an active SNOM emitter where a deposited layer of fluorescent or even electrofluorescent material is used to illuminate the sub-wavelength aperture. In this way active nano-light sources based on organic light emitting diodes (OLED) technology can be realized in future.

The silicon dioxide aperture tip introduced above is a quite interesting base structure for completely different applications. Figure 4.14b shows for example a miniaturized electrode that might be used as working electrode in scanning electrochemical microscopy (SECM). Another application we are currently working on is the electrical and optical measurement of the pH value on an ultrasmall scale. Possible applications fields are, for example, the locally resolved in situ investigation of proton pumping bacteria (e.g., pseudomonas putida, pseudomnonas fluorescence) or the pH value measurement in extremely small chemical microreactors.

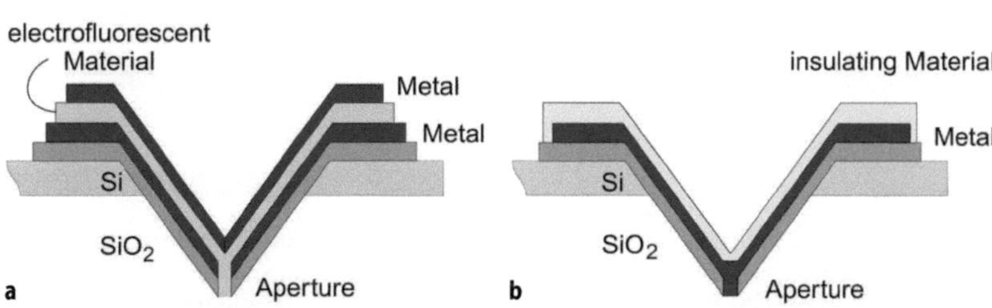

Fig. 4.14. Proposed probes obtained by modification of the bare aperture tip: **a** an active near-field emitting tip based on OLED technology, **b** ultrasmall sample electrode for scanning electrochemical microscopy (SECM)

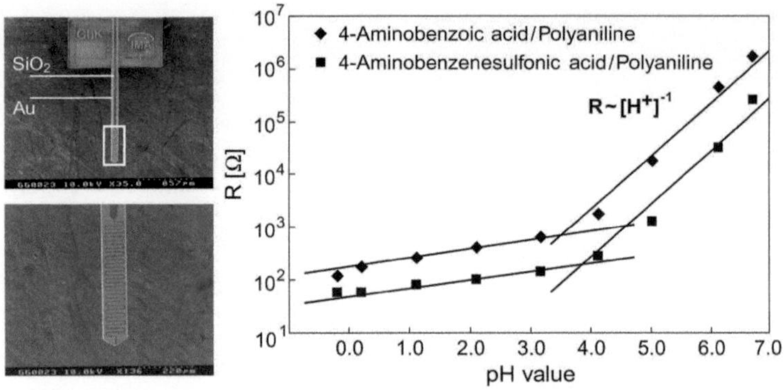

Fig. 4.15. SEM images of an interdigitated polyaniline-based resistive pH value sensor on a cantilever probe (upper image). The resistance characteristics of the pH probe obtained in phosphoric acid of varying pH-value shows two almost linear regions of different sensitivity (with courtesy of T. Kurzenknabe, M. Müller-Wiegand, G. Georgiev, N. Nicoloso, E. Oesterschulze)

For the latter application interdigitated sensors electrochemically coated with polyaniline [160–162] have already been realized on thermally oxidized silicon cantilever probes as shown in the left part of Fig. 4.15. The geometry of the probe is compatible with typical dimensions of microtiter plates. Polyaniline as well as some of its derivatives have been utilized and investigated as pH-value-sensitive materials [163–165]. The electrically measured characteristics of the interdigitated probe as a function of pH value are given in the right part of Fig. 4.15 and demonstrate its feasibility. A further decrease of probe dimensions however is hindered owing to technological as well as experimental limitations (e.g., the realization of ultrasmall electrodes separated by less than 100 nm or so).

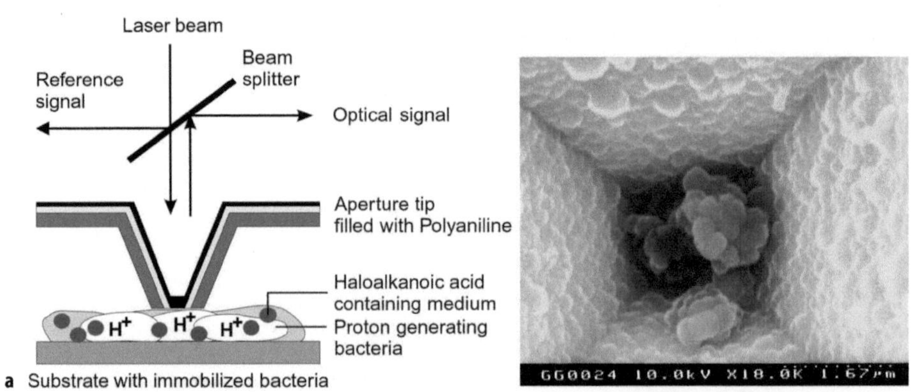

Fig. 4.16a, b. Ultrasmall optically operated pH-value probe based on the hollow aperture tip in Fig. 9. The optical active polyaniline film was electrochemically deposited onto a Au/Ti layer into the tip (in collaboration with F. Schmidt, Universität Kassel, Mikrobiologie (Kassel) Germany. **a** Optical measurement set-up and cross-section through the modified tip. **b** SEM image revealing the globular structure of the thin polyaniline film inside the hollow tip (with courtesy of T. Kurzenknabe, M. Müller-Wiegand, G. Georgiev, N. Nicoloso, E. Oesterschulze)

To overcome these kinds of limitations of electrical-based pH-value-sensitive probes, we currently propose to exploit the pH-dependant optical absorption of polyaniline [166–169] for high-resolution imaging of the pH -value using the aperture tip in Fig. 4.9. For this purpose we recently succeeded in electrochemically depositing polyaniline into the silicon dioxide aperture that was first coated with thin layers of Ti/Au. Figure 4.16b shows an example of a deposited polyaniline film in the aperture tip. Protons entering through the aperture and diffusing into the polyaniline film cause a change in the absorbance of the latter which is optically detected. As shown in Fig. 4.16a the absorbance can be determined by focusing a laser beam into the pyramidal tip and measuring the reflected beam while monitoring the original one as reference. First preliminary results indicate the feasibility of this novel approach.

4.3.2.3 *Active Near-Field Optical Probes*

A completely different approach of a light emitting active near-field probe for SNOM amid SFM applications is presented in Fig. 4.17. It comprises a conventional monolithic GaAs cantilever probe with a sharpened tip and an integrated laser diode that illuminates the tip from the rear. Light emission is restricted by a sub-wavelength-sized aperture in the 100-nm thin metal layer at the tips apex. The high refractive index of GaAs ($n = 3.8$ at a wavelength of 980 nm) is advantageous because it reduces the transmission intensity losses of the tapered tip due to the cut-off effect [128]. The compact geometry of the VCSELs allows one to accomplish near-field cantilever probes with indistinguishable dimensions to conventional SFM cantilevers. Furthermore, our probe design is capable of exploiting the beam of the VCSEL emitted in the opposite direction for the beam deflection measurement in the SFM mode thereby omitting any external light source. Thus these kinds of probes are predestined to be operated in a parallel arrangement that is advantageous, for example, for a very compact set-up of high-density near-field optical data storage devices [170] or near-field optical lithography.

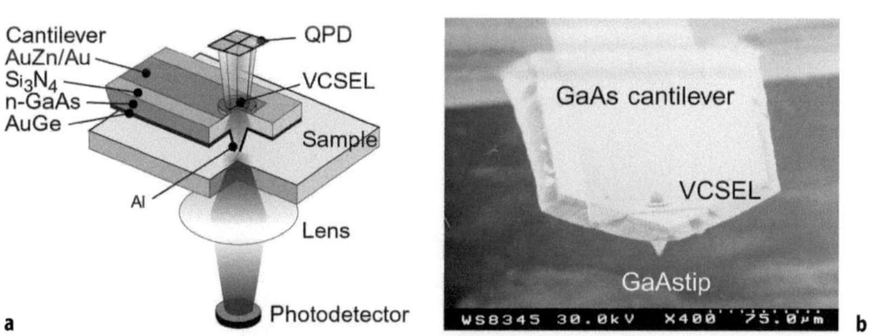

Fig. 4.17. **a** Schematic view and **b** SEM image of the first active GaAs near-field probe with an integrated VCSEL operating at a wavelength of about 977 nm [127, 128]

4.4 Probes for Ultrafast Scanning Probe Microscopy

In ultrafast scanning probe microscopy (USPM) voltage contrast of electronic devices is detected with a pump/probe experiment [171, 172]. An optical train of ultrashort pulses emitted from a Ti-sapphire laser is converted into an electrical pulse using a photoconductive switch [173–175]. This pulse is sent to the device under test (DUT). Simultaneously, the same optical pulses are used to turn on and off a photoconductive switch [176, 177] integrated into a GaAs SFM cantilever to measure a high-speed electrical signal using photoconductive sampling technique [178]. A SEM image of an appropriate sampling probe is shown in the lower part of Fig. 4.18. It carries a microwave strip line on top of a thin low-temperature-grown (LT) GaAs layer that was deposited onto a cantilever made from an insulating GaAs substrate [179]. Illuminating the gap between two of the adjacent microwave electrodes with ultrashort optical pulses allows one to sample the voltage that is picked up with the very strip line on the GaAs cantilever from the DUT. The time resolution is achieved by varying the time delay between the synchronized pump and probe beam [180].

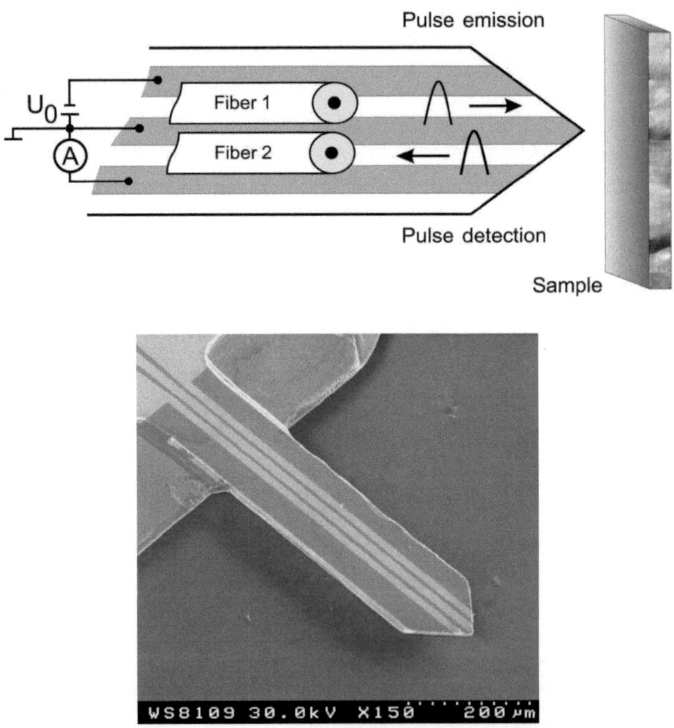

Fig. 4.18. Scheme of a scanning microwave microscope based on a GaAs cantilever with two adjacent photoconductive switches used for launching and subsequent detection of ultrafast signals that interact locally with the sample surface. SEM image of a realized USPM cantilever with microstrip line [179]

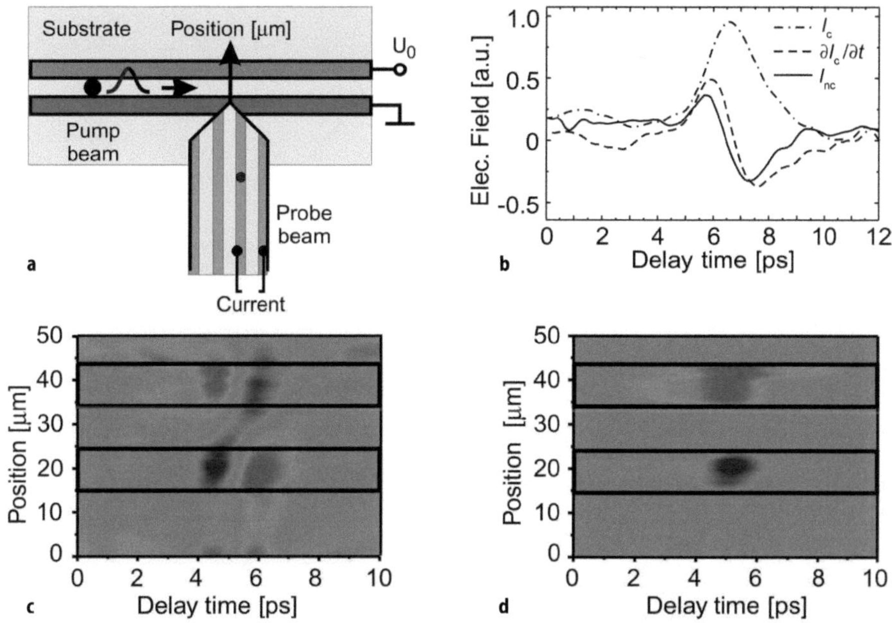

Fig. 4.19. **a** Spatio/temporal sampling of the field distribution of pulses launched on a microwave strip line with a photoconductive switch integrated onto a GaAs cantilever. **b** The resolved electric field signal on the grounded electrode of the transmission line: in contact mode I_c, in noncontact mode I_{nc}, and the numerically differentiated contact mode signal. **c, d** Spatio/temporal measurements of the field distribution of the stripline mode in contact and noncontact mode, respectively [179–182]

A typical result of the investigation of a microwave strip line is given in Fig. 4.19. The experiment was conducted as shown inn Fig. 4.19a. The temporal signal shown in Fig. 4.19b was picked up on a fixed position on the ground electrode and reveals a pulse width of the strip line mode of about 1.2 ps [179, 181]. Figure 4.19c and d show similar results of a spatio/temporal investigation where the cantilever was scanned across the strip line in the contact and noncontact mode to obtain the complete pulse shape, respectively. Obviously, the differentiated signal is obtained in noncontact mode owing to the differentiating behavior of the gap capacitance between the cantilever tip and sample surface [181]. Operating the cantilever probe in the noncontact mode should thus allow the high-resolution investigation of the local dielectric function in a microwave microscope as schematically shown in the upper part of Fig. 4.18.

4.4.1 Improved Sampling Technique

It has to be pointed out that for the conventional photoconductive sampling technique as described above, the carrier lifetime limits the temporal resolution of the system to 1–2 ps [183, 184]. Usually materials with reduced carrier lifetimes such

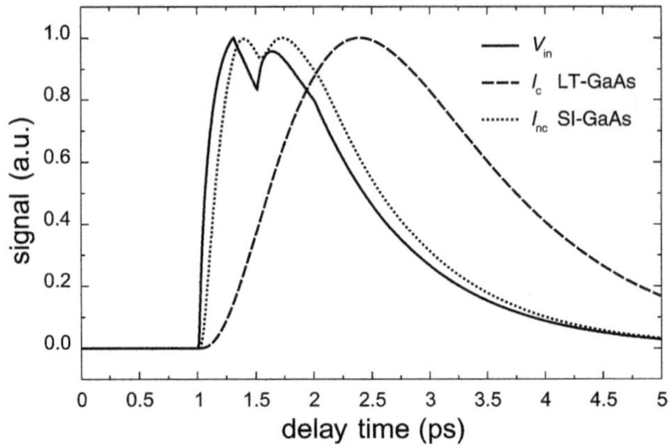

Fig. 4.20. The original signal V_{in} on the strip line is a superposition of two 500-fs-long pulses delayed by 300 fs. Additionally shown are the signals obtained for the conventional sampling with fast LT-GaAs material in the contact mode (*dashed line*) and for the novel technique with long carrier lifetime material in the noncontact mode (*dotted line*) [182]

as low-temperature-grown GaAs (LTGaAs) or ion implanted semiconducting layers are utilized [185–189]. However, the introduction of a high defect density by such techniques usually results in low carrier mobility, poor stability, and a lack of fabrication processes of other devices.

To get rid of these material restrictions, we proposed an alternative sampling method. The idea of the improved sampling concept relies on the mathematical theorem that the differentiation of the Heavyside function $\Theta\,(t – t_0)$ results in Diracs delta-function:

$$\delta\,(t - t_0) = \frac{d\Theta\,(t - t_0)}{dt} \tag{4.8}$$

That is if the photoconductive switch is formed using a material with a longer carrier lifetime, such as semi-insulating GaAs (SI-GaAs), than the transfer function of the photoconductive switch considering a timescale of a few picoseconds corresponds to a step-like function. If this photoconductive switch is integrated into the GaAs cantilever and the latter is operated in the noncontact mode, sampled signals will be differentiated and thus the time resolution should be improved substantially. First theoretical simulations confirm the working principle as can be seen in Fig. 4.20 [182]. To demonstrate the high efficiency of this improved technique a superposition of two pulses delayed by 300 fs was sampled. For the conventional sampling technique with LTGaAs the obtained signal in the contact mode and for the novel technique with SI-GaAs the noncontact signals are plotted together in Fig. 4.20. As can be seen, the high temporal resolution of the new technique allows one to resolve the two superimposed pulses, whereas the conventional method, even with a switch response as short as 600 fs, is not capable of recovering the double peak. This concept may be used in future for ultrafast pulse generation and sampling.

References

1. Binnig G, Rohrer H, Gerber C, Weibel E (1981) Tunneling through a controllable vacuum gap. Appl Phys Lett 40:178–180
2. Binnig G, Rohrer H (1982) Scanning tunneling microscopy. Helv Phys Acta 55:726
3. Binnig G, Rohrer H, Gerber C, Weibel E (1982) Surface studies by scanning tunneling microscopy. Phys Rev Lett 49:57–61
4. Binnig G, Quate CF, Gerber C (1986) Atomic force microscope. Phys Rev Lett 56:930–933, 1986
5. Wiesendanger R (1994) Scanning probe microscopy and spectroscopy. Cambridge University Press
6. Eigler DM, Schweizer BK (1990) Positioning single atoms with a scanning tunneling microscope. Nature 344:524–526
7. Crommie MF, Lutz CP, Eigler DM (1993) Confinement of electrons to quantum corrals on a metal surface. Science 262:218–220
8. Martin Y, Williams CC, Wickramasinghe HK (1987) Atomic force microscopy-force mapping and profiling on a sub 100-Å scale 1. Appl Phys 61:4723–4729
9. Sarid D (1991) Scanning-force-microscopy. Oxford University Press, New York
10. Albrecht TR, Grutter P, Horne D, Rugar D (1991) Frequency modulation detection using high-q cantilevers for enhanced force microscope sensitivity. J Appl Phys 69:668–673
11. Meyer B, Heinzelmann H (1995) Scanning force microscopy. In: Wiesenndanger R, Güntherodt HJ (eds) Scanning tunneling microscopy II. Springer, Berlin Heidelberg New York, pp 99–149
12. Anczykowski B, Krüger D, Fuchs H (1996) Cantilever dynamics in quasinon-contact force microscopy: spectroscopic aspects. Phys Rev B 53:15485–15488
13. Schwarz UD, Zwörner O, Köster P, Wiesendanger R (1997) Quantitative analysis of the frictional properties of solid materials at low loads II. MICA and germanium sulfide. Phys Rev B 56:6997–7000
14. Giessibl FJ, Bielefeldt H (2000) Physical interpretation of frequency-modulation atomic force microscopy. Phys Rev B 61:9968–9971
15. Hölscher H, Gotsmann B, Allers W, Schwarz UD, Fuchs H, Wiesendanger R (2001) Measurement of conservative and dissipative tip-sample interaction forces with a dynamic force microscope using the frequency modulation technique. Phys Rev B 64:075402
16. Schulz M, Blachnik R (1982). Landolt-Börnstein, vol 111/17a. Springer, Berlin Heidelberg New York, pp 61–83
17. Blakemore JS (1982) Semiconducting and other major properties of gallium arsenide. J Appl Phys 53:R123–R183
18. von Münch (1982) Landolt-Börnstein, vol III/17a. Springer, Berlin Heidelberg New York, pp 36–42
19. Heuberger A (1991) Mikromechanik. Springer, Berlin Heidelberg New York
20. Johansson S, Ericson F, Schweitz J (1989) Influence of surface coatings on elasticity, residual stresses, and fracture properties of silicon microelements. J Appl Phys 65:122–128
21. Lai J, Perazzo T, Shi Z, Majumdar A (1997) Optimization and performance of high-resolution micro-optomechanical thermal sensors. Sens Actuators A 58:113–119
22. Hazel JL, Tsukruk VV (1999) Spring constants of composite ceramic/gold cantilevers for scanning probe microscopy. Thin Solid Films 339:249–257
23. Madou M (1997) Fundamentals of microfabrication. CRC Press LLC
24. Oesterschulze E (2001) Advances in imaging and electron physics, vol 118, chap recent developments of probes for scanning probe microscopy. Academic Press, pp 129–206
25. Wolf S, Tauber RN (1986) Silicon processing for the VLSI Era, vol 1. Lattice Press, California
26. Katz LB (1990) VLSI Technology, chap 3. McGraw-Hill
27. Wolter O, Bayer T, Greschner J (1991). Micromachined silicon sensors for scanning force microscopy. J Vac Sci Technol B 9:1353–1357

28. Ravi TS, Marcus RB (1991) Oxidation sharpening of silicon tips. J Vac Sci Technol B9: 2733–2737
29. Zhang Y, Zhang Y (1996) Formation of single tips of oxidation-sharpened Si. Appl Phys Lett 69:4260–4261
30. Marcus RB, Ravi TS, Gmitter T, Chin K, Liu D, Orvis WJ, Ciarlo DR, Hunt CE, Trujillo J (1990) Formation of silicon tips with <1 nm radius. Appl Phys Lett 56:236–238
31. Barth W, Debski T, Abedinov N, Ivanov T, Heerlein H, Volland B, Gotszalk T, Rangelow IW, Torkar K, Fritzenwallner F, Grabiec P, Studzinska K, Kostic I, Hudek P (2001) Evaluation and fabrication of AFM array for ESA-midas/rosetta space mission. Microelectron Eng 57–58:825–831
32. Collins SD (1997) Etch stop techniques for micromachning. J Electrochem Soc 144: 2242–2262
33. Nakano S, Ogiso H, Yabe A (1999) Advanced micromachine fabrication using ion-implanted layers. Nuclear instruments and methods in physics research. Mater Sci Eng B 155:79–84
34. Yang J, Ono T, Esashi M (2000) Mechanical behaviour of ultrathin microcantilevers. Sens Actuators 82:102–107
35. Itoh J, Tohma Y, Kanemamu S, Shimizu K (1995) Fabrication of an ultrasharp and high-aspect-ratio microprobe with a silicon-on-insulator wafer for scanning force microscopy. J Vac Sci Technol B 13:331–334
36. Hosaka S, Etoh K, Kikukawa A, Koyanagi H (2000) Megahertz silicon atomic force microscopy (AFM) cantilever and high-speed readout in AFM-based recording. J Vac Sci Technol B 18:94–99
37. MaCarthy J, Pei Z, Becker M, Atteridge D (2000) FIB micromachined submicron thickness cantilevers for the study of thin film properties. Thin Solid Films 358:146–151
38. Albrecht TR, Akamine S, Carver TB, Quate CF (1990) Microfabrication of cantilever styli for the atomic force microscope. J Vac Sci Technol A 8:3386–3396
39. Spindt CA, Bordie I, Humphrey L, Westerberg BR (1996) Physical properties of thin-film field emission cathodes with molybdenum cones. J Appl Phys 47:5248–5262
40. Mihalcea C, Scholz W, Werner S, Münster S, Oesterschulze E, Kassing R (1996) Multi-purpose sensor tips for scanning near-field microscopy. Appl Phys Lett 68:3531–3533
41. Hantschel T, Pape U, Slesazeck S, Niedermann P, Vandervorst W (2000) Mounting of moulded AFM probes by soldering. Proc SPIE 4175:62–73
42. Scholz W, Albert D, Malavé A, Werner S, Mihalcea C, Kulisch W, Oesterschulze E (1997) Fabrication of monolithic diamond probes for scanning probe microscopy applications. In: Micromachining and Imaging, SPIE vol 3009–09, pp 61–71
43. Tortonese M, Yamada H, Barett RC, Quate CF (1991) Atomic force microscopy using a piezoresistive cantilever. IEEE 91CH2817–5:448
44. Minne SC, Manahis SR, Quate CF (1995) Parallel atomic force microscopy using cantilevers with integrated piezoresistive sensors and integrated piezoelectric actuators. Appl Phys Lett 67:3918–3920
45. Jumpertz R, von den Hart A, Ohlsson O, Saumenbach S, Schelten J (1998) Piezoresistive sensors on AFM cantilevers with atomic resolution. Microelectron Eng 41/42:441–444
46. Volodin A, van Haesendonck C (1998) Low temperature force microscopy based on piezoresistive cantilevers operating at a higher flexural mode. Appl Phys A 66:S305–S308
47. Su Y, Brunnschweiler A, Evans AGR, Ensell G (1999) Piezomesistive silicon V-AFM cantilevers for high-speed imaging. Sens Actuators 76:139–144
48. Brugger J, Despont M, Rossel C, Rothuizen H, Vettiger P, Willemin M (1999) Microfabricated ultrasensitive piezoresistive cantilevers for torque magnetometry. Sens Actuators 73: 235–242
49. Gotszalk T, Grabiec P, Rangelow I (2000) Piezoresitive sensors for scanning probe microscopy. Ultramicroscopy 82:39–48
50. Erlandsson R, McClelland GM, Mate CM, Chiang S (1988) Atomic force microscopy using optical interferometry. J Vac Sci Technol A 6:266
51. Meyer G, Amer M (1988) Novel optical approach to atomic force microscopy. Appl Phys Lett 53:1045

52. Rugar D, Mamin HJ, Guethner P (1989) Improved fiber-optic interferometer for atomic force microscopy. Appl Phys Lett 55:2588
53. Putman CAJ, de Grooth BG, van Hulst N, Greve J (1991) A theoretical comparison between interferometnic and optical beam deflection technique for the measurement of cantilever displacement in AFM. Ultramicroscopy 42–44:1509–1513
54. Ruf A, Abraham M, Diebel J, Ehrfeld W, Güthner P, Lacier M, Mayr K, Reinhardt J (1997) Integrated Fabry-Perot distance control for atomic force microscopy. J Vac Sci Technol B 15:579585–
55. Göddenhenrich T, Lembe H, Hartmann U, Heiden C (1990) Force microscope with capactive displacement detection. J Vac Sci Technol A 8:383
56. Itoh T, Suga T (1994) Scanning force microscope using a piezoelectric microcantilever. J Vac Sci Technol B 12:1581–1585
57. Giessibl FJ (2000) Atomic resolution on Si (111) (7×7) by noncontact atomic force microscopy with a force sensor based on a quartz tuning fork. Appl Phys Lett 76:1470–1472
58. Minne SC, Yarahioglu G, Manahis SR, Adams JD, Zesch J, Atalar A, Quate CF (1998) Automated parallel high-speed atomic force microscopy. Appl Phys Lett 72(18):2340–2342
59. Minne SC, Flueckiger P, Soh HT, Quate CF (1995) Atomic force microscope lithography using amorphous silicon as a resist and advances in parallel operation. J Vac Sci Technol B 13:1380–1385
60. Minne SC, Manahis SR, Atalar A, Quate CF (1996) Independent parallel lithography using the atomic force microscope. J Vac Sci Technol B 14:2456–2461
61. Wilder K, Soh HT, Minne SC, Manahis SR, Quate CF (1997) Cantilever arrays for lithography. Naval Res Rev XXIX:35–48
62. Chui BW, Stowe TD, Kenny TW, Mamin HJ, Terris BD, Rugar D (1996) Low-stiffness silicon cantilevers for thermal writing and piezo-resistive feadback with the atomic force microscope. Appl Phys Lett 69:2767
63. Vettiger P, Despont M, Drechsler U, During U, Häberle W, Lutwyche MI, Rothuizen HB, Stutz R, Widmer R, Binning GK (2000) The millipede – more than one thousand tips for future AFM data storage. IBM J Res Develop 44:323–340
64. Despont M, Brugger J, Drechsler U, During U, Häberle W, Lutwyche M, Rothuizen H, Stutz R, Widmer R, Binning G, Rohrer H, Vettiger P (2000) VLSI-NEMS chip for parallel AFM data storage. Sens Actuators 80:100–107
65. Garcia N, Levanyuk AP, Minyukov SA, Binh TV (1995) Estimations for the characteristics of GHz range nanocantilevers: eigenfrequencies and quality factors. Surface Sci 328: 337–342
66. Walters DA, Cleveland JP, Thomson NH, Hansma PK, Wendmann MA, Gurley G, Elings V (1996) Short cantilevers for atomic force microscopy. Rev Sci Instr 67:3583–3590
67. Stowe TD, Yasumura K, Kenny TW (1997) Attonewton force detection using ultrathin silicon cantilevers. Appl Phys Lett 71:288–290
68. Wago O, Zuger K, Wegener R, Kendrick R, Yannoni CS, Rugain D (1997) Magnetic resonance force detection and spectroscopy of electron spins in phosphorous-doped silicon. Rev Sci Instr 68:1823–1826
69. Paloczi GT, Smith BL, Hansma PK, Walters DA (1998) Rapid imaging of calcite crystal growth using atomic force microscopy with small cantilevers. Appl Phys Lett 73:1658–1660
70. Kawakatsu H, Toshiyoshi H, Saya D, Fukushima K, Fujita H (2000) Strength measurement and calculations on silicon-based nanometric oscillators for scanning force microscopy operating in the gigahertz range. Appl Surf Sci 157:320–325
71. Saya D, Fukushima K, Toshiyoshi H, Fujita H, Hashiguchi G, Kawakatsu H (2000) Fabrication of silicon-based filiform-necked nanometmic oscillator. Jpn J Appl Phys 39:3793–3798
72. Bergen R, Lang HP,Gerber C, Gimzewski JK, Fabian JH, Scandella L, Meyer B, Guntherodt HJ (1998) Micromechanical thermogravimetry. Chem Phys Lett 294:363–369
73. Lang HP, Berger R, Battiston FM, Ramseyer JP, Meyer C, Andreolli B, Brugger J, Vettiger P, Despont M, Mezzacasa T, Scandehla L, Güntherodt HJ, Gerber C, Gimzewski JK (1998) A chemical sensor based on a micromechanical cantilever for the identification of gases and vapors. Appl Phys A 66:S61–S64

74. Bachels T, Schäfer R (1999) Formation enthalpies of Sn clusters: a calorimetric investigation. Chem Phys Lett 300:177–182
75. Jung MY, Kim DW, Choi SS, Kang CJ, Kuk Y (1999) Characterization of bimetallic cantilever for chemical sensor applications. Jpn J Appl Phys 38:6555–6557
76. Lang HP, Baller MK, Bergen R, Gerber C, Gimzewski JK, Battiston FM, Fornaro P, Ramseyer JP, Meyer B, Guntherodt HJ (1999) An artificial nose based on a micromechanical cantilever array. Anal Chim Acta 393:59–65
77. Baller MK, Lang HP, Fritz J, Gerber C, Gimzewski JK, Drechslem U, Rothuizen H, Despont M, Vettiger P, Battiston FM, Ramseyer JP, Fornaro P, Meyer B, Gunthemodt HJ (2000) A cantilever array-based artificial nose. Ultramicroscopy 82:1–9
78. Boisen A, Thaysen J, Jensenius H, Hansen O (2000) Environmental sensors based on micromachined cantilevers with integrated mead-out. Ultramicroscopy 82:11–16
79. Barnes JR, Stephenson RJ, Welland MB, Gerber C, Gimzewski JK (1994) Photothermal spectroscopy with femtojoule sensitivity using a micromechanical device. Nature 372:79–81
80. Berger R, Delamarche B, Lang HP, Gerber C, Gimzewski JK, Meyer B, Guntherodt HJ (1997) Surface stress in the self-assembly of alkanethiols on gold. Nature 276:2021–2024
81. Akama Y, Nishimura B, Sakai A (1990) New scanning tunneling microscopy tip for measuring surface topography. J Vac Sci Technol A 8:429–433
82. Broers AN, Molzen WW, Cuomo JJ, Wittels ND (1996) Electron beam fabrication of 80-A metal structures. Appl Phys Lett 29:596–598
83. Okayama S, Komuro M, Mitzutani W, Tokumoto H, Okano M, Shimizu K, Kobayashi Y, Matsumoto F, Wakiyama S, Shigeno M, Sakai F, Fujiwara S, Kitamura O, Quo M, Kajimura K (1988) Observation of microfabricated patterns by scanning tunneling microscopy. J Vac Sci Technol A 6:440–444
84. Ichihashi T, Matsui S (1988) In situ observation of electron beam induced chemical vapor deposition by transmission electron microscopy. J Vac Sci Technol B 6:1869–1872
85. Dai H, Hafner JH, Rinzler AG, Colbert DT, Smalley RB (1996) Nanotubes as nanoprobes in scanning probe microscopy. Nature 384:147–150
86. Iiiji S (1991) Helical microtubules of graphitic carbon. Nature 354:56–58
87. Dresselhaus MS, Dresselhaus G, Eklund PC (1996) Science of fullerenes and carbon nanotubes. Academic Press
88. Harris TD, Gershoni D, Grober RD, Pfeiffer L, West P, Chand N (1996) Near-field optical spectroscopy of single quantum wire. Appl Phys Lett 68:988–990
89. Nagy G, Scarmozzino R, Osgood H, Dai RM, Smalley RB, Michaels CA, Flynn GW, McLane GF (1998) Carbon nanotube tipped atomic force microscopy for measurement of <100 nm etch morphology on semiconductors. Appl Phys Lett 73:529–531
90. Wong SS, Woolley AT, Odon TW, Huang JL, Kim P, Vezenov DV, Lieber CM (1998) Single-walled carbon nanotube probes for high-resolution nanostructure imaging. Appl Phys Lett 73:3465–3467
91. Dai H, Franklin N, Han J (1998) Exploiting the properties of carbon nanotubes for nanolithography. Appl Phys Lett 73:1508–1510
92. Teacy MM, Ebbesen TW, Gibson JM (1996) Exceptionally high Young's modulus observed for individual carbon nanotubes. Nature 381:678–680
93. Wong BW, Sheehan PB, Lieber CM (1992) Nanobeam mechanics: elasticity, strength, and toughness of nanorods and nanotubes. Science 277:1971–1975
94. Salvetat JP, Briggs AD, Bonard JM, Basca RR, Kuhik AJ, Stöckhi T, Burnham NA, Forro L (1999) Elastic and shear moduli of single-walled carbon nanotube ropes. Phys Rev Lett 82:944–947
95. Falvo MR, Chary GJ, Taylor RM, Chin V, Brooks FP, Washburn S, Superfine R (1997) Bending and buckling of carbon nanotubes under large strain. Nature 389:582–584
96. Nardehhi MB, Yakobsen BI, Bernhole J (1998) Brittle and ductile behavior in carbon nanotubes. Phys Rev Lett 81:4656–4659
97. Falvo MR, Chary GJ, Paulson S, Taylor RM, Chin V, Brooks FP, Washburn S, Superfine R (1999) Nanomanipulation experiments exploring frictional and mechanical properties of carbon nanotubes. Micros Microanal 4:504–512

98. Ru CQ (2000) Effective bending stiffness of carbon nanotubes. Phys Rev B 62:9973–9976
99. Wong SS, Joselevich B, Woolley AT, Cheung CC, Lieber CM (1998) Covalently functionalized nanotubes as nanometre-sized probes in chemistry and biology. Nature 394:52–55
100. Terrones M, Hsu WK, Schilder A, Terrones H, Grobert N, Hare JP, Zhu YQ, Schwoerer M, Prassides K, Kroto HW, Walton DRM (1998) Novel nanotubes and encapsulated nanowires. Appl Phys A 66:307–317
101. Nishijima H, Kamo S, Akita S, Nakayama Y, Hohmura KI, Yoshimura SH, Takeyasu K (1999) Carbon-nanotube tips for scanning probe microscopy: preparation by a controlled process and observation of deoxyribonucleic acid. Appl Phys Lett 74:4061–4063
102. Akita S, Nishijima H, Nakayama Y, Tokumasu F, Takeyasu K (1999) Carbon nanotube tips for a scanning probe microscope: their fabrication and properties. J Phys D Appl Phys 32:1044–1048
103. Barwich V, Bammerhin M, Baratoff A, Bennewitz R, Guggisberg M, Loppacher C, Pfeiffer O, Meyer B, Gunthemodt HJ, Salvetat JP, Bonard JM, Forro L (2000) Carbon nanotubes as tips in non-contact SFM. Appl Surf Sci 157:269–273
104. Kulisch W (1999) Deposition of diamond-like superhard materials, vol 157. Tracts in modern physics. Springer, Berlin Heidelberg New York
105. Kulisch W, Malavé A, Lippold G, Scholz W, Mihalcea C, Oesterschulze E (1997) Fabrication of integrated diamond cantilevers with tips for SPM applications. Diamond Relat Mater 6:906
106. Malavé A, Oesterschulze E, Kulisch W, Trenkler T, Hantschel T, Vandervorst W (1999) Diamond tips and cantilevers for the characterization of semiconductor devices. Diamond Relat Mater 8:283–287
107. Hantschel T, Trenkler T, Vandervorst W, Malavé A, Büchel D, Kulisch W, Oesterschulze E (1999) Tip-on-tip: a novel AFM tip configuration for the electrical characterization of semiconductor devices. Microelectr Eng 46:113–116
108. Binning G, Rohrer H (1996) Scanning tunneling microscopy. IBM J Res Develop 30:355
109. Martin O, Drake B, Hansma PK (1987) Atomic force microscopy of liquid-covered surfaces: atomic resolution images. Appl Phys Lett 51:484–486
110. Visser BP, Gerritsen JW, Van Enckevort WJP, Van Kempen H (1992) Tip for scanning tunneling microscopy made of monocrystaline, semiconducting, chemical vapour deposited diamond. Appl Phys Lett 60:3232–3234
111. Kang WP, Davidson JL, Howell M, Bhuva B, Kinser DL, Kerns DV (1996) Micropatterned polycrystalline diamond field emitter vacuum diode arrays. J Vac Sci Technol B 14:2068–2071
112. Germann GJ, McClelland GM, Mitsuda Y, Buck M, Seki H (1990) Diamond force microscope tips by chemical vapor deposition. Rev Sci Inst 63:4053–4055
113. Liu N, Ma Z, Chu X, Hu T, Xue Z, Jiang X, Pang S (1994) Fabrication of diamond tips by the microwave plasma chemical vapour deposition. J Vac Sci Technol B 12:1712–1715
114. Niedermann P, Hänni W, Blanc N, Christoph R, Burger J (1996) Chemical vapour deposition diamond for tips in nanoprobe experiments. J Vac Sci Technol A 14:1233–1236
115. Oesterschulze E, Scholz W, Mihalcea C, Albert D, Sobisch B, Kulisch W (1997) Fabrication of small diamond tips for scanning probe microscopy application. Appl Phys Lett 70:435–437
116. Niedermann P, Hänni W, Morel D, Perret A, Skinner N, Indermühle PF, de Rooij NF, Buffat PA (1998) CVD diamond probes for nanotechnology. Appl Phys A 66:S31–S34
117. Mihalcea C, Scholz W, Malavé A, Albert D, Kulisch W, Oesterschulze E (1998) Fabrication of monolithic diamond probes for scanning probe microscopy applications. Appl Phys A 66:S87–S90
118. Trenkler T, Hantschel T, Stephenson R, De Wolf P, Vandervorst W, Hellemans L, Malavé A, Büchel D, Oesterschulze E, Kulisch W, Niedermann P, Sulzbach T, Ohlsson O (2000) Evaluating probes for 'electrical' atomic force microscopy. J Vac Sci Technol B 18:418–427
119. Beuret C, Akiyama T, Staufer U, de Rooij NF, Niedermann P, Hänni W (1998) Conical diamond tips realized by a double-molding process for high-resolution profilometry and atomic force microscopy applications. Appl Phys Lett 76:1621–1623

120. Malavé A, Ludolph K, Leinhos T, Lehrer C, Frey L, Oesterschulze E (2001) All-diamond probes for scanning probe microscopy applications realized by a proximity lithography process. Appl Phys A (in press)
121. Oesterschulze E, Malavé A, Keyser UF, Haug RJ (2001) Diamond cantilevers with integrated tip for nanomachining. Diamond Relat Mater (in press)
122. Yuan G, Jin Y, Jin C, Zhang B, Song H, Ning Y, Zhou T, Jiang H, Li S, Tian Y, Gu C (1998) Growth of diamond on silicon tips. J Crystal Growth 186:382–385
123. Howes MJ, Morgan DV (eds) Gallium arsenide – materials, devices, and circuits. Wiley
124. Prins MWJ, Groenveld RHM, Abraham DL, Van Kempen H (1995) Near-field magneto-optical imaging in scanning tunneling microscopy. Appl Phys Lett 66:1141–1143
125. Heisig S, Oesterschulze E (1998) Gallium arsenide probes for scanning near-field probe microscopy. Appl Phys A 66:385–390
126. Prins MWJ, van den Wielen MCMM, R. Jansen, Abraham DL, Van Kempen H (1994) Photoamperic probes in scanning tunneling microscopy. Appl Phys Lett 64:1207–1209
127. Heisig S, Rudow O, Oesterschulze E (2000) Optical active gallium arsenide cantilever probes for combined scanning near-field optical microscopy and scanning force microscopy. J Vac Sci Technol B 18:1134–1137
128. Heisig S, Rudow O, Oesterschulze E (2000) Scanning near-field optical microscopy in the near-infrared using light emitting cantilever probes. Appl Phys Lett 77:1071–1073
129. Goodman JW (1989) Introduction to Fourier optics. McGraw-Hill
130. Hecht B Optik, 3rd edn. Addison-Wesley
131. Fischer UC (1998) Scanning near-field optical microscopy. In: Wiesendanger R (ed) Scanning probe microscopy. Springer, Berlin Heidelberg New York, pp 161–209
132. Novotny L (1996) Light propagation and light confinement in near-field optics. PhD thesis, Swiss Federal Institute of Technology, Zürich
133. Fillard JP (1996) Near field optics and nanoscopy. World Scientific
134. Paesler MA, Moyer PJ (1996) Near-field optics – theory, instrumentation, and applications. Wiley
135. Ohtsu M (1998) Near-field nano/atom optics and technology. Springer, Berlin Heidelberg New York
136. Vollkopf A, Rudow O, Oesterschulze E (2001) Technology to reduce the aperture size of microfabricated aperture SNOM tips. J Electrochem Soc 148:G587–G591
137. Ruiter AG, Moers MHP, van Hulst NF, de Boer M (1996) Microfabrication of near-field optical probes. J Vac Sci Technol B 14:597–601
138. Ruiter AGT, Moers MHP, Jalocha A, van Hulst NF (1995) Development of an integrated NSOM probe. Ultramicroscopy 61:139–143
139. Mihalcea C, Vollkopf A, Oesterschulze E (2000) Reproducible large area microfabrication of sub 100 nm apertures on hollow tips. J Electrochem Soc 147:1970
140. Deal BB, Grove AS (1965) General relationship for the thermal oxidation of silicon. J Appl Phys 36:3770–3778
141. EerNisse BP (1979) Stress in thermal SiO_2 during growth. Appl Phys Lett 35:810
142. Wilson LO, Marcus RB (1987) Oxidation of curved silicon surfaces. J Electrochem Soc 134:481–491
143. Kobeda B, Irene BA (1988) SiO_2 film stress distribution during thermal oxidation of Si. J Vac Sci Technol B 6:574–578
144. Kao DH, McVittie JP, Nix WD, Krishna CS (1987) Two-dimensional thermal oxidation of silicon – I. Experiments. IEEE Trans Electron Dev BD-34:1008–1017
145. Kao DH, McVittie JP, Nix WD, Krishna CS (1988) Two-dimensional thermal oxidation of silicon–Il. Modeling stress effects in wet oxides. IEEE Trans Electron Dev BD-35:25–37
146. Law ME, Tasch A (1998) FLOOPS; Florida object oriented processing simulator. University of Florida and University of Texas, at Austin, http://www.swamp.tec.ufl.edu
147. Senez V, Collard D, Baccus B (1994) Analysis and application of a viscoelastic model for silicon oxidation. J Appl Phys 76:3285–3296
148. Hu SM (1998) Effect of process parameters on stress development in two-dimensional oxidation. J Appl Phys 64:323–330

149. Rafferty CS, Borucki L, Dutton RW (1989) Plastic flow during thermal oxidation of silicon. Appl Phys Lett 54:1516–1518
150. Lee MB, Kourogi M, Yatsui T, Tsutsuin T, Atoda N, Ohtsu M (1999) Silicon planar-apertured probe array for high-density near-field optical data storage. Appl Opt 38:3566–3571
151. Lee MB, Atoda N, Tsutsui K, Ohtsu M (1999) Nanometric aperture arrays fabricated by wet and dry etching of silicon for near-field optical storage application. J Vac Sci Technol B 17:2462–2466
152. Kim YJ, Kurihara K, K. Suzuki, Nomura M, Mitsugi S, Chiba M, Goto K (2000) Fabrication of micro-pyramidal probe array with aperture for near field optical memory applications. Jpn J Appl Phys 39:L1538–L1541
153. Grober RD, Schoelkopf RJ, Prober DB (1997) Optical antenna: towards a unity efficiency near-field optical probe. Appl Phys Lett 70:1354–1356
154. Keilmann F (1991) Scanning tip for optical radiation. US patent 4,994,818
155. Fischer UC, Zapletal M (1992) The concept of a coaxial tip as a probe for scanning near field optical microscopy and steps towards a realisation. Ultramicroscopy 42–44:393–398
156. Leinhos T, Rudow O, Stopka M, Vollkopf A, Oesterschulze E (1999) Coaxial probes for scanning near-field microscopy. J Microsc 194:349–352
157. Grober RD, Schoelkopf RJ, Prober DB (1997) High efficiency near-field electromagnetic probe having a bowtie antenna structure. US patent 5,696,372
158. Oesterschulze E, Georgiev G, Vollkopf A, Rudow O (2001) Transmission line probe on base of a bow-tie antenna. J Microscopy 202:39–44
159. Rudow O, Vollkopf A, Müller-Wiegand M, Georgiev G, Oesterschulze E (2001) Theoretical investigations of a coaxial probe concept for scanning near-field optical microscopy. Opt Commun 189:187–192
160. Diaz AF, Logan JA (1980) Electroactive polyaniline films. J Electroanal Chem 111:111–114
161. Wei Y, Tang X, Su Yn (1989) Autoacceleration and kinetics of electrochemical polymerization of polyaniline. J Phys Chem 93:4878–4881
162. Duic L, Mandic Z (1992) Counter-ion and pH effect on the electrochemical synthesis of polyaniline. J Electroanal Chem 335:207–221
163. Chiang JC, MacDiarmid AG (1986) Polyanilin: protonic doping of the emeraldine form to the metallic regime. Synth Metals 13:193–205
164. Talaie A (1997) Conducting polymer based pH detector: a new outlook to pH sensing technology. Polymer 38:1145–1150
165. Talaie A, Lee JY, Lee YK, Jang J, Romagnoli JA, Taguchi T, Maeder B (2000) Dynamic sensing using intelligent composite: an investigation to development of new pH sensors and electrochromic devices. Thin Solid Films 363:163–166
166. Pringsheim B, Terpetschnig B, Wolfbeis OS (1997) Optical sensing of pH using thin films of substituded polyanilines. Anal Chim Acta 357:247–252
167. Grummt UW, Pron A, Zagorska M, Lefrant S (1997) Polyaniline based optical pH sensor. Anal Chim Acta 357:253–259
168. Ge Z, Brown CW, Sun L, Yang SC (1993) Fiberoptic pH sensor based on evanescent-wave absorption-spectroscopy. Anal Chem 65:2335–2338
169. Jin Z, Su Y, Duan Y (2000) An improved optical pH sensor based on polyaniline. Sens Actuators B 71:118–122
170. Shinada S, Koyama F, Nishiyama N, Arai M, Goto K, Iga K (1999) Fabrication of microapertur surface emitting laser for near field optical data storage. Jpn J Appl Phys 38:L1327–L1329
171. Weiss S, Ogletree DF, Botkin D, Salmeron M, Chemala DS (1993) Ultrafast scanning probe microscopy. Appl Phys Lett 63:2567
172. Botkin DA (1995) Ultrafast tunneling microscopy. PhD thesis, UC Berkeley
173. Ketchen MB, Grischkowsky D, Chen CC, Chi CC, Duling IN, Halas NJ, Halbout JM, Kash JA, Li GP (1986) Generation of subpicosecond electrical pulses on coplanar transmission lines. Appl Phys Lett 48:751–753
174. Heiliger HM, Pfeiffer T, Roskos HG, Kurz H (1996) External photoconductive switches as generators and detectors of picosecond electric transients. Microelectron Eng 31:415–426

175. Jensen RJ, Keil UD, Hvam JH (1997) Spatio-temporal imaging of voltage pulses with an ultrafast scanning tunneling microscope. Appl Phys Lett 70:2762–2764
176. Auston DH (1975) Picosecond optoelectronic switching and gatinng in silicon. Appl Phys Lett 26:101–103
177. Kroekel D, Grischkowsky D, Ketchen MB (1989) Subpicosecond electrical pulse generation using photoconductinve switches with long carrier lifetimes. Appl Phys Lett 54:1046–1047
178. Kim J, Williamson S, Nees J, Wakana SI, Whitaker J (1993) Photoconductive sampling probe with 2.3 ps temporal resolution and 4 μV sensitivity. Appl Phys Lett 62:2268–2270
179. Steffens WM, Heisig S, Keil U, Oesterschulze E (1999) Spatio-temporal imaging of voltage pulses with a laser gated photoconductive sampling probe. Appl Phys B 69:455–458
180. Keil UD, Jensen JR, Hvam JM (1998) Transient measurements with an ultrafast scanning tunneling microscope. Appl Phys A 66:S23–S26
181. Steffens WM, Oesterschulze E (1999) Atomic force microscope cantilever for voltage probe with ultrafast the resolution. Electron Lett 35:1106–1108
182. Oesterschulze E, Steffens WM (2001) Cantilever probes for spatio-temporal imaging of voltage pulses with an ultrafast scanning probe microscope. J Vac Sci Technol B 19:107–110
183. Keil UD, Jensen JR, Hvam JM (1997) Fiber coupled ultrafast tunneling microscope. J Appl Phys 81:2929–2934
184. Steffens W (1999) Detektion von ultrakurzen elektrischen Signalen mit hoher Ortsauflösung. PhD thesis, Universität Gesamthochschule Kassel
185. Gupta S, Frankel MY, Valdmanis JA, Whitaker GA, Mourou JF (1991) Subpicosecond carrier lifetime in GaAs grown by molecular beam epitaxy at low temperatues. Appl Phys Lett 59:3276–3278
186. Harmon BS, Melloch MR, Woodall JM, Nolte DD, Otsuka N, Chang CL (1991) Carrier lifetime versus anneal in low temperature grown GaAs. Appl Phys Lett 63:2248–2250
187. Smith FW (1992) Device applications of low-temperature-grown GaAs. In: Witt GL, Calawa AR, Mishra UK, Weber ER (eds) Low temperature (LT) GaAs and related materials. Material Res Soc, pp 3–11
188. Liu X, Prasad A, Chen WM, Kurpiewski A, Stoschek, Liliental-Weber Z, Weber BR (1994) Mechanism responsible for the semi-insulating properties of low-temperature-grown GaAs. Appl Phys Lett 59:3276–3278
189. Keil UD, Jensen JR, Hvam JM (1998) Transient measurements with an ultrafast scanning tunneling microscope on semiconductor surfaces. Appl Phys Lett 72:1644–1646

5 Tip Characterization for Dimensional Nanometrology

John S. Villarrubia

5.1 Introduction

Technological trends are requiring dimensional metrology of features smaller than a micrometer in a number of industries. In some cases (e.g., pharmaceuticals and other biological or medical applications) the nanometer size scale is determined by the fact that function is driven by chemistry, and the size scale is therefore that of molecules. In other cases (semiconductor electronics, data storage) sizes of components are driven to the smallest manufacturable because performance (speed, data density) improves with miniaturization. Advanced materials, ceramics, and composites increasingly may benefit from the assessment of grain size and other aspects of nanostructure.

Dimensional quantities typically of interest include pitch (the distance between similarly shaped features, often used to set the scale of an image), width, height, and roughness. According to its current roadmap [1] the semiconductor industry will soon be manufacturing sub-50-nm transistor gates with a targeted process variation of less than ±5 nm (3 standard deviations). Topographic features used to store data in optical disks require metrology at a similar size scale [2]. Surface microroughness may determine gate oxide integrity in integrated circuits, and is a critical input for models of friction and wear [3].

While optical and electron beam-based measurement technologies generally have a speed advantage over SPM (scanning probe microscopy), SPM has important advantages of its own. Resolution is typically very high. Even individual atoms may sometimes be resolved. Unlike the alternatives, SPM images are inherently three-dimensional. This is important for height and roughness determinations. An advantage that is sometimes overlooked is the relatively straightforward interaction mechanism between instrument and sample. Size measurements using optical or e-beam methods are complicated by the fact that the probe-specimen interaction is often material dependent. For example, lines of the same width in different materials will have different optical and electron scattering properties and therefore produce different images in an optical microscope or scanning electron microscope (SEM). Except for very soft materials, this seems less likely to be an issue for contact or near-contact modes of topographic SPM imaging. Thus, even when other measurement technologies are used on the production line because of their speed, SPM may have a critical role to play behind the scenes in primary calibrations.

Dimensional metrology that pushes the limits of a microscopy necessarily concerns itself with image artifacts. Some artifacts in SPM have their origins in noise, motion errors of the tip and/or sample stage (for example, due to nonlinearities of the piezoelectric actuators used for scanning or bending of the tip or cantilever), or effects due to the finite size of the tip. Noise is typically addressed by using filtering or averaging methods. Motion errors may be minimized by the use of interferometry or calibrated capacitance gauges to traceably measure the position of the tip relative to the sample [4, 5].

In this chapter we focus mainly on tip size effects, which can be partially corrected with a good estimate of the tip shape. Firstly we review tip characterization followed by detailed descriptions of the two main methods. The treatment will be practical. The aim is to acquaint the reader with important measurement issues associated with the tip and to make accessible in practical form recent advances in tip and sample reconstruction techniques. To that end, results will be presented with plausibility arguments and graphical illustrations of their meaning. Detailed derivations will be left to the references. In addition to a general introduction to the ideas, references are given to computer codes, and recipes or procedures are provided for the application of those codes to typical measurements. We devote attention to such questions as: How should one measure the images and other inputs required for tip characterization? Given the inputs, how does one obtain the tip estimate? What can go wrong with this process? How does one recognize the sources and symptoms of error, and how might they be controlled?

5.2 Tip Characterization Overview

In any microscope, there is some level at which the image differs from the sample due to non-ideal instrument behavior. An important source of instrumental artifacts in the SPM is the tip. The ideal tip would be perfectly sharp, but all real tips have non-vanishing size. The measured image is a dilation of this non-ideal tip with the specimen. Dilation and its dual, erosion, are sometimes confused with convolution and deconvolution, to which they are analogous. Convolution and deconvolution usually refer to linear operations that are strict inverses of each other. Dilation and erosion, as we will see, are nonlinear and are not, strictly speaking, invertible.

Good measurement practice dictates that one estimate, report and, if possible, compensate for tip effects whenever they might have an important influence on the quantity being measured [6]. Tip shape often significantly affects width measurements or the depth of narrow depressions, as illustrated in Fig. 5.1. The valleys between peaks in a sinusoidal surface with short wavelength and/or high amplitude are also inaccessible to the tip, analogously to the v-shaped depression in the figure. Because the broad range of spatial frequencies contained in rough surfaces often includes some of these, tip shape is important for surface roughness determinations.

Since 1990 methods of tip measurement have been developed based upon deducing the shape of the tip from images of special specimens called "tip characterizers". The ability to substitute a characterizer for the unknown specimen allows

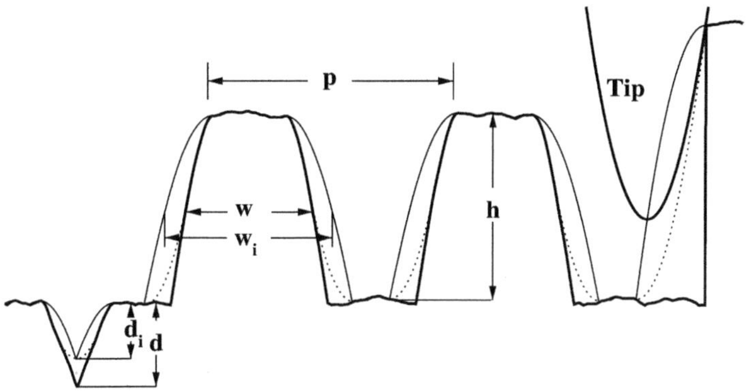

Fig. 5.1. Imaging distortions due to finite-sized tip. Shown are the specimen surface (*thick line*), image (*thinner line*), line of deepest tip penetration (*dotted*), and tip (*labeled*). Note that the pitch (p) and the height (h) of well-separated bumps are the same whether measured from the true surface or the image, while their widths (e.g., w_i versus w) and the depth (d_i versus d) of the v-groove on the *left* are more strongly distorted

routine checking of the tip. There are two methods depending upon whether the characterizer shape is known or not.

When the tip characterizer's geometry is known its influence can be accounted for. In a reconstruction process analogous to deconvolution, the characterizer's influence upon the image is removed, leaving information about the tip shape. This method has appeared in a variety of forms. The simplest are formulated only for characterizer shapes falling within a specified class, usually spheres. The equations for reconstruction are then derived by simple geometrical arguments particular to that class of shapes. Examples of such class-specific reconstruction have been given in the context of characterizers made from biomolecules [7], colloidal spheres [8–10], and polystyrene latex balls [11, 12]. Completely general solutions, of which these class-specific solutions represent special cases, had been derived somewhat earlier in the form of differential equations [13] or Legendre transforms [14]. Though completely general, reconstruction from noisy images using these methods could be problematical because of the need for numerical derivatives. Later, an equally general but more convenient method was developed. The method was described independently by a number of different authors in varying mathematical or algorithmic forms [15–24], but whether stated explicitly or not, all of these appear to be mathematically equivalent to erosion.

The second method places bounds on the tip even when the characterizer geometry is not independently known. It relies upon characterizers with one or more sharp features that image all or part of the tip. In its simplest form, a spike [25–29], an edge [30, 31], or other suitably sharp object is used to deliberately produce an image of the tip (e.g., Fig. 5.2). The tip's shape is then approximated by its image. This "direct imaging" or "inverse imaging", in which the sample produces an image of the tip, is a special case of the more general "blind reconstruction" that was developed independently at about the same time [19, 22–24, 32]. In its more general form, blind reconstruction permits the combination of information from mul-

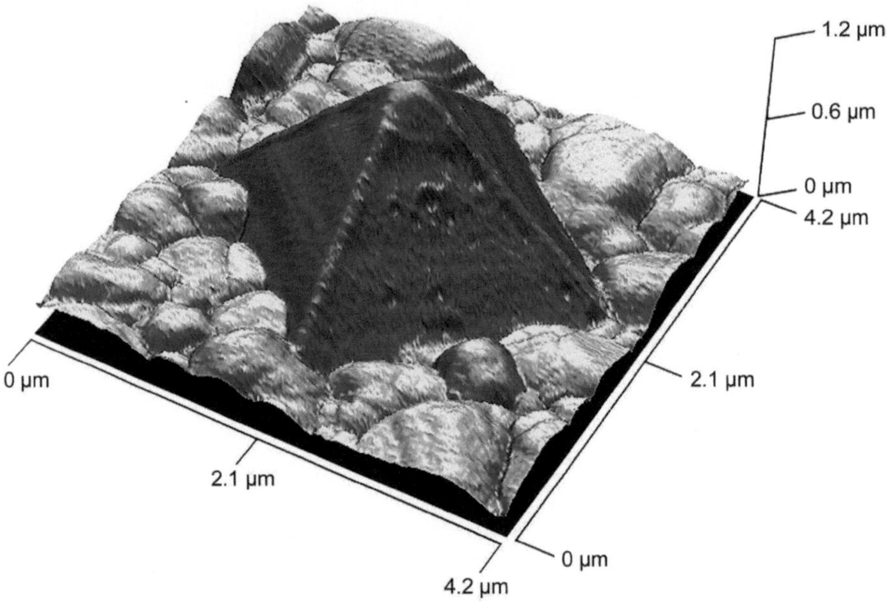

Fig. 5.2. Example of a tip image produced by a pin-like surface structure on a vapor-deposited copper film (Reprinted from Surface Science, 323, F. Atamny and A. Baiker, "Direct imaging of the tip shape by AFM", p L314, Copyright 1995, with permission from Elsevier Science)

tiple complementary tip images, some of which may be so poor that they are individually hardly recognizable as images.

One limitation of all methods that rely upon tip characterizers is that they provide only an outer bound on the tip shape. Only parts of the tip that actually touch the characterizer can be reconstructed. This need not be a problem if the characterizer is designed such that all parts of the tip that touch the unknown specimen also touch the tip characterizer.

Tip measurement using a characterizer has several advantages over alternatives. Independent imaging techniques, like SEM or TEM (transmission electron microscopy), are usually more time-consuming, and require additional handling of the tip with consequent risk of damage or contamination. This disadvantage is heightened by the fact that SPM tips wear or are damaged in use and therefore require frequent measurement, perhaps as often as before and after each image in careful work. SEM and TEM also have measurement limitations of their own. For example, they produce two-dimensional images, not the three-dimensional tip model required for detailed surface reconstruction. Nonetheless, these methods have their place for inspecting and qualifying tips, where only rough size information is required, or in cases when time and care are warranted, as in an occasional independent check of tip characterizer results [33].

5.3 Measuring Tip Shape with a Known Tip Characterizer

This section describes a method to analyze an SPM image of a tip characterizer to obtain the tip's detailed geometry. In order to use this procedure, the actual shape of the tip characterizer must be known with an uncertainty at least as small as required for the final tip determination. If it is not practical to obtain the actual characterizer geometry, blind estimation of the tip shape, described later, is an alternative.

In the next section, we discuss in general terms how this kind of tip reconstruction works. The following section provides a recipe for obtaining the image and doing the calculation as well as some remarks on interpreting the results. This is followed by a summary of the procedure.

5.3.1 Dilation and Erosion

In SPM, the image can be modeled by a dilation, $I = S \oplus P$ [15, 16, 19–24, 32]. In the most general case, I and S are full solid representations of the image and specimen, and P is the reflection of the tip through the origin. Currently available algorithms require us to restrict consideration to objects that can be represented by single-valued functions, that is, by the top surfaces of I, S, and P, which are respectively $i(x,y)$, $s(x,y)$, and $p(x,y)$. (This restriction is the same one implicitly made by the general practice of representing data by a 2D array of heights, as in a grayscale image.) Because P is the reflection of the tip through the origin, p is related to the tip's surface, t, by $p(x,y) = -t(-x,-y)$. In terms of these functions, the dilation operation has a particularly simple meaning:

$$i(x, y) = \max_{(u,v)} [s(x - u, y - v) + p(u, v)] \tag{5.1}$$

The max operation is taken over all (u,v) in the horizontal plane. A graphical interpretation of dilation is shown in Fig. 5.3a. The max operation of Eq. (5.1) forms the outer envelope of various translates of s and p with respect to each other, in each of which the apex of p is in contact with a point on s. Some of these translates are shown in the figure.

We may think of the rectangular and spherical parts of the hypothetical sample in Fig. 5.3a as possible tip characterizers. If we have measured $i(x,y)$ and we know these characterizer shapes, then the best estimate, P_{r}, of the tip shape is given by $P_{\mathrm{r}} = I \ominus S$ where $\ominus$ denotes erosion. In terms of the surfaces defining the tops of these objects,

$$p_r(x, y) = \min_{(u,v)} [i(x + u, y + v) - s(u, v)] \tag{5.2}$$

Erosion is graphically similar to dilation, except that the lower bound of translates of the inverted characterizer form an outer envelope for the inverted tip. Figures 5.3b and 5.3c demonstrate this operation. Figure 5.3b erodes the rectangular characterizer from the corresponding part of the image, while Fig. 5.3c does the same for the spherical characterizer. Erosion is not strictly the inverse of dilation. This is because there are areas of the characterizer that are not contacted by the tip and, like-

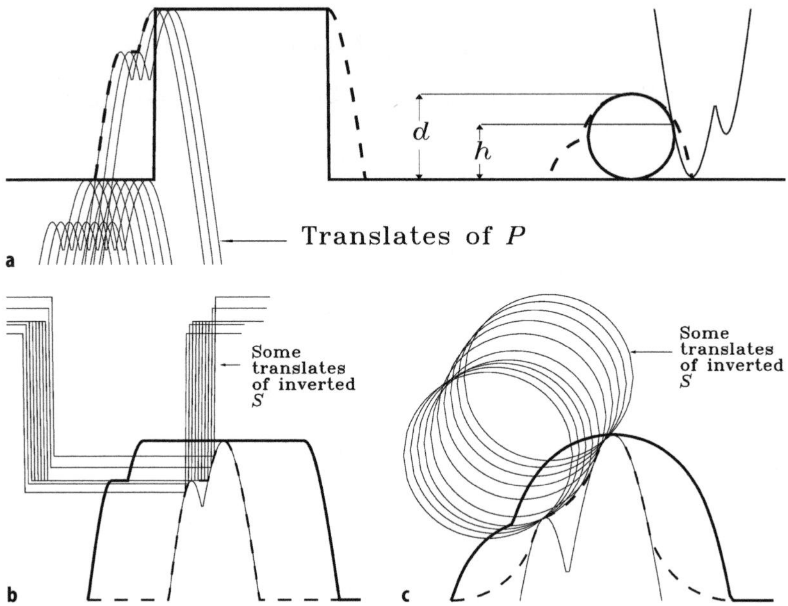

Fig. 5.3. Illustrations of dilation and erosion. **a** The dilation of rectangular and circular cross section characterizer geometries (*thick continuous curve*) by a tip (*right*) can be formed by the maximum envelope (*dashed line*) of translates of the reflected tip, some of which are shown. The apex of translates must fall on the characterizer surface. Reconstruction of the tip by erosion is illustrated in **b** and **c** for the two characterizer shapes. The image is shown by the *thick continuous lines*, the reconstructed tip by the *dashed lines*, and the actual tip by the *thin continuous lines*

wise, areas of the tip not contacted by the characterizer. The image bounds such areas but contains no detailed information about them. For this reason dilation is not strictly invertible, and p_r is not everywhere equal to p. Instead, $p_r \geq p$ with the equality applying for those points of the tip that touched the characterizer and the inequality elsewhere. This can be observed in the figure, where neither characterizer is sharp enough to reconstruct the narrow depression between the tip's primary and secondary maxima.

5.3.2 Reconstructing the Tip

5.3.2.1 *Obtaining the Inputs*

Equation (5.2) describes how to reconstruct the tip shape. We need only calculate p_r. In practice, we generally represent the functions i, s, and p_r by arrays of heights (grayscale images). Their arguments (x, y, etc.) take on only discrete values. Clearly, the inputs required are an image and an image-like array representation of the actual tip characterizer geometry.

5.3.2.2 *The Image*

The image, $i(x,y)$, required by Eq. (5.2) is obtained experimentally by scanning a tip characterizer with the tip to be measured. The reconstructed tip will have the same pixel resolution as the image, so the pixel spacing should be chosen with this in mind.

This procedure assumes that the image really is a dilation of the tip characterizer and tip. This is only approximately true in practice. Deviations from the model are produced by noise, scanner nonlinearities, finite feedback loop response time, etc. Good tip reconstruction requires that these effects be small compared to the tip size. If this is not already the case, some additional data acquisition precautions may be required, for example, appropriate isolation and filtering (for noise), slow enough scan speed relative to the feedback loop response time, and software or hardware linearization of the scan. If post-acquisition digital filtering of the data is employed to reduce noise, it is advisable to filter the image prior to calculating the erosion (rather than calculating, and then filtering, p_r), and median filters may be preferable to area averaging or low pass filters [32].

A related issue arises over the coordinate system. Both $i(x,y)$ and $s(x,y)$ must be described in the same coordinate system. If the x, y, and z motion axes of the instrument are not truly orthogonal or if their calibration is incorrect, orthogonality or scale errors are introduced into the image. Such errors are unlikely to be intentionally duplicated in the model, $s(x,y)$, of the characterizer. For example, a spherical tip characterizer could appear elongated along one axis if the instrument is out of calibration only along that axis. If the model of the characterizer is spherical, it will not have the same asymmetry. This results in a characterization error comparable in magnitude to the shape errors introduced by the faulty coordinate system. The remedy is to properly calibrate the scale and orthogonality of the instrument.

5.3.2.3 *The Tip Characterizer*

The input, $s(x,y)$, is an array of heights constructed on the computer based on the model for the characterizer geometry. It should be to the same scale in x, y, and z as the image.

A tip characterizer will provide good reconstruction only for those parts of the tip it touches. This generally dictates that the height of the characterizer be at least as large as the relief on the samples ultimately to be measured. Height alone is not sufficient, however, since a tall but gently sloped characterizer may never touch the tip far from its apex. In Fig. 5.3a, for example, the highest contact point for the spherical characterizer is at a height, h, that is less than the diameter, d, of the sphere. The amount of the tip that is touched depends upon the slope as well as the height. For this reason, it is an advantage for a characterizer to have tall steep features. If it is necessary to measure narrow depressions in a tip (for example, to accurately measure a double tip) it is also necessary for the characterizer to have a sharp-topped feature.

Finally, it is necessary to have an accurate model of the characterizer. Here, "accurate" means having uncertainties everywhere small compared to the desired uncertainty in the calculated tip. The need for a model of the characterizer gives an advantage to shapes that are simple and symmetrical. For example, if it is known that the characterizer is spherical, it may not be necessary to have an independent measurement of it at all. Since the measured height of an isolated sphere is not affected by the tip size, this height could be used as the sphere's diameter, which would then be used to model an appropriate $s(x,y)$. (Note, however, that it is not enough to know that the characterizer particle is spherical in vacuum or solution, where the forces on it are small and/or symmetrical. Spherical particles on a substrate can deform significantly depending upon the work of adhesion [34].) The requirement to know the characterizer also gives an advantage to shapes that are small, since shape variability tends to increase with size. Clearly, some of the requirements of this and the previous paragraph are in tension with one another, and an appropriate compromise is required.

5.3.2.4 *Doing the Calculation*

Dilation and erosion are standard mathematical routines available in a variety of commercial mathematical software packages. (Note that some packages distinguish between "binary" and "grayscale" versions. Equations (5.1) and (5.2) correspond to the grayscale variety. Some may also restrict the data type of the inputs or produce incorrect values near the edges of the result.) Alternatively, published codes [32] can be used to create a program for calculating dilation and erosion. Use the inputs determined above to calculate p_r.

5.3.3 Interpreting and Checking the Result

The erosion formula of Eq. (5.2) returns a single output, p_r, which is a reflected tip. The estimate of the actual tip geometry is found if needed by reflecting p_r through the origin, that is, by forming $-p_r(-x,-y)$. This estimate is, in principle, an outer bound on the tip shape, equal to the real tip only for those areas of the tip that made contact with the characterizer. In practice, of course, experimental errors can cause

a 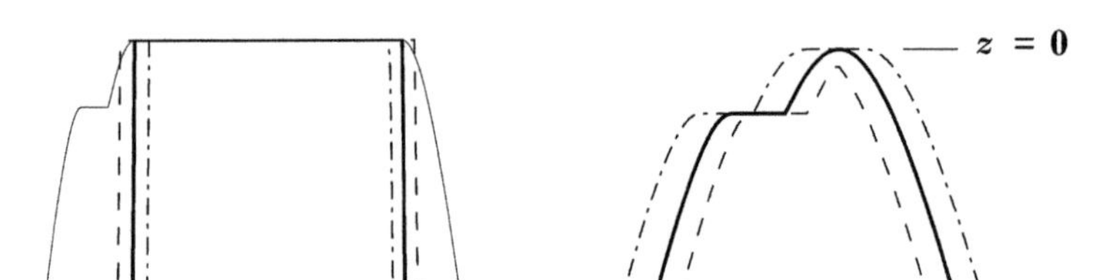 b

Fig. 5.4. Effect of characterizer estimation errors. **a** An image (*thin continuous line*) and three characterizer models. The true model is indicated by the *thick continuous line*, while the *dashed* and *dash-dot lines* show width over- and underestimates by 10%. In **b** are shown (expanded scale) the resulting reconstructed inverted tips. The line type for each tip is the same as for the corresponding characterizer model in **a**. There exists a test (see discussion in text) for avoiding overestimation of the characterizer the estimate to deviate from the true tip even for those areas that made contact (e.g., characterizer estimation errors in Fig. 5.4).

The careful analyst can avoid some characterizer estimation errors because they produce recognizable symptoms in the result. For example, in Fig. 5.4a the dashed characterizer model is wider at the top than the measured image (continuous line). This is a physical impossibility. Since tips have finite width, the image can only be blunter than the sample, never sharper. In the corresponding reconstructed (dashed) tip in Fig. 5.4b, the symptom of this error is that the apex is at $z < 0$. In general, it is possible to catch such consistency errors by verifying that $S \oplus P_r = I$. That is, dilation of the characterizer model by the reconstructed tip should reproduce the measured image. This equality may fail to be exactly followed due to noise or other instrumental artifacts in the image, but if the discrepancy is much larger than those effects, it is a sign that assumptions about the characterizer geometry are inconsistent with the measurement. Not every error in S is inconsistent with the image, so some errors may not be caught this way. For example, the underestimated characterizer in Fig. 5.4 produces an overly broad tip that is nevertheless consistent with the measured image.

5.3.4 Summary of the Method

To measure a tip using the known characterizer method:

1. Choose a characterizer of known geometry.
2. Take an image, $i(x,y)$, of the characterizer using the unknown tip, using care to minimize non-tip-related instrumental artifacts.
3. Create a corresponding height map, $s(x,y)$, of the known characterizer, taking care to use the same coordinate scale as the image.
4. Reconstruct the reflected tip shape, $p_r(x,y)$. Obtain the unreflected tip shape by forming $t(x,y) = -p_r(-x,-y)$ if desired.
5. Check the result by dilating S with P_r using Eq. (5.1). The result should be equal to the measured image, apart from (presumably small) differences due to noise and non-tip-related instrumental artifacts. If the answer does not check, re-evaluate the geometry of the characterizer.

In careful work, it is advisable to measure the tip shape both before and after imaging the unknown samples, in order to insure that the tip did not change during imaging.

5.4 Measuring Tip Shape without a Known Characterizer

If the tip characterizer is known, it is easy to show [19, 32] that the best procedure to use is tip reconstruction by erosion $P_r = I \ominus S$, where I is the image, S the known characterizer, and P_r the reconstructed outer bound on the tip, reflected through the origin) as described above. A known characterizer is one for which a three-dimensional height map with errors small compared to the tip size is available. This level of accuracy is often difficult to achieve in practice. In such cases, it is still possible with a good characterizer to determine a tight outer bound on the tip shape using the blind reconstruction procedure described here. It is based upon the iteration equation,

$$P_{i+1} = \bigcap_{x \in I} \{(I - x) \oplus [P_i \cap (x - I)]\} \cap P_i \tag{5.3}$$

An equivalent expression was derived in ref. [19], where a version in terms of min and max functions, can also be found. For our purposes we need not delve in too much detail into the mathematical intricacies of this equation. The main thing to notice it is that, unlike the previous tip reconstruction Eq. (5.2), Eq. (5.3) makes no reference to the tip characterizer, S. The only items referenced are the image, I, of the characterizer, points (x) contained within this image, and an estimate, P_i, of an outer bound for the tip on the ith iteration. To use the equation, one needs a starting, possibly crude, outer bound, P_0, for the 0th iteration. This bound then gets refined through successive iterations until it becomes the reconstructed tip shape. The necessary inputs can all be determined without knowledge of the tip characterizer's true geometry.

Our discussion of blind reconstruction is divided into several sections. We first describe the principle upon which it works. We then explain where to find computer code to do the required calculations, how to determine appropriate inputs for the algorithm, and how to interpret and use its output. We then provide some extensions to the basic recipe followed by a summary of the procedure. Finally, we describe an experimental test of the method.

5.4.1 Why Does it Work?

From one point of view it is surprising to be able to reconstruct the tip shape without independent knowledge of the tip characterizer. The image is a dilation of the sample by the tip $I = S \oplus P$. Solving for P without knowing S appears to solve a single equation with two unknowns.

From another point of view it is not surprising at all. Most practitioners of SPM have at one time or another seen obvious tip images in a scan of a sample with sharp features. We have already seen a deliberately arranged example in Fig. 5.2. This situation is depicted schematically in Fig. 5.5a. The sharp sample feature produces an inverted image of the tip. Suppose we cut this tip image out of the larger data file, invert it, and call it our tip estimate, as has sometimes been done [25–31]. This estimate has structure derived in part from the tip and in part from the underlying surface feature. If the surface feature were a spike with width, $w = 0$, our result would be equal to the actual tip shape. Wider features lead to wider tip estimates. Since the true w must be greater than zero our estimate is larger than the actual tip. It represents an outer bound. This outer bound is obtained without knowledge of the underlying surface feature, so it represents a rudimentary form of blind reconstruction. The procedure we will describe here represents a quantification or formalization of this intuitive insight.

The real power of blind reconstruction arises when we observe that other parts of the image supply independent outer bounds (Fig. 5.5b). For each point on the image, even those that are not local maxima, one can derive a corresponding outer

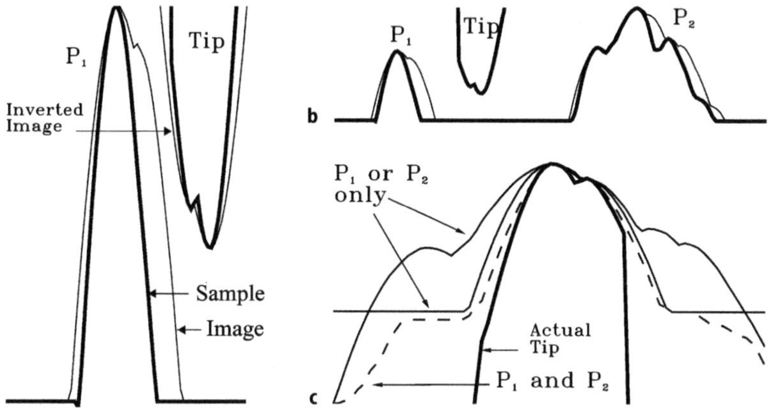

Fig. 5.5. **a** Rudimentary blind reconstruction of a tip using a single image maximum. The thin-lined tip is simply an inversion of the image peak, P_1. **b** The image (*thin line*) of the sample (*thick line*) may contain more than one peak, labeled P_1 and P_2. **c** Reconstructed tips are compared to the actual tip. The *dashed line* is a full blind reconstruction obtained by forcing a single tip to be consistent with both peaks. The *thin solid lines* each use only one of the peaks

bound on the tip [19]. These outer bounds represent mixtures of the *same* tip geometry with a variety of *different* surface structures. Only the part of the structure that is, in some appropriate sense, common to all of the independent bounds can conceivably belong to the tip (Fig. 5.5c). A candidate tip larger than this must be rejected on the grounds that it is inconsistent with the image. That is, there exists no physically realizable surface that would produce the observed image with that candidate tip. Mathematical details are available elsewhere [19, 22, 32].

5.4.2 Reconstructing the Tip

5.4.2.1 *How to Choose Inputs*

The blind reconstruction procedure requires three inputs. These are 1) an image of a tip characterizer, 2) a starting outer bound on the tip shape, and 3) a threshold value that tells the algorithm how to discriminate between non-tip-related instrumental artifacts (e.g., noise) and real shape information in the image.

5.4.2.1.1 *Choosing a Tip Characterizer*

One of the strengths of blind reconstruction is that it produces a valid outer bound on the tip from an image of any tip characterizer acquired with the tip. It does not follow, however, that any tip characterizer is as good as any other. An appropriate analogy is using an SPM image as an estimate of the sample. Any tip will produce an image that is an outer bound on the sample, but the bound produced by a sharp

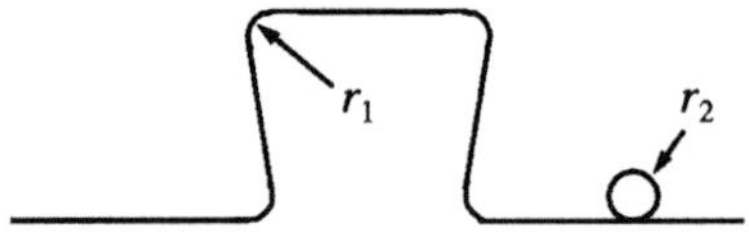

Fig. 5.6. Desirable features in a tip characterizer include tall steep-walled features and small radii at corners and tops

tip is tighter than the bound produced by a blunt one. Similarly, in tip reconstruction any characterizer will produce an outer bound on the tip, but a better characterizer produces a tighter bound.

Figure 5.6 illustrates desirable features of a characterizer. It combines a tall wide feature with steep sidewalls and sharp edges with a shorter feature with small radius of curvature on top. The steep (or undercut) sides of the taller feature are intended to increase the maximum distance from the tip's apex at which contact is made. The distance required depends upon the unknown specimen that the tip will measure. To be safe, the distance should be greater than or equal to the relief (maximum minus minimum height) in the image of the unknown. The characterizer's radius of curvature at points that contact the tip should be as small as possible – these ultimately determine the size of characterization errors for the points on the tip that they contact. Once again, radii to be encountered in the specimen one ultimately wants to measure set the appropriate scale. Other geometries, for example sharp-topped edges (i.e., a razor blade geometry) in a variety of orientations or simply a very rough surface with random high-curvature features, may be appropriate depending upon the application.

5.4.2.1.2
Acquiring a Characterizer Image

Blind reconstruction extracts tip shape information from an image. We therefore want the image to be as free of spurious (non-geometrical) information as possible. That is, we want to minimize noise, failure of the feedback loop to track the topography, hysteresis and creep in the piezoelectric scanner, and anything else that produces image features unrepresentative of dilation of the surface with the tip. The goal is for these effects to produce height errors small compared to the distortion caused by the finite size of the tip. For some instruments and samples, these unwanted effects might be small enough without taking special precautions. If this is not the case, strategies include appropriate isolation and filtering (for noise), proper choice of feedback parameters and scan speed slow enough relative to the feedback response time (for good feedback response), and software or hardware linearization of the scan (to correct hysteresis and creep).

The pixel resolution in the reconstructed tip will be the same as that for the image. If n pixels nm^{-1} are required to adequately describe the tip, the image must be acquired with n pixels nm^{-1}.

It is common practice to flatten SPM images. This is a background subtraction that changes the average height of a scan line with respect to its neighbors, generally by making all lines have the same average height. This practice is intended to remove spurious line-to-line variations in height that are really due to scanner misbehavior instead of actual topography. It is a good thing to do when there is reason to believe topographic variations in average height truly are small, as for instance when imaging a flat or uniformly rough surface with scan distances long enough to make a good statistical sampling. In such a case, flattening removes spurious sharp edges between lines that would otherwise be interpreted as requiring similar sharpness in the tip. It is a bad thing to do when the real variations in average line height are not small, as for example in an image of a tall object or a depression on a uniform substrate. In such a case, flattening *adds* spurious sharp edges. It would be better not to flatten, or to calculate correction parameters based on parts of the image that really are flat (e.g., the substrate only).

In addition to flattening line-to-line height variations as just discussed, background subtraction is often used to remove tilt before displaying an image. It is worth noting that the correct method to remove tilt is a rotation [35]. Subtracting a tilted surface is not the same as rotation, and introduces a geometrical distortion, a cosine error. This is second order in tilt angle, so usually unimportant. However, if for any reason (perhaps deliberate) the characterizer has large tilt, one should not do background subtraction prior to tip reconstruction. Do the tip reconstruction first, then background subtract (if desired) for display purposes only.

It is sometimes useful to filter images for further noise reduction. Keep in mind that Fourier low pass filters or smoothing filters of the area-averaging type will smooth not only spurious sharp features, but also actual ones that contain most of the tip information. Median filtering is a good alternative. Filtering the image followed by tip reconstruction does not in general yield the same result as tip reconstruction followed by filtering the result. Filtering first is preferred.

5.4.2.1.3
Initial Outer Bound

Blind reconstruction is an iterative procedure. An initial outer bound, P_0, is iteratively refined to determine a shape that is the union of all tips that are both a) smaller than the starting outer bound and b) consistent with the image. P_0 is therefore one of the inputs.

P_0 is specified as a rectangular array of height values with the same lateral and vertical scaling as the image. The user must decide both the lateral size of this array (number of pixels and lines) and the height values with which to populate it. Figure 5.7 illustrates how to choose the lateral size. The tip contacts the characterizer at points whose lateral coordinates form a bounded (though not necessarily connected) region in the x–y plane. The size of the array must be chosen large enough to contain this "footprint". Of course, we don't know the tip shape, or its footprint, a priori. Therefore we estimate it, then enlarge the result by a suitable safety factor. The estimate can be based on the nominal tip shape and sample relief. For example, a parabolic tip, $t(x,y) = (x^2+y^2)/(2r)$, with radius r contacts a vertical wall (the worst case) of height, h, at a distance $\sqrt{2rh}$ from the apex, giving a footprint of diameter at most $2\sqrt{2rh}$. A suitable allowance for uncertainty in r might be to double or triple this dimension. Alternatively, one can choose a lateral size by observing the lateral extent of the taller maxima in the image. However, if the tip is double (or worse), multiple image maxima may arise from a single sur-

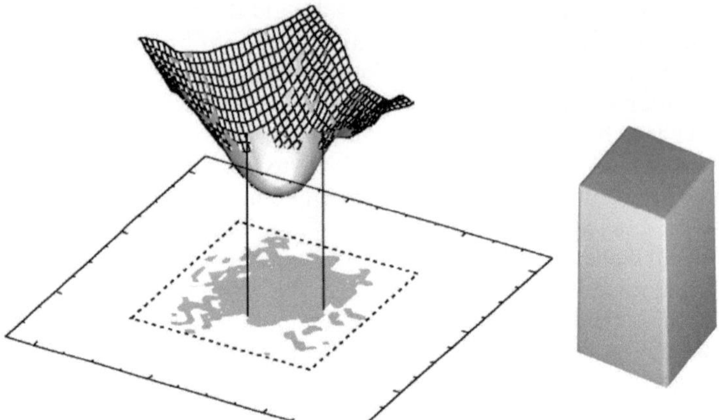

a b

Fig. 5.7. **a** An SPM tip is here shown rendered partly as a solid and partly as a wire mesh. The *solid region* indicates those parts of the tip that actually make contact with a particular surface (not shown). A projection of the contact points into the horizontal plane comprises the tip's "footprint", shown in *gray*. A rectangular bounding box is indicated by the *dashed line*. **b** A common choice of initial outer bound is a rectangular pillar. Its reflection through the origin (as required by morphology algorithms) is shown here. The bottom is at $-\infty$, the top face at $z = 0$, and the size of the face is large enough to bound the tip's footprint; that is, it is equal to or larger than the dashed bounding box in *a*

face feature. If this possibility is suspected, a rectangle large enough to encompass several image maxima must be chosen.

Choosing the dimension larger than necessary increases the computation time, which scales in the worst case as N^2, N being the number of pixels in P_0. Choosing it dramatically larger than necessary (e.g., equal to the image size) also degrades the reconstruction result, but this is not a significant issue for more reasonable choices. Choosing it smaller than the true tip footprint makes the array too small to hold an outer bound on the tip shape – the algorithm may produce a result dramatically smaller than the actual tip.

In order to populate the P_0 array with appropriate non-zero height values it is necessary to pay attention to conventions for tip description. We have alluded above to the fact that the reflection of a tip through the origin plays an important role in the mathematics of imaging. The software in ref. [32] adopts the convention that all tips, including P_0 and the reconstruction result, are described in this inverted space. An array describing an uninverted tip must be reflected left to right and top to bottom, then all heights multiplied by -1. Additionally, since we do not want to consider as different two tips that differ by nothing other than a translation, all tips are required to have their apex at $z = 0$. The (x,y) location of this apex must be supplied. Other implementations of blind reconstruction may adopt different conventions. Two easy choices of height values that meet the above conventions are commonly used. The first is all zeros, as shown in Fig. 5.7b. The second is obtained by cutting out an array of the appropriate size centered on one of the peaks in the characterizer image and translating it so its maximum z is 0. In this case, it is convenient that the image maxima are already inverted tip images, so need no further inversion.

5.4.2.1.4
Threshold Parameter

As already mentioned, there are many instrumental artifacts that can lead to sharp features unrelated to the tip. If we nonetheless treat the image as though all features arise from tip/sample dilation, we usually get an anomalously sharp tip reconstruction. If the non-tip artifacts are small compared to the tip ones, we can discriminate against them by using a threshold. The threshold is a height, as shown in Fig. 5.8, that defines a separation between tip and non-tip effects. Figure 5.8 depicts a clean separation. There are substantial tip effects, but no non-tip ones, above the threshold. In such a case, good tip reconstruction is possible.

Choosing an appropriate threshold might seem to require us to know the size of instrumental artifacts that are not yet well studied experimentally. The suggested solution is to try the calculation with a range of thresholds (Fig. 5.9). A threshold that is too small produces a sharp result with an unphysical discontinuity in the slope at the apex (curve 1 in the figure). As the threshold is increased, the result usually changes little at first (curves 2 through 4). Then, one reaches a transition point (between curves 4 and 5) at which the result jumps suddenly from the unphysical one to one that is physically reasonable, and which then changes only gradually (curves 5 to 10) as the threshold continues to increase. Our working hypothesis is that the best choice of threshold is one just above this transition region. Picking out the transition may be aided by plotting the derivative of the tip volume (or area for a 2D tip as in the figure) as a function of threshold value. This hypothesis appears to work in simulations [32] and in an experimental test [33]. To set the scale of thresholds to try, it is useful first to bracket the correct threshold

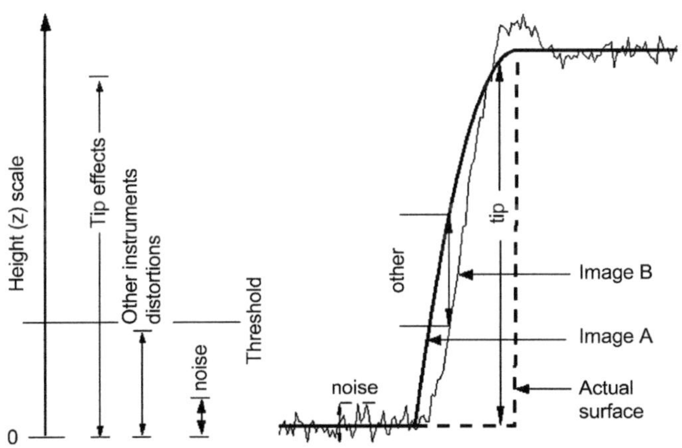

Fig. 5.8. The concept of a threshold to separate tip effects from other instrumental distortions. At the right is shown a schematic surface (*dashed*). Image *A* assumes ideal instrument behavior except for the tip size. Image *B* includes noise and other instrumental artifacts, here represented by a finite time constant that causes delayed instrument response followed by overshoot. Magnitudes of noise, tip effects, and other instrumental effects are shown by the *double-headed arrows*, then transferred for comparison to the common z-scale on the *left*. The indicated threshold value defines a region (above the threshold) where only tip effects are operative

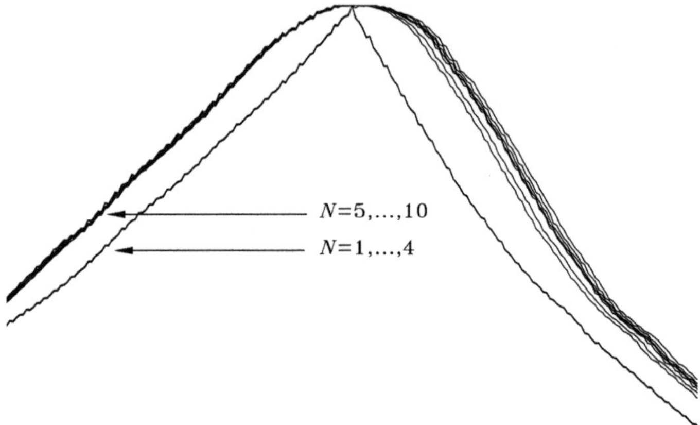

Fig. 5.9. Choosing the right threshold: the 10 tip profiles, some of which lie on top of each other, represent reconstruction results for thresholds uniformly spaced from 20% to 165% of the optimum value. Increasing *N* corresponds to increasing threshold. The transition from the unphysically sharp result occurs over a small range of thresholds. From ref. [32]

with one that is obviously too small, as described above, and one that is too large. The symptom of the latter case is a reconstruction with a plateau at its apex. To save time during this early trial and error phase, calculations can be limited to a subset of the image, provided it is suitably representative. The method described here employs a single threshold, the determination of which is consequently relatively straightforward. It is also possible to use more than one threshold, for example, one for the fast-scan and one for the slow-scan direction [36], or to accommodate spatially varying instrument distortions [37]. These methods have the disadvantage of a larger number of paramcters to optimize, but they provide flexibility that might be important in some situations.

Occasionally, one finds images for which the transition is not sharp, and there appears to be no satisfactory choice of threshold. It is likely that these represent images in which non-tip instrumental artifacts extend over the full range of the tip ones, or even beyond. Clean separation of the regimes is impossible. Tip reconstruction, whether blind or otherwise, is not likely to be possible from such images because they depart significantly from the dilation model of imaging that all tip reconstruction by tip characterizer methods currently employ.

5.4.2.2
How to do the Calculation

Reference [32] contains a library of routines, in C, to perform morphological operations, including dilation, erosion, and blind reconstruction. The codes, with inputs as determined above, may be used to calculate the tip shape.

5.4.2.3 Interpretation, Limitations, and Use of the Result

The result, P_R, of the calculation is an outer bound. In the convention of ref. [32] the outer bound is expressed as an inverted tip. It is not expected to be close to the true (inverted) tip shape at distances from the apex greater than the relief on the characterizer. It is expected to be close for those parts of the tip that touched high curvature parts of the characterizer. Any coordinate system errors in the image due to poor scale calibration or scanner non-orthogonality will be replicated in the tip description.

Sometimes a tip collides with the surface and changes in the middle of an image. If this happens, the assumption that a single tip shape produced all of the observed features is not valid, and the tip shape returned by blind reconstruction will generally be incorrect. Repeating the measurement and requiring the results to be consistent can guard against this possibility. Fortunately, collisions are usually obvious in the image – features appear to change discontinuously from one scan line to the next.

P_R is used to reconstruct the surface of an unknown sample, S, from its image, I, by calculating $S_R = I \ominus P_R$. Since P_R is an outer bound on the tip, the resulting S_R is an inner bound on those parts of S that were touched by the tip. Since I is an outer bound on S, dimensions of parts of S touched by the tip are bracketed by I and S_R (Fig. 5.10). If the tip and the characterizer are both sharp, these brackets may be quite tight. This is an important result for width measurements, since without an estimate of the tip a protrusion's width might be anything between zero and the width of its image.

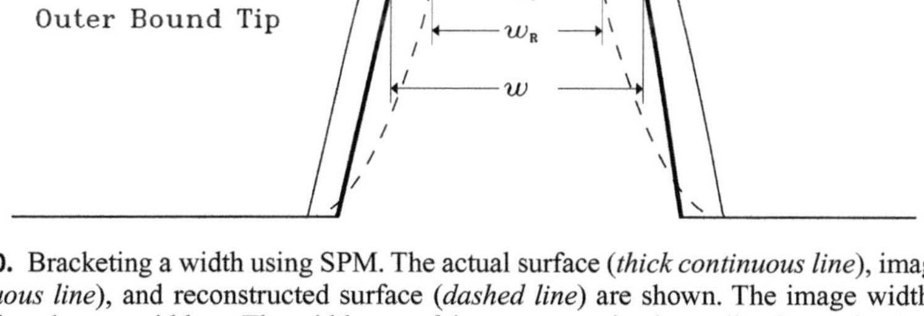

Fig. 5.10. Bracketing a width using SPM. The actual surface (*thick continuous line*), image (*thin continuous line*), and reconstructed surface (*dashed line*) are shown. The image width, w_I, is larger than the true width, w. The width, w_R, of the reconstruction is smaller than w for those parts of the sample with which the tip made contact. From ref. [19]

5.4.3 Other Modes of Operation

5.4.3.1 *Reconstructing a Surface Self-Consistently*

Since it is not necessary to know the characterizer geometry for blind reconstruction, the question arises whether we may dispense altogether with the separate characterizer. Why not image the unknown specimen, reconstruct the tip shape from the image, then erode that tip from the image to reconstruct the unknown specimen's surface?

There are some advantages to such self-consistent reconstruction. By eliminating the separate characterizer, an approach of the tip to a new surface between the time the tip is characterized and when it is actually used is no longer needed. Since the approach often risks damaging the tip by bumping it into the surface, fewer approaches reduces the risk that the tip is changed between characterization and use.

However, there are limitations to this procedure that make it inapplicable to some samples. What is really happening when a sample is used as its own characterizer is that the sharpest sample features characterize the tip with errors comparable to their size [22]. When that tip is then eroded from the image to reconstruct the surface, those same sample features are underestimated by an amount comparable to their size – that is, an error of $\approx$100% in width or radius! Therefore, these smallest, sharpest features are not well characterized by the self-consistent procedure. If these are the features of interest, a separate sharper characterizer is needed. On the other hand, if the interesting features on the unknown are the moderate to large ones, and other features of the sample meet the requirements of a good characterizer as described earlier, the self-consistent procedure will be applicable. It may sometimes be arranged for this to be the case. For example, if the sample is something that is deposited on a substrate, using a tip characterizer as the substrate is sometimes an option – or the tip characterizer itself may consist of small particles that can be deposited on the sample.

5.4.3.2 *Tips Consistent with Multiple Images*

As is evident from the above description, blind reconstruction determines tip shapes consistent with all of the features in the image. It is a simple extension of this concept to determine a tip shape consistent with two images. One simply performs a tip reconstruction on the first image, uses the result as the starting outer bound (P_0) for the second image, returns with that result to the first image, and so on iteratively until there is no further change. The generalization to more than two images is straightforward. This might be useful, for example, in combining the best features of tip characterizers from different suppliers. There is an additional issue to be aware of when doing this. The images must all have been acquired with the same tip. For example, with cantilevered instruments it is necessary to insure that the cantilever angle was close to the same for all images.

5.4.3.3 *Combining Blind Reconstruction with the Known Characterizer Method*

As already mentioned, if the characterizer is known, one determines a reconstructed tip by $P_r = I \ominus S$. P_r is an outer bound on the tip shape, good where the tip touched S and not as good elsewhere. Most candidates for known characterizers are small, so P_r would be a tight bound only near the tip's apex. However, there is no reason not to use P_r as the starting outer bound for blind reconstruction [22, 32]. In this way, we can make full use of all available information for a characterizer that is partly known.

5.4.4 Summary of the Method

To measure a tip using blind reconstruction:

1. Choose a characterizer.
2. Take an image of the characterizer using the unknown tip, using care to minimize non-tip-related instrumental artifacts.
3. Choose a starting outer bound for the tip shape.
4. Choose a low value (e.g., zero) for the threshold parameter. This forms the bottom of a threshold interval.
5. Use the above inputs to determine a (reflected) tip shape using published software [32]. This result should be too sharp.
6. By trial and error, find a threshold value high enough to produce a calculated tip that is too blunt (e.g., has a plateau at the apex or is obviously blunter than some protrusions visible in the image). This forms the top of the threshold interval.
7. Having thus bracketed the correct threshold, calculate tip shapes for thresholds regularly spaced within the interval. The resulting tips should transition sharply from the overly sharp result.
8. Choose a tip from just above this transition region. This is the estimated (reflected) tip shape. If desired, reflect this result to form an unreflected description of the tip.

In careful work, it is advisable to measure the tip shape both before and after imaging unknown samples, in order to insure that the tip did not change during imaging.

5.4.5 Experimental Test of Blind Reconstruction

The use of SPM itself to determine the shape of the tip through the use of a tip characterizer has the advantage that, since SPM is a three-dimensional imaging technique, the characterization comes with just the sort of three-dimensional estimate that is needed in order to be useful in surface reconstruction. More traditional two-dimensional microscopies such as optical microscopy or scanning electron microscopy (SEM), in which the image contrast is provided by something other than the sample's height, do not directly lead to a three-dimensional tip geometry. Rather, they produce a profile, or projection of the tip into the plane normal to the observation axis. Their probes are no longer tips in mechanical contact (or near contact) with the sample. Rather, they are beams of photons or electrons directed at the sample from a distance. These probes interact with the sample according to their own rules, giving rise to imaging artifacts analogous to the tip artifacts to which SPM is subject. The size scale at which these artifacts appear is determined by the physics of the particular interaction. For obvious reasons, this size scale is of the same order as the resolution.

All of this creates a problem for the experimentalist who wishes to provide a validation of the use of tip characterizers for measuring tips. Such a validation ought to take the form of a comparison between the measured tip shape and the true shape. Since we are not given a priori knowledge of the true shape, the best we can generally manage is to replace the true shape in the comparison with a good reference measurement. However, since SPM is three-dimensional with an intrinsic resolution higher than that of most microscopies, it is difficult to come up with a suitable reference measurement.

One approach is to perform the measurements on larger tips, so the inaccuracies of the reference technique are small compared to the tip size. If a tip characterizer with high relief is used, the tip can be measured for a distance from its apex that is also large compared to the reference technique's uncertainties. Dongmo et al. [33] performed such a test using coated diamond stylus tips and SEM as the reference technique. Good agreement was obtained as shown in Fig. 5.11, which shows blind reconstruction results superimposed upon the SEM images. Far from

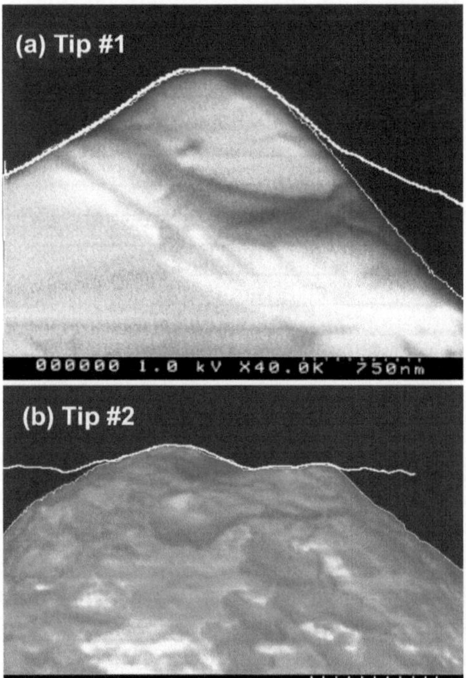

Fig. 5.11. Comparison of SEM and blind reconstruction results. Superimposed on each SEM image are the edges as determined from that image (*thin line*) and by blind reconstruction (*thick line*). **a** *Tip #1*, with single maximum. **b** *Tip #2*, a broken tip with more complicated structure. From ref. [33]

Fig. 5.12. Comparison of the SEM profile (*thin line*) and the profile obtained by the blind reconstruction (*thick*) in the case of **a** *Tip #1* and **b** *Tip #2*. The *shaded area* represents plus or minus twice the estimated combined standard uncertainty of the measurement. From ref. [33]

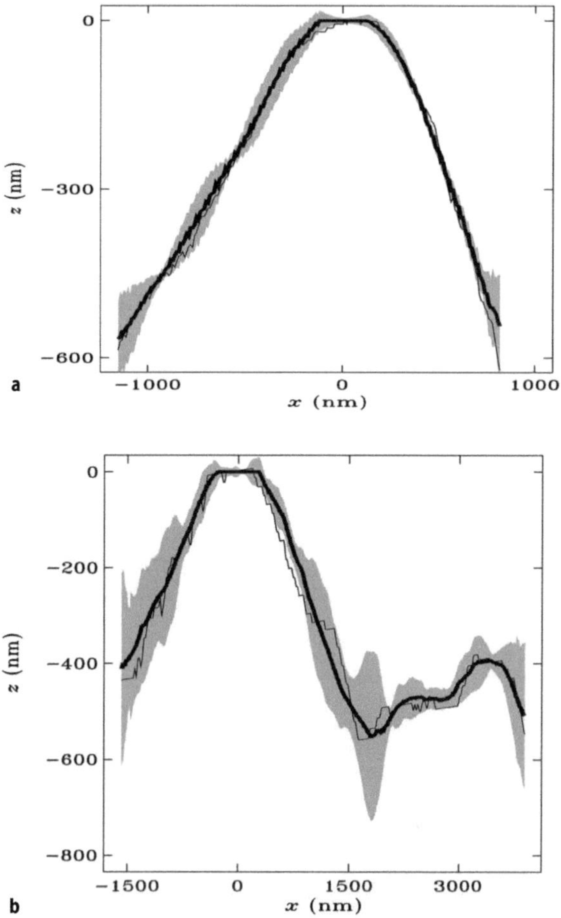

the apex, where the tip was untouched by the characterizer, the bound placed by blind reconstruction departs significantly from the actual tip, as expected. Within a few hundred nanometers of the apex lie parts of the tip that could be reached by the characterizer. Figure 5.12 shows good agreement of the SEM and blind reconstruction results in this region relative to their combined uncertainties.

Reconstruction of a standard pyramidal SiN tip was also done in the same study. The uncertainties in the SEM due to charging (if the tip were not coated with a conducting film) or due to the coating (if it were) would have been comparable to the tip's size, so an independent SEM measurement was not reported. However, the reconstructed tip's radius was 30 nm, in conformity with the manufacturer's claims for such a tip, and it was a 4-fold symmetric pyramid with face angles that agreed well with the crystallographically determined angles of the pit from which it was formed. It exhibited wear on two adjacent faces. When a tip is lifted for the retrace, as that one was, it only contacts the surface while scanning in one direction. One therefore expects most of the wear on one face (if one face happens to be lined up exactly perpendicular to the scan direction) or two adjacent ones (if the alignment is anything other than that). Thus, the conformity of the reconstruction with expectations is remarkable.

References

1. International technology roadmap for semiconductors (2001) Semiconductor Industry Association, Metrology section, Table 98a, p 8 (http://public.itrs.net/Files/2002 Update/2001ITRS/Home.htm)
2. Chernoff DA, Burkhead DL (1999) J Vac Sci Technol A 17:1457
3. Majumdar A, Bhushan B (1995) In: Bhushan B (ed) Handbook of micro/nano tribology. CRC Press, Boca Raton
4. Dixson R, Köning R, Vorburger TV, Fu J, Tsai VW (1998) Proc SPIE 3332:420
5. Kramar J, Jun J, Penzes W, Scire F, Teague C, Villarrubia J, Amatucci E, Gilsinn D (1999) Proc SPIE 3677:1017
6. ASTM E1813–96. Measuring and reporting probe tip shape in scanning probe microscopy
7. Thundat T, Zheng X-Y, Sharp SL, Allison DP, Warmack RJ, Joy CD, Ferrell TL (1992) Scan Microsc 6:903
8. Xu S, Arnsdorf MF (1994) J Microsc 173:199
9. Vesenka J, Miller R, Henderson E (1994) Rev Sci Instrum 65:2249
10. Derose JA, Revel JP (1999) J Microsc (Oxford) 195:1
11. Odin C, Aimé JP, El Kaakour Z, Bouhacina T (1994) Surf Sci 317:321
12. Van Cleef M, Holt SA, Watson GS, Myhra S (1996) J Microsc 181:2
13. Reiss G, Schneider F, Vancea J, Hoffmann H (1990) Appl Phys Lett 57:867
14. Keller D (1991) Surf Sci 253:353
15. Gallarda H, Jain R (1991) Proc SPIE 1464:459
16. Pingali GS, Jain R (1992) Proc IEEE workshop on applications of computer vision, p 282
17. Keller DJ, Franke FS (1993) Surf Sci 294:409
18. Markiewicz P, Goh MC (1994) Langmuir 10:5
19. Villarrubia JS (1994) Surf Sci 321:287
20. Bonnet N, Dongmo S, Vautrot P, Troyon M (1994) Microsc Microanal Microstruct 5:477
21. Wilson DL, Kump KS, Eppell SJ, Marchant RE (1995) Langmuir 11:265
22. Villarrubia JS (1996) J Vac Sci Technol B 14:1518
23. Dongmo S, Troyon M, Vautrot P, Delain E, Bonnet N (1996) J Vac Sci Technol B 14:1552
24. Williams PM, Shakesheff KM, Davies MC, Jackson DE, Roberts CJ, Tendler SJB (1996) J Vac Sci Technol B 14:1557
25. Montelius L, Tegenfeldt JO, van Heeren P (1994) J Vac Sci Technol B 12:2222
26. Bogdanov AL, Erts D, Nilsson B, Olin H (1994) J Vac Sci Technol B 12:3681
27. Atamny F, Baiker A (1995) Surf Sci 323:L314
28. Bao GW, Li SFY (1998) Talanta 45:751
29. Seah MP, Spencer SJ, Cumpson PJ, Johnstone JE (1999) Appl Surf Sci 144–145:151
30. Siedle P, Butt HJ, Bamberg E, Wang DN, Kuhlbrandt W, Zach J, Haider M (1993) Inst Phys Conf Ser 130:361
31. Martin Y, Wickramasinghe HK (1995) J Vac Sci Technol B 13:2335
32. Villarrubia JS (1997) J Res Natl Inst Stand Technol 102:425 (http://nvl.nist.gov/pub/nistpubs/jres/102/4/j24vil.pdf)
33. Dongmo S, Villarrubia JS, Jones SN, Renegar TB, Postek MT, Song JF (2000) Ultramicrosc 85:141
34. Mahoney W, Schaefer DM, Patil A, Andres RP, Reifenberger R (1994) Surf Sci 316:383
35. Pingali GS, Jain R (1996) In: Sanz JLC (ed) Image technology. Springer, Berlin Heidelberg New York, p 519
36. Todd BA, Eppell SJ (2001) Surf Sci 491:473
37. Williams PM, Davies MC, Roberts CJ, Tendler SJB (1998) Appl Phys A: Mater Sci Process 66:S911

Part II
Characterization

6 Micro/Nanotribology Studies Using Scanning Probe Microscopy

Bharat Bhushan

6.1 Introduction

The mechanisms and dynamics of the interactions of two contacting solids during relative motion, ranging from atomic to microscale, need to be understood in order to develop fundamental understanding of adhesion, friction, wear, indentation, and lubrication processes. At most solid-solid interfaces of technological relevance, contact occurs at many asperities. Consequently the importance of investigating single asperity contacts in studies of the fundamental micro/nanomechanical and micro/nanotribological properties of surfaces and interfaces has long been recognized. The recent emergence and proliferation of proximal probes, in particular scanning probe microscopies (the scanning tunneling microscope and the atomic force microscope), and the surface force apparatus, and of computational techniques for simulating tip-surface interactions and interfacial properties, has allowed systematic investigations of interfacial problems with high resolution as well as ways and means for modifying and manipulating nanoscale structures. These advances have led to the appearance of the new field of micro/nanotribology, which pertains to experimental and theoretical investigations of interfacial processes on scales ranging from the atomic and molecular to the microscale, occurring during adhesion, friction, scratching, wear, indentation, and thin-film lubrication at sliding surfaces [1–12].

The micro/nanotribological studies are needed to develop fundamental understanding of interfacial phenomena on a small scale and to study interfacial phenomena in micro/nanostructures used in magnetic storage systems, micro/nanoelectromechanical systems (MEMS/NEMS), and other applications [1, 5–7, 13, 14]. Friction and wear of lightly loaded micro/nanocomponents are highly dependent on the surface interactions (few atomic layers). These structures are generally lubricated with molecularly thin films. Micro/nanotribological studies are also valuable in fundamental understanding of interfacial phenomena in macrostructures to provide a bridge between science and engineering.

The surface force apparatus (SFA), the scanning tunneling microscopes (STM), atomic force and friction force microscopes (AFM and FFM) are widely used in micro/nanotribological studies. Typical operating parameters are compared in Table 6.1. The SFA was developed in 1968 and is commonly employed to study both static and dynamic properties of molecularly thin films sandwiched between two molecularly smooth surfaces. The STM, developed in 1981, allows imaging of electrically conducting surfaces with atomic resolution, and has been used for

Table 6.1. Comparison of typical operating parameters in SFA, STM and AFM/FFM used for micro/nanotribological studies

Operating parameter	SFA	STM[a]	AFM/FFM
Radius of mating surface/tip	≈ 10 mm	5–100 nm	5–100 nm
Radius of contact area	10–40 μm	N/A	0.05–0.5 nm
Normal load	10–100 mN	N/A	<0.1 nN–500 nN
Sliding velocity	0.001–100 μm s^{-1}	0.02–2 μm s^{-1} (scan size ≈ 1 nm × 1 nm to 125 μm × 125 μm; scan rate <1–122 Hz)	0.02–2 μm s^{-1} (scan size ≈ 1 nm × 1 nm to 125 nm × 125 nm; scan rate) <1–122 Hz
Sample limitations	Typically atomically-smooth, optically transparent mica; opaque ceramic, smooth surfaces can also be used	Electrically conducting samples	None

[a] Can only be used for atomic-scale imaging.

imaging of clean surfaces as well as of lubricant molecules. The introduction of the atomic force microscope in 1985 provided a method for measuring ultrasmall forces between a probe tip and an engineering (electrically conducting or insulating) surface, and has been used for topographical measurements of surfaces on the nanoscale, as well as for adhesion and electrostatic force measurements. Subsequent modifications of the AFM led to the development of the friction force microscope (FFM), designed for atomic-scale and microscale studies of friction. This instrument measures forces in the scanning direction. The AFM is also being used for investigations of scratching, wear, indentation, detection of transfer of material, boundary lubrication, and fabrication and machining. Meanwhile, significant progress in understanding the fundamental nature of bonding and interactions in materials, combined with advances in computer-based modeling and simulation methods, has allowed theoretical studies of complex interfacial phenomena with high resolution in space and time. Such simulations provide insights into atomic-scale energetics, structure, dynamics, thermodynamics, transport and rheological aspects of tribological processes.

The nature of interactions between two surfaces brought close together, and those between two surfaces in contact as they are separated, have been studied experimentally with the surface force apparatus. This has led to a basic understanding of the normal forces between surfaces, and the way in which these are modified by the presence of a thin liquid or a polymer film. The frictional properties of such systems have been studied by moving the surfaces laterally, and such experiments have provided insights into the molecular-scale operation of lubricants such as thin liquid or polymer films. Complementary to these studies are those in which the AFM or FFM is used to provide a model asperity in contact with a solid or lubricated surface. These experiments have demonstrated that the relationship between friction and surface roughness is not always simple or obvious. AFM studies have also revealed much about the nanoscale nature of intimate contact during wear and indentation.

In this chapter, we present a review of significant aspects of micro/nanotribological studies conducted using AFM/FFM.

6.2 Friction and Adhesion

6.2.1 Atomic-Scale Friction

To study friction mechanisms on an atomic scale, a well-characterized freshly cleaved surface of highly-oriented pyrolytic graphite (HOPG) has been studied by Mate et al. [15] and Ruan and Bhushan [16]. For the friction calibration technique, see Bhushan [2]. The atomic-scale friction force of HOPG exhibited the same periodicity as that of the corresponding topography (Fig. 6.1a), but the peaks in friction and those in topography are displaced relative to each other (Fig. 6.1b). A Fourier expansion of the interatomic potential was used to calculate the conservative interatomic forces between atoms of the FFM tip and those of the graphite surface. Maxima in the interatomic forces in the normal and lateral directions do not occur at the same location, which explains the observed shift between the peaks in the lateral force and those in the corresponding topography. Furthermore, the observed local variations in friction force were explained by variation in the intrinsic lateral force between the sample and the FFM tip resulting from an atomic-scale stick-slip process [15, 16].

6.2.2 Microscale Friction

Local variations in the microscale friction of cleaved graphite are observed, Fig. 6.2. These arise from structural changes that occur during the cleaving process [17]. The cleaved HOPG surface is largely atomically smooth but exhibits line-shaped regions in which the coefficient of friction is more than an order of magnitude larger. Transmission electron microscopy indicates that the line-shaped regions consist of graphite planes of different orientation, as well as of amorphous carbon. Differences in friction have also been observed for multiphase ceramic materials [18]. Figure 6.3 shows the surface roughness and friction force maps of Al_2O_3-TiC (70–30 wt %). TiC grains have a Knoop hardness of about 2800 kg mm^{-2}; therefore, they do not polish as much and result in a slightly higher elevation (about 2–3 nm higher than that of Al_2O_3 grains). TiC grains exhibit higher friction force than Al_2O_3 grains. The coefficients of friction of TiC and Al_2O_3 grains are 0.034 and 0.026, respectively, and the coefficient of friction of Al_2O_3-TiC composite is 0.03. Local variation in friction force also arises from the scratches present on the Al_2O_3-TiC surface. Meyer et al. [19] also used FFM to measure structural variations of organic mono and multilayer films. All of these measurements suggest that the FFM can be used for structural mapping of the surfaces. FFM measurements

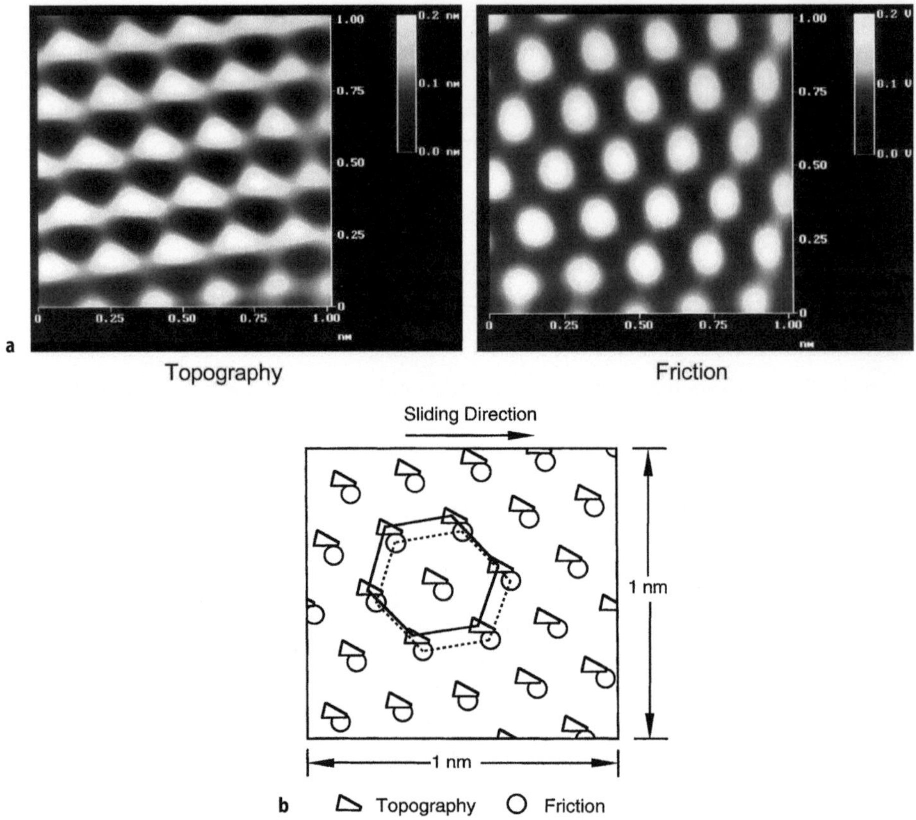

Fig. 6.1. **a** Gray-scale plots of surface topography and friction force maps of a 1 nm × 1 nm area of freshly cleaved HOPG, showing the atomic-scale variation of topography and friction, and **b** schematic of superimposed topography and friction maps from **a**; the *symbols* correspond to maxima. Note the spatial shift between the two plots [16]

can be used to map chemical variations, as indicated by the use of the FFM with a modified probe tip to map the spatial arrangement of chemical functional groups in mixed organic monolayer films [20]. Here, sample regions that had stronger interactions with the functionalized probe tip exhibited larger friction.

Local variations in the microscale friction of nominally rough, homogeneous surfaces can be significant, and are seen to depend on the local surface slope rather than the surface height distribution. This dependence was first reported by Bhushan and Ruan [21], Bhushan et al. [22], and Bhushan [23] and later discussed in more detail by Koinkar and Bhushan [24] and Sundararajan and Bhushan [25]. In order to show elegantly any correlation between local values of friction and surface roughness, surface roughness and friction force maps of a silicon grid with square pits were obtained (Fig. 6.4) [25]. Figure 6.4 shows the surface roughness map, the slopes of the roughness map taken along the sliding direction (surface slope map) and the friction force map for various samples. There is a strong correlation be-

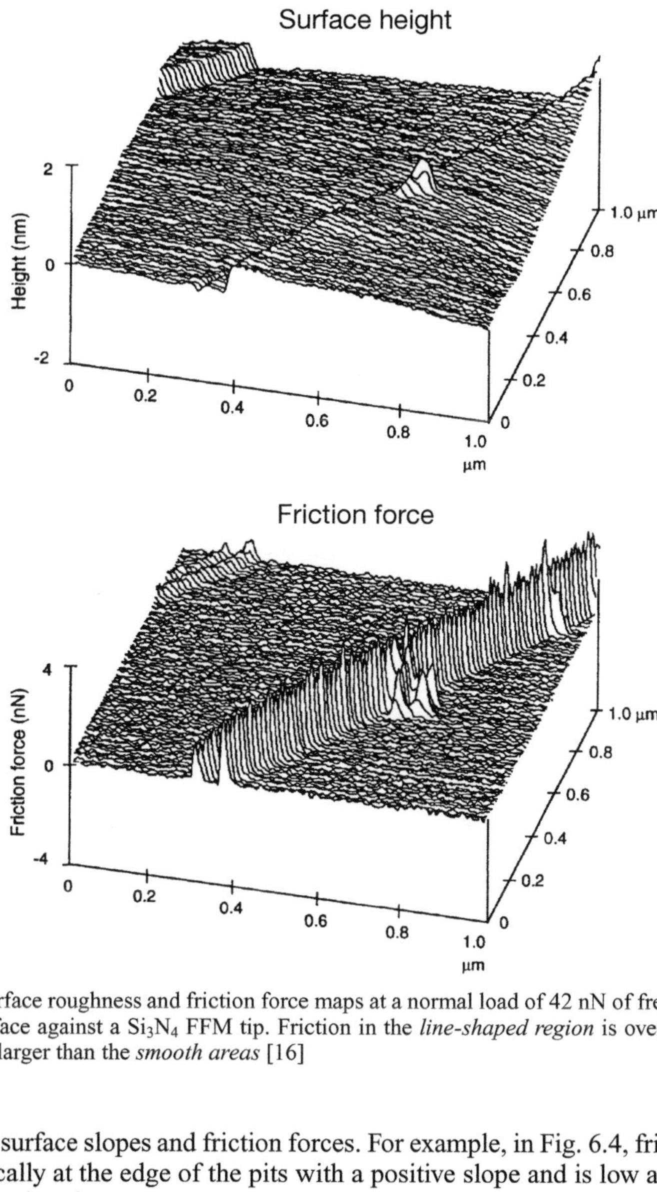

Fig. 6.2. Surface roughness and friction force maps at a normal load of 42 nN of freshly cleaved HOPG surface against a Si_3N_4 FFM tip. Friction in the *line-shaped region* is over an order of magnitude larger than the *smooth areas* [16]

tween the surface slopes and friction forces. For example, in Fig. 6.4, friction force is high locally at the edge of the pits with a positive slope and is low at the edges with a negative slope.

We now examine the mechanism of microscale friction, which may explain the resemblance between the slope of surface roughness maps and the corresponding friction force maps [2, 3, 16, 17, 21, 22, 24, 25]. A so-called ratchet mechanism is believed to be the dominant mechanism for the local variations in the friction force map, which suggests a direct correspondence between the coefficient of friction and the surface slope. With the tip sliding over the leading (ascending) edge of an asperity, the surface slope is positive; it is negative during sliding over the trailing

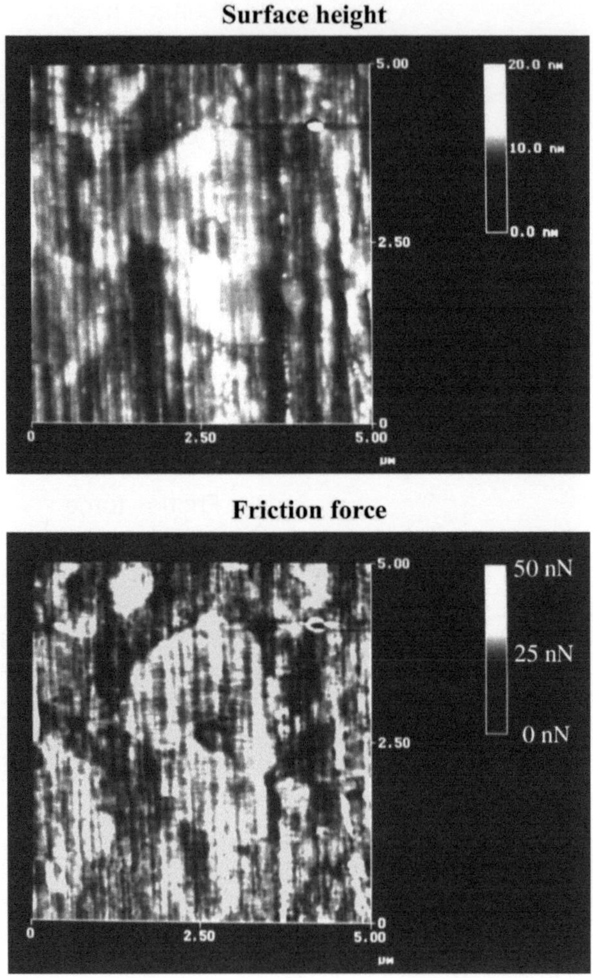

Fig. 6.3. Gray-scale surface roughness ($\sigma = 0.80$ nm) and friction force maps (mean = 7.0 nN, $\sigma = 0.90$ nN) for Al_2O_3-TiC (70–30 wt %) at a normal load of 138 nN [18]

(descending) edge of an asperity. Thus, measured friction is high at the leading edge of asperities and low at the trailing edge. In addition to the slope effect, the collision of the tip when encountering an asperity with a positive slope produces additional torsion of the cantilever beam leading to higher measured friction force. When encountering an asperity with the same negative slope, however, there is no collision effect and hence no effect on torsion. This effect also contributes to the difference in friction forces when the tip scans up and down on the same topography feature. The ratchet mechanism and the collision effects thus semiquantitatively explain the correlation between the slopes of the roughness maps and friction force maps observed in Fig. 6.4.

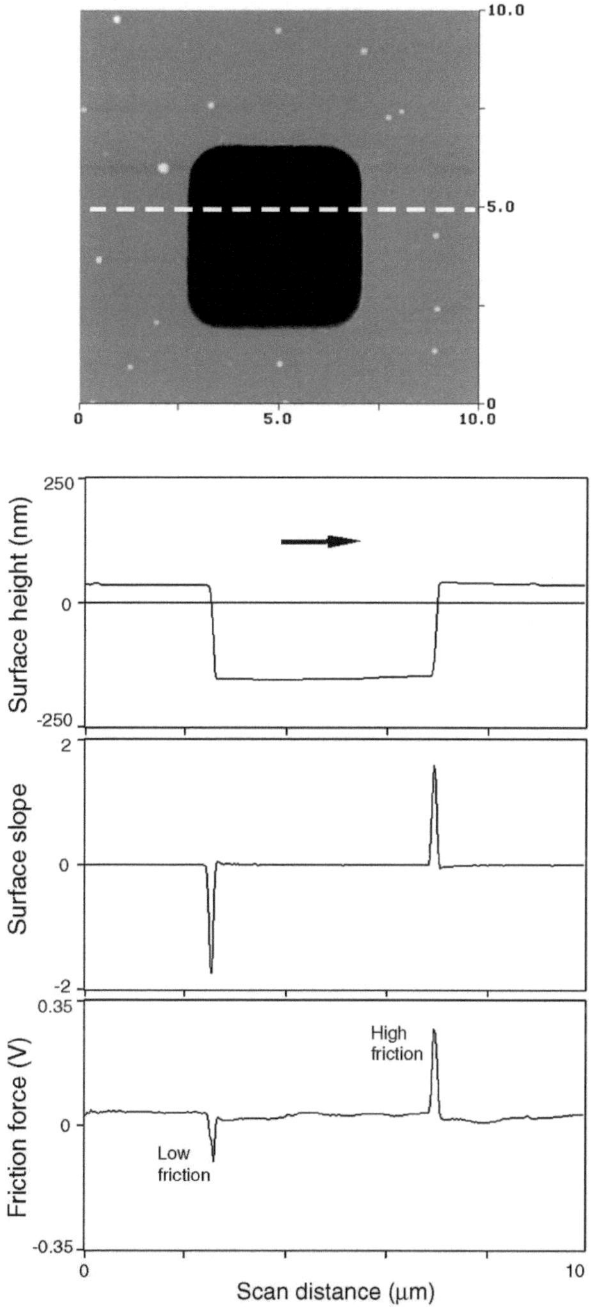

Fig. 6.4. Surface roughness map, surface slope map taken in the sample sliding direction (*horizontal axis*), and friction force map for a silicon grid (with 5-μm square pits of depth 180 nm and a pitch of 10 μm) [25]

6.2.3 Directionality Effect on Microfriction

During friction measurements, the friction force data from both the forward (trace) and backward (retrace) scans are useful in understanding the origins of the observed friction forces. Magnitudes of material-induced effects are independent of the scanning direction, whereas topography-induced effects are different between forward and backward scanning directions. Since the sign of the friction force changes as the scanning direction is reversed (because of the reversal of torque applied to the end of the tip), addition of the friction force data of the forward and backward scan eliminates the material-induced effects while topography-induced

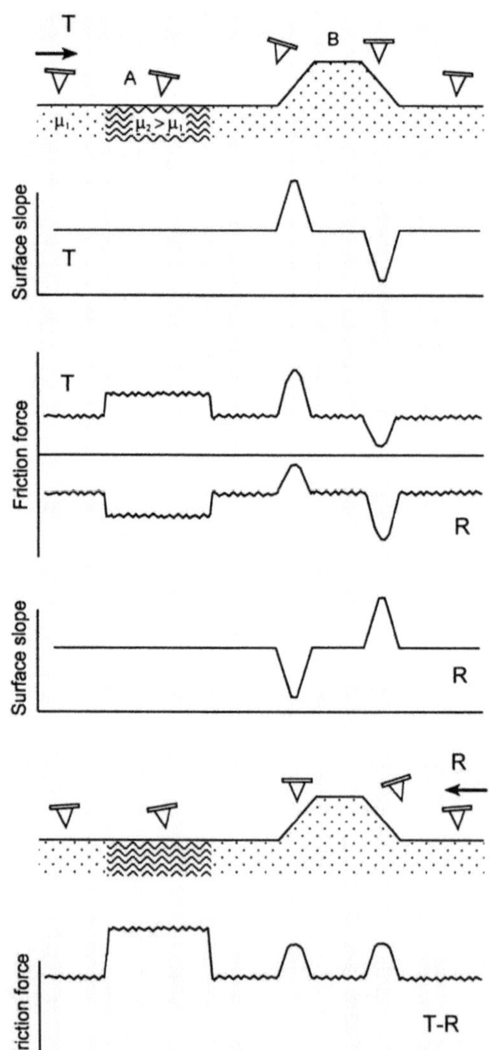

Fig. 6.5. Schematic illustration showing the effect of an asperity (making an angle θ with the horizontal plane) on the surface in contact with the tip on local friction in the presence of adhesive friction mechanism. W and F are the normal and friction forces, respectively, and S and N are the force components along and perpendicular to the local surface of the sample at the contact point, respectively

effects still remain. Subtraction of the data between forward and backward scans does not eliminate either effect, Fig. 6.5 [25].

Owing to the reversal of the sign of the retrace (R) friction force with respect to the trace (T) data, the friction force variations due to topography are in the same direction (peaks in trace corresponds to peaks in retrace). However, the magnitudes of the peaks in trace and retrace at a given location are different. An increase in the friction force experienced by the tip when scanning up a sharp change in topography is more than the decrease in the friction force experienced when scanning down the same topography change, partly because of collision effects discussed earlier. Asperities on engineering surfaces are asymmetrical which also affect the magnitude of friction force in the two directions. Asymmetry in tip shape may also have an effect on the directionality effect of friction. We will note later that the magnitudes of surface slopes are virtually identical; therefore, the tip shape asymmetry should not have much effect. Because of the differences in the magnitude of friction forces in the two directions, subtracting the two friction data yields a residual peak.

In order to facilitate comparison of directionality effect on friction, it is important to take into account the sign change of the surface slope and friction force in the trace and retrace directions. Figure 6.6 shows surface height, surface slope, and friction force data for a silicon grid in the trace and retrace directions. The correlation between surface slope and friction forces is clear. The third column in the figures shows retrace slope and friction data with an inverted sign (–retrace). Now we can compare trace data with –retrace data. It is clear that the friction experienced by the tip is dependent upon the scanning direction because of surface topography. In addition to the effect of topographical changes discussed earlier, during surface-finishing processes, material can be transferred preferentially onto one side of the asperities, which also causes asymmetry and direction dependence. Reduction in local variations and in directionality of friction properties requires careful optimization of surface roughness distributions and of surface-finishing processes.

The directionality as a result of surface asperities effect will be also manifested in macroscopic friction data, that is, the coefficient of friction may be different in one sliding direction than in the other. Asymmetrical shape of asperities accentuates this effect. The frictional directionality can also exist in materials with particles having a preferred orientation. The directionality effect in friction on a macroscale is observed in some magnetic tapes. In a macroscale test, a 12.7-mm-wide polymeric magnetic tape was wrapped over an aluminum drum and slid in a reciprocating motion with a normal load of 0.5 N and a sliding speed of about 60 mm s^{-1} [1]. The coefficient of friction as a function of sliding distance in either direction is shown in Fig. 6.7. We note that the coefficient of friction on a macroscale for this tape is different in different directions. Directionality in friction is sometimes observed on the macroscale; on the microscale this is the norm [2, 14]. On the macroscale, the effect of surface asperities normally is averaged out over a large number of contacting asperities.

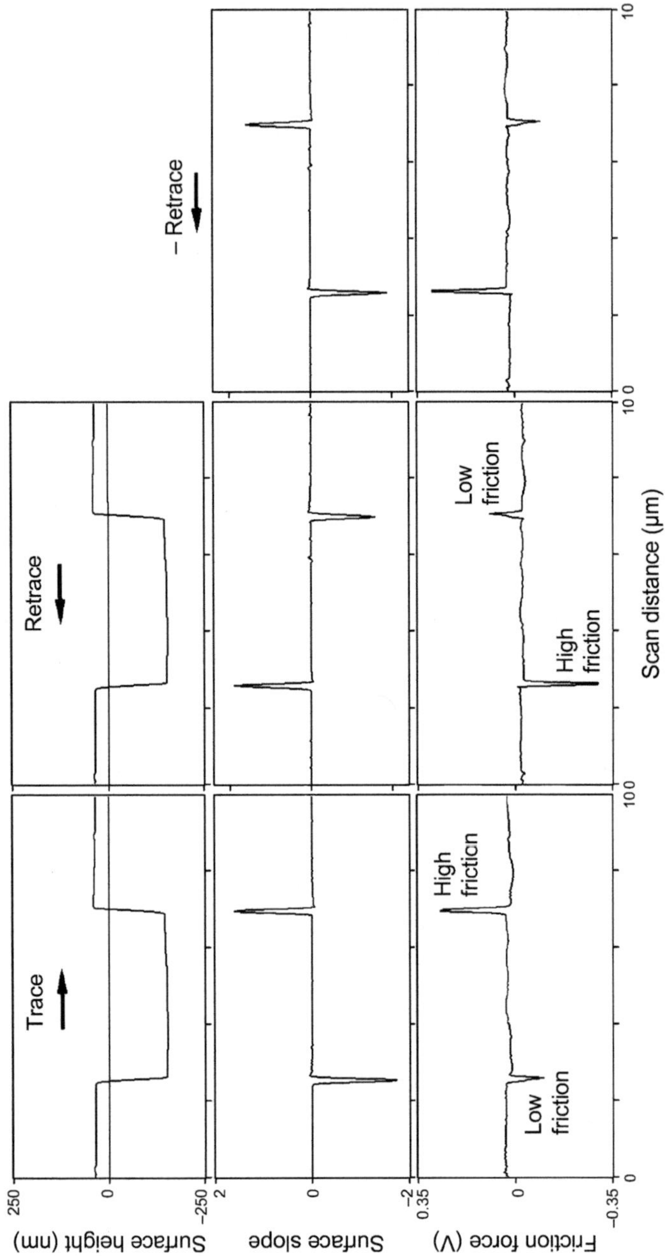

Fig. 6.6. Two-dimensional profiles of surface heights, surface slopes, and friction forces for scans across the silicon grid pit. *Arrows* indicate the tip sliding direction [25]

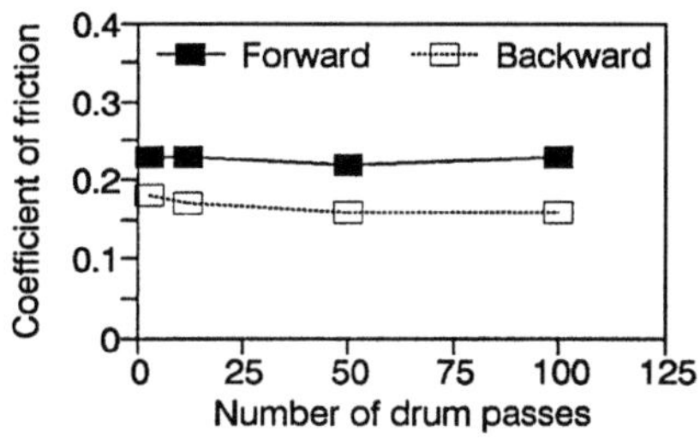

Fig. 6.7. Coefficient of macroscale friction as a function of drum passes for a polymeric magnetic tape sliding over an aluminum drum in a reciprocating mode in both directions. Normal load = 0.5 N over 12.7-mm-wide tape, sliding speed = 60 mm s^{-1} [23]

6.2.4 Velocity Dependence on Microfriction

AFM/FFM experiments are generally conducted at relative velocities as high as few tens of μm s^{-1}. To simulate applications, it is of interest to conduct friction experiments at higher velocities. High velocities can be achieved by mounting either the sample or the cantilever beam on a shear wave transducer (ultrasonic transducer) to produce surface oscillations at MHz frequencies [26, 27]. The velocities of the order of a few mm s^{-1} can thus be achieved. The effect of in-plane and out-of-plane sample vibration amplitude on the coefficient of friction is shown in Fig. 6.8. Vibration of a sample at ultrasonic frequencies (>20 kHz) can substantially reduce the coefficient of friction, known as ultrasonic lubrication or sonolubrication. When the surface is vibrated in-plane, classical hydrodynamic lubrication develops hydrodynamic pressure, which supports the tip and reduces friction. When the surface is vibrated out-of-plane, a lift-off caused by the squeeze-film lubrication (a form of hydrodynamic lubrication), reduces friction.

Fig. 6.8a,b. Reduction of coefficient of friction, measured at a normal load of 100 nN and average tip separation as a function of surface amplitude on a single-crystal silicon subjected to **a** in-plane and **b** out-of-plane vibrations at about 1 MHz against a silicon nitride tip [26]

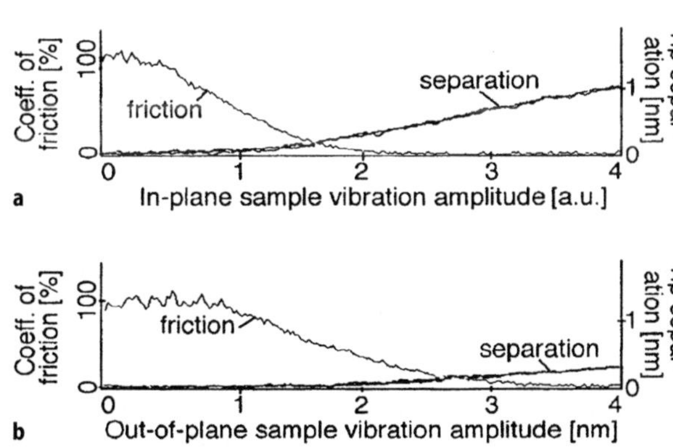

6.2.5 Effect of Tip Radii and Humidity on Adhesion and Friction

The tip radius and relative humidity affect adhesion and friction for dry and lubricated surfaces [28, 29]. Figure 6.10 shows the variation of single-point adhesive force measurements as a function of tip radius on a Si(100) sample for several humidities. The adhesive force data are also plotted as a function of relative humidity for several tip radii. The general trend at humidities up to the ambient is that a 50-nm radius Si_3N_4 tip exhibits a slightly lower adhesive force as compared to the other microtips of larger radii; in the latter case, values are similar. Thus for the microtips, there is no appreciable variation in adhesive force with tip radius at a given humidity up to the ambient. The adhesive force increases as relative humidity increases for all tips.

Sources of adhesive force between a tip and a sample surface are van der Waals attraction and meniscus formation [3, 8]. Relative magnitudes of the forces from the two sources are dependent upon various factors including the distance between the tip and the sample surface, their surface roughnesses, their hydrophobicity, and relative humidity [30]. For most surfaces with some roughness, meniscus contribution dominates at moderate to high humidities. The trend in adhesive forces as a function of tip radii and relative humidity, in Fig. 6.9, can be explained by the presence of meniscus forces, which arise from capillary condensation of water vapor from the environment forming meniscus bridges. If enough liquid is present to form a meniscus bridge, the meniscus force should increase with an increase in tip radius (proportional to tip radius for a spherical tip) and should be independent of the relative humidity or water film thickness. In addition, an increase in tip radius in a dry environment results in increased contact area leading to higher values of van der Waals forces. However, if nanoasperities on the tip and the sample are considered then the number of contacting and near-contacting asperities forming meniscus bridges increases with an increase of humidity leading to an increase in meniscus forces. This explains the trends observed in Fig. 6.9. From the data, the tip radius has little effect on the adhesive forces at low humidities but increases with tip radius at high humidity. Adhesive force also increases with an increase in humidity for all tips. This observation suggests that thickness of the liquid film at low humidities is insufficient to form continuous meniscus bridges to affect adhesive forces in the case of all tips.

Figure 6.9 also shows the variation in coefficient of friction as a function of tip radius at a given humidity, and as a function of relative humidity for a given tip radius for Si(100). It can be observed that for 0% RH, the coefficient of friction is about the same for the tip radii except for the largest tip, which shows a higher value. At all other humidities, the trend consistently shows that the coefficient of friction increases with tip radius. An increase in friction with tip radius at low to moderate humidities arises from increased contact area (higher van der Waals forces) and higher values of shear forces required for larger contact area. At high humidities, similar to adhesive force data, an increase with tip radius occurs because of both contact area and meniscus effects. Although AFM/FFM measurements are able to measure the combined effect of the contribution of van der Waals and meniscus forces towards friction force or adhesive force, it is difficult to mea-

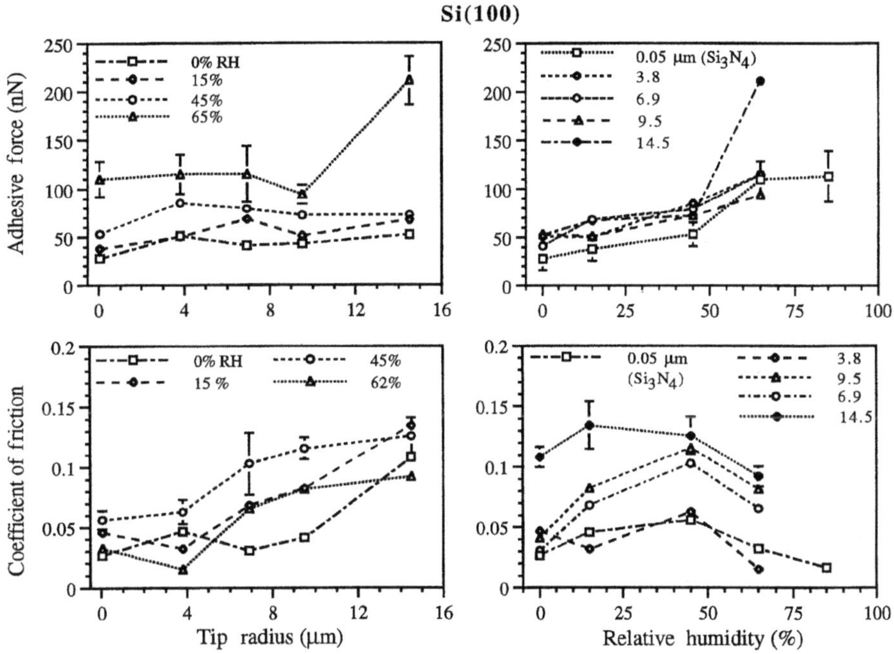

Fig. 6.9. Adhesive force and coefficient of friction as a function of tip radius at several humidities and as a function of relative humidity at several tip radii on Si(100) [28]

sure their individual contributions separately. It can be seen that for all tips, the coefficient of friction increases with humidity to about ambient, beyond which it starts to decrease. The initial increase in the coefficient of friction with humidity arises from the fact that the thickness of the water film increases with an increase in the humidity, which results in a larger number of nanoasperities forming meniscus bridges and leads to higher friction (larger shear force). The same trend is expected with the microtips beyond 65% RH. This is attributed to the fact that at higher humidities, the adsorbed water film on the surface acts as a lubricant between the two surfaces. Thus the interface is changed at higher humidities, resulting in lower shear strength and hence lower friction force and coefficient of friction.

6.2.6 Scale Dependence on Friction

Table 6.2 shows the coefficient of friction measured for various materials on nano/micro and macroscales. It is clearly observed that friction values are scale dependent. The values on the nano/microscale are much lower than those on the macroscale. There can be the following four and possibly more differences in the operating conditions responsible for these differences. First, the contact stresses at AFM conditions, in spite of small tip radii, generally do not exceed the sample hardness that minimizes plastic deformation. Average contact stresses in macro-

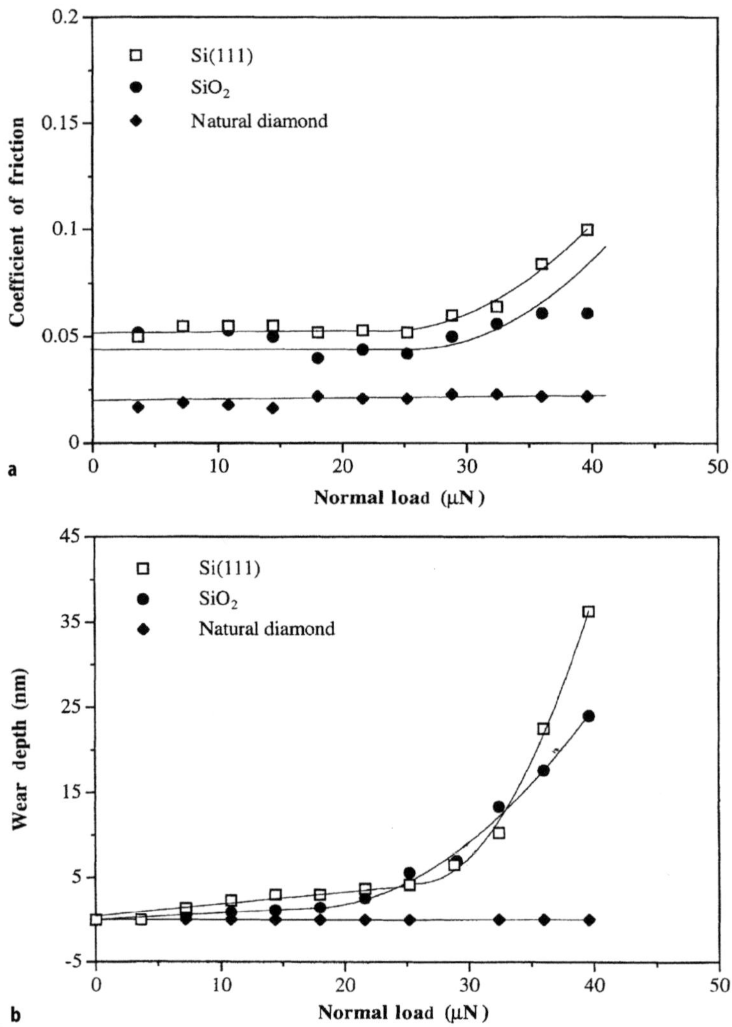

Fig. 6.10. **a** Coefficient of friction as a function of normal load and **b** corresponding wear depth as a function of normal load for silicon, SiO_2 coating and natural diamond. *Inflections* in the curves for silicon and SiO_2 correspond to the contact stresses equal to the hardnesses of these materials [32]

contacts are generally lower than that in AFM contacts; however, a large number of asperities come in contact which go through plastic deformation. Second, when measured for the small contact areas and very low loads used in microscale studies, indentation hardness and modulus of elasticity are higher than at the macroscale, as will be discussed later [9, 31]. Lack of plastic deformation and improved mechanical properties reduce the degree of wear and friction. Third, the small apparent area of contact reduces the number of particles trapped at the interface, and thus minimizes the third-body plowing contribution to the friction force [9]. As

Table 6.2. Surface roughness (standard deviation of surface heights or σ) and coefficients of frictionon nano- and macroscales of various samples in air

Material	σ (nm)	Coefficient of nanoscale friction versus Si_3N_4 tip [a]	Coefficient of macroscale friction versus Si_3N_4 ball [b]
Graphite (HOPG)	0.09	0.006	0.1
Natural diamond	2.3	0.04	0.2
Si(100)	0.14	0.07	0.4

[a] Tip radius of about 50 nm in the load range of 10–150 nN (2.5–6.1 GPa) and a scanning speed of 0.5 nm s^{-1} and scan area of 1 nm × 1 nm for HOPG and a scanning speed of 4 µm s^{-1} and scan area of 1 µm × 1 µm for diamond and Si(100).
[b] Ball radius of 3 mm at a normal load of 1 N (0.6 GPa) and average sliding speed of 0.8 mm s^{-1}.

a fourth and final difference, we have seen in the previous section that coefficient of friction increases with an increase in the AFM tip radius. AFM data are taken with a sharp tip, whereas asperities coming in contact in macroscale tests range from nanoasperities to much larger asperities, which may be responsible for larger values of friction force on the macroscale.

To demonstrate the load dependence on the coefficient on friction, stiff cantilevers were used to conduct friction experiments at high loads, Fig. 6.10 [32]. At higher loads (with contact stresses exceeding the hardness of the softer material), as anticipated, the coefficient of friction for microscale measurements increases towards values comparable with those obtained from macroscale measurements, and surface damage also increases. At high loads, plowing is a dominant contributor to the friction force. Based on these results, Amontons' law of friction, which states that the coefficient of friction is independent of apparent area and normal load, does not hold for microscale measurements. These findings also suggest that microcomponents sliding under lightly loaded conditions should experience ultralow friction and near-zero wear [9].

6.3 Scratching, Wear, and Local Deformation

6.3.1 Microscale Scratching

The AFM can be used to investigate how surface materials can be moved or removed on micro to nanoscales, for example, in scratching and wear [2] (where these things are undesirable), and nanofabrication/nanomachining (where they are desirable). Scratching can be performed at random loads to determine the scratch resistance of materials and coatings. The coefficient of friction is measured during scratching and the load at which the coefficient of friction increases rapidly is known as the "critical load," which is a measure of scratch resistance. In addition, post-scratch imaging can be performed in situ with the AFM in tapping mode to study failure mechanisms. Figure 6.11 shows data from a scratch test on Si(100)

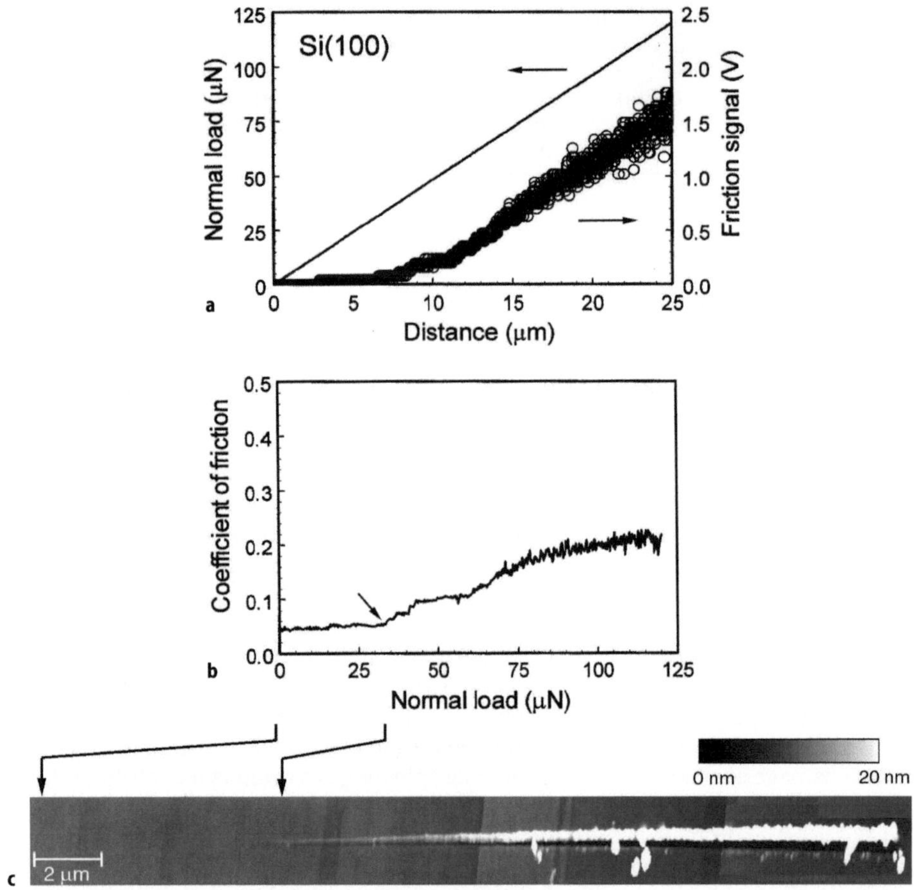

Fig. 6.11. **a** Applied normal load and friction signal measured during the microscratch experiment on Si(100) as a function of scratch distance, **b** friction data plotted in the form of coefficient of friction as a function of normal load, and **c** AFM surface height image of scratch obtained in tapping mode [33]

with a scratch length of 25 μm and a scratching velocity of 0.5 μm s^{-1}. At the beginning of the scratch, the coefficient of friction is 0.04 which is a typical value for silicon. At about 35 μN (indicated by the arrow in the figure), there is a sharp increase in the coefficient of friction, which indicates the critical load. Beyond the critical load, the coefficient of friction continues to increase steadily. In the post-scratch image, we note that at the critical load, a clear groove starts to form. This implies that Si(100) was damaged by plowing at the critical load, associated with the plastic flow of the material. At and after the critical load, small and uniform debris is observed and the amount of debris increases with increasing normal load. Sundararajan and Bhushan [33] have also used this technique to measure the scratch resistance of diamond-like carbon coatings ranging in thickness from 3.5 to 20 nm.

6.3.2 Microscale Wear

By scanning the sample in two dimensions with the AFM, wear scars are generated on the surface. Figure 6.12 shows the effect of normal load on wear depth. We note that wear depth is very small below 20 μN of normal load [34, 35]. A normal load of 20 μN corresponds to contact stresses comparable to the hardness of the silicon. Primarily, elastic deformation at loads below 20 μN is responsible for low wear [32].

To understand wear mechanisms, evolution of wear can be studied using AFM. Figure 6.13 shows evolution of wear marks of a DLC-coated disk sample. The data illustrate how the microwear profile for a load of 20 μN develops as a function of the number of scanning cycles [22]. Wear is not uniform, but is initiated at the nanoscratches. Surface defects (with high surface energy) present at nanoscratches act as initiation sites for wear. Coating deposition also may not be uniform on and near nanoscratches, which may lead to coating delamination. Thus, scratch-free surfaces will be relatively resistant to wear.

Wear precursors (precursors to measurable wear) can be studied by making surface potential measurements [36–38]. The contact potential difference or simply surface potential between two surfaces depends on a variety of parameters such as electronic work function, adsorption, and oxide layers. The surface potential map of an interface gives a measure of changes in the work function, which is sensitive to both physical and chemical conditions of the surfaces including structural and chemical changes. Before material is actually removed in a wear process, the surface experiences stresses that result in surface and subsurface changes of structure and/or chemistry. These can cause changes in the measured potential of a surface. An AFM tip allows mapping of surface potential with nanoscale resolution. Surface height and change in surface potential maps of a polished single-crystal aluminum (100) sample abraded using a diamond tip at loads of 1 μN and 9 μN are shown in Fig. 6.14a (note that the sign of the change in surface potential is reversed here from that in ref. [38]). It is evident that both abraded regions show a large potential contrast (≈0.17 V), with respect to the non-abraded area. The black region in the lower-right-hand part of the topography scan shows a step that was created during the polishing phase. There is no potential contrast between the high region and the low region of the sample, indicating that the technique is independent of

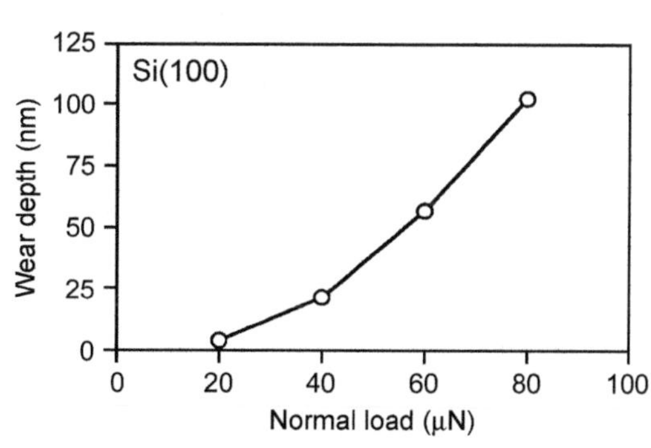

Fig. 6.12. Wear depth as a function of normal load for Si(100) after one cycle [35]

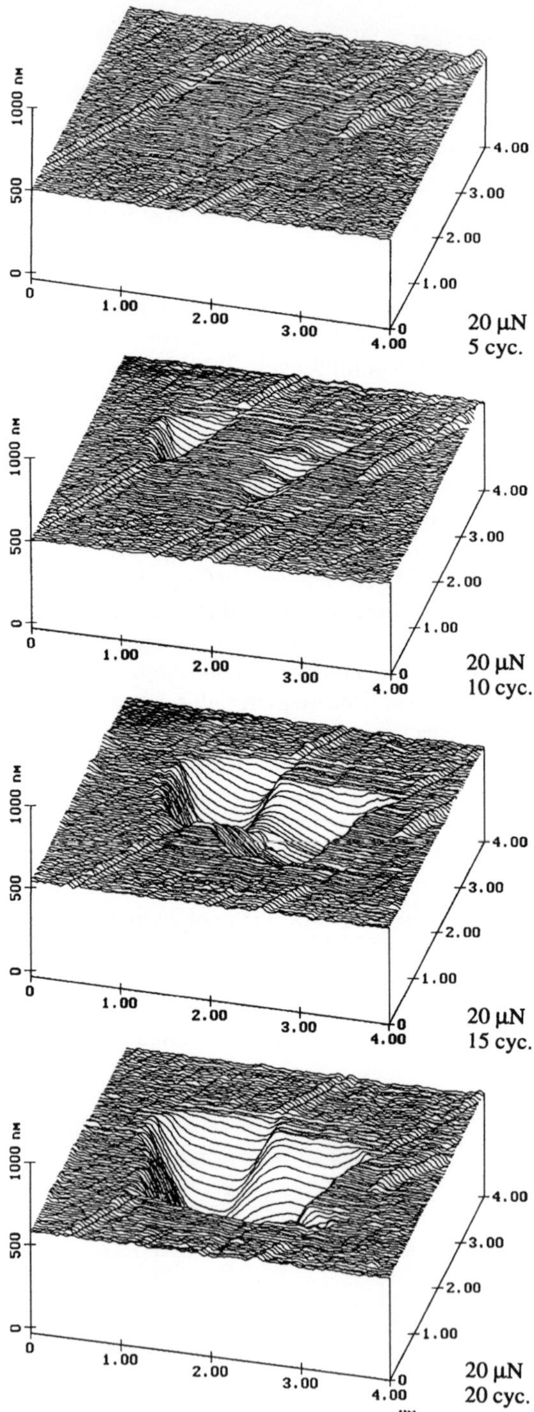

Fig. 6.13. Surface plots of diamond-like carbon-coated thin-film disk showing the worn region; the normal load and number of test cycles are indicated [22]

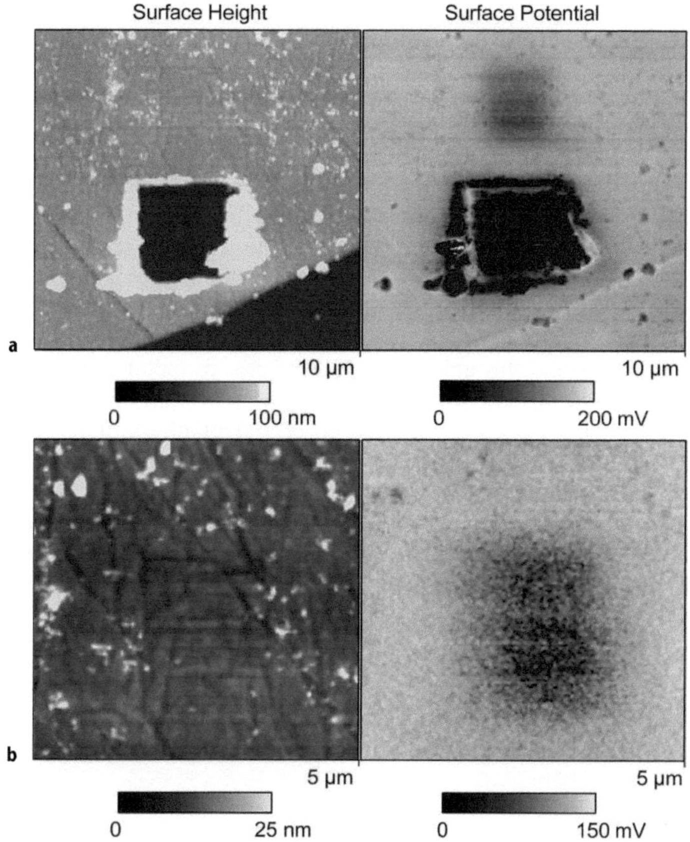

Fig. 6.14. **a** Surface height and change in surface potential maps of wear regions generated at 1 µN (*top*) and 9 µN (*bottom*) on a single-crystal aluminum sample showing bright contrast in the surface potential map on the worn regions. **b** Close-up of upper (low load) wear region [38]

surface height. Figure 6.14b shows a close-up scan of the upper (low load) wear region in Fig. 6.14a. Notice that while there is no detectable change in the surface topography, there is nonetheless, a large change in the potential of the surface in the worn region. Indeed the wear mark of Fig. 6.14b might not be visible at all in the topography map was it not for the noted absence of wear debris generated nearby and then swept off during the low-load scan. Thus, even in the case of zero wear (no measurable deformation of the surface using AFM), there can be a significant change in the surface potential inside the wear mark which is useful for the study of wear precursors. It is believed that the removal of thin contaminant layer including the natural oxide layer gives rise to the initial change in surface potential. The structural changes, which precede generation of wear debris and/or measurable wear scars, occur under ultralow loads in the top few nanometers of the sample, and are primarily responsible for the subsequent changes in surface potential.

6.3.3 In Situ Characterization of Local Deformation

In situ surface characterization of local deformation of materials and thin films can be carried out using an AFM. Failure mechanisms of polymeric thin films have been studied [39, 40]. Figure 6.15 shows topographical images of the three magnetic tapes, with thicknesses ranging from about 7 to 8 µm, and the polyethylene terephthalate (PET) substrate of 6-µm thickness, after being strained to 3.75%, which is well beyond the elastic limit of the substrate. They reported that cracking of the coatings started at about 1% strain for all tapes much before the substrate starts to yield at about 2% strain. Metal-particle (MP) tape develops short and numerous cracks perpendicular to the direction of loading. In tapes with metallic coating, the cracks extend throughout the tape width. In metal evaporated (ME) tape with DLC coating, there is a bulge in the coating around the primary cracks that are initiated when the substrate is still elastic, like crack 'A' in the figure. The white band on the right side of the figure is the bulge of another crack. The secondary cracks like 'B' and 'C' are generated at higher strains and are straighter compared to the primary cracks. In ME tape which has a Co-O film on PET substrate, with a thickness ratio of 0.03, both with and without DLC coating, no difference is observed in the rate of growth between primary and secondary cracks.

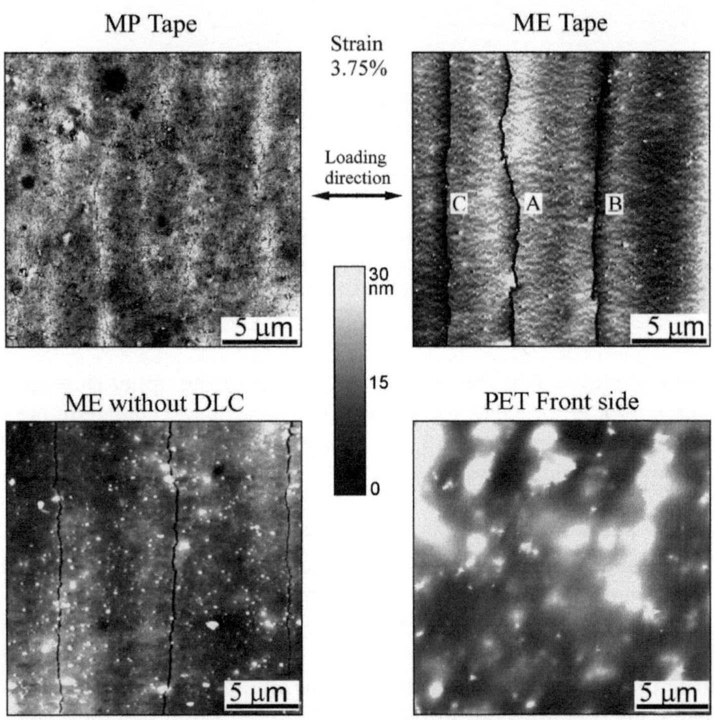

Fig. 6.15. Comparison of the crack morphologies at 3.75% strain in three magnetic tapes and PET substrate. *Cracks B* and *C*, nucleated at higher strains are more linear than *crack A* [40]

The failure is cohesive with no bulging of the coating. This seems to suggest that the DLC coating has residual stresses that relax when the coating cracks, causing delamination. Since the stresses are already relaxed, the secondary crack does not delaminate. The presence of the residual stress is confirmed by the fact that a free-standing ME tape curls up (in a cylindrical form with its axis perpendicular to the tape length) with a radius of curvature of about 6 mm and the ME tape without DLC does not curl. The magnetic coating side of PET substrate is much smoother at smaller scan lengths. However, in 20-µm scans, it has lot of bulging outs, which appear as white spots in the figure. These spots change shape even while scanning the samples in tapping mode at very low contact forces.

In situ surface characterization of unstretched and stretched films has been used to measure Poisson's ratio of polymeric thin films [41]. Uniaxial tension is applied by the tensile stage. Surface height profiles obtained from the AFM images of unstretched and stretched samples are used to monitor the changes in displacements of the polymer films in the longitudinal and lateral direction simultaneously.

6.4 Indentation

Mechanical properties such as hardness and Young's modulus of elasticity can be determined on micro to picoscales using the AFM [21, 22, 42, 43] and a depth-sensing indentation system used in conjunction with an AFM [31, 44–46].

6.4.1 Nanoscale Indentation

To make accurate measurements of hardness at shallow depths, a depth-sensing indentation system is used [31]. Figure 6.16 shows the load-displacement curves at different peak loads for Si(100). Loading/unloading curves often exhibit discontinuities, particularly at high loads. Discontinuities, or referred to as pop-ins, during initial part of the loading part of the curve mark a sharp transition from pure elastic loading to plastic deformation of specimen surface, and thus correspond to an initial yield point. The sharp discontinuities in displacement observed during unloading part of the curves are believed to be due to formation of lateral cracks, which form at the base of the median crack, which results in the surface of the specimen being thrust upward. Load-displacement data at residual depths as low as about 1 nm can be obtained, and the indentation hardness of surface films has been measured for Si(100) [31, 44–46]. The hardness of silicon on the nanoscale is found to be higher than on the microscale, Fig. 6.17. Microhardness has also been reported to be higher than that on the millimeter scale by several investigators. The data reported to date show that hardness exhibits size effect. According to the strain gradient plasticity theory advanced by Fleck et al. [47], large strain gradients inherent in small indentations lead to accumulation of geometrically necessary dislocations that cause enhanced hardening. In addition, the decrease in hardness with an increase in indentation depth can possibly be rationalized on the basis that as the volume of deformed material increases, there is a higher probability of encountering material defects.

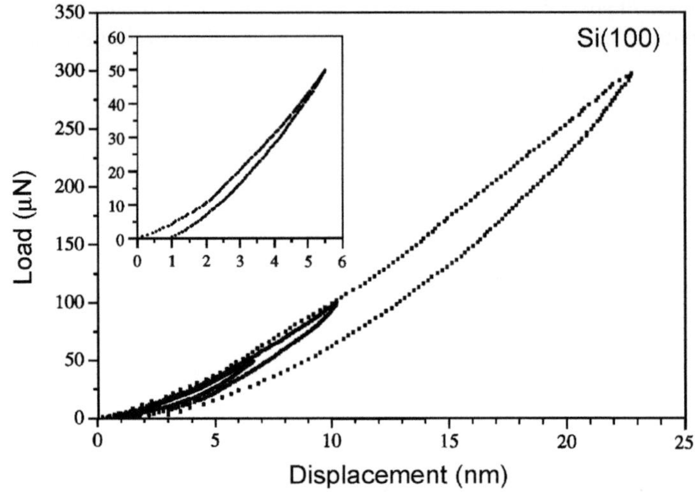

Fig. 6.16. Load-displacement curves at various peak loads for Si(100) [31]

Fig. 6.17. Indentation hardness as a function of residual indentation depth for Si(100) [31]

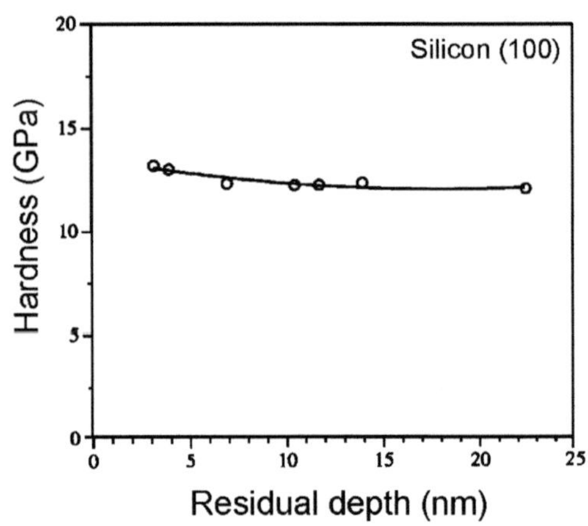

Bhushan and Koinkar [42] have used AFM measurements to show that ion implantation of silicon surfaces increases their hardness and thus their wear resistance. Formation of surface alloy films with improved mechanical properties by ion implantation is of growing technological importance as a means of improving the mechanical properties of materials. Hardness of 20-nm-thick diamond-like carbon films has been measured by Kulkarni and Bhushan [46].

The creep and strain-rate effects (viscoelastic effects) of ceramics can be studied using a depth-sensing indentation system. Bhushan et al. [31] and Kulkarni and Bhushan [44–46] have reported that ceramics exhibit significant plasticity and creep on the nanoscale. Figure 6.18 shows the load-displacement curves for single-crystal silicon at various peak loads held at 180 s. To demonstrate the creep effects, the load-displacement curves for a 500-μN peak load held at 0 and 30 s are also shown

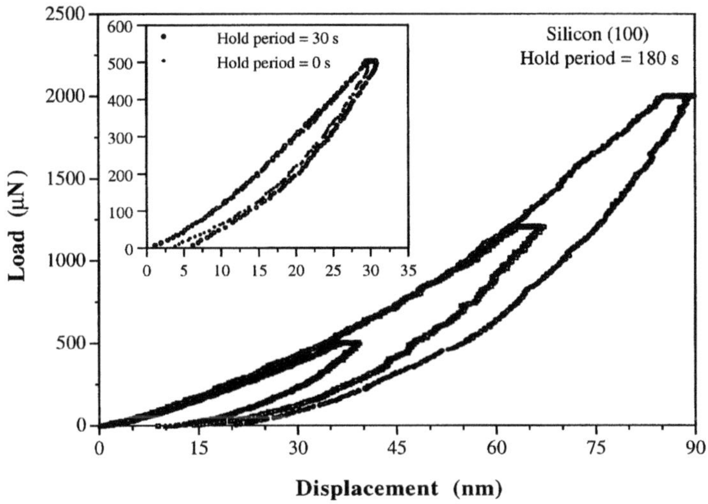

Fig. 6.18. Creep behavior of Si(100) [31]

as an inset. Note that significant creep occurs at room temperature. The mechanism of dislocation glide plasticity is believed to dominate the indentation creep process on the macroscale. The viscoelastic effects observed here for silicon at ambient temperature could arise from size effects mentioned earlier. Most likely, creep and strain rate experiments are being conducted on the hydrated films present on the silicon surface in ambient environment and these films are expected to be viscoelastic.

The Young's modulus of elasticity is calculated from the slope of the indentation curve during unloading.

6.5 Boundary Lubrication

6.5.1 Perfluoropolyether Lubricants

The classical approach to lubrication uses freely supported multimolecular layers of liquid lubricants [8, 14, 48, 49]. The liquid lubricants are sometimes chemically bonded to improve their wear resistance [8, 14, 48]. Partially chemically bonded, molecularly thick perfluoropolyether (PFPE) films are used for lubrication of magnetic storage media [14]. These are considered as potential candidate lubricants for micro/nanoelectromechanical systems (MEMS/NEMS).

For boundary lubrication studies, friction, adhesion, and durability experiments have been performed on virgin Si (100) surfaces and silicon surfaces lubricated with two PFPE lubricants – Z-15 (with $-CF_3$ nonpolar end groups) and Z-DOL (with –OH polar end groups) [50–52]. Z-DOL film was thermally bonded at 150 °C for 30 min and the unbonded fraction was removed by a solvent (BW) [14]. The thickness of Z-15 and Z-DOL films were 2 · 8 nm and 2 · 3 nm, respectively.

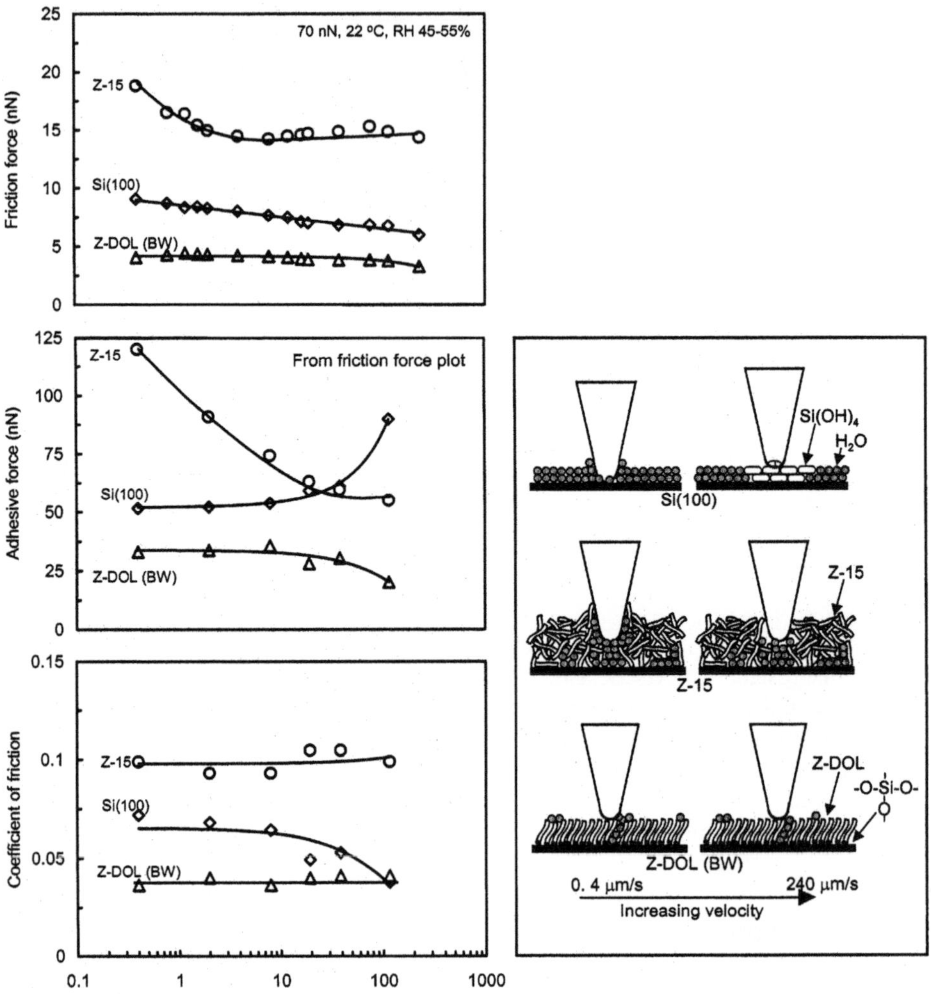

Fig. 6.19. The influence of velocity on the friction force, adhesive force, and coefficient of friction of Si(100) and Z-15 and Z-DOL (BW) films at 70 nN, in ambient air. The schematic (*right*) shows the change of surface composition (by tribochemical reaction) and formation of meniscus while increasing the velocity [52]

To study the velocity effect on friction and adhesion, the variation of friction force, adhesive force, and coefficient of friction of Si(100), Z-15 and Z-DOL(BW) as a function of velocity are summarized in Fig. 6.19. It indicates that, for silicon wafer, the friction force decreases logarithmically with increasing velocity. For Z-15, the friction force decreases with increasing velocity up to 10 μm s^{-1}, after which it remains almost constant. The velocity has a much smaller effect on the friction force of Z-DOL (BW): it reduced slightly only at very high velocity. Figure 6.19 also indicates that the adhesive force of Si(100) is increased when the velocity is higher than 10 µm s^{-1}. The adhesive force of Z-15 is reduced dramatically with a velocity increase up to 20 µm s^{-1}, after which it is reduced slightly. The adhesive force of Z-DOL (BW) is also decreased at high velocity. In the testing range of velocity, only the coefficient of friction of Si(100) decreases with velocity, but the coefficients of friction of Z-15 and Z-DOL (BW) almost remain constant. This implies that the friction mechanisms of Z-15 and Z-DOL (BW) do not change with the variation of velocity.

The mechanisms of the effect of velocity on the adhesion and friction are explained based on schematics shown in Fig. 6.19 (right). For Si(100), tribochemical reaction plays a major role. Although, at high velocity, the meniscus is broken and does not have enough time to rebuild, the contact stresses and high velocity lead to tribochemical reactions of Si(100) wafer which has a native oxide (SiO_2) and Si_3N_4 tip with water molecules and forms $Si(OH)_4$. The $Si(OH)_4$ is removed and continuously replenished during sliding. The $Si(OH)_4$ layer between the tip and Si(100) surface is known to be of low shear strength and causes a decrease in friction force and coefficient of friction [8, 48]. The chemical bonds of Si–OH between the tip and Si(100) surface induce large adhesive force. For Z-15 film, at high velocity the meniscus formed by condensed water and Z-15 molecules is broken and does not have enough time to rebuild; therefore, the adhesive force and consequently friction force is reduced. The friction mechanism for Z-15 film is shearing the same viscous liquid even at high velocity range; thus the coefficient of friction of Z-15 does not change with velocity. For Z-DOL (BW) film, the surface can adsorb few water molecules under ambient conditions, and at high velocity these molecules are displaced which is responsible for a slight decrease in friction force and adhesive force. Koinkar and Bhushan [50, 51] have suggested that in the case of samples with mobile films, such as condensed water and Z-15 films, alignment of liquid molecules (shear thinning) is responsible for the drop in friction force with an increase in scanning velocity. This could be another reason for the decrease in friction force for Si(100) and Z-15 film with velocity in this study.

To study the relative humidity effect on friction and adhesion, the variation of friction force, adhesive force, and coefficient of friction of Si(100), Z-15, and Z-DOL (BW) as a function of relative humidity are shown in Fig. 6.20. It shows that for Si(100) and Z-15 film, the friction force increases with a relative humidity increase up to 45%, and then it shows a slight decrease with further increase in the relative humidity. Z-DOL (BW) has a smaller friction force than Si(100) and Z-15 in the whole testing range. Its friction force shows a relative apparent increase when the relative humidity is higher than 45%. For Si(100), Z-15, and Z-DOL (BW), their adhesive forces increase with relative humidity. Their coefficients of friction increase with relative humidity up to 45%, after which they decrease with further increasing of the relative humidity. It is also observed that the humidity effect on Si(100) really depends on the history of the Si(100) sample. As the surface of Si(100) wafer readily adsorb water in air, without any pre-treatment the Si(100) used in our study almost reaches its saturate stage of adsorbed water, and is responsible for less effect during increasing relative humidity. However, once the Si(100) wafer was thermally treated by baking at 150 °C for 1 h, a bigger effect was observed.

The schematic (right) in Fig. 6.20 shows that Si(100), because its high free surface energy, can adsorb more water molecules while increasing relative humidity.

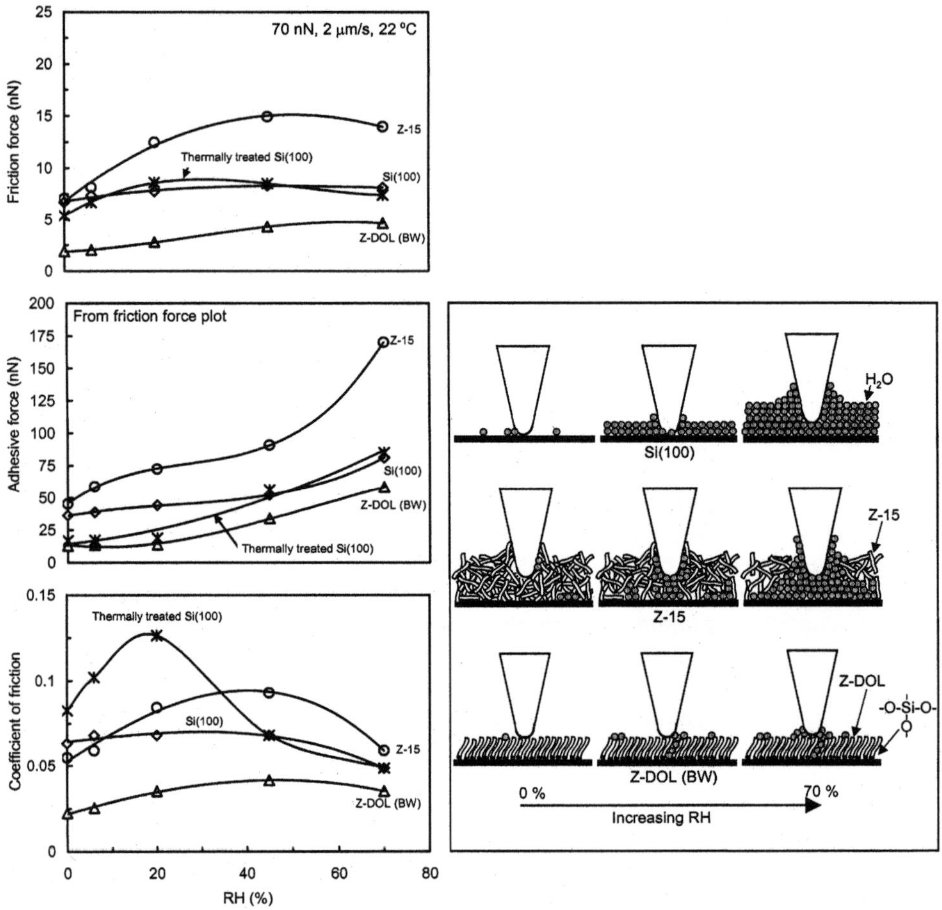

Fig. 6.20. The influence of relative humidity on the friction force, adhesive force, and coefficient of friction of Si(100) and Z-15 and Z-DOL (BW) films at 70 nN, 2 μm s^{-1}, and in 22 °C air. Schematic (*right*) shows the change of meniscus while increasing the relative humidity. In this figure, the thermally treated Si(100) represents the Si(100) wafer that was baked at 150 °C for 1 h in an oven (in order to remove the adsorbed water) just before it was placed in the 0% RH chamber [52]

As discussed earlier, for Z-15 film in the humid environment, the condensed water from the humid environment competes with the lubricant film present on the sample surface, and interaction of the liquid lubricant film with the silicon substrate is weakened and a boundary layer of the liquid lubricant forms puddles. This dewetting allows water molecules to be adsorbed on the Si(100) substrate mixed with Z-15 molecules [50, 51]. Obviously, more water molecules can be adsorbed on Z-15 surface while increasing relative humidity. The more adsorbed water molecules in the case of Si(100), along with lubricant molecules in Z-15 film case, form a bigger water meniscus which leads to an increase of friction force, adhesive force, and coefficient of friction of Si(100) and Z-15 with humidity. But at

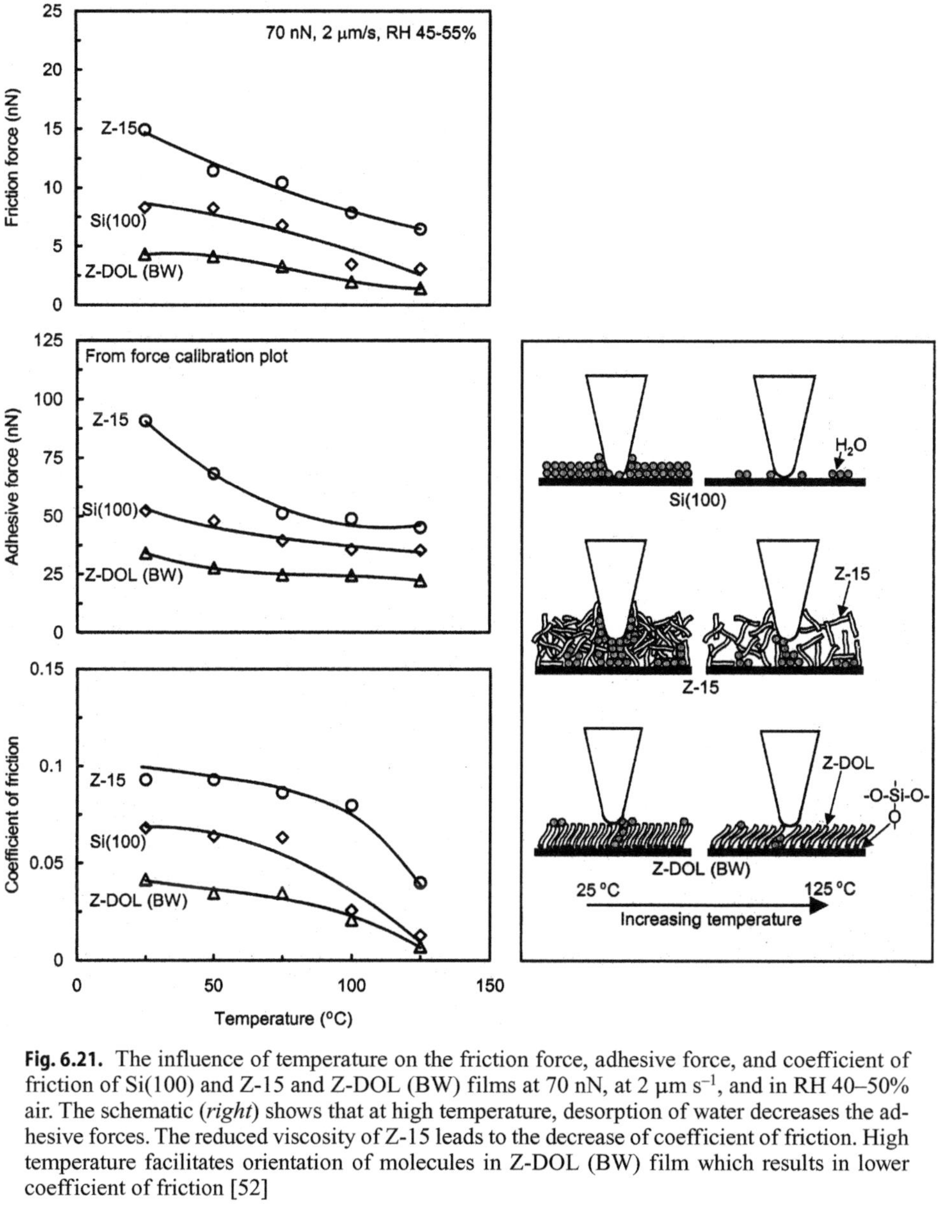

Fig. 6.21. The influence of temperature on the friction force, adhesive force, and coefficient of friction of Si(100) and Z-15 and Z-DOL (BW) films at 70 nN, at 2 μm s^{-1}, and in RH 40–50% air. The schematic (*right*) shows that at high temperature, desorption of water decreases the adhesive forces. The reduced viscosity of Z-15 leads to the decrease of coefficient of friction. High temperature facilitates orientation of molecules in Z-DOL (BW) film which results in lower coefficient of friction [52]

very high humidity of 70%, large quantities of adsorbed water can form a continuous water layer that separate the tip and sample surface, and act as a kind of lubricant, which causes a decrease in the friction force and coefficient of friction. For Z-DOL (BW) film, because of their hydrophobic surface properties, water molecules can be adsorbed and cause an increase in the adhesive force and friction force only at high humidity (RH ≥ 45%).

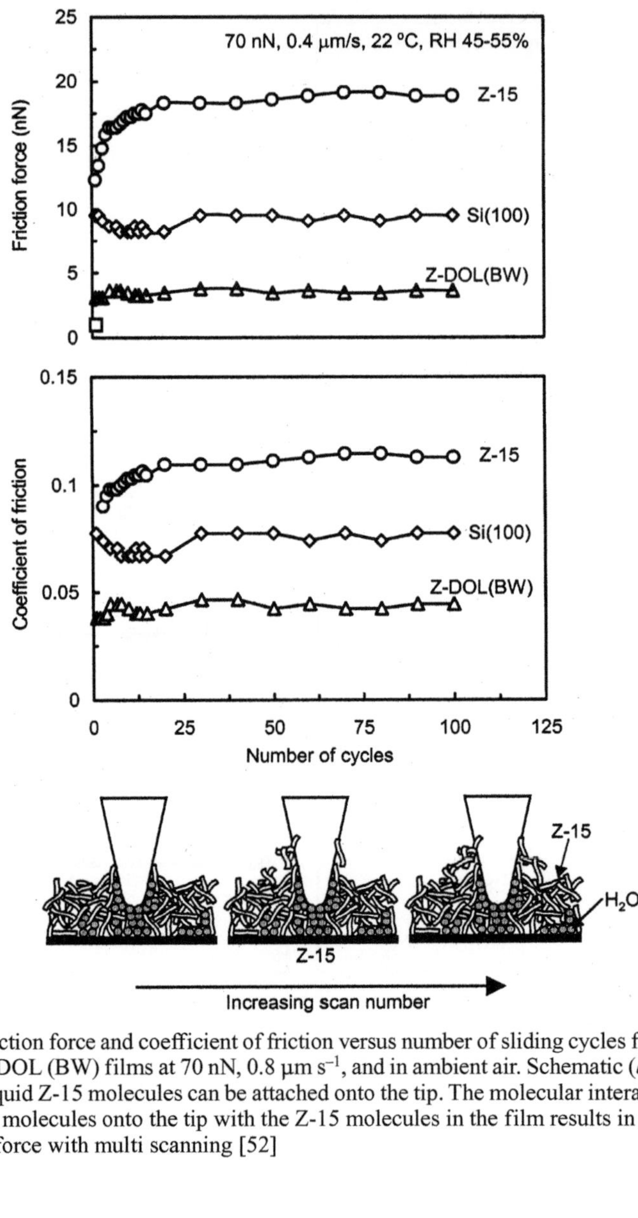

Fig. 6.22. Friction force and coefficient of friction versus number of sliding cycles for Si(100) and Z-15 and Z-DOL (BW) films at 70 nN, 0.8 μm s^{-1}, and in ambient air. Schematic (*bottom*) shows that some liquid Z-15 molecules can be attached onto the tip. The molecular interaction between the attached molecules onto the tip with the Z-15 molecules in the film results in an increase of the friction force with multi scanning [52]

To study the temperature effect on friction and adhesion, the variation of friction force, adhesive force, and coefficient of friction of Si(100), Z-15, and Z-DOL (BW) as a function of temperature are summarized in Fig. 6.21. It shows that the increasing temperature causes a decrease of friction force, adhesive force, and coefficient of friction of Si(100), Z-15, and Z-DOL (BW). The schematic (right) in Fig. 6.21 indicates that at high temperature, desorption of water leads to the decrease of friction force, adhesive forces, and coefficient of friction for all of the samples. For Z-15 film, the reduction of viscosity at high temperature also has a contribution to the decrease of friction force and coefficient of friction. In the case of Z-DOL (BW) film, molecules more easily oriented at high temperature, which may be partly responsible for the low friction force and coefficient of friction.

To study the durability of lubricant films on the nanoscale, the friction of Si(100), Z-15, and Z-DOL (BW) as a function of the number of scanning cycles are shown in Fig. 6.22. As observed earlier, friction force and coefficient of friction of Z-15 is higher than that of Si(100) with the lowest values for Z-DOL(BW). During cycling, friction force and coefficient of friction of Si(100) show a slight decrease during initial few cycles then remain constant. This is related to the removal of the top adsorbed layer. In the case of Z-15 film, the friction force and coefficient of friction show an increase during the initial few cycles and then approach to higher and stable values. This is believed to be caused by the attachment of the Z-15 molecules onto the tip. The molecular interaction between these attached molecules to the tip and molecules on the film surface is responsible for an increase in the friction. But after several scans, this molecular interaction reaches the equilibrium and after that friction force and coefficient of friction remain constant. In the case of Z-DOL (BW) film, the friction force and coefficient of friction start out low and remain low during the entire test for 100 cycles. It suggests that Z-DOL (BW) molecules do not get attached or displaced as readily as Z-15.

6.5.2 Self-Assembled Monolayers

For lubrication of MEMS/NEMS, another effective approach involves the deposition of organized and dense molecular layers of long-chain molecules. Two common methods to produce monolayers and thin films are the Langmuir-Blodgett (L-B) deposition and self-assembled monolayers (SAMs) by chemical grafting of molecules [53]. LB films are bonded to the substrate by weak van der Waals attraction, while SAMs are chemically bonded via covalent bonds to the substrate. Because of the choice of chain length and terminal linking group that SAMs offer, they hold great promise for boundary lubrication of MEMS/NEMS.

A number of studies on the tribological properties of various SAMs have been conducted [53–57]. It has been reported that SAMs with high-compliance long carbon chains exhibit low friction; chain compliance is desirable for low friction. Based on Bhushan and Liu [55], the friction mechanism of SAMs is explained by a so-called molecular spring model (Fig. 6.23). According to this model, the chemically adsorbed self-assembled molecules on a substrate are just like assembled molecular springs anchored to the substrate. An asperity sliding on the surface of

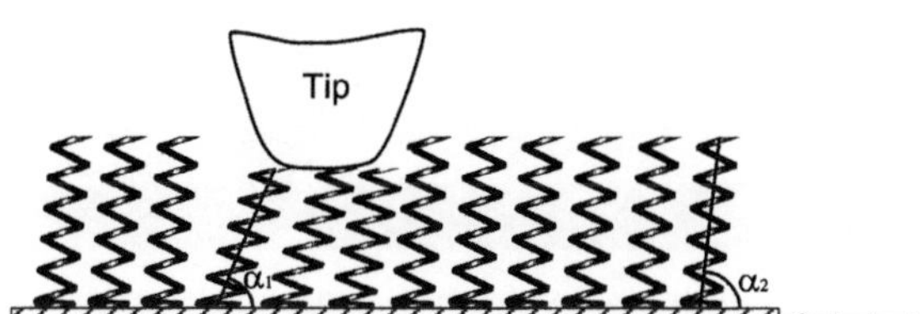

Fig. 6.23. Molecular spring model of SAMs. In this figure, $\alpha_1 < \alpha_2$, which is caused by the further orientation under the normal load applied by an asperity tip [55]

SAMs is like a tip sliding on the top of "molecular springs or brush". The molecular spring assembly has compliant features and can experience orientation and compression under load. The orientation of the molecular springs or brush under normal load reduces the shearing force at the interface, which in turn reduces the friction force. The SAMs with high-compliance long carbon chains also exhibit the best wear resistance [55, 57].

6.5.3 Lubricant Film Thickness Measurements

Liquid film thickness measurement of thin lubricant films (of the order of 10 nm or thicker) with nanometer lateral resolution can be made with the AFM [2, 58]. The lubricant thickness is obtained by measuring the force on the tip as it approaches, contacts, and pushes through the liquid film and ultimately contacts the substrate. The distance between the sharp snap-in (owing to the formation of a liquid meniscus between the film and the tip) at the liquid surface and the hard repulsion at the substrate surface is a measure of the liquid film thickness.

Lubricant film thickness mapping of ultrathin films (of the order of 2 nm) can be obtained using friction force microscopy [50] and adhesive force mapping [29]. Figure 6.24 shows gray-scale plots of the surface topography and friction force obtained simultaneously for unbonded Demnum-type PFPE lubricant film on silicon. The friction force plot shows well-distinguished low-friction and high-friction regions roughly corresponding to high and low regions in surface topography (thick and thin lubricant regions). A uniformly lubricated sample does not show such a variation in the friction. Friction force imaging can thus be used to measure the lubricant uniformity on the sample surface, which cannot be identified by surface topography alone. Figure 6.25 shows the gray-scale plots of the adhesive force dis-

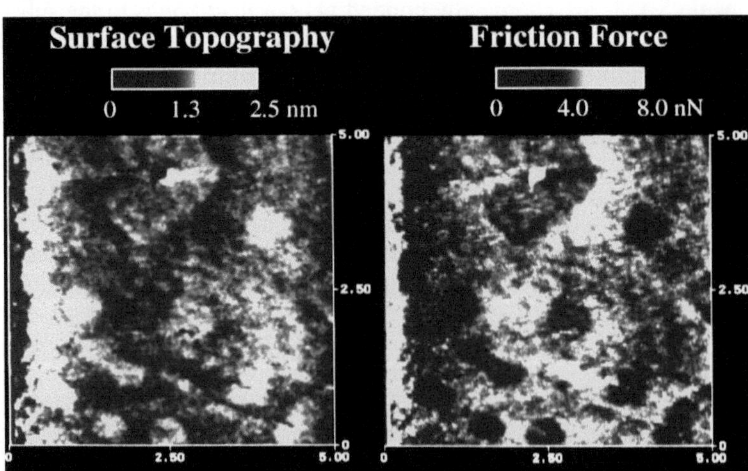

Fig. 6.24. Gray-scale plots of the surface topography and friction force obtained simultaneously for unbonded Demnum type perfluoropolyether lubricant film on silicon [50]

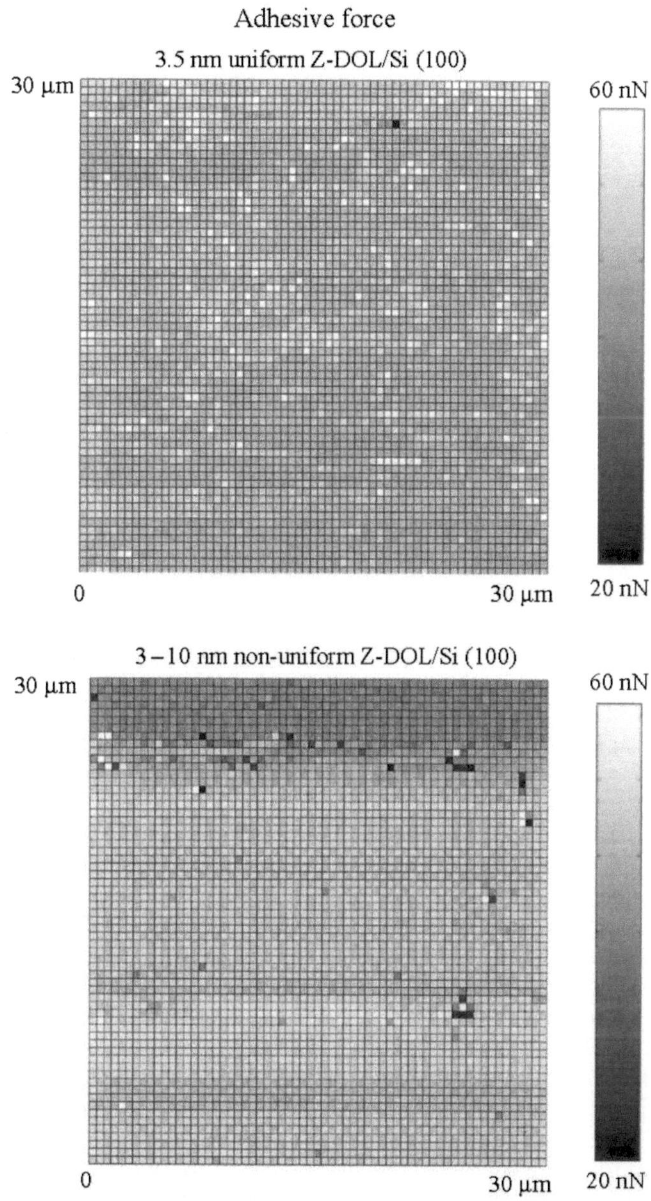

Fig. 6.25. Gray-scale plots of the adhesive force distribution of a uniformly-coated, 3.5-nm-thick unbonded PFPE film on silicon and 3- to 10-nm-thick unbonded PFPE film on silicon that was deliberately coated nonuniformly by vibrating the sample during the coating process [29]

tribution for silicon samples coated uniformly and nonuniformly with Z-DOL-type PFPE lubricant. A region that has adhesive force distinctly different from the other region for the nonuniformly coated sample can be clearly seen. This implies that the liquid film thickness is nonuniform giving rise to a difference in the meniscus forces.

6.6 Conclusions

At most solid-solid interfaces of technological relevance, contact occurs at many asperities. A sharp AFM/FFM tip sliding on a surface simulates just one such contact. However, asperities come in all shapes and sizes. The effect of radius of a single asperity (tip) on the friction/adhesion performance can be studied using tips of different radii. AFM/FFM are used to study various tribological phenomena, which include surface roughness, adhesion, friction, scratching, wear, indentation, detection of material transfer, and boundary lubrication. Measurement of atomic-scale friction of a freshly cleaved highly oriented pyrolytic graphite exhibits the same periodicity as that of the corresponding topography. However, the peaks in friction and those in the corresponding topography are displaced relative to each other. Variations in atomic-scale friction and the observed displacement can be explained by the variation in interatomic forces in the normal and lateral directions. Local variations in microscale friction occur and are found to correspond to the local slopes, suggesting that a ratchet mechanism and collision effects are responsible for this variation. Directionality in the friction is observed on both micro and macroscales, which results from the surface roughness and surface preparation. Anisotropy in surface roughness accentuates this effect. Microscale friction is generally found to be smaller than the macrofriction as there is less plowing contribution in microscale measurements. Microscale friction is load dependent and friction values increase with an increase in the normal load, approaching the macrofriction at contact stresses higher than the hardness of the softer material. The tip radius also has an effect on the adhesion and friction.

The mechanism of material removal on the microscale has also been studied. Wear rate for single-crystal silicon is negligible below 20 µN and is much higher and remains approximately constant at higher loads. Elastic deformation at low loads is responsible for negligible wear. Most of the wear debris is loose. Evolution of wear has also been studied using AFM. Wear is initiated at nanoscratches. For a sliding interface requiring near-zero friction and wear, contact stresses should be below the hardness of the softer material to minimize plastic deformation and surfaces should be free of nanoscratches. Furthermore, wear precursors can be detected at early stages of wear by using surface potential measurements. It is found that even in the case of zero wear (no measurable deformation of the surface using AFM); there can be a significant change in the surface potential inside the wear mark which is useful for study of wear precursors. In situ surface characterization of local deformation of materials is studied to better understand failure mechanisms.

Modified AFM can be used to obtain load-displacement curves and for measurement of nanoindentation hardness and Young's modulus of elasticity, with depth of indentation as low as 1 nm. Hardness of ceramics on nanoscales is found to be higher than that on the microscale. Ceramics exhibit significant plasticity and creep on the nanoscale. Scratching and indentation on nanoscales are powerful ways to screen for adhesion and resistance to deformation of ultrathin films. Detection of material transfer on the nanoscale is possible with AFM.

Boundary lubrication studies and measurement of lubricant film thickness with a lateral resolution on the nanoscale can be conducted using AFM. Chemically bonded lubricant films and self-assembled monolayers are superior in friction and wear resistance.

Investigations of wear, scratching, and indentation on nanoscales using the AFM can provide insights into failure mechanisms of materials. Coefficients of friction, wear rates, and mechanical properties such as hardness have been found to be different on the nanoscale than on the macroscale; generally, coefficients of friction and wear rates on micro and nanoscales are smaller, whereas hardness is greater. Therefore, micro/nanotribological studies may help define the regimes for ultralow friction and near-zero wear. These studies also provide insight into molecular origins of adhesion, friction, and lubrication mechanisms.

References

1. Bhushan B (1997) Micro/nanotribology and its applications, vol E330. Kluwer Academic, Dordrecht, Netherlands
2. Bhushan B (1999) Handbook of micro/nanotribology, 2nd edn. CRC Press, Boca Raton, Florida
3. Bhushan B (1999) Nanoscale tribophysics and tribomechanics. Wear 225–229:465–492
4. Bhushan B (1999) Wear and mechanical characterization on micro- to picoscales using AFM. Int Mat Rev 44:105–117
5. Bhushan B (2001) Modern tribology handbook, vol 1: principles of tribology. CRC Press, Boca Raton, Florida
6. Bhushan B (2001) Fundamentals of tribology and bridging the gap between the macro- and micro/nanoscales. NATO science series II, vol 10. Kluwer Academic, Dordrecht, Netherlands
7. Bhushan B (2001) Nano- to microscale wear and mechanical characterization studies using scanning probe microscopy. Wear 251:1105–1123
8. Bhushan B (2002) Introduction to tribology. Wiley, New York
9. Bhushan B, Israelachvili JN, Landman U (1995) Nanotribology: friction, wear and lubrication at the atomic scale. Nature 374:607–616
10. Guntherodt HJ, Anselmetti D, Meyer E (1995) Forces in scanning probe methods, Vol E286. Kluwer Academic, Dordrecht, Netherlands
11. Persson BNJ, Tosatti E (1996) Physics of sliding friction, vol E311. Kluwer Academic, Dordrecht, Netherlands
12. Singer IL, Pollock HM (1992) Fundamentals of friction: macroscopic and microscopic processes, vol E220. Kluwer Academic, Dordrecht, Netherlands
13. Bhushan B (1998) Tribology issues and opportunities in MEMS. Kluwer Academic, Dordrecht, Netherlands
14. Bhushan B (1996) Tribology and mechanics of magnetic storage devices, 2nd edn. Springer, Berlin Heidelberg New York
15. Mate CM, McClelland GM, Erlandsson R, Chiang S (1987) Atomic-scale friction of a tungsten tip on a graphite surface. Phys Rev Lett 59:1942–1945
16. Ruan J, Bhushan B (1994) Atomic-scale and microscale friction of graphite and diamond using friction force microscopy. J Appl Phys 76:5022–5035
17. Ruan J, Bhushan B (1994) Frictional behavior of highly oriented pyrolytic graphite. J Appl Phys 76:8117–8120

18. Koinkar VN, Bhushan B (1996) Microtribological studies of Al_2O_3-TiC, polycrystalline and single-crystal Mn-Zn Ferrite and SiC head slider materials. Wear 202:110–122
19. Meyer E, Overney R, Luthi R, Brodbeck D, Howald L, Frommer J, Guntherodt HJ, Wolter O, Fujihira M, Takano T, Gotoh Y (1992) Friction force microscopy of mixed Langmuir-Blodgett films. Thin Solid Films 220:132–137
20. Frisbie CD, Rozsnyai LF, Noy A, Wrighton MS, Lieber CM (1994) Functional group imaging by chemical force microscopy. Science 265:2071–2074
21. Bhushan B, Ruan J (1994) Atomic-scale friction measurements using friction force microscopy: part II – application to magnetic media. ASME J Trib 116:389–396
22. Bhushan B, Koinkar VN, Ruan J (1994) Microtribology of magnetic media. Proc Inst Mech Eng Part J J Eng Tribol 208:17–29
23. Bhushan B (1995) Micro/Nanotribology and its applications to magnetic storage devices and MEMS. Tribol Int 28:85–95
24. Koinkar VN, Bhushan B (1997) Effect of scan size and surface roughness on microscale friction measurements. J Appl Phys 81:2472–2479
25. Sundararajan S, Bhushan B (2000) Topography-induced contributions to friction forces measured using an atomic force/friction force microscope. J Appl Phys 88:4825–4831
26. Scherer V, Arnold W, Bhushan B (1998) Active friction control using ultrasonic vibration. In: Bhushan B (ed) Tribology issues and opportunities in MEMS. Kluwer Academic, Dordrecht, Netherlands, pp 463–469
27. Scherer V, Arnold W, Bhushan B (1999) Lateral force microscopy using acoustic friction force microscopy. Surf Interface Anal 27:578–587
28. Bhushan B, Sundararajan S (1998) Micro/nanoscale friction and wear mechanisms of thin films using atomic force and friction force microscopy. Acta Mater 46:3793–3804
29. Bhushan B, Dandavate C (2000) Thin-film friction and adhesion studies using atomic force microscopy. J Appl Phys 87:1201–1210
30. Stifter T, Marti O, Bhushan B (2000) Theoretical investigation of the distance dependence of capillary and van der Waals forces in scanning force microscopy. Phys Rev B 62:13667–13673
31. Bhushan B, Kulkarni AV, Bonin W, Wyrobek JT (1996) Nano/picoindentation measurement using a capacitance transducer system in atomic force microscopy. Philos Mag 74:1117–1128
32. Bhushan B, Kulkarni AV (1996) Effect of normal load on microscale friction measurements. Thin Solid Films 278:49–56; 293:333
33. Sundararajan S, Bhushan B (2001) Development of a continuous microscratch technique in an atomic force microscope and its application to study scratch resistance of ultra-thin hard amorphous carbon coatings. J Mater Res 16:75–84
34. Koinkar VN, Bhushan B (1997) Scanning and transmission electron microscopies of single-crystal silicon microworn/machined using atomic force microscopy. J Mater Res 12:3219–3224
35. Zhao X, Bhushan B (1998) Material removal mechanism of single-crystal silicon on nanoscale and at ultralow loads. Wear 223:66–78
36. Bhushan B, Goldade AV (2000) Measurements and analysis of surface potential change during wear of single crystal silicon (100) at ultralow loads using Kelvin probe microscopy. Appl Surf Sci 157:373–381
37. Bhushan B, Goldade AV (2000) Kelvin probe microscopy measurements of surface potential change under wear at low loads. Wear 244:104–117
38. DeVecchio D, Bhushan B (1998) Use of a nanoscale Kelvin probe for detecting wear precursors. Rev Sci Instrum 69:3618–3624
39. Bobji MS, Bhushan B (2001) Atomic force microscopic study of the micro-cracking of magnetic thin films under tension. Scripta Mater 44:39–42
40. Bobji MS, Bhushan B (2001) In-situ microscopic surface characterization studies of polymeric thin films during tensile deformation using atomic force microscopy. J Mater Res 16:844–855
41. Bhushan B, Mokashi PS, Ma T (2003) A new technique to measure Poisson's ratio of ultra-thin polymeric films using atomic force microscopy. Rev Sci Instrum 74:1043–1047

42. Bhushan B, Koinkar VN (1994) Tribological studies of silicon for magnetic recording applications. J Appl Phys 75:5741–5746
43. Bhushan B, Koinkar VN (1994) Nanoindentation hardness measurements using atomic force microscopy. Appl Phys Lett 64:1653–1655
44. Kulkarni AV, Bhushan B (1996) Nanoscale mechanical property measurements using modified atomic force microscopy. Thin Solid Films 290–291:206–210
45. Kulkarni AV, Bhushan B (1996) Nano/picoindentation measurements on single-crystal aluminum using modified atomic force microscopy. Mater Lett 29:221–227
46. Kulkarni AV, Bhushan B (1997) Nanoindentation measurement of amorphous carbon coatings. J Mater Res 12:2707–2714
47. Fleck NA, Muller GM, Ashby MF, Hutchinson JW (1994) Strain gradient plasticity: theory and experiment. Acta Metall Mater 42:475–487
48. Bhushan B (1999) Principles and applications of tribology. Wiley, New York
49. Bowden FP, Tabor D (1950) The friction and lubrication of solids, part 1. Clarendon Press, Oxford, UK
50. Koinkar VN, Bhushan B (1996) Micro/nanoscale studies of boundary layers of liquid lubricants for magnetic disks. J Appl Phys 79:8071–8075
51. Koinkar VN, Bhushan B (1996) Microtribological studies of unlubricated and lubricated surfaces using atomic force/friction force microscopy. J Vac Sci Technol A 14:2378–2391
52. Liu H, Bhushan B (2003) Nanotribological characterizations of molecularly-thick lubricant films for applications to MEMS/NEMS. Ultramicroscopy 97:321–340
53. Bhushan B (2001) Self-assembled monolayers for controlling hydrophobicity and/or friction and wear. In: Bhushan B (ed) Modern tribology handbook, vol 2. Materials, coatings, and industrial applications. CRC Press, Boca Raton, Florida, pp 909–929
54. Bhushan G, Kulkarni AV, Koinkar VN, Boehm M, Odoni L, Martelet C, Belin M (1995) Microtribological characterization of self-assembled and Langmuir-Blodgett monolayers by atomic and trictin force microscopy. Langmuir 11:3189–3198
55. Bhushan B, Liu H (2001) Nanotribological properties and mechanisms of alkylthiol and biphenyl thiol self-assembled monolayers studied by AFM, Phys Rev B 63:245412–1 to –11
56. Liu H, Bhushan B (2001) Investigation of the adhesion, friction, and wear properties of biphenyl thiol self-assembled monolayers by atomic force microscopy. J Vac Sci Technol A 19:1234–1240
57. Liu H, Bhushan B (2002) Investigation of nanotribological properties of self-assembled monolayers with alkyl and biphenyl spacer chains. Ultramicroscopy 91:185–202
58. Bhushan B, Blackman GS (1991) Atomic force microscopy of magnetic rigid disks and sliders and its applications to tribology. ASME J Tribol 113:452–458

7 Visualization of Polymer Structures with Atomic Force Microscopy

Sergei Magonov

7.1 Introduction

In last two decades, microscopic characterization of materials has significantly advanced with the inventions of scanning tunneling microscopy (STM) and atomic force microscopy (AFM) [1, 2]. In STM, tunneling current between a sharp metallic probe placed in close proximity to a conducting surface is used as a probing interaction. Tunneling current in the range of nanoamperes originates when a bias voltage is applied between this probe and the conducting sample. Rastering of the probe is performed over the surface at separations small enough for stable detection of tunneling current between these two electrodes. This is realized with high accuracy using a three-dimensional piezoceramic actuator. In scanning tunneling microscopes, a feedback mechanism keeps the tip-sample current constant in every surface location by adjusting the vertical tip-sample separation. Changes in the applied voltage to the piezoactuator, which are needed to adjust the tip-sample separation, are presented in the height image. This image, to a first approximation, reproduces surface topography of the sample. Atomic-scale resolution, routinely achieved in STM, has made it an invaluable addition to the family of microscopic techniques. Another remarkable feature of such microscopes is their ability to examine samples not only in an ultrahigh (UHV) vacuum but also at ambient conditions and even under liquids. At present, STM has become a mature technique that is widely applied to the visualization of atomic structures and atomic-scale processes on different substrates, especially in UHV conditions.

A major limitation of STM applications to various materials is the need for sample conductivity. There have been successful but limited attempts to broaden STM capabilities to studies of non-conducting samples. This capability was demonstrated by imaging of organic layers lying immediately on conducting substrates. STM images of a number of low molecular weight liquid crystals, normal alkanes and alkylthiols on different substrates (graphite, Au, Pt, etc) show molecular-scale patterns in which individual atoms, chemical groups, and conformations of organic molecules can be recognized (Fig. 7.1a) [3]. This happens due to the admixture of electron contributions of atoms belonging to a non-conducting layer to a partial electron density map dominated by contributions of atoms of the conducting substrate. Although such STM images provide structural information about a limited number of organic layers, these results can be very useful for understanding structural motifs of their epitaxy on different substrates.

Fig. 7.1. **a** STM image of normal alkane $C_{36}H_{74}$ on graphite, **b** AFM contact mode image of 2,4-hexadienylene-bis(*p*-fluorobenzenesulfonate) crystal

A search for a more universal tip-sample interaction that could be applied to surface imaging led to the invention of atomic force microscopy (AFM) [2]. In this technique, the repulsive force between a sharp probe fixed on a sensitive cantilever and the sample is used for imaging. As in the case of STM, the piezoelectric actuator adjusts the probe-sample distance to keep the tip-sample force constant during scanning, and the height image shows sample topography. Tunneling current and repulsive force strongly depend on the tip-sample separation. This explains extremely high-resolution imaging achieved in STM and AFM, which are the main scanning probe techniques. In the AFM contact mode, the cantilever's bending caused by tip-sample force is measured with the deflection of a laser beam from the cantilever's backside. In this mode, the probe maintains permanent contact with the sample during scanning and it exercises normal and lateral force interactions. A feedback control applied during imaging ensures that the probe scans over the surface keeping the interaction force at the constant level. Atomic-scale resolution is routinely achieved in STM on various samples, yet it is a more delicate issue for AFM. In many cases, atomic-scale patterns that are obtained in AFM contact mode on flat faces of crystalline compounds are examples of lattice imaging. A regular atomic-scale pattern, which was recorded on the *bc* place of a monomer crystal of 2,4-hexadienylenebis(*p*-fluorobenzenesulfonate) in shown in Fig. 7.1b. The pattern is characterized by the repeat distances corresponding to the crystallographic unit cell and shows a small surface reconstruction [3]. The normal and lateral force sensing by the AFM probe is periodically changing in registry with the surface periodic lattice when the probe apex slides on the crystalline sample [4]. In this case, a tip-sample contact area might be larger than atomic size, and, therefore, single atomic-scale defects cannot be detected.

An introduction of commercial Si_3N_4 and Si probes was the key event leading to the broad use of AFM. First applications, which were performed in the contact mode, were very informative. In addition to high-resolution imaging of surfaces, AFM has been effectively used for evaluation of sample mechanical properties by modulating the tip-sample force interaction. It has been also recognized that by recording lateral tip-sample force interactions this technique can be applied to friction measurements. Unfortunately, lateral forces applied to soft samples might induce a strong shearing deformation and sample damage. This appears to be the case even when soft cantilevers are applied in contact mode in attempts to reduce the tip-sample force interactions. Further force minimization has been achieved by performing measurements under liquid [5]. In ambient-condition studies, a capillary force tends to attract the probe towards the sample thus limiting the force control. By immersing the sample and the probe into a liquid one can eliminate this effect and minimize the force below the 1-nN level. Yet for many polymer samples, imaging under liquid should be considered only as an optional, but not routine operation. All these issues have limited contact-mode AFM studies of polymer materials.

This situation changed drastically with the appearance of AFM oscillatory mode known as tapping mode [6]. In this mode, the probe oscillates vertically and, during imaging, it contacts the sample surface intermittently preventing shearing of a sample surface. The damping of the cantilever amplitude to a pre-selected level is used for feedback control while imaging in tapping mode. Such an operation is gentler than in contact mode, despite the fact that stiffer probes are used in tapping mode. Stiffer probes are needed to overcome the capillary force in each cycle of the probe oscillation. In tapping mode measurements (at ambient conditions or under liquid) the image resolution is defined by the smallest tip-sample contact area (1–3 nm) that can be achieved by adjusting the tip force. High-resolution imaging in tapping mode is demonstrated in Fig. 7.2a–d. In oscillatory modes, atomic-scale resolution can be obtained only in ultrahigh vacuum [7], where the contamination surface liquid layer is completely eliminated and the tip-sample force interactions can be controlled at a level that minimizes tip-sample contact to the atomic-size area. Under these conditions, true atomic-scale resolution has been achieved on a number of samples. Despite some limitations of image resolution, there is no doubt that the introduction of tapping mode has revolutionized AFM applications because a broad range of samples and materials has become accessible for ambient studies.

Further progress in AFM is related to a broad use of phase imaging in tapping mode. Variations in the phase of the oscillating probe while interacting with the sample are related to mechanical and adhesion properties of a particular location. The study of these variations has made phase imaging very important for compositional analysis of polymeric materials. Though tapping mode imaging has become the most frequently applied AFM mode, one should not underestimate the potential of other modes. In many cases the combined use of several AFM modes is quit informative.

Development of AFM imaging techniques would not have been possible without advances in local probing of various sample properties (mechanical, adhesive, electric, thermal, etc.). There are several techniques that address such measurements. Mechanical properties of different samples, including polymer materials can be studied by AFM nanoindentation [8]. Electrical properties of polymer materials are examined with electric force and surface potential microscopy. A combination of thermal and optical detection with AFM is also realized. Though qualitative differences of the above properties are useful for compositional mapping, quantitative local measurements have been a challenge since the appearance of AFM. Among the problems to resolve are preparation of specialized probes, op-

Fig. 7.2. a,b Tapping mode images of normal alkanes $C_{36}H_{74}$ and $C_{60}H_{122}$ on graphite. *Insert* in *a* shows 2D-FFT power spectra of the image in *a*. **c,d** Tapping mode images of single macromolecules of 5th generation liquid crystalline carbosilane dendrimer on Si wafer. *Insert* between *c* and *d* shows chemical structure of these macromolecules

timization of instrumentation for particular measurements, and a lack of appropriate theoretical analysis of microscopic tip-sample force interactions.

Initially introduced as an accessory to STM, AFM has become the leading scanning probe technique primarily due to its applicability to a broad range of samples and due to its penetration into different industries. Solving problems of real world samples is a proven AFM capability, which is now well recognized in semiconductor, data storage, plastic and rubber industries, and in biotechnology. Measurements of surface roughness of Si wafers, imaging of sub-micron technological profiles, such as those common for CD and DVD, and compositional mapping of multicomponent polymer materials are only few of the important industrial applications successfully realized with AFM.

In general, AFM of polymer materials is a broad field, which includes, besides imaging, local studies of mechanical, electric, magnetic, thermal, and optical properties. In this chapter, the main emphasize will be on visualization of polymer structures. The chapter consists of six sections. The most essential experimental aspects of AFM applications to polymers will be discussed in second section. This section contains a description of operational and imaging modes, procedures for tip force optimization, peculiarities of imaging in different environments and at different temperatures, and various sample preparation techniques. All these features are indispensable constituents of successful AFM analysis of polymer samples. The next three sections then include practical examples of AFM imaging of individual macromolecules, morphology studies of semicrystalline polymer samples, and compositional mapping of heterogeneous polymer materials. Future developments in AFM are discussed in the final section.

The third section deals with visualization of single macromolecules on various substrates and self-assembled structures in the first layers adsorbed on substrate. This area is "hot" because of a recently increased interest in nanometer-scale science and technology. Synthesis of new polymers with well-controlled molecular weight and chain branching and well-defined end group functionality, and the development of technological procedures for nanoscale patterning of substrates are only a few of the efforts leading to a build-up of a variety of functional molecular assemblies. AFM is the most suitable tool for observation of single macromolecules on selected surfaces and is widely used for examination of the molecular weight distribution of polymers, polymer conformation, and their thermal motion. Different aspects of imaging single macromolecules and their self-assembly will be mentioned below. It will also be demonstrated that AFM is developing into a versatile manipulator of nanoscale objects.

Studies of semicrystalline polymers are presented in the 4th section. Such polymers are characterized by crystallinity, size of lamellae and lamellar architecture at micron and sub-micron scales, as well as by the type of morphology at larger scales (tens and hundreds of microns). A key issue is the high-resolution visualization of polymer lamellae, which is the basic construction element of semicrystalline morphology with dimensions in the 5–20 nm range. Lamellae are recognized by their specific shape, which is enhanced by the image contrast differentiating sample locations with different density. There are strong indications that AFM senses smaller variations in the polymer density than those detectable by small-angle X-ray scattering (SAXS). Another important question concerning AFM studies of semicrystalline polymers deals with the similarities and differences between surface and bulk morphology. Sample etching and microtomy, which allow AFM observation of bulk morphology, are helpful in this respect. New capabilities in studies of semicrystalline polymers are opened with the introduction of the high-temperature AFM accessory. It can be used to monitor structural changes of crystalline polymers associated with melting/crystallization and glass transitions in situ.

Compositional mapping of heterogeneous polymer samples is described in the 5th section. Copolymers, polymer blends, paints, rubber materials, and composite materials are heterogeneous polymer materials, which include two or more components with different chemical structures. Such materials represent the most important type of industrial samples analyzed with AFM. Visualization of different components of the samples is based on recognition of individual components by specific shape/morphology and by image contrast related to differences of the components' properties (mechanical, adhesive, electrical, etc.). Compositional imaging with AFM does not require selective chemical staining common for such analysis with transmission electron microscopy (TEM). It is worth noting that the contrast of AFM images related to different components is not completely understood due to complexity of the tip-sample interactions, especially, in case of polymers – materials with non-linear viscoelasticity. Therefore, in many cases a researcher should work out a general recipe for compositional mapping of a particular family of materials. To realize this in practice, a researcher might find it useful to examine model samples prepared by varying a composition of a particular blend or multicomponent material. The helpful recipe should indicate the best-suited AFM mode and experimental conditions to get the most pronounced image contrast of different components. As long as distribution of components can be mapped, it does not matter what imaging mode and what imaging conditions are. Two examples of such recipes will be discussed below.

In the final section, further developments of AFM and its applications to polymers will be discussed.

7.2 Basic Aspects of AFM Imaging of Polymers

7.2.1 Instrumentation and Operational Modes

The AFM images presented in this chapter were obtained with a commercial scanning probe microscope Nanoscope IIIa MultiMode (Digital Instruments, Santa Barbara, CA). The microscopes from this manufacturer are the most common in research laboratories worldwide, and they give access to a wide spectrum of operation modes developed in recent years. The contact mode and tapping mode are the main imaging modes. Piezoelectric scanners, which are the most common actuators in scanning probe microscopes, allow lateral imaging on areas up to 200 micron×200 microns and vertical profiling up to 10 microns. An optical microscope combined with AFM stage is useful for a navigation of the AFM probe to a sample location of interest. The measurements can be conducted at ambient conditions and also under liquids and at elevated temperatures (up to 250 °C). All these conditions are useful for studies of polymers. Digital Instruments' Dimension scanning probe microscope equipped with a new controller – Nanoscope IV, a close-loop head and NanoMan software was applied for AFM experiments, which include a tip-assisted manipulation of single macromolecules.

Microfabricated probes made of Si_3N_4 or etched from Si wafers are most common for AFM measurements. The probes consist of a rectangular or triangular cantilever with a sharp pyramidal tip at the end. The tip apex is sharper in Si probes where it is often in the range 10–20 nm. Various thin coatings are applied to the probes in order to increase optical reflectivity and electric conductivity or provide magnetic sensitivity. Unfortunately, the coating can make the tip apex less sharp. For studies of polymeric samples, a broad range of the probes can be used. This is related to the fact that stiffness of polymeric materials varies in a broad range and matching of the probe stiffness to that of a polymer sample or of different components in it might be quite useful for imaging purposes. The probe choice also depends on a particular type of operation mode and environmental conditions. For imaging in air, Si probes with stiffness of 3–5 N m^{-1} are most useful, especially when high-resolution and low-force imaging is required. For compositional imaging, which is commonly conducted at elevated tip forces, Si probes with stiffness 30–40 N m^{-1} will be a good choice. Imaging under liquids can be done with soft Si probes (0.3–1 N m^{-1}). Commercial Si probes from 4–5 main manufacturers have quite comparable characteristics.

At the beginning of the discussion of AFM operational modes, local measurements using force curves will be shortly introduced. Force curves, which are available in main imaging modes (contact mode and tapping mode), are graphs reflecting the cantilever deflection changes versus z-travel of the scanner (DvZ) in the contact mode and the cantilever amplitude or phase changes versus z-travel of the scanner (AvZ and PvZ) in the tapping mode. These graphs are used for adjustment of the tip-sample force interactions, for calibration of the cantilever's sensitivity, and also for local mechanical measurements. In such measurements, which are typically performed at frequencies between 0.1 Hz and 100 kHz, one can examine sample stiffness, adhesion, or effects of long-range forces such as electrostatic interactions. By varying the z-excursion the probe is either brought into contact with a sample and induces its deformation or it is retracted from the sample thus exercising tip-sample adhesion and long-range forces. At high loads, the sample can be deformed plastically with a formation of local pits, whose dimensions reflect the sample mechanical properties. Such measurements are the subject of AFM-related nanoindentation, which is often applied to polymers [8]. When a bit of a sample is stuck either physically or chemically to the probe, then the probe retraction from the surface causes stretching of this material. In some cases, the force curves reflect fracture events during this stretching, which are assigned to rupture of individual bonds [9]. For compositional mapping, families of the force curves obtained at multiple locations on the surface of heterogeneous materials can be presented as images in the mode known as force volume [10]. Unfortunately, such measurements consume much more time than imaging in other modes that can be applied for the same purpose.

The contact mode, which was briefly described in the introduction, remains an important AFM mode in many applications. The results of the contact mode measurements are typically presented by height and deflection images. The height image presents variations of a vertical position of the scanner needed to keep the tip-sample force constant during imaging. In other words, this image reproduces the sample surface topography under the assumption that tip-sample force is weak enough not to cause a sample deformation. The deflection (or error signal) image, which reflects instantaneous changes of the cantilever deflection when the probe is brought to new surface location during scanning, resembles a map of local derivatives of height profile lines. This image distinctively shows edges of surface features, which are barely seen in the height image. Therefore, the deflection image helps to recognize surface steps, crystal boundaries, fibers, and other fine structures. Typically, both images are arrays of data, which are collected in a line-by-line fashion in 512×512 surface locations. Grey-scale or color-scale images are used to reflect lateral variations of height or other data recorded during imaging.

The contact mode should be applied to soft materials such as polymers with a special care even when there is no visible sample damage. The level of the tip force applied to the sample in this mode is measured via the cantilever bending or set-point deflection. At the beginning of imaging, a probe is pre-loaded on a sample to an operator-chosen set-point level, and this might cause an undesirable sample deformation. As a result, height image will not represent surface topography correctly. To prevent this, one should perform imaging at different set-point deflection levels to be sure that the change of the tip force does not induce height image variations. When this is the case then one might perform lateral force imaging and force modulation experiments, which can provide additional information about a particular sample.

In lateral force operation, the height image is accompanied by the lateral force image, which shows changes of the lateral photodetector signal at different sample locations, primarily due to variations in friction. The contrast of such an image can be related to presence of different components of a heterogeneous sample. Yet the interpretation of the lateral force images can be non-trivial due to "contaminations" caused by topography, tip-force level, and adhesion. There are a number of approaches to get valuable information about sample friction from lateral force measurements.

Force modulation imaging was introduced in the contact mode for identification of hard and soft locations in composite samples [11]. In addition to rastering the probe at the constant deflection level, the probe or the sample are vibrated vertically at the resonance frequency of the Z-actuator. When the tip contacts a hard location the cantilever AC amplitude becomes large. Thus the contrast of the amplitude image correlates with the local stiffness. Such an image is recorded simultaneously with the height image, which represents the sample topography. A pre-loading of the probe on the sample limits applications of force modulation to polymers. To overcome this limitation, the analog of the force modulation has been developed to use with tapping mode.

In the tapping-mode AFM, the probe oscillation is induced at its resonance frequency (50–300 KHz) and the change in amplitude, which is caused by the tip-sample force interactions, is applied to feedback in order to track the surface topography. In general, the height image reflects sample topography. The amplitude image, which is often recorded together with height image, is the analog of the deflection (error) image of contact mode and it emphasizes fine surface features. The phase changes in tapping mode are presented in the phase image, which is generally recorded simultaneously with the height image. The experimental results accumulated during last 5 years show that phase images are extremely useful in AFM analysis of polymers. First, in studies of topographic features, phase images (similar to amplitude images) complement height images in contrasting their minor structural details. Second, high sensitivity of phase contrast to material properties allows compositional mapping of multicomponent polymer systems. Therefore, for many samples, height image, which shows topographic features, is quite different from the phase image, which indicates the different components.

There are a couple of useful extensions of the tapping mode based on the two-pass operation. In EFM and magnetic force microscopy (MFM), a probe with electric or ferromagnetic coating is used to study surface electric or magnetic properties in the following way. For each scan, the height profile is recorded in the first run and, in the second run, the probe is moved along this height profile while being lifted 5–50 nm over the surface. During the second pass, the changes of the probe resonant frequency and phase, which are caused by long-distance forces between the probe and sample locations with different electric or magnetic properties, are registered. The height and frequency (phase) images based on the data obtained in this procedure represent surface topography and a map of electric or magnetic properties, respectively.

Force modulation measurements, which were initially performed in contact mode, can be also conducted in tapping mode. To do so, one might use the 2-pass method: the first topographic pass is accompanied by the second pass, in which the probe is engaged into a sample (negative lift). During the second pass, the probe oscillates at the resonance frequency of a piezoactuator (typically, 8–10 kHz) and the cantilever deflections, which are larger on stiff sample locations and smaller on soft sample locations, reflect mechanical properties of the sample. Force modulation in tapping mode is more applicable to polymeric materials compared to one in contact mode. Nevertheless it is more destructive than imaging in conventional tapping mode.

7.2.2
Sample Preparation

In AFM studies of polymeric samples, researchers might pursue different goals, for example, examination of surface roughness of photoresist, visualization of surface structures of polymer Langmuir-Blodgett films, or examination of carbon black distribution in rubber samples. Depending on a particular application, polymeric samples can be studied either directly without any sample preparation or it is the only way to observe structures of interest.

AFM studies of thin and ultrathin (thickness below 100 nm) polymer films are straightforward for samples prepared by spin casting of their solution on a flat substrate or by dipping the substrate into solution. Among commonly used substrates are polycrystalline silicon, glass, gold films with large crystal terraces, and atomically flat surfaces of mica, graphite or inorganic layered materials. When polymer concentration in solution approaches 10^{-5} wt%, dispersions of nm-scale particles and individual macromolecules can be obtained. Increase of the polymer concentration leads to films of different thicknesses and interplay between macromolecules' interactions with a substrate and intermolecular interactions within the films can be studied by visualization conformations and packing of macromolecules in such samples.

Surfaces of polymer coatings, paints, films, sheets, blocks, or pellets produced in this way can be studied without any preparation if size of surface corrugations does not exceed 5–6 microns, the typical vertical range of scanners in commercial microscopes. A typical example of the surface of a commercial polymer film is shown in Fig. 7.3a. When a polymer layer to be examined is on a curved surface (fiber, cylinder, etc.), then only a top part of such a sample is accessible for AFM imaging. A special accessory for tilting or rotating the sample is needed to expose different surface areas to the study.

In many cases, the structures of interest are in the bulk of the polymer. For example, the impact strength of a polymer blend is determined mainly by the bulk microstructure, not the surface structure. Usually an ultramicrotome should be used to section the sample and the image in Fig. 7.3b shows a cross-section of multilayer

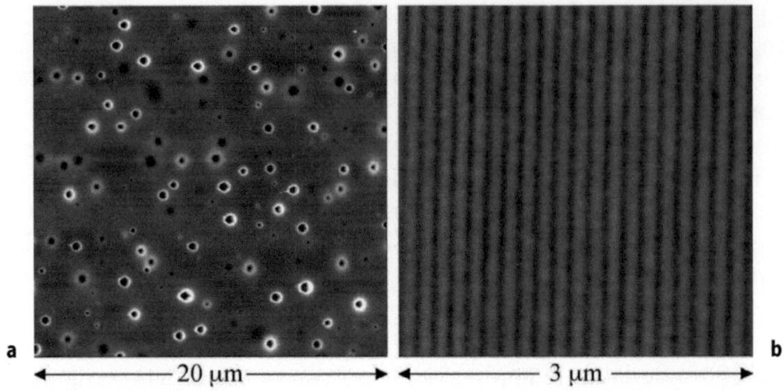

Fig. 7.3a,b. Height images of Nafion membrane and cross-section of multilayer films consisting of alternative layers of PET and PMMA

polymer film, which was obtained on the microtomed surface. Such a procedure needs a skilled operator yet the preparation of the flat faced-off block, which is used for AFM studies, is much easier than making ultrathin polymer sections for TEM analysis. A use of a diamond knife is essential for making flat polymer surfaces. Cutting temperature also requires optimization, and the preparation is often done at low temperature below glass transition of a polymer sample or its components. There is also no need for staining of the sample for AFM examination, whereas this procedure is essential for compositional imaging with TEM. In some instances, it might be helpful to use different etching procedures (by chemical means or plasma) for the removal of non-specific surface layers or amorphous material of semicrystalline polymers.

7.2.3 Imaging in Tapping Mode

At present, the tapping mode is more often applied for imaging polymers than the contact mode because it almost eliminates lateral forces. Additionally, the short duration of the tip-sample contact in tapping mode limits the development of inelastic deformation in polymer materials. These useful features of tapping do not mean that tip-sample force interactions should be completely neglected. In reality, they are more challenging for understanding and evaluation compared to the tip-sample forces in contact mode. Several aspects of the control of the tip-sample force interactions in tapping mode will be discussed now. This control is useful for optimization of imaging and for a rational image analysis. The main idea is to show that one should be extremely careful when interpreting height and phase images, especially for heterogeneous polymer samples, so that erroneous conclusions are avoided.

In contrast to the contact mode, in which cantilever deflection is a direct measure of tip-sample force, the amplitude of an oscillating probe does not describe tip-sample forces completely. In addition to the cantilever amplitude, its frequency and phase change substantially in response to the tip-sample forces. Monitoring of frequency and phase changes helps to determine whether an overall force exerted on a probe is attractive or repulsive, and to compare the overall force in different sample locations or on different sample components.

For rational analysis of AFM images it is quite important to conduct experiments in a reproducible fashion to limit the number of parameters influencing the image contrast. Several issues should be taken into consideration.

At the beginning of the experiment, the resonant frequency of the probe is determined in the immediate vicinity of the sample surface and the driving frequency is usually chosen at the resonant frequency [12]. An operator chooses the amplitude of the free-oscillating cantilever (A_0) by varying the voltage applied to the piezo-element, which is positioned close to the fixed end of the probe. It is typically in the 1–150 nm range. Engagement of the probe into a sample surface is typically performed at a set-point amplitude (A_{sp}) that ensures the minimal amplitude damping and minimal force interactions (i.e., $A_{sp} \approx 0.9\,A_0$). After this, scanning of the probe over the sample surface can be performed at different A_{sp}, which is usually varied in the 0.9–0.2 A_0 range. Both these parameters define the level of the tip-sample force interactions as well as stiffness of the cantilever. In the initial applications of tapping mode, it was found that with increasing A_0 and decreasing A_{sp} the level of the tip-sample interactions increases [13].

The effect of A_{sp} on contrast of height and phase images is demonstrated by a practical example taken from imaging of a binary latex blend, Fig. 7.4a–d. This latex blend consists of particles of the same size but made of two different acrylic polymers with different glass transitions ($T_g = 20$ °C and 30 °C). The ratio of these polymers is 1:3 with the minor component having higher T_g. The images of this sample, which were recorded with $A_0 = 20$ nm and at different A_{sp} values show pronounced height and phase variations. It is worth noting that these changes are completely reversible. When the tip-sample force interactions were minimized by choosing $A_{sp} = 0.88\,A_0$, the height and phase images exhibit a similar pattern related to surface corrugations on the micron scale, Fig. 7.4a. Most likely a low-tip-force level led to imaging of the surface top layer, which might consist of a contamination material formed on this hydrophilic surface. A small decrease of A_{sp} to 0.83 A_0 led to the appearance of some nanometer-scale details mostly seen in the phase image, Fig. 7.4b. Individual latex particles became visible in the height and phase images after further A_{sp} decrease, Fig. 7.4c. In these images one sees that the sample indeed consists of latex particles of identical size. Therefore, with an increase in the tip force the probe penetrates through the contamination layer and reveals topography of the latex blend. Further lowering of A_{sp} to 0.4 A_0 induced changes in the height and phase images that are related to different mechanical properties of components because at ambient conditions one polymer is in a hard, glassy state and the other is in a soft, rubbery state. In the height image, individual particles of the soft polymer are barely distinguished in the height image (Fig. 7.4d), whereas particles of the hard polymer are clearly seen. The differences in the phase image are even more pronounced: stiffer particles exhibit brighter contrast than their surroundings, which consist of softer particles. In addition, the phase image reveals bright, needle-like structures that can be assigned to small crystals of a surfactant used in the polymerization process.

These observations as well as previous results teach us that imaging of polymer samples can be performed at low-tip-sample force interactions (light tapping) and

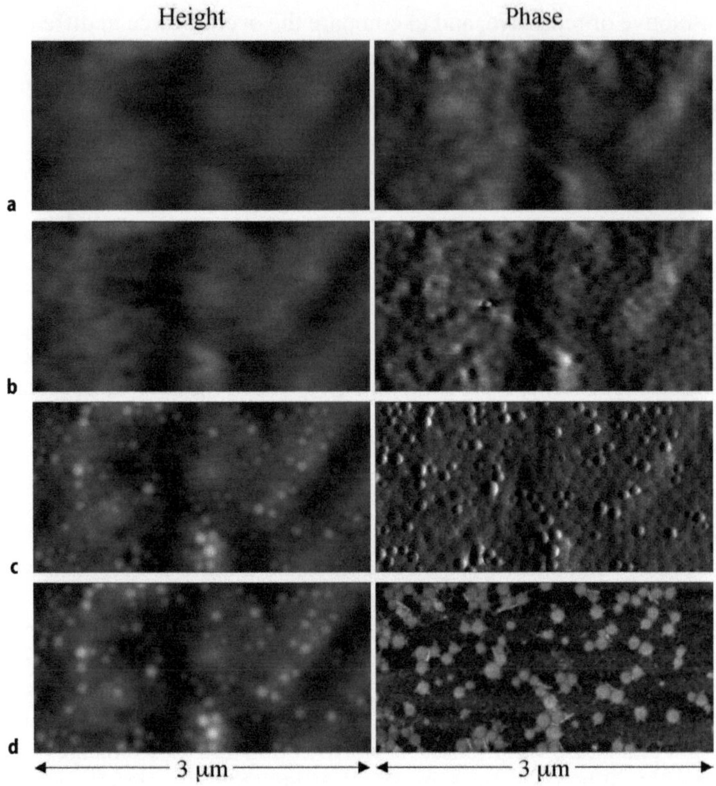

Fig. 7.4a–d. Height (*left*) and phase (*right*) images of binary latex blend obtained at $A_0 = 20$ nm and $A_{sp} = 0.88$ A$_0$ (**a**), $A_{sp} = 0.83\,A_0$ (**b**), $A_{sp} = 0.80\,A_0$ (**c**), $A_{sp} = 0.45\,A_0$ (**d**)

at elevated forces (hard tapping) [14]. At light tapping, the probe interacts primarily with top surface layers and reveals either an overall contamination layer or individual structures of material forming its surface. Light tapping imaging of the latex blend at $A_0 = 0.88\,A_0$ can even be named non-contact imaging with respect to the latex particles and contact imaging with respect to the contamination layer. The contrast of the phase image obtained at $A_{sp} = 0.4\,A_0$ distinctively points out different components of the latex blend and can be used for quantitative evaluation of composition of multicomponent polymer materials. In essence, these are the conditions of hard tapping because depressions of soft particles are also seen in the height image. Strong variations in height images at different A_{sp} have been recorded in studies of a melt-blended sample of two polyethylenes (PE) with a large difference in density (0.86 g cm^{-3} and 0.92 g cm^{-3}) [14]. Analysis of the height changes, which are related to tip-induced depressions of low-density locations, shows how to estimate tip-sample forces at different A_0 and A_{sp}. Unfortunately, the height contrast is noticeable only when mechanical properties or densities of the components of the examined sample are substantially different. This happens quite rarely and in most cases the phase contrast is more sensitive to the sample heterogeneity. The

a

b

30 μm 30 μm

Fig. 7.5a,b. Height (*left*) and phase (*right*) images of microlayer PE film, which consists of alternating layers of polymers with different branching density. Narrow layers – high-density PE ($\rho = 0.92$ g cm^{-3}), wide layers low-density PE ($\rho = 0.86$ g cm^{-3}). Images were obtained with Si probe (stiffness 40 N m^{-1}), $A_0 = 80$ nm and $A_{sp} = 40$ nm

images and cross-sectional profiles in Figs. 7.5 and 7.6, which were obtained on a multilayer PE sample, demonstrate this. If compared to the melt-blended sample, the multilayer material with the same PE components exhibits different mechanical behavior. There were no height variations when A_{sp} was reduced to $0.5\,A_0$. The phase contrast behaves differently, and the high-density layers are seen much brighter than the low-density layers, Fig. 7.5a,b. This is also noticeable in the height and phase cross-section profiles and images, Fig. 7.6a,b. The described measurements were performed at the resonance frequency of a Si probe with stiffness 40 N m^{-1}, which is typical for tapping mode studies. One might increase an effective stiffness of the probe using its second fundamental frequency, which is over 6 times higher than the first one, for imaging. In this case, pronounced phase changes were observed at high A_{sp}, Fig. 7.6c–d.

The trend found in studies of the multilayer PE is common for blends of PE and other polyolefins when imaging is conducted using probes that have a stiffness 40–50 N m^{-1} at large A_0 (80–120 nm) and $A_{sp} = 0.5$–$0.4\,A_0$. Behavior of the phase contrast is different when softer probes are applied. Differences in adhesive properties of components influence phase contrast more substantially when imaging is performed at low A_0.

Such an empirical approach can be helpful in studies of heterogeneous samples despite the fact that there are a number of theoretical attempts to describe the behavior of the phase in tapping mode measurements. A general correlation between phase behavior and energy dissipated in the tip-sample junction in tapping mode has been established [14–17]. Yet this result does not necessarily help in answering the following question: What material properties are mostly responsible for the energy dissipation? Adhesion, capillary effects, and indentation forces are involved in the interplay leading to energy dissipation of the probe interacting with a sample. So far, it is not clear how to quantify individual contributions of these interactions. Additional difficulties related to polymer samples are caused by a non-linearity of viscoelastic behavior of these materials in the tip-sample contact.

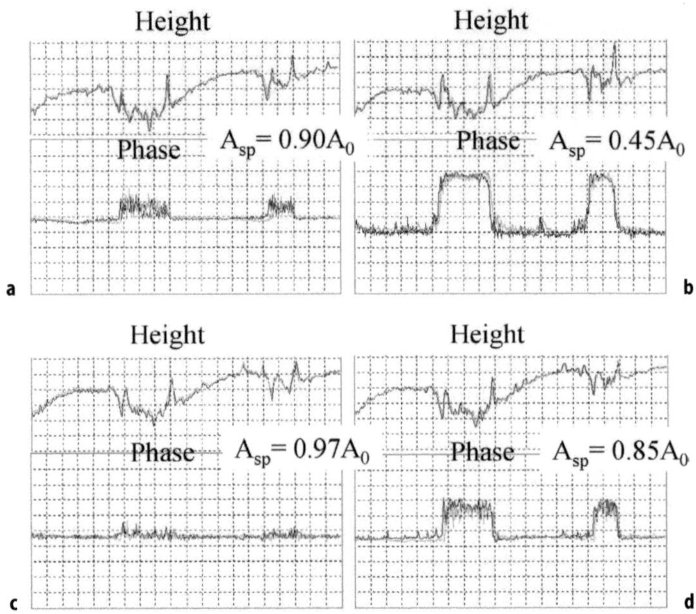

Fig. 7.6. a,b Height (*top*) and phase (*bottom*) profiles taken along microtomed surface of multilayer PE at A_0 = 80 nm and different A_{sp}. Probe was oscillating at its main resonant frequency. **c,d** Height (*top*) and phase (*bottom*) profiles taken along microtomed surface of multilayer PE at A_0 = 80 nm and different A_{sp}. Probe was oscillating at its second fundamental resonant frequency, which is approximately 6 times higher than the main resonant frequency

One additional remark should be made concerning tapping mode imaging of heterogeneous polymer samples such as smooth films and layers on different substrates. The feedback mechanism used in the microscopes is based on the recording of amplitude changes at the resonance frequency of the free-oscillating probe. Unfortunately, heterogeneous locations might exhibit different adhesion causing different shifts of the cantilever resonant frequency at these locations. The related drop of the amplitude (at the "old" resonant frequency) will lead to different heights of these locations that do not exist in reality. Therefore, patches of a polymer monolayer might appear in height image lower than their substrate or one will see surface corrugations of different blocks in block copolymers films, whereas their surface is smooth. Luckily, these "faked" corrugations are only a few nanometers in height. Therefore, they "contaminate" only images of a limited number of samples. To avoid these problems one should attempt to obtain height images in light tapping, when the cantilever phase is practically not changed from its value for a free-oscillating cantilever. The use of a small A_0 value and soft probes will be advantageous for this purpose.

7.2.4 Studies at Elevated Temperatures and in Different Environments

To date, the majority of AFM measurements of polymers have been performed at ambient conditions. A recently introduced heating accessory allows imaging at elevated temperatures up to 250 °C [18]. This capability further enhances AFM applications to polymers because it allows in situ monitoring of structural transformations during thermal phase transitions [19] and optimization of imaging by adjusting the observation temperature for a better discrimination of sample components in compositional imaging. [20]. Several problems have been solved to ensure stable imaging in tapping mode at elevated temperatures. They include cooling of a piezoscanner, the dual heating of a sample location using sample and probe heaters, monitoring of the surface temperature with 1 °C accuracy by using the temperature dependence of the probe resonant frequency, and environmental control of the sample by purging with inert gases. Imaging at elevated temperature is not much different from that at ambient conditions. Two issues should be mentioned. Because of unsolved problems with different expansions of the microscope components in the vertical direction, the probe should be retracted from the sample during sample heating when the sample temperature is changed more than 10 °C. After the sample and the microscope settle to new temperature (it takes 2–3 min), driving frequency should be re-tuned to accommodate the temperature-related shift of the resonant frequency of the probe. Also for imaging of polymers close to their melting temperature one might use larger oscillating amplitudes of the probe in order to overcome adhesive tip-sample interactions. Several examples of temperature studies will be discussed below.

AFM imaging in different gas atmospheres is conducted rarely however it might have many undiscovered applications. For example, an increase in air humidity allows one to observe the unfolding of hydrophilic PMMA segments in PS-PMMA block copolymer sub-monolayer films on mica [21]. Inert atmosphere is required for heating of some polymer samples to avoid their oxidation. This is best achieved by purging with He gas, which is also an excellent heat transfer agent. In addition, controlling air substitution with He near the sample is straightforward thanks to a pronounced shift of the cantilever resonance to high frequencies and a substantial increase in the cantilever quality factor.

AFM imaging of polymers under liquid is also not frequently reported. There have been attempts to use this environment for imaging at lower tip-sample forces than are achievable in air. Such realization will allow imaging of much softer top layer structures than studies in air. Indeed, this possibility was demonstrated in imaging of PE samples in the contact and tapping modes. Another function of the liquid environment could be directed towards selective swelling or etching of components in block copolymers, blends, and other heterogeneous polymer samples. This capability is demonstrated by images in Fig. 7.7a,b. A microphase pattern in thin films of PS-*b*-PVP block copolymer is barely seen in the height and phase images under ambient conditions and also under water. Immersion of this material into methanol or acidic water (pH 2.2) reveals the microphase separation due to surface changes caused by selective swelling of PVP blocks. The images show well-defined hexagonal ordering in this material developed after long-term annealing at high temperatures.

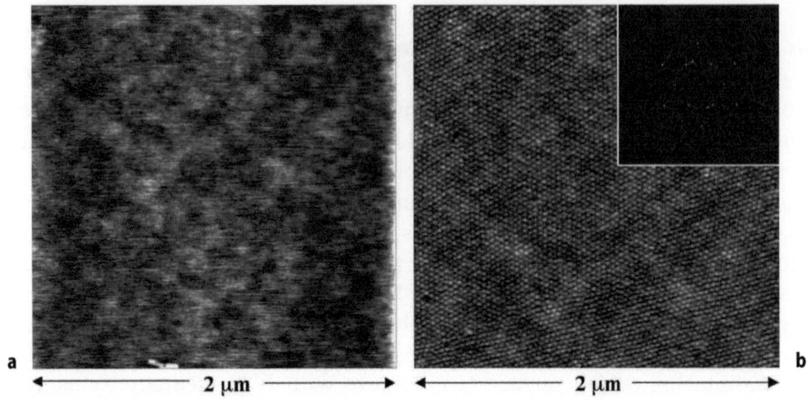

Fig. 7.7a,b. Height images of PS-*b*-PVP block copolymer on Si substrate, which were recorded in air and under methanol/water (50/50) mixture

7.3 Observations of Single Macromolecules and Their Self-Assemblies

AFM gives researchers a unique capability for visualization of single macromolecules, their conformations, and packing motifs in single layers. In the first STM and AFM applications, imaging of single DNA macromolecules was a challenging subject. The fixation of DNA to mica was a key issue for the successful imaging of DNA in the contact mode [22]. This is less important in tapping mode and, nowadays, imaging of DNA on mica in air or in water using tapping mode has become a routine experiment. Thanks in part to successful studies of biological macromolecules with AFM, imaging of individual synthetic macromolecules has been brought into a focus. First, such imaging has been demonstrated on single-layer assemblies of polystyrene-*block*-polymethylmethacrylate prepared by the Langmuir-Blodgett technique on water and transferred onto mica [21]. Later, these applications were further extended to visualization of spherically shaped star-burst macromolecules [23–25], cylindrically shaped macromolecular brushes [26], and polymers with mini-dendritic groups [27].

7.3.1 Chain Molecules on Substrates

Individual polymer molecules have been observed in AFM images as isolated species as well as in flat arrays and layers on various substrates (mica, graphite, Si wafer). On a flat substrate, individual carbosilane dendrimers are seen as spheres of 3–5 nm in size depending on the generation number Fig. 7.2c–d. Carbosilane dendrimers with alkylcyanobiphenyls as terminal groups are characterized by strong intermolecular interactions, which dominate the interplay with the adsorbate-substrate interactions during formation of layers and bulk material [25]. Visualization of chain macromolecules is more intriguing. Solvent-free molecules might adopt either coiled or extended conformations, which can also collapse.

These questions are addressed in a number of AFM studies of various polymers with mini-dendritic groups performed in recent years.

Linear polystyrene (PS) and polymethylmethacrylate (PMMA) with large side chains formed by mini-dendritic groups were the first polymers of this kind to be examined with AFM [27, 28]. Spreading polymer macromolecules on a flat substrate by spin casting of their dilute solutions or by dipping the substrate into such solutions is a typical procedure for AFM studies of such polymer macromolecules and their self-assemblies. In contrast to ordinary chain macromolecules, chains of polymer brushes, and polymers with mini-dendritic side groups prefer an extended conformation due to steric repulsions of densely grafted side chains. Therefore, in AFM images they are seen as worm-like objects several nanometers in diameter. The contour length of these worm-like species can be considered as a measure of their molecular weight, and for a number of polymers the related histograms have been compared with molecular weight distribution estimated from GPC and light scattering data [26]. It is remarkable that a polymer with mini-dendritic side groups with alkyl chains exhibits preferential orientations along surface lattice directions of graphite. Such epitaxy helps unravel the macromolecules on the substrate and ensures their mechanical stability with respect to the tip force in AFM experiments. A couple of sketches and images of polystyrene with the mini-dendritic groups in Fig. 7.8a–d illustrates this discussion [29]. The chemical structure of the monomer unit is shown in Fig. 7.8a. Terminal groups contain 14 alkyl groups, which provide this molecule with a hydrophobic jacket. In bulk, the polymer formed mesomorphic columnar order, and the diameter of individual chains in this state is around 5 nm. A sketch in Fig. 7.8b presents a possible orientation of the polymer chains on the surface of graphite. The spacing of the graphite lattice and a repeat distance along the alkyl chain in the all-*trans* conformation are close to 0.25 nm, and this is believed to be a reason for epitaxial orientation of alkyl chains along the main directions of graphite. As will be shown below, this is not entirely true. Nevertheless, such a model helps to explain the ordering of the polymer chains seen in AFM images. Single macromolecules of the PS with mini-dendritic groups are seen as stretched worms in the image in Fig. 7.8c. They are approximately 6-nm wide, which is close to the molecule diameter in the bulk state. Yet the height of the structures is much smaller (approximately 1.5 nm). The only reasonable explanation of such discrepancy is a tip-induced deformation of the molecules. Segments of individual chains are oriented preferentially along the directions consistent with 3-fold symmetry of the substrate. This is more pronounced in Fig. 7.8d, which shows a graphite surface with a higher concentration of the polymer chains. One also sees a tendency of individual chains to aggregate and form ordered domains, which is most likely related to hydrophobic interactions between the chain jackets. It has been noticed that thermal annealing of the samples enhances this aggregation. Quantitative estimates of the contour length of the PS with mini-dendritic side groups on graphite, which have been made from AFM images, gave very different numbers to those expected from molecular weight measurements performed with size-exclusive chromatography. These data assume a strong longitudinal collapse of the polymer chains in solvent-free state. The data for the PMMA with mini-dendritic groups suggest only negligible collapse [30].

a b c d

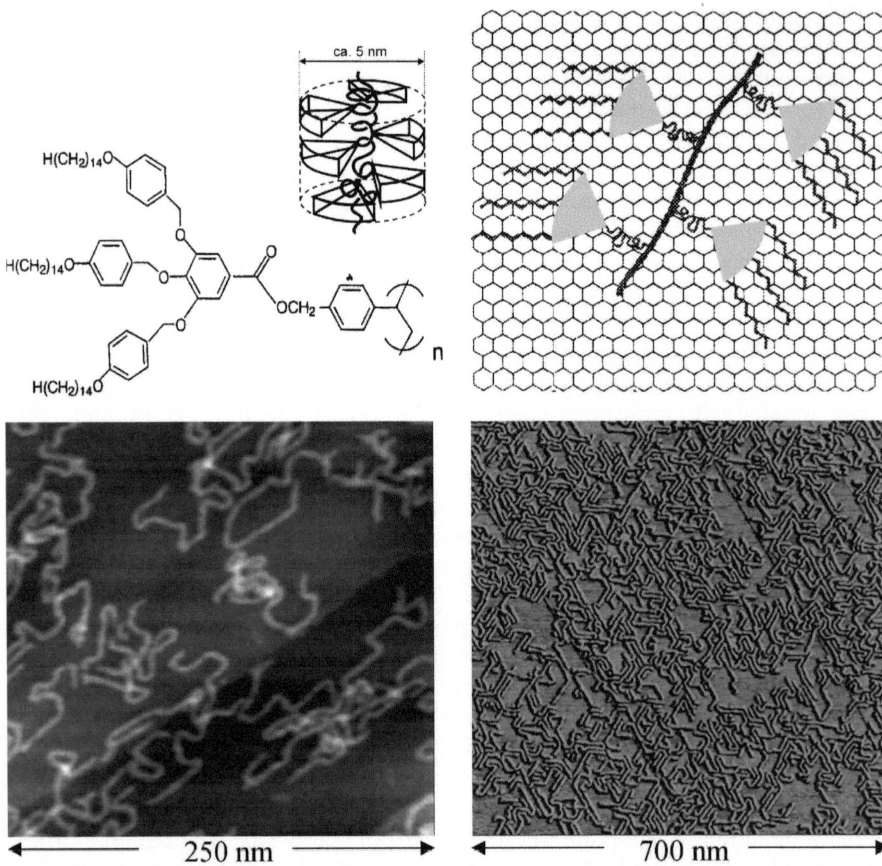

Fig. 7.8. a,b Illustrations of chemical structure of PS with mini-dendritic groups and their possible epitaxy on graphite lattice. **c,d** Images of chains of PS with mini-dendritic groups on graphite with different coverage [29]

In the search for new polymers whose self-organization favors the formation of columnar mesophases with different functionality, the type of mini-dendritic groups and chemistry of chain backbones has been varied extensively. AFM imaging was applied for visualization of single chains of polymers with 7-oxanonbornene units in the backbone and different dendrons as side-chains, Fig. 7.9a–d. Molecular weight of these polymers is also drastically different. The images of the first polymer in Fig. 7.9a,b show similar features to those for the PS with mini-dendritic groups. Individual macromolecules are seen extended along three main directions. Analysis of their height profiles shows that the chain width is in the 5.5–6.5 nm range (as measured at half of the height peak) and height is around 0.7 nm. These data are "contaminated" by a convolution of the tip apex shape during profiling of the macromolecules, whose width is comparable with the tip contact area, and by possible tip-induced deformation. After thermal annealing, the macromolecules aggregate into large domains whose height is slightly larger

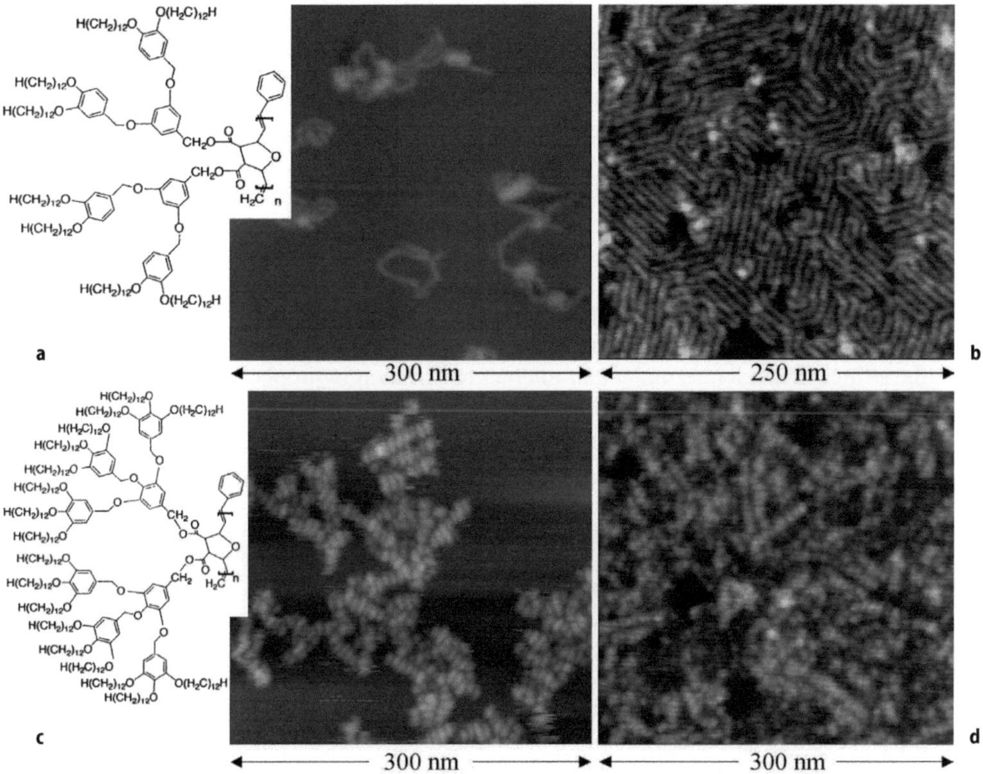

Fig. 7.9a–d. Height images of poly(oxanonbornene)s with of mini-dendritic groups of different kind (chemical structures are shown in the *inserts*) on graphite. Images in *a* and *c* were obtained on surface locations with a low coverage; images in *b* and *d* were obtained on surface locations with a high coverage

($\approx$ 0.8 nm) than that of the individual molecules. Measurements of a repeat distance between the tops of the closely packed individual macromolecules inside the domain are more reliable for defining the macromolecule width. Such measurements for the image in Fig. 7.9b gave a repeat distance of 5.5 nm. This number is still higher than the 4.1-nm diameter of the chains in bulk hexagonal columnar phase estimated by X-ray diffraction [28]. Most likely, the difference reflects variations in two-dimensional and three-dimensional packing of these macromolecules.

A number of bright spots on the single macromolecules in Fig. 7.9a and at different locations of the domain in Fig. 7.9b are worthy of notice. They might be caused by structural defects such as chain intersections, "hair spins", and others discussed in the analysis of AFM images of the layers of PS with mini-dendritic groups on graphite and mica. It was mentioned that these defects are more common for mica, where absence of epitaxy suggests more complex interactions defining chain conformation on the atomically flat substrate [27].

Most of the AFM studies of polymers with mini-dendritic groups [26–29] have been carried out with relatively large macromolecules whose contour length is of the order of hundreds of nanometers. It is rather important to learn if shorter macromolecules with mini-dendritic groups interact with graphite strong enough to be examined with AFM. The poly(oxanonbornene) with the larger dendrons has a small polymerization degree of 15. AFM images of spin-cast samples of this polymer on graphite are shown in Fig. 7.9c–d. The first image shows a network of strings formed from extended beads. The beads are approximately 6 nm in width and their length varies in the 12–18 nm range. Occasionally, single beads have been seen. Large domains have been observed after thermal annealing of the sample, Fig. 7.9d. One can distinguish a number of linear strings of beads, and a dominant number of beads seen through the domain are organized into the strings. A partial epitaxy of the strings to graphite is distinguishable in both images. It is reasonable to assign the individual beads to single poly(oxanonbornene) chains, which form the strings through intermolecular interactions of their hydrophobic jackets and epitaxy to graphite. Thermal annealing seems to enhance intermolecular interactions between the individual macromolecules.

7.3.2 Self-Assembly of Individual Macromolecules

It is natural to expand visualization of single macromolecules to studies of their self-assembly, which results from interplay of different interaction forces. As has already been shown above, AFM is extremely useful for such studies. This will be demonstrated further in studies of polymers with mini-dendritic groups on graphite. Macromolecules with other types of backbone, poly(oxazoline), and poly(acetylene carbazole)s will be considered for this purpose.

The cylindrical shape of macromolecules of poly(oxazoline)s with mini-dendron, Fig. 7.10a, leads to their self-organization into one-, two- and three-dimensional lattices [31, 32]. TEM studies performed on films of these polymers reveal homeotropic and planar aligned assemblies of supramolecular cylinders formed of the macromolecules. The one-dimensional and two-dimensional assemblies of the poly(oxazolines) with alkyl chains of different length and with different degrees of polymerization have been observed on graphite with AFM. Small domains, which consisted of 5–10 rods, have been seen in the sample prepared by spin casting on graphite, Fig. 7.10b. The rods are 10–50 nm in length. Thermal annealing of the sample has induced the formation of large domains with close-packed rods extending on for hundreds of nanometers. A close look at one such domain is given in Fig. 7.10c. The spacing between the rods in small and large domains is approximately 4.5 nm, which is slightly larger than the diameter of the polymer macromolecules in the bulk (3.7 nm), which was measured with X-ray diffraction. The rod length exceeds the expected longitudinal dimension of the poly(oxazoline), and the rods most likely present the one-dimensional macromolecule self-assemblies, which are formed via hydrophilic recognition of their poly(ethyleneimine) core. This packing into the two-dimensional domains can be explained by hydrophilic interactions of their jackets.

The suggested motifs for self-organization of the poly(oxazoline)s are supported by AFM images of another polymer, in which the alkyl side chains are slightly longer. Images of the samples of this polymer with different molecular weight are

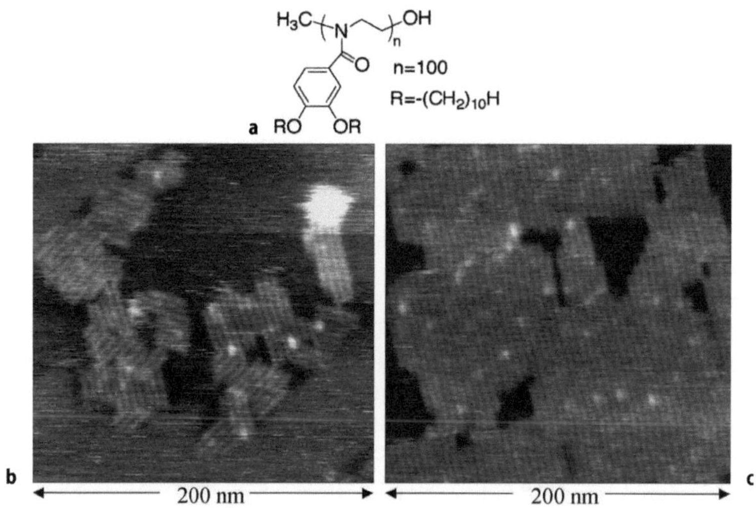

Fig. 7.10. **a** Sketch of chemical structure of a repeat unit of poly(oxazoline) with mini-dendritic groups. **b,c** Images of this polymer after spin casting and after thermal annealing, respectively

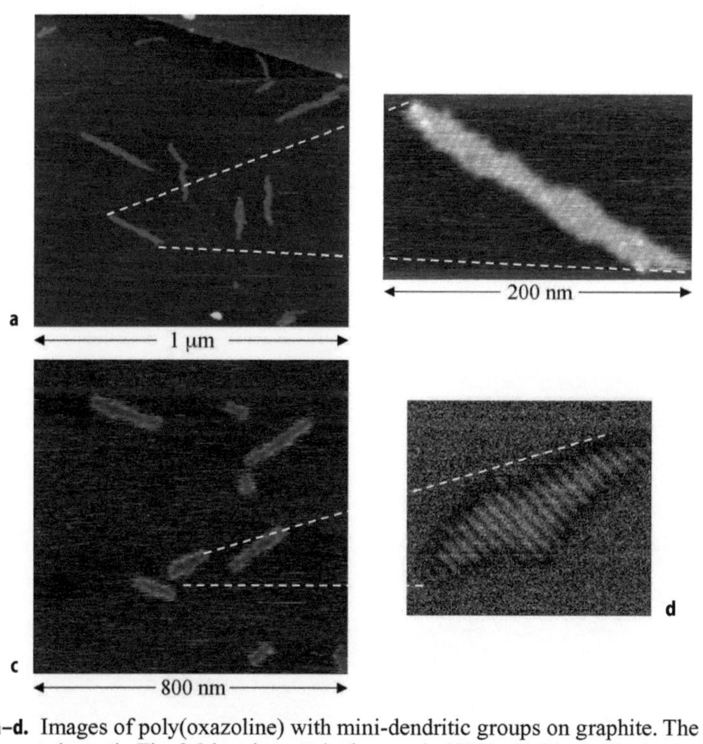

Fig. 7.11a–d. Images of poly(oxazoline) with mini-dendritic groups on graphite. The polymer is similar to one shown in Fig. 3.2 but the terminal group is $(CH_2)_{12}H$. The image in *a* was obtained for the polymer with degree of polymerization 100, the image in *b* for polymer with polymerization degree 500

shown in Fig. 7.11a–d. Domains with rod-like substructure, which are oriented along the main directions of the substrate, have been found in samples of these polymers, Fig. 7.11a,c. A pitch of 4.7 nm was determined from the high-resolution images in Fig. 7.11b,d. This value is consistent with the 4.0-nm diameters of these macromolecules in bulk. The domains are more regular in width for polymers with low molecular weight, which might be consistent with its narrower polydispersity. Yet the number-averaged contour width of the rods in the domain (23.6 nm) shown in Fig. 7.11b, is almost twice as large as that of the estimated contour length of the macromolecules. Therefore, the rods are formed from a couple of the extended macromolecules.

Synthetic polyacetylenes with carbazole groups offers new capabilities for tuning self-assembly of macromolecules via stereoisomerization and chirality, optimization of photoconductivity and chemical sensing, and better processability. AFM has also been applied to the characterization of such polymers [32]. Images of a sub-monolayer of one of these polymers on graphite are shown in Fig. 7.12a,b. This patch seen in Fig. 7.12a is formed from rods that are characterized by hexagonal packing consistent with the symmetry of the underlying graphite. An averaged distance between neighboring rods is around 6 nm, whereas the lattice parameter of the hexagonal columnar phase in the bulk is 4.7 nm. At some locations, one can identify individual lamellar blocks (indicated as #1 and #2) with an averaged width of approximately 25 nm, which is two to three times larger than an estimated con-

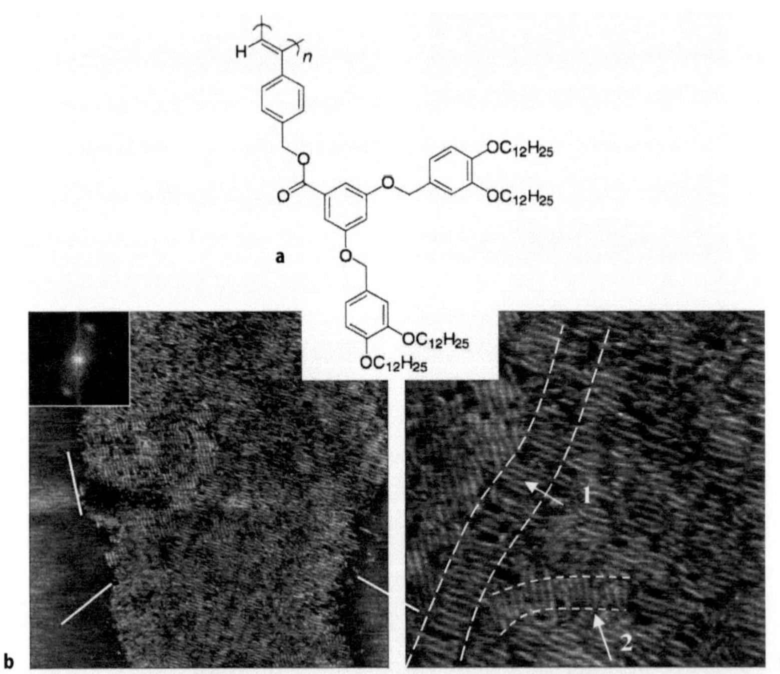

Fig. 7.12. a Sketch of chemical structure of repeat unit of poly(acetylene carbazole) with mini-dendritic groups. **b,c** Images of the polymer domain on groups on graphite

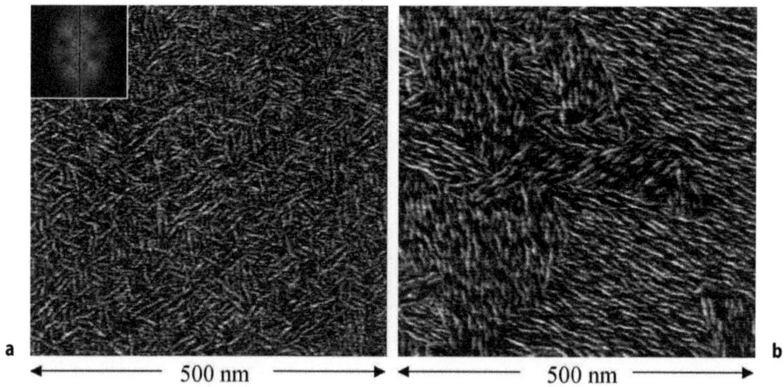

Fig. 7.13a,b. Images of poly(acetylene carbazole) with mini-dendritic groups on graphite of as-received layer (**a**) and after annealing at 50 °C (**b**)

tour length of this macromolecule (9.4 nm). This implies one-dimensional ordering of the macromolecules into the rods.

AFM phase images of the surface region of a spin-cast sample, which is completely covered by the polymer layer, are shown in Fig. 7.13a,b. This layer is characterized by an interwoven pattern formed by numerous rods 4–5 nm in width and 30–50 nm in length. The epitaxy of the polymer layer to graphite is evident yet the interactions with the substrate are not very strong. After annealing at 50 °C, structure of this layer underwent distinctive changes related to the formation of sub-micron domains with different orientation of elongated linear structures composed of smaller rods, Fig. 7.13b. The macromolecular order of this domain can be described as a two-dimensional nematic phase. Intermolecular interactions within the layer are much stronger than the interactions with the substrate that led to the liquid crystalline order in this polymer.

7.3.3 Thermal Motion, Adhesion, and Manipulation

Recent progress in AFM instrumentation that allows imaging at elevated temperatures and reproducible positioning of the probe with nanometer accuracy opens broad possibilities for studies of dynamic behavior of single macromolecules. It is important to consider several results of monitoring of thermal motion of polymers with mini-dendritic groups and probe-assisted manipulation of these objects. The first example is taken from AFM studies of thermal motion of one of the polymers shown in Fig. 7.14a. Single molecules of this polymer have been examined at elevated temperatures on three different substrates: graphite, mica, and WSe_2 [33]. WSe_2 is an inorganic layered crystal with an atomically flat surface formed by top surface Se atoms, which are arranged in a closed hexagonal lattice with 0.32-nm repeat distance.

An AFM image of the individual polymer chains on WSe_2 shows that extended macromolecules adopt orientations reflecting the 3-fold symmetry of this substrate, Fig. 7.14b. It appears that the general symmetry of the flat substrate, and not the

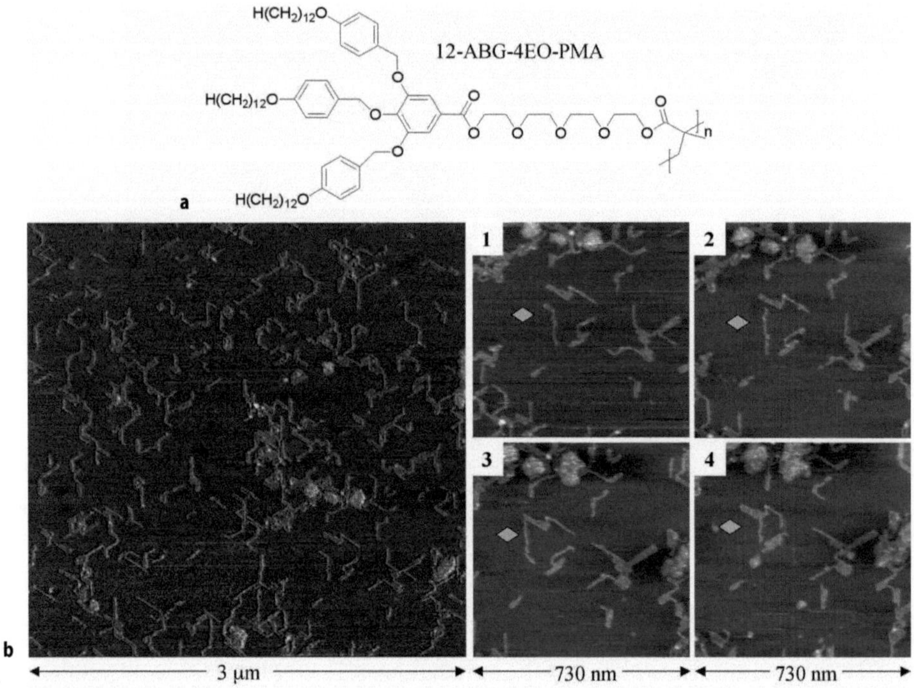

Fig. 7.14. **a** Sketch of chemical structure of PMMA with mini-dendritic groups. **b** Image of the polymer molecules on WSe_2. **c** Sequence of images obtained at 40 °C with 10-min time interval between the images

match between the repeat distances of terminal alkyl groups in the all-*trans* conformation and the substrate, is responsible for the observed epitaxy. A sequence of images 1–4 in Fig. 7.14c was captured with intervals of approximately 10 min between images after the sample was heated to 40 °C. In the center of these images one sees a number of molecules changing their relative positions with each other. The macromolecule at the right of the diamond-like marks has moved gradually up to join a couple of others, which form a zigzag pattern. During this translation, the chain kept its parts oriented along the directions of 3-fold symmetry. At present, the observed phenomenon of the macromolecule translation along the substrate is not fully understood. One might suggest that a weakened adhesion of the macromolecules to WSe_2, which is at least partially due to their increased thermal motion, and a periodical disturbance by the tip force have promoted this process. Due to the epitaxy to graphite, temporarily stable surface locations of extended macromolecules are along the energetically favorable substrate directions. Hydrophobic interactions with other macromolecules facilitate the chains aggregation. It is worth noting that similar motions of the macromolecules were detected on graphite surface at slightly elevated temperatures of 30–40 °C. The situation was quite the opposite for the macromolecules deposited on mica. At this substrate, there were no changes of the macromolecules' shape on heating up to 170 °C [33].

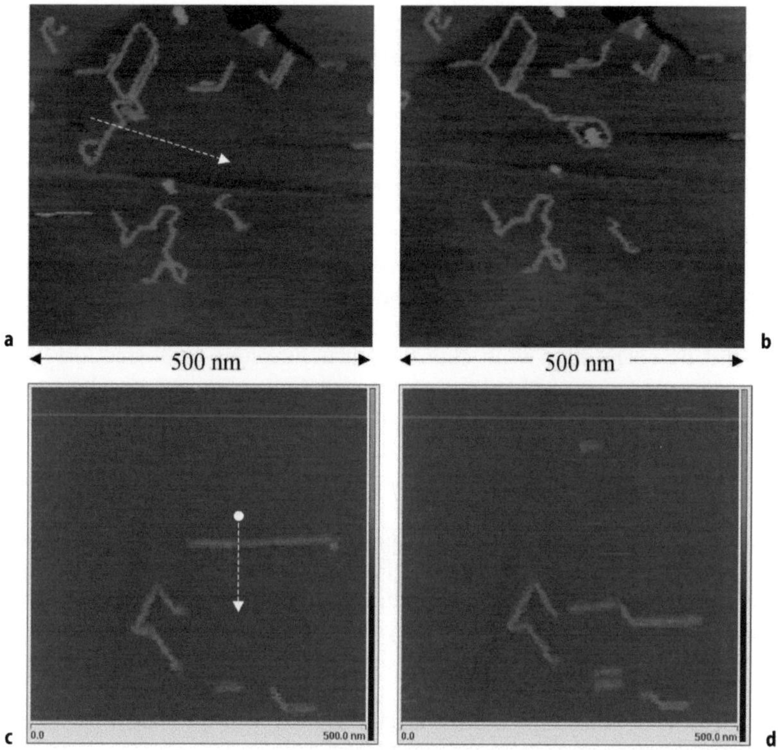

Fig. 7.15. Height images showing the results of tip-assisted transport of single molecules of PMMA with mini-dendritic groups on graphite **a,b** and WSe_2 **c,d**. Motions of the tip are indicated with *dotted lines*

Observations of macromolecule motion at elevated temperatures on three selected substrates indicate differences of the polymer adhesion to these substrates. The results are consistent with the data of tip-assisted manipulation [33], which are presented in Fig. 7.15a–d. In these experiments, an AFM probe was brought to the location of interest, moved to the close vicinity of the surface, and then transported along an operator-defined trajectory. These trajectories are shown by dotted lines in the images in Fig. 7.15a,c. The images taken after these actions are shown in Fig. 7.15b,d, respectively. These images show that the tip has moved the macromolecules on both substrates. In the case of the polymer molecule on graphite, its motion was affected by the attachment to the other macromolecule at the top, Fig. 7.15b. The translation of the single-lying stretched macromolecule on WSe_2 has proceeded with a kink formation at the location touched by the probe. The interaction with the substrate has determined the macromolecules' orientations in their initial and final positions. All attempts to move the same macromolecules lying on mica have been unsuccessful: the probe cut the molecules into parts.

A more elaborate description of these results and their analysis will be presented elsewhere. Here, it is appropriate to mention that a combination of AFM studies of macromolecules at different temperatures with the tip-assisted manipulation can provide unique information about adhesion of individual polymer molecules to selected substrates.

7.4 Studies of Semicrystalline Polymers

A large number of polymers exhibit various degrees of crystallinity that define mechanical and thermal behavior of these materials. For years, studies of polymer crystallinity were performed with a large variety of physical and chemical methods. Yet there are many open questions concerning the fundamental nature of this property and its correlation with chemical structure and conformation of polymer chains, sample size, material thermal history, etc. In contrast to low molecular weight compounds, polymer crystallization in most cases leads to only partially ordered material, in which crystalline structures coexist with an amorphous polymer. Studies of semicrystalline polymers with AFM have been conducted since the advent of this method [34]. Initially, the high resolution of AFM motivated researchers to study single polymer crystals [35–37] and morphology of different polymer samples. With the progress of AFM applications and better understanding of the sensitivity of AFM to local material properties, the use of this method in characterization of semicrystalline polymers has increased substantially. It has become clear that AFM complements not only other microscopic methods but also diffraction methods in elucidation of polymer morphology and lamellar structure. In this section, several examples illustrating the capabilities of AFM for studies of semicrystalline polymers will be discussed.

7.4.1 Single Crystals of Normal Alkanes and Polyethylene

Crystallization of PE, a polymer with a simple chemical structure of its repeat unit, has been explored in more detail than crystallization of any other polymer. Normal alkanes are ideal model compounds of PE, and AFM studies of alkanes with different chain length are helpful for exploring such important processes as folding and unfolding of polymer chains. Epitaxial layers of $C_{36}H_{74}$ and $C_{60}H_{122}$ on graphite are shown in Fig. 7.2a,b. Recent high-temperature AFM studies have revealed that epitaxy of extended-chain $C_{60}H_{122}$ molecules to graphite exhibits at temperatures up to 145 °C, a temperature that is much higher than the melting temperature of bulk crystals of this alkane [40]. Similar behavior was found for long and ultralong alkanes $C_{122}H_{246}$, $C_{242}H_{486}$, and $C_{390}H_{782}$ [41, 42]. Images of layers of these alkanes on graphite are presented in Fig. 7.16a–c. They demonstrate ribbons with a width (14 nm for $C_{122}H_{246}$, 29 nm for $C_{242}H_{486}$, and 45 nm for $C_{390}H_{782}$) comparable with the length of the extended chains of these alkanes. This is consistent with the fact that the all-*trans* conformation of alkane chains is energetically favorable. The image of $C_{122}H_{246}$ sample at room temperature shows domains consisting of the flat-lying lamellae whose orientations reflect the substrate symmetry, Fig. 7.16a. These domains form a topmost layer of multilayer epitaxy on graphite. When the samples of the long and ultralong alkane layers on graphite are

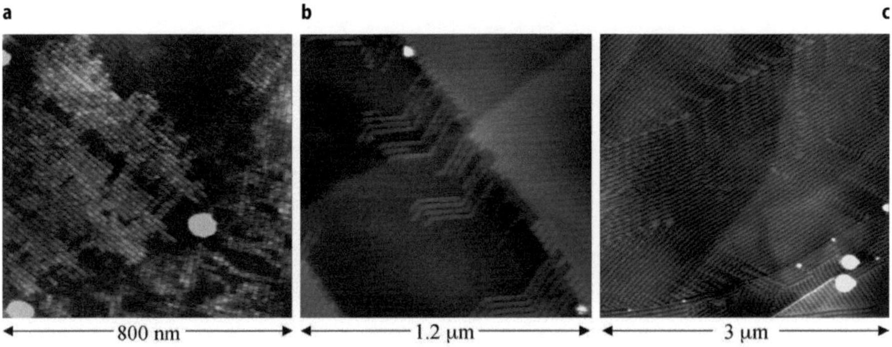

Fig. 7.16. **a** height image of ultralong alkane $C_{242}H_{486}$ layers on graphite at 120 °C; **b** height image of long alkane $C_{122}H_{246}$ layers of graphite at room temperature; and **c** phase image of ultralong alkane $C_{390}H_{782}$ layer on graphite at 170 °C

heated to temperatures in the 160–190 °C range the epitaxial order disappears. On cooling it re-appears in a layer-by-layer fashion. The image in Fig. 7.16b shows a complete first layer and a part of the second layer of $C_{242}H_{486}$ lamellae. This is one of a series of images recorded at 120 °C during the epitaxial lamellae growth. Note that in contrast to the $C_{122}H_{246}$ layer, the individual $C_{242}H_{486}$ lamellae form multiple kinks in registry with the substrate symmetry. The kinks appear and disappear with temperature changes even in the range of a few degrees, thus reflecting interplay between many factors governing chain packing is such layers. Local kink waves have been seen in the first $C_{390}H_{782}$ layers on the substrate at temperatures (170 °C) close to the complete "melting" of the epitaxy, Fig. 7.16c. There is a definite analogy between the kinks observed in the epitaxial alkane layers chain tilt in single crystals of alkanes and PE. Therefore, the studies of these local defects in layers formed from flat-lying lamellae and their temperature behavior currently in progress will be useful for understanding effects of chain tilt in polymer crystals.

Generally, the crystallization of polymers is characterized by multiple folding of chain macromolecules to form two-dimensional crystalline lamellae. This process is well defined in dilute polymer solutions where only few chain entanglements are present. Individual lozenge-shaped lamellae of PE grown in solutions are the best-explored polymer crystals [38, 42]. These crystals, which form initially as hollow pyramids, collapse into two-dimensional platelets while being deposited on the substrate. The length of a straight segment of polymer chains defines the lamellar thickness, which is usually in the 10–12 nm range. Single crystals of PE have been examined in AFM images with great structural detail on such topics as sectors, screw dislocations, and overgrowth [43]. The single PE crystal shown in the image in Fig. 7.17a exhibits a lozenge shape with the sector boundaries aligned along the diagonals. Studies of single crystals of long alkanes have shown that crystals of $C_{150}H_{302}$ could be crystallized with chains folded in two, whereas crystals of $C_{102}H_{206}$ could only be obtained in extended-chain form. Images of single crystals of $C_{122}H_{246}$, $C_{242}H_{486}$ and $C_{390}H_{782}$ are collected in Fig. 7.17b–d. Crystals of long alkane $C_{122}H_{246}$ have a platelet shape and their height (14 nm) is comparable with the length of the extended $C_{122}H_{246}$ chain. Crystals of the ultralong alkanes,

Fig. 7.17. **a** Phase image of single crystal of PE; **b–d** height images of single crystals of long and ultralong alkanes, $C_{122}H_{246}$, $C_{242}H_{486}$, and $C_{390}H_{782}$, respectively

Fig. 7.17c–d, exhibit the lozenge shape and those of $C_{390}H_{782}$ also show sectorization boundaries similar to those of single PE crystals. The height of the crystals (10 nm for $C_{242}H_{486}$ and 9.5 nm for $C_{390}H_{782}$) is consistent with the length of the alkanes in the folded-in-three chain conformation for $C_{242}H_{486}$ and the folded-in-five conformation for $C_{390}H_{782}$. Therefore, the ultralong alkanes have "polymeric-like" character that is further confirmed by the comparative temperature studies of these crystals and single crystals of PE [41].

AFM imaging of single PE crystals at elevated temperatures reveals lamellar thickening [44] caused by unfolding of individual chains from a kinetically favorable folded state to the energetically favorable extended-chain conformation, Fig. 7.18a. AFM observations of crystal annealing reveal formation of holes surrounded by thickened regions, which is consistent with early TEM data [38]. Thus, the sensitivity of AFM to height measurements makes it possible to monitor chain unfolding in situ with exceptional precision. As shown in the TEM studies [38], annealing of single PE crystals containing traces of solvent (xylene) proceeds differently with the reorientation of chains, which become aligned along the crystal surface. A similar phenomenon was observed in high-temperature AFM examination of single PE crystals on Si substrate that were not completely dried (i.e., free) of traces of xylene used as the solvent. Annealing of such PE crystals leads to the formation of holes and stacks of edge-on lamellae, Fig. 7.18b. The latter implies the polymer chain reorientation and a strong guess that local recrystallization takes

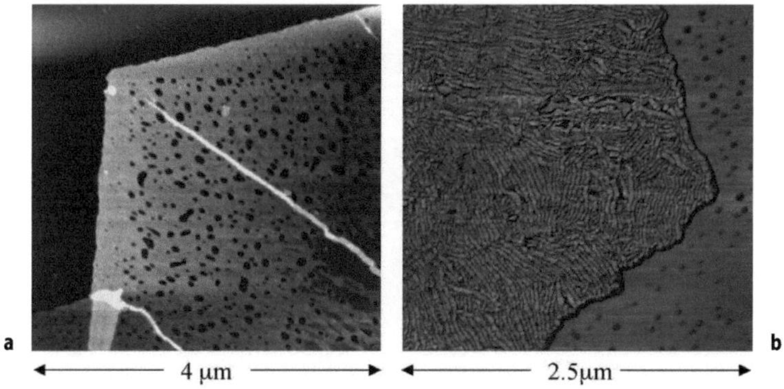

Fig. 7.18. **a** Height image of a part of dry PE crystal during annealing at 115 °C. **b** Phase image of a part of wet PE crystal after annealing at 105 °C

place during annealing of the "wet" PE crystals. These observations are in line with the results of AFM monitoring of annealing of dry and wet crystals of the ultralong alkanes [41]. The strong influence of solvent traces on the temperature behavior of polymer crystals should be in the forefront of the mind of a researcher dealing with structural and temperatures studies of polymer samples where the removal of solvent traces might be a difficult and time-consuming process.

7.4.2
Morphology of Semicrystalline Polymers on Surfaces and in Bulk

Polymer crystallization from melt and from concentrated solutions is a much more complex phenomenon compared to crystallization in diluted solutions, and their mechanisms are far from being completely understood. It is documented that crystal growth from melt leads to the formation of characteristic morphological patterns, among which spherulitic morphology is probably the best known.

Spherulitic morphology has been observed on the surface of ultrathin film (20 nm in thickness) of LDPE prepared by dipping a piece of Si wafer into a hot polymer solution in xylene, Fig. 7.19a. Most of the lamellae, which are seen in the high-resolution images in Fig. 7.19c–d, are oriented edge-on with the respect to the substrate, and only a few of them exhibit a flat-lying orientation. The phase image emphasizes the granular structure of the lamellar edges and the flat-lying surfaces. The grains of the lamellar edges are 10–15 nm in diameter, and those seen on the surface of flat-lying lamellae are 6–8 nm in diameter. The granular structure is similar to that reported in the recent AFM studies of a number of crystalline materials [45, 46]. The hypothesis that these grains with folded chain interior and less-ordered exterior are the building units of crystalline architecture will be further discussed below.

Most likely, the architecture of the dip-crystallized LDPE layer is a result of competition between the rate of crystallization and the speed of pulling. Though the metastable character of this morphology can be expected, only AFM provides a direct visualization of how this morphology transforms to a more stable one and how fast these changes occur on the micron and sub-micron scale [47]. The image

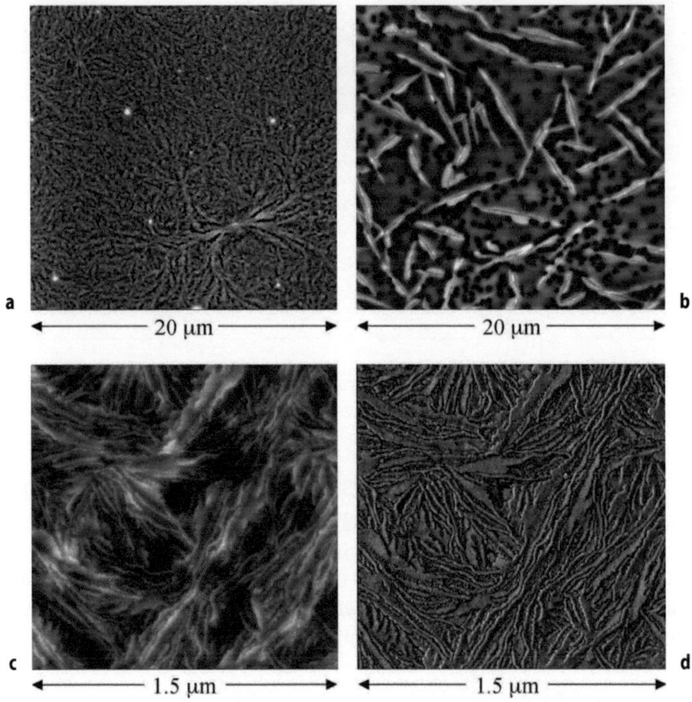

Fig. 7.19. Images of ultra thin LDPE film as-received by dipping (**a**), (**c**), (**d**) and after annealing at 90 °C (**b**)

in Fig. 7.19b shows the film morphology after heating to 90 °C, which is below melting temperature of this polymer measured by DSC. This morphology is characterized by lamellar platelets surrounded by an amorphous layer with a large number of small holes, which appear as a result of partial dewetting. The AFM images have documented the appearance of this morphology in the result of recrystallization. Melting of the initial morphology was accompanied by the growth of lamellar platelets [47]. This recrystallization is quite similar to one suggested for the wet crystals of PE and ultralong alkanes.

Another demonstration of the morphological changes that accompany the melting of semicrystalline polymers is given in AFM images in Fig. 7.20a–c. The images were obtained at different temperatures of an ultrathin film of isotactic polypropylene (iPP) deposited on a Si substrate by spin casting of hot xylene solution. Initial morphology of this film is characterized by single crystals, which are embedded into a network of tiny fibrils, Fig. 7.20a. High stereoregularity of this iPP is the likely reason for the single crystal formation. The fibrilar network is less thermally stable that the single crystals, and it disappears after heating to 120 °C, Fig. 7.20b. A complete melting of the single crystals was observed at temperatures higher than 160 °C, Fig. 7.20c.

It is well established that crystalline morphology strongly depends on the stereoregularity of polymer chains. Linear PE crystallizes as high-density mater-

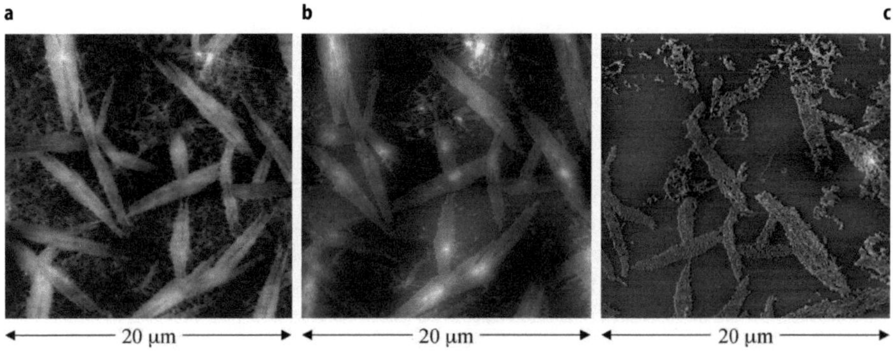

Fig. 7.20a–c. Images of ultrathin film of iPP at different temperatures: 25 °C (**a**), 120 °C (**b**) and 160 °C (**c**)

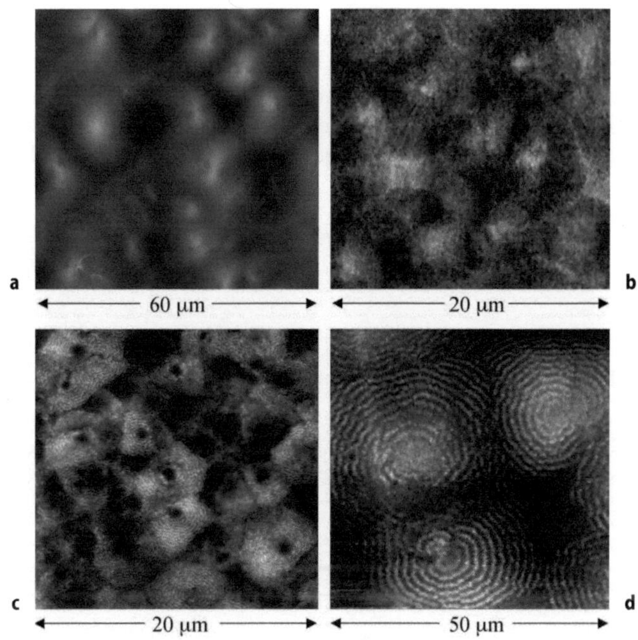

Fig. 7.21. **a** Image of surface of melt-crystallized LDPE film. **b,c** Images of microtomed surface of bulk LDPE samples before and after permanganate etching, respectively. **d** Image of microtomed surface of LLDPE/HDPE blend after permanganate etching

ial (HDPE) with high crystallinity, whereas the presence of branches of different lengths (starting with methyl) gives low-density material, LDPE. It also has lower crystallinity, which is reflected in the morphology of LDPE. Poorly developed spherulites, also known as sheaf-like patters are seen on the surface of a melt-crystallized LDPE sample, Fig. 7.21a. Differences in polymer morphology on surface and in the bulk are also expected. An access to bulk morphology is provided by cutting bulk samples with an ultramicrotome. However this procedure might also destroy some material, which will appear on the microtomed surface, Fig. 7.21b. In some cases, chemical etching of semicrystalline samples is applied to remove amorphous polymer and thus bulk crystalline morphology is opened for visualization. Indeed, AFM images of a microtomed LDPE surface subjected to permanganate etching clearly show numerous spherulites with banded structure, Fig. 7.21c. Extensive folding into lamellae expected for linear macromolecular chains will give material with higher crystallinity. This is characteristic for a blend of HDPE with linear low-density PE (LLDPE), which also is characterized by linear structure of its chains. Indeed, the image of HDPE/LLDPE blend obtained on a microtomed and etched surface of this blend, Fig. 7.21d, shows large banded spherulites. They occupy a larger part of the surface area seen in the image.

7.4.3
Lamellar Structure of Melt-Crystallized Polymers

High-resolution images in Fig. 7.22a–d show lamellar structures of polymer samples with different crystallinity. Height variations of the surface of banded spherulite correlate with the change of lamellar orientation. Elevations are formed of stacks of lamellae that are lying almost flat, where depressed locations are formed of stacks of lamellae standing edge-on, Fig. 7.22a. Of special concern are

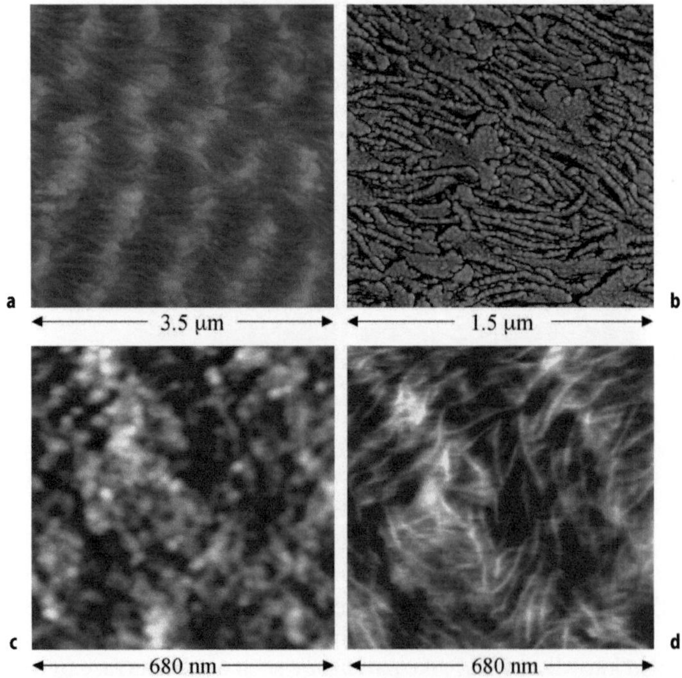

Fig. 7.22. a,b Height and phase images of lamellar surface of a banded spherulite of the LLDPE/HDPE blend. **c,d** Height images of surface of melt-crystallized LDPE film obtained in light and hard tapping, respectively

granular structures, which are seen at lamellar edges and also on the surface of flat-lying lamellae in Fig. 7.22b. During the past 20 years, there have been a number of electron microscopy observations of similar structures, which were questionable [45, 46]. At present, numerous AFM images of semicrystalline samples have confirmed the existence of granular structures, yet their nature might be diverse and it requires further studies [47, 48].

A surface of the sheaf-like crystals of LDPE consisting of grains is seen in the high-resolution image in Fig. 7.22c. With an increase the tip-sample force, the image pattern is different; it shows slightly curved fibrils, Fig. 7.22d. The described changes are fully reversible, and grainy morphology is restored when the tip-sample force is lowered again. As mentioned above, this suggests that the fibrils consist of a more rigid core and a rubbery-like exterior [47]. In this case the granular morphology is characteristic of the polymer organization at the surface. Traces of granular structures can also be found in the bulk, yet their nature is not necessarily the same.

Surface structure of two more types were found on melt-crystallized samples of HDPE and a metallocene-based PE (mPE) with a high number of octene branches and low density, Fig. 7.23a,b. The structures visible in the first image represent edges of HDPE lamellae that might degenerate into fibrils due to a lack of polymer material on the surface, Fig. 7.23a. Granular substructure is seen on the lamellar edges. The surface of the mPE sample looks quite different as seen from the image in Fig. 7.23b, which was obtained in light tapping. In contrast to HDPE, the mPE polyethylene exhibits mostly grainy surface features. When imaged in hard tapping, the same location looks different, and short fibrils are seen everywhere. Therefore, there is some analogy of the granular structure to one observed on the surface of LDPE, yet the crystallinity of mPE is very small.

So far, most of the AFM studies of semicrystalline polymers were illustrated by images of different PE and PP. High-contrast imaging of lamellar structures of these polymers is facilitated by the fact that at room temperature amorphous material in their samples is in a rubbery-like state. This is not necessarily the case for other polymers. For example, AFM images of lamellar structures of syndiotactic PS (sPS) provide high-contrast patterns at temperatures above its $T_g \approx 100$ °C, Fig. 7.23c–d. The presented images reveal differences in lamellar structure of α- and β-crystalline modifications of sPS. It is worth emphasizing that AFM images show distinctively lamellar structure of α-crystalline modifications of sPS, despite small variations of densities of crystalline and amorphous materials in this polymer. In this sense, AFM complements SAXS, which has difficulties in examining crystalline structure of α-phase sPS. Visualization of lamellar structures with AFM also can be successfully applied to justify real-space models derived from SAXS data, which are obtained in reciprocal space. For example, formation of lamellar stacks is one possible model that has been derived from SAXS data of PET crystallization and has been unambiguously proved by AFM images obtained during in situ visualization of the crystallization of this polymer [19]. One should not forget that AFM images provide a local two-dimensional view of lamellar structures where SAXS present an averaged three-dimensional data. Therefore, only a combination of these techniques can provide a reliable description of crystalline architecture. A quantitative analysis of AFM images is important for this description.

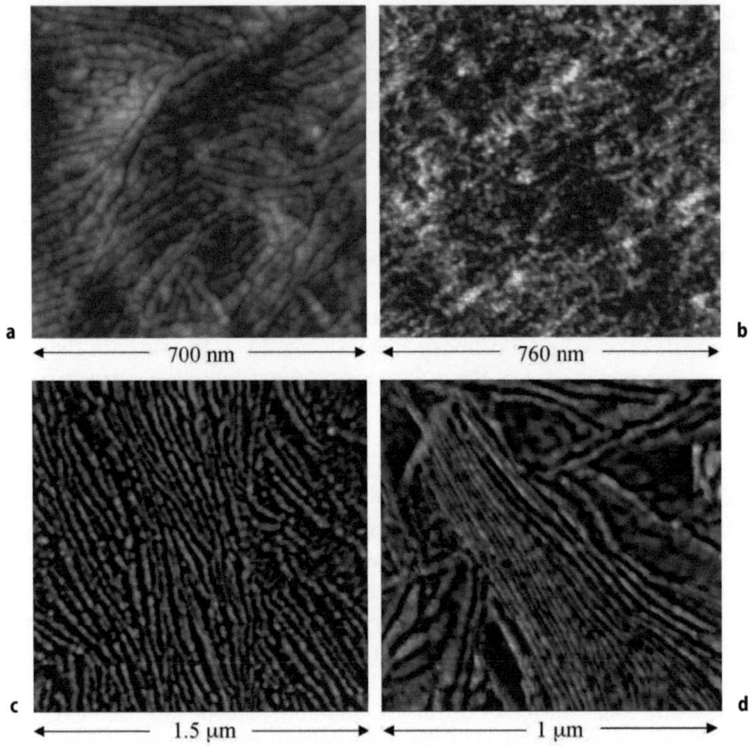

Fig. 7.23a–d. Images of lamellar surface of melt-crystallized HDPE film (**a**), melt-crystallized film of mPE ($\rho = 0.86$ g cm^{-3}) (**b**), and lamellar surface of sPS in crystalline α-modification (**c**), and β-modification (**d**). Images in *a* and *b* were obtained at room temperature, images in *c* and *d* at 120 °C

First efforts to apply correlation analysis routinely used in SAXS for quantification of AFM images of lamellar structures were made recently [49]. Such treatment is applied to the image locations, where stacks of edge-on standing lamellae are seen. If the contrast of lamellae is high enough (images obtained in hard tapping might be the best for this) then the correlation analysis gives not only a pitch along this stack similar to long period determined from SAXS data, but also an averaged lamellar core width. These data can be used for evaluation of linear crystallinity in the samples of semicrystalline polymers [50].

In addition to semicrystalline polymers, there are a number of so-called mesomorphic polymers that in addition to crystalline and amorphous states exhibit intermediate states (mesophases) with some degree of order. AFM has been applied to study structural changes in the –50 °C to +50 °C range, which covers transitions from crystalline phase to a mesophase and finally to isotropic state, for different crystalline modifications of polydiethylsiloxane (PDES) [51]. This is one of only few AFM studies performed at sub-zero temperatures. The AFM images of mesomorphic β-phase PDES are shown in Fig. 7.24a,b. Extended rods of approximately 500 nm in width are stretched from the bottom to the top in Fig. 7.24a. They pre-

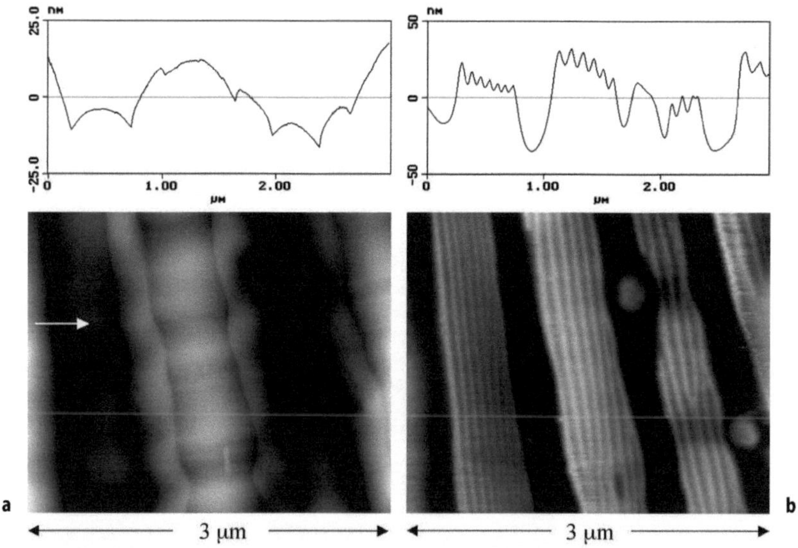

Fig. 7.24a,b. Height images of PDES at room temperature, which were obtained in light and hard tapping, respectively. Cross-section profiles taken along the direction indicated with an *arrow* are shown *above* the images

sent large lamellae of the mesophases of β-PDES. Imaging at elevated force has revealed that these lamellae consist of a rigid core (Fig. 7.24b) that is wrapped in ill-ordered polymer layers [51]. The AFM probe penetrates through this exterior and reveals its core structure. After the tip force is lowered, the lamellae restore their original shape. This is one of the examples of sub-surface imaging that can be realized for semicrystalline polymers with top surface being in a rubbery-like state.

7.5 Compositional Imaging of Heterogeneous Polymer Systems

Compositional AFM imaging of latex and PE blends was demonstrated above. Several examples of AFM imaging of heterogeneous samples of block copolymer, polymer blends, and multicomponent rubbery-like materials are presented below. The results are obtained in studies of thin films and bulk samples, whose surfaces were prepared with a cryo-ultramicrotome. It will be shown that that the contrast related to different components can show up not only in phase images but also in height images.

7.5.1 Morphology of Block Copolymers

Morphology of block copolymers prepared by bonding blocks of chemically different polymers in a regular fashion reflects micro-phase separation on the scales defined by a length of individual blocks. Visualization of the block copolymer morphology with TEM in most cases requires a selective staining of the components,

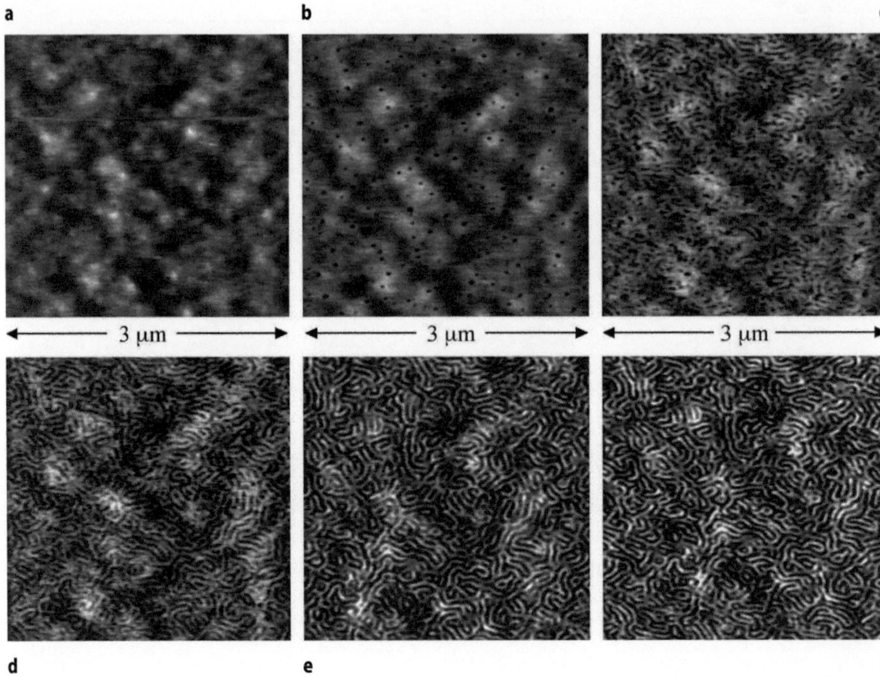

Fig. 7.25a–f. Height images of annealed film of triblock block copolymer polystyrene-*b*-polybutadiene-*b*-polystyrene at different A_{sp}/A_0 ratios: 0.91, 0.88, 0.84, 0.81, 0.74, 0.64. $A_0 = 20$ nm

whereas in AFM differences in mechanical properties of the components can be sufficient for this purpose. Ambient AFM studies of many block copolymer samples provide images with a pronounced contrast when one component is in glassy state and another in a rubbery-like state. Of course, such imaging should be performed at elevated tip-sample forces. This is illustrated by AFM height images of triblock copolymer of polystyrene-*block*-polybutadiene-*block*-polystyrene (SBS) films in Fig. 7.25a–f. Under light tapping conditions, the image reveals a slightly corrugated topography of the top layer. Note that the topmost layer is formed of polybutadiene blocks, which are rubbery at room temperature and have lower surface energy than polystyrene blocks. With the tip-force increase at lower A_{sp}, the height image changes drastically. This is related to a gradual tip penetration into the material until it reaches a microphase-separated morphology formed by alternating PS and PB cylinders. The glassy PS cylinders are typically brighter than rubbery PB cylinders. A similar sequence of patterns is recorded in the phase images at different A_{sp} [52]. The contrast between the components is strong at temperatures of around 100 °C where softening of PS block occurs. For some block copolymers, high-contrast imaging can be achieved at temperatures above the glass transition of one component but below the glass transition of another [53]. Note that in AFM studies of materials with a rubbery top layer that the probe might be able to detect rigid structures hidden below the surface at depths of hundreds of nanometers.

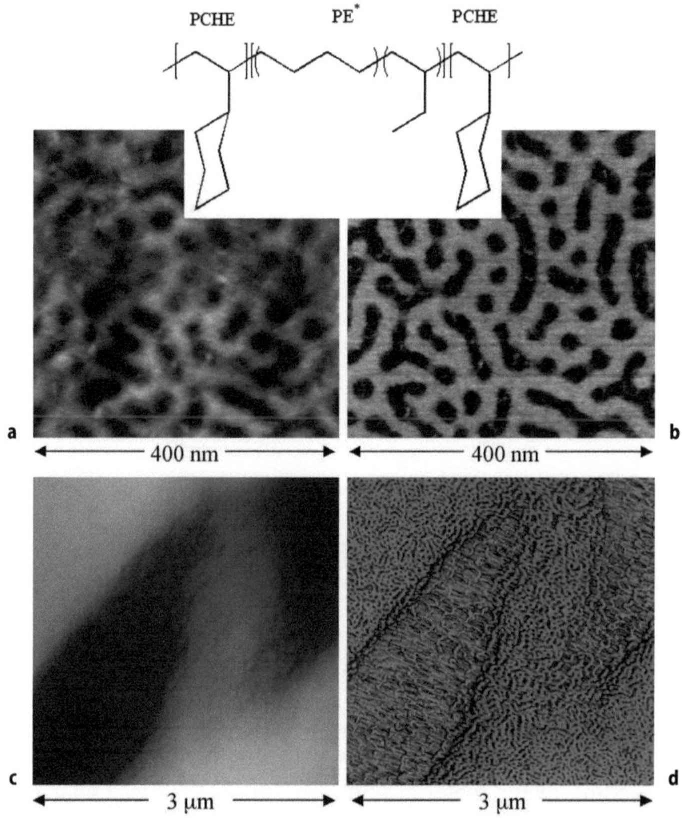

Fig. 7.26. **a,b** Height and phase images of microphase separated morphology of triblock copolymer PCHE-*b*-PE-*b*-PCHE film. **c,d** Height and phase images of the film location where two crazes met. Chemical structure of this block copolymer is shown in *insert* to the images in *a* and *b*

A new kind of block copolymers of poly(cyclohexylethylene)-poly(ethylene)-poly(cyclohexylethylene) (PCHE-PE-PCHE) were synthesized by catalytic hydrogenation of SBS, see the insert in Fig. 7.26a,b. These novel materials are characterized by micro-phase separation of semicrystalline PE blocks and glassy PCHE blocks and their morphology is seen in the AFM images in Fig. 7.26b. The semicrystalline nature of PE blocks makes them seen darker in the phase images. Tiny bright features of 4–5 nm in width distinguishable inside the PE blocks, present small lamellae of the crystalline polymer. AFM has substantially complemented TEM in visualization of structural changes of PCHE-*b*-PE-*b*-PCHE block copolymer caused by stretching [54]. The images of neighboring crazes formed in thin film of this block copolymer are shown in Fig. 7.26c–d. The height image shows surface depression at the crazes, whereas the phase image demonstrates local changes of the nanoscale structure. Unfolding of PE blocks into nanofibrils is clearly seen inside the crazed material. The nanofibrils, which are oriented perpendicular to the direction of craze growth, alternate with glassy domains extended

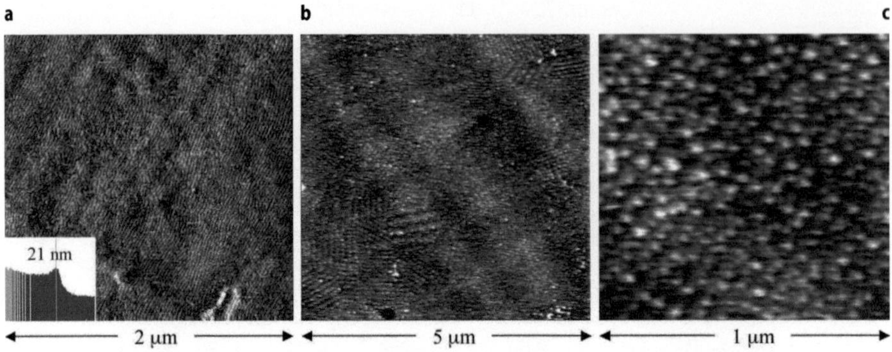

Fig. 7.27. **a** Phase image of microtomed surface of SEBS block copolymer with the 2D-FFT power spectra shown in the *insert*. **b,c** Height images of microtomed surface of SEBS/oil (60/40) mixture

along the craze in a wavy fashion. Further details of AFM characterization of crazes in PCHE/PE block copolymers will be given in a separate paper [55].

In a number of practical applications, properties of block copolymers are modified using oil as filler. This causes changes in the morphology of triblock copolymers, which can be examined with AFM. Microphase separation of pure triblock copolymer poly(styrene)-*b*-poly(ethylene-butylene)-*b*-poly(styrene) (SEBS) is characterized by a lamellar arrangement with a repeat distance of 21 nm, Fig. 7.27a. Due to partial damage (this surface was prepared by cryo-microtomy) the periodic pattern of this surface is less pronounced than that of spin-cast films of block copolymers. Loading of the material with 40% oil has led to a change from the lamellae-type morphology to micellar morphology. On a large surface area, Fig. 7.27b, one sees differently oriented domains formed of periodic rows of elevated features. The local order is not perfect but can be characterized by a structural factor of 53 nm. In other terms, the separation between elevated locations that can be assigned to micelle cores is substantially increased compared to the spacing between lamellae in the pure triblock copolymer. Most likely, this is due to the incorporation of oil into the ethylene-butylene phase, which becomes substantially greater. The micelle cores are formed of harder PS blocks, which is a minor component in this material [56].

7.5.2 Studies of Polymer Blends

In contrast to block copolymers, polymer blends are composed of chemically different polymers that are not chemically bonded to each other. Interactions between different components define their degree of miscibility, and phase separation is a common phenomenon for polymer blends. Dynamics of phase separation in polymer blends have been examined with various microscopic and scattering techniques. The latter provides averaged information in reciprocal space, whereas the former provides information in real space. AFM is a quite suitable microscopic technique for this purpose due to enhanced resolution and a larger experimental

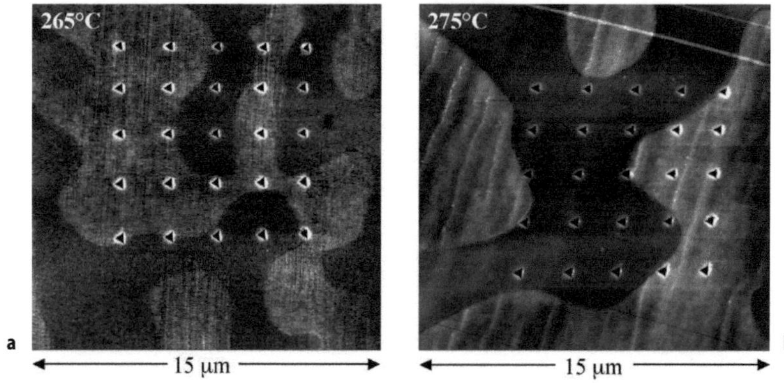

Fig. 7.28a,b. Height images of microtomed surfaces of cross-sections of TMPC/PS (50/50) films annealed at different temperatures for 500 s

spatial window. It provides real-space observation of different coarsening processes and allows direct quantification of phase volume fractions and computation of domain interface curvatures of various morphologies.

Initially, AFM was been frequently applied to studies of phase separations in thin and ultrathin films on various substrates [57–59]. Later, this technique was used for examination of spinodal decomposition of tetramethyl bisphenol A polycarbonate (TMPC) and PS mixtures in bulk, at both critical and off-critical compositions [60, 61]. Relatively thick TMPC/PS films were quenched after different times of annealing times at temperatures just above a lower critical solution temperature. The fractured surfaces of the phase-separated thick films exhibit a nanometer-scale surface topography that is resolvable by AFM, Fig. 7.28a,b. By comparing the AFM images with optical micrographs the identity of the topographic patterns and phase-separated morphology was established. Most likely, the height variations stemmed from a differential plastic deformation of the phases. The higher domains have been assigned to those enriched in PS, that is, the component with lower molecular weight. This assignment was supported by the nanoindentation results. The pits produced by the constant-force indenting exhibit well-pronounced pile-ups on the brighter domains consistent with their more plastic character, Fig. 7.28a,b. Different structural factors of these two morphologies are also noticeable from the AFM images.

In the study of TMPC/PS blends with different compositions, phase separation was induced by isothermal annealing at various temperatures within 240–275 °C and annealing times up to 1,000 s [60, 61]. The surface profiles, revealed in height images, were analyzed according to their in-plane spinodal morphology and their (perpendicular) height distribution. 2D-FFT analysis of the images revealed that the structural factors range between 200 nm and 12 microns. Average height profiles between two plateaus were argued to be a function of the composition difference between coexisting phases.

Correlation between morphology of polymer blends and their mechanical properties is one of the key issues for understanding performance of polymeric materials and their optimization. Different compatabilizers that might improve blend-

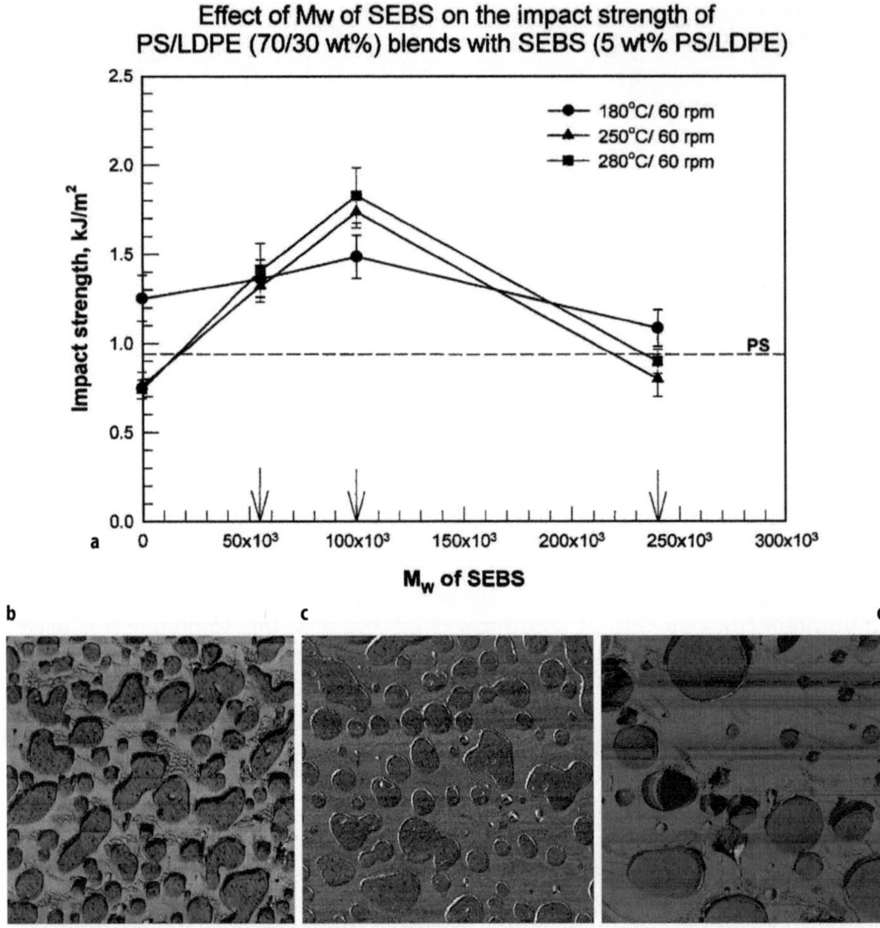

Fig. 7.29. **a** Dependence of the impact strength of PS/LDPE (70/30 wt%) blends with SEBS (5 wt% PS/LDPE) on molecular weight of SEBS. **b–d** Phase images of microtomed surfaces of PS/LDPE blends with SEBS of molecular weight 100 K, 50 K and 250 K, respectively. Samples and the mechanical data – courtesy of J. Higgins (Imperial College, London, UK)

ing and adhesion between different components are added to polymer mixtures in small quantities during extrusion. Block copolymers whose different blocks are miscible with different components of a polymer blend are potential candidates for this function. The effect of SEBS block copolymer on the impact strength of LDPE/PS blend is seen from the graph in Fig. 7.29a. Best mechanical performance was achieved by adding SEBS with 100 000 g/mol molecular weight. The AFM images in Fig. 7.29b–d show three different morphologies for the samples with SEBS of different molecular weight. The images were obtained on the microtomed surfaces of the bulk samples. The dark contrast of the phase images belongs to LDPE domains, which are more compliant than glassy PS. A similar assignment was made in the phase images of triblock copolymer PCHE/PS. Finally, this interpre-

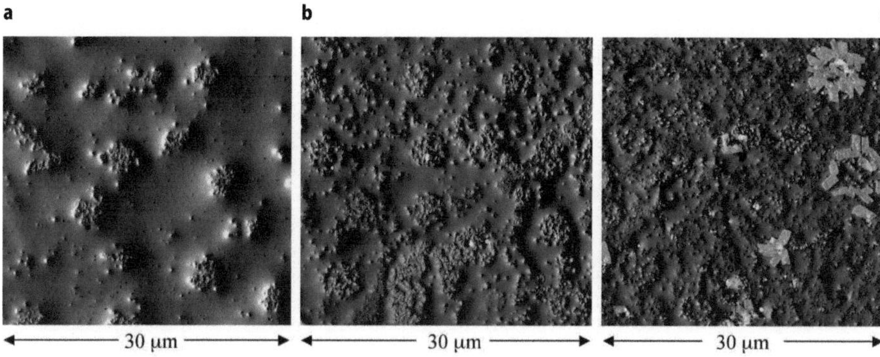

Fig. 7.30a–c. Phase images of microtomed surface of EPDM rubbers subjected to sulfur-based curing. Concentration of the curing agent is 0.5%wt in (**a**), 1%wt in (**b**) and 1.5% in (**c**)

tation was proved by high-magnification images of the dark domains (not shown here) that revealed lamellar structures of these domains. The comparison of the morphologies with the mechanical data shows that a narrower size distribution of LDPE domains and their more homogeneous distribution inside the matrix are in favor of higher impact strength of the blend. It is worth noting that identification of traces of minor component – SEBS in the AFM images of the blends either in LDPE or at the interfaces – has not been successful. One reason is disturbance of the interfaces by the microtome knife during sample preparation. In the worse performed material, some LDPE domains were removed from the matrix by the cutting due to poor adhesion, Fig. 7.29d.

Compositional imaging of multicomponent polymer materials is more challenging than for the above-described block copolymers and blends. Several examples of compositional imaging of rubbery materials and thermoplastic vulcanizates are collected below. The sensitivity of AFM to variations of cross-linking density has been demonstrated in studies of rubber blends, including NR and EPDM [62]. The phase images showing morphological changes following cross-linking of EPDM rubber with sulfur as the cure agent are presented in Fig. 7.30a–c [63]. The curative's concentration varies from 0.5% to 1.5%. The cross-linking resulted in growth of domains consisting of harder locations inside the rubber. With an increase in the curative's concentration from 0.5% to 1.0%, the surface area covered by domains enlarged from 20% to 40%, Fig. 7.30a,b. The cross-linked regions cover most of the surface area in the sample with 1.5% curative, Fig. 7.30c. Bright platelets also seen in this image are probably sulfur crystals formed in excess of the cross-linking system.

When in polymer blend rubber is mixed with plastics, compositional imaging and identification of the components can be achieved in different AFM modes. The images in Fig. 7.31a,b show the morphology of iPP/EPDM blend. The depression locations seen in the height image exhibit a darker contrast in the phase image. Therefore, the domains can be assigned to soft rubbery regions, which are surrounded by hard iPP matrix. The force curves obtained in contact mode presented in Fig. 7.31c supports this assignment. A slope of the DvZ curve measured on the

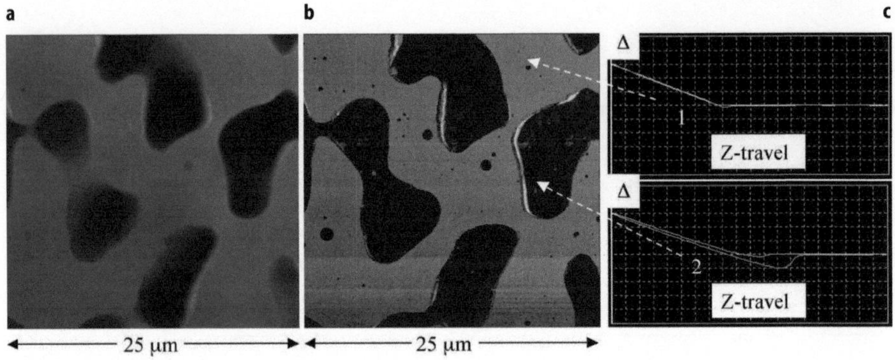

Fig. 7.31. a,b Height and phase images of microtomed surface of iPP/EPDM blend. **c** Deflection-vs-Z-travel *curves 1* and *2* recorded on iPP and EPDM locations, respectively. *Dotted lines* indicate these locations

matrix is steeper than a slope of the DvZ curve recorded on the rubber domain. A map of DvZ curves provides the same pattern as the phase image in Fig. 7.31b. This is also true for an amplitude image obtained in force modulation mode.

Morphology of thermoplastic vulcanizate formed by mixing of iPP with EPDM and carbon black particles is more complex and strongly depends on mixing conditions. The height and phase images of this material are shown in Fig. 7.32a,b. The matrix has been assigned to iPP filled with carbon black (CB) particles. This was justified by observation of lamellar structure found in some locations less filled with CB. The latter are seen as 30–40 nm bright spots scattered over the matrix. There is a partial aggregation of CB. Domains of different size that are embedded into the matrix present EPDM rubber. Variations of phase contrast seen in these domains are related to locations with different cross-linking density. Segregation of conducting CB particles in the matrix favors a formation of percolation network inside this material. The phase image obtained in EFM mode, Fig. 7.32c, exhibits a darker contrast for the matrix material. It is an indication that these regions are

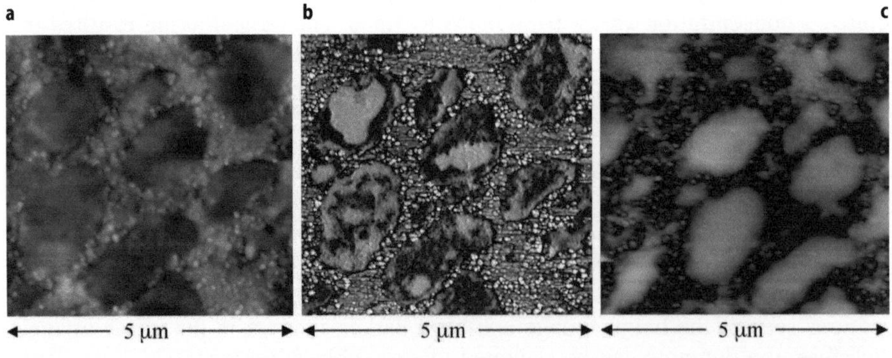

Fig. 7.32. a,b Height and phase images of thermoplastic vulcanizate, consisting of iPP matrix filled with carbon black particles and domains of EPDM rubber. **c** Phase images of the same location as in *a,b* obtained in electric force microscopy

electrically conducting where the brighter contrast of the rubber domains is consistent with absence of CB particles in these locations.

In general, the analysis of AFM images of rubbery-like materials might be quite complicated because of the large number of components involved. Therefore, empirical recipes obtained from studies of model systems will be quite helpful. One such recipe that was worked out for multicomponent rubber materials [64] indicates that when phase imaging is performed with $A_0 = 60$ nm and $A_{sp} = 0.4 A_0$, then brightness of the contrast in phase images will be the highest on fillers (CB, silica, etc.) and will drop in the following sequence: plastics, butyl rubber, EPDM, SBR, BR, NR, oil. One should also take into account that effective stiffness of rubbery-like components depends on the size of such domains. AFM probe might feel a small rubbery-like inclusion in a rigid matrix as a stiffer material compared to a large rubber-like domain.

7.6 Future Outlook

During the last decade AFM has been recognized as a useful characterization tool for the analysis of polymers. High-resolution imaging and compositional mapping are the key features for the majority of applications. The examples collected in this chapter support this conclusion. Further expansion of AFM experimental data is expected with a broader use of temperature accessories. So far, imaging can be performed routinely at elevated temperatures, whereas studies at low temperatures are not common. No doubt that researchers will welcome the development of instrumentation, which will allow imaging at low temperatures. Temperature variations as well as environmental control and imaging under liquids will substantially broaden and deepen AFM possibilities in studies of polymers. Improved control of tip-sample forces and the ability to operate at very low force levels is another desirable feature. It will allow imaging of extremely soft surface features, which could not been "seen" at tip-force levels used in current studies. Lowering of tip force might potentially increase image resolution, which will also benefit from use of sharper probes.

Progress is also expected in local measurements of mechanical properties. There are a number of different approaches (force modulation, oscillatory shearing, nanoindentation, etc.), which are applied for this purpose. These techniques provide semi-quantitative data in a relatively narrow frequency range, mostly around resonant frequencies of the piezoactuators or probes. Although such measurements are extremely important they should be extended to a broader frequency range over several decades of frequencies where molecular motion and relaxation of macromolecules take place. Such studies of polymer viscoelastic behavior have become important not only for macroscopic samples but also for functional plastic and rubbery-like structures and elements with dimensions in the millimeter and micron range. Measurements at those scales should be performed with AFM-related techniques. The demand for such studies will further increase with development of nanotechnology.

The probing of materials with AFM is basically realized through mechanical interactions. Therefore, only indirect correlation between these measurements and the chemical nature of a sample or its constituents can be derived from such studies. This does not fulfill the need of compositional analysis with chemical identification. The latest achievements in this field are primarily related with soft X-ray spectromicroscopy, which provide chemically sensitive mapping with resolution of tens of nanometers [65]. Yet this technique requires the use of synchrotron radiation sources. Conventional methods like FTIR-microscopy provide chemical mapping only with resolution of several microns. Recently, a combination of AFM and infrared spectrometry has been successfully applied to collecting IR spectra of polymers from the tip-sample junction [66]. In perspective, such a combination might become an extremely powerful mapping technique with lateral resolution in the tens of nanometers range.

References

1. Binnig G, Rohrer H, Gerber C, Weibel E (1982) Phys Rev Lett 49:57
2. Binnig G, Quate C, Gerber C (1986) Phys Rev Lett 56:930
3. Magonov SN, Whangbo M-H (1996) Surface analysis with STM and AFM. VCH. Weinheim
4. Landman U, Luedke WD, Nitzan A (1989) Surf Sci Let 10:L177
5. Weinsenhorn AL, Maivald P, Butt H-J, Hansma PK (1992) 45:11226
6. Zhong Q, Innis D, Kjoller K, Elings VB (1993) Surf Sci Let (1993) 290:L688
7. Sugawara J, Ohta M, Ueyama H, Morita S (1995) Science 270:1646
8. VanLandingham MR, McKnight SH, Palmese GR, Elings JR, Huang X, Bogetti TA, Eduljee RF, Gillespie JW Jr (1997) J Adhesion 64:31
9. Grandbois M, Beyer M, Rief M, Clausen-Schaumann H, Gaub HE (1999) Science 283:1727
10. Heinz WF, Hoh JH (1999) Trends Biotechnol 17:143
11. Maivald P, Butt H-J, Gould SAC, Prater CB, Drake B, Gurley JA, Elings VB, Hansma PK (1991) Nanotechnology 2:13
12. When the resonant frequency of the probe is defined away from the sample surface, its retuning in the immediate vicinity of the sample surface is needed. It will eliminate a frequency shift of the probe caused by damping effect of air squeezed near the surface
13. Hoeper R, Gesang T, Possart W, Hennemann O-D, Boseck S (1995) Ultramicroscopy 60:17
14. Magonov SN (2000) Atomic force microscopy in analysis of polymers. In: Meyers RA (ed) Encyclopedia of analytical chemistry. Wiley, Chichester, p 7432
15. Cleveland JP, Anczykowski B, Schmid AE, Elings VB (1998) Appl Phys Lett 72:2613
16. Wang L (1998) Appl Phys Lett 73:3781
17. Bar G, Brandsch R, Whangbo M-H (1998) Surf Sci Lett 411:L802
18. Ivanov D, Daniels R, Magonov S (2001) Exploring the high temperature AFM and its use for studies polymers. Application note. Digital Instruments/Veeco Metrology Group, Santa Barbara CA, USA
19. Ivanov DA, Amalou Z, Magonov SN (2001) Macromolecules 34:8944
20. Fasolka MJ, Mayes AM, Magonov S (2001) Ultramicroscopy 90:21
21. Kumaki J, Nishikawa Y, Hashimoto T (1996) JACS 118:3321
22. Lyubchenko YL, Jacobs BL, Lindsay SM (1992) Nucleic Acid Res 20:3983
23. Percec V, Ahn C-H, Ungar G, Yeardley DSP, Möller M, Sheiko S (1998) Nature 391:161
24. Frey H, Lach C, Lorenz K (1998) Adv Mater 10:279
25. Ponomarenko SA, Boiko NI, Shibaev VP, Magonov SN (2000) Langmuir 16:5487
26. Gerle M, Fischer K, Schmidt M, Roos S, Mueller AHE, Sheiko SS, Prokhorova SA, Möller M (1999) Macromolecules 32:2629
27. Prokhorova SA, Sheiko SS, Ahn C-H, Percec V, Möller M (1999) Macromolecules 32:2653
28. Prokhorova SA, Sheiko SS, Mourran A, Azumi R, Beginn U, Zipp G, Ahn C-H, Holerca MN, Percec V, Möller M (2000) Langmuir 16:6862
29. Sheiko SS, Möller M (2001) Chem Revs 101:4099
30. Prokhorova SA, Sheiko SS, Moeller M, Ahn C-H, Percec V (1998) Macromol Rapid Commun 19:366

31. Percec V, Holerca MN, Magonov SN, Yeardley DJP, Ungar G, Duan H, Hudson SD (2001) Biomacromolecules 2:706
32. Percec V, Obata M, Rudick JC, De BB, Glodde M, Bera TK, Magonov SN, Balagurusamy VSK, Heiney PA (2002) J Polym Sci Polym Chem Ed 40:3509
33. Tartsch B, Magonov SN, Möller M (2003) Adv Mater (in press)
34. Patil R, Kim S-J, Smith E, Reneker DH, Weisenhorn AL (1991) Polym Commun 64:117
35. Patil R, Reneker DH (1994) Polymer 35:1909
36. Nie H-Y, Motomatsu M, Mizutani W, Tokumoto H (1996) Polymer 37:183
37. Boyd RD, Badyal JPS (1999) Adv Mater 9:895
38. Geil PH (1993) Polymer single crystals. Wiley
39. Barham P (1993) Crystallization and morphology of semicrystalline polymers. In: Cahn RW, Haasen P, Kramer EJ, Thomas EL (eds) Materials science and technology, structure and properties of polymers. VCH, vol 12, p 153
40. Magonov SN, Yerina NA (2003) Langmuir 19:500
41. Magonov SN, Yerina NA, Ungar G, Reneker DH, Ivanov DA (2003) Macromolecules 36: 5637
42. Wunderlich B (1973) Macromolecular physics, vol 1, (1976) vol 2, (1980) vol 3. Academic Press
43. Craemer K, Wawkuschewski A, Domb A, Cantow H-J, Magonov SN (1995) Polym Bull 35:457
44. Tian M, Loos J (2001) J Polym Sci Phys Ed 39:763
45. Kanig G (1991) Colloid Polym Sci 269:1118
46. Michler GH (1992) Kunstoff-Mikromechanik. Carl Hanser Verlag, p 187
47. Magonov SN, Godovsky YuK (1999) Amer Lab 31:52
48. Hugel T, Strobl G, Thomann R (1999) Acta Polym 50:214
49. Basire C, Ivanov DA (2000) Phys Rev Lett 85:5587
50. Ivanov DA, Magonov S (2002) Atomic force microscopy studies of semicrystalline polymers at variable temperature. In: Sommer J-C, Reiter G (eds) Polymer crystallization: observations, concepts and interpretations. Springer, Berlin Heidelberg New York
51. Godovsky YuK, Papkov VS, Magonov SN (2001) Macromolecules 34:976
52. Magonov SN, Cleveland J, Elings V, Denley R, Whangbo M-H (1997) Surf Sci 85:201
53. Fasolka MJ, Mayes AM, Magonov S (2001) Ultramicroscopy 90:21
54. Ruokolainen J, Fredrickson GH, Kramer EJ, Ryu CY, Hahn SF, Magonov SN (2002) Macromolecules 35:9391
55. Ryu CY, Ruokolainen J, Fredrickson GH, Kramer EJ, Hahn SF, Magonov SN, in preparation
56. Watanabe H, Kotaka T (1983) Macromolecules 16:1783
57. Krausch G, Dai CA, Kramer EJ, Bates FS (1993) Phys Rev Lett 71:3669
58. Tanaka K, Yoon J-S, Takahara A, Kajiyama T (1995) Macromolecules 28:934
59. Karim A, Slawecki TM, Kumar SK, Douglas JF, Satija SK, Han CC, Russell TP, Liu Y, Overney R, Sokolov J, Rafailovich MH (1998) Macromolecules 31:857
60. Cabral JT, Higgins JS, McLeish TCB, Strausser S, Magonov SN (2001) Macromolecules 34:3748
61. Cabral JT, Higgins JS, Yerina NA, Magonov SN (2002) Macromolecules 35:1950
62. Galuska AA, Poulter RR, McElrath KO (1997) Surf Interface Anal 25:418
63. Yerina NA, Magonov SN (2003) Rubber Ind Technol (in press)
64. Galuska AA, personal communication
65. Koprinarov I, Hitchcock AP, Li WH, Heng YM, Stoever HDH (2001) Macromolecules 34: 4424
66. Hammiche A, Pollock HM, Reading M, Claybourn M, Turner PH, Jewkes K (1999) Appl Spectroscopy 53:810

8 Displacement and Strain Field Measurements from SPM Images

Jürgen Keller · Dietmar Vogel · Andreas Schubert · Bernd Michel

8.1 Introduction

Rapid miniaturization of microelectronics, microtechnology, and nanotechnology products has revealed new challenges corresponding to thermo-mechanical reliability. Besides accelerated testing of products and components and numerical finite element analysis, mainly deformation measurement methods are looked for. They allow us to understand response of components to environmental and functional thermo-mechanical loading and are part of advanced reliability studies.

The manufacturing, handling, and control of microscopic and nanoscopic objects require the quantification of their geometrical, kinematic, and mechanical properties. While the measurement of geometrical and size data is more easily accessible by scanning force microscopy (SFM) and related methods, kinematic and mechanical characterization is a general problem for micro and nano objects and devices. Displacements and their derivatives are two basic properties to be measured for kinematic and mechanical description. Until now only a few methods exist to obtain accessible quantified field data for these tiny regions.

Most of the published experimental studies on thermally and/or mechanically induced deformation are qualitative and semi-quantitative analyses or make use of surface profiles depicted from topography scans only [1–4]. For that purpose different kinds of SFM imaging have been used [1, 5]. Among the published quantitative approaches two promising techniques exist – Moiré [6–8] and image correlation-based methods [9–11]. Both techniques utilize load state images captured by scanning force microscopes.

Published work on Moiré measurement reports rather single point displacement or strain measurement [6]. Moiré fringes from crosshatched grids are taken to calculate the mentioned quantities averaging over the whole SFM scan field. In this case one can also employ scans of moderate line number depending on the desired accuracy. Nevertheless, the exploitation of advanced phase shifting algorithms allows one to compute displacement maps from a couple of SFM scans shifted relative to each other [8]. However, this technique is not yet applicable to thermal measurements [6] and needs further development. Especially the influence of smallest scan drifts during capturing of multiple scans for one measurement has to be analyzed in more detail. Another difficulty regarding the Moiré approach is the time-consuming object grid preparation.

In contrast to Moiré measurements, correlation-type measurements are based on higher pixel resolution SFM scans. They allow one to measure displacements and strains with moderate spatial resolution within the SFM scan area [11, 12]. Digital image correlation (DIC) is the technique currently used by most of the authors measuring object deformations from SFM images. Thus, the present chapter is restricted to the detailed description of this approach.

Both DIC and Moiré techniques applied to SFM imaging have their roots in related optical methods. Technology needs to measure in smaller regions have driven developments to transfer optical methods to higher-resolution imaging equipment. Starting from optical Moiré [13], Moiré interferometry [13], and coherent speckle measurement techniques [14], first attempts were made to establish Moiré and DIC measurement methods in scanning electron microscopy (SEM) [15]. Later on they were extended to scanning force microscopy (SFM) [11]. Because SEM as well as SFM imaging are scanning methods, image capture principally takes more time than for optical imaging. This prerequisite and the high magnification of SEM and SFM images have put new issues on the agenda. Namely, stability and accuracy problems had to be solved in the past in order to establish these techniques.

Considering DIC techniques SFM images are captured subsequently for different object states. Mechanical and/or thermal loading is carried out by special loading stages developed for SFM and SEM application. Locally applied cross-correlation algorithms are utilized to compute displacement fields and the corresponding first-order derivatives from SFM images [16]. In this chapter we will first describe the underlying basic principles of digital image correlation. We will then discuss the specific application of DIC to SFM micrographs. This is followed by two examples of the utilization of SFM-based DIC for thermomechanical analysis on microcomponents and nanocomponents. Finally we give an idea of how to use DIC with SFM micrographs to measure kinematic properties of moving solid-state objects.

8.2 Basics of Digital Image Correlation

Correlation analysis on gray scale images can be realized as a *field measuring and characterization method* making use of digital image or pattern acquisition and subsequent digital image processing. Then a complete set of very local image pattern is tracked between two or more object states represented by different digitized images. One of the most outstanding advantages of field measuring methods in connection with digital image processing is the possibility to obtain full two-dimensional field information instead of only point-wise data. Such methods are increasingly applied to problems of displacement and deformation analysis as well as to positioning or repositioning problems.

The method of correlation analysis on gray scale image pattern offers a lot of interesting new possibilities for applications in many areas of materials science and production technology. With the description of materials deformation due to thermal and mechanical loading elastic properties such as Poisson's ratio and the coefficient of thermal expansion (CTE) are obtained by correlation techniques [17]. Other fields of application are the evaluation of system response of complete structures and components including material interfaces. Furthermore, the method can be applied to positioning and repositioning purposes as well as to the detection of microstructural changes in optically perceptible surfaces.

Compared to other displacement and strain measurement techniques such as laser interferometric or Moiré methods digital image correlation has several advantages:

- In many cases only relatively simple low-cost hardware is required (optical measurements) or already existing microscopic tools like SEM and SFM can be utilized without any changes.
- Once implemented in a well-designed software code, the correlation analysis of gray scale images is user friendly and easy to understand in the measuring and post-processing process.
- For optical micrographs no special preparation of the objects under investigation is needed at all. For low magnifications, demands on the experimental environment (e.g., vibration isolation, stability of environmental parameters like temperature) are lower than in laser-based measurements.
- According to its nature the method possesses an excellent downscaling capability. By using microscopic imaging principles, very small objects can also be investigated. Therefore, correlation analysis of gray scale images is predestinated for qualitative and quantitative characterization of micromechanical and nanomaterial properties.

8.2.1 Cross-Correlation Algorithms on Gray Scale Images

Digital image correlation methods on gray scale images were established by several research groups. Examples from different fields of applications can be found in various publications [9, 16, 18–23].

Early attempts to acquire high-resolution SEM images for subsequent deformation analysis have been made by Davidson [21] rescanning SEM photographs. Modern SEMs allow one to capture digital images and to apply correlation algorithms directly to them. This approach has been chosen by different research labs and is described in several publications [16, 22, 23]. The authors have developed and refined different tools and equipment in order to apply SEM images for deformation analysis on thermomechanically loaded electronics packages. The respective technique was established as microDAC, which means micro deformation analysis by means of correlation algorithms [22].

The microDAC technique, by definition, is a method of digital image processing. Digitized micrographs of the analyzed objects in at least two or more different states (e.g., before and during mechanical or thermal loading) have to be obtained by means of an appropriate imaging technique. Generally, the utilized cross-correlation algorithms can be applied to micrographs extracted from very different sources. Digitized photographs or video sequences but also images from, for example, optical microscopy, SEM, LSM, or SPM are suitable for the application of digital image correlation. The basic idea of the underlying mathematical algorithms follows from the fact that images of different kinds commonly allow one to record local and unique object patterns, within the more global object shape and structure. These patterns are maintained, if the objects are stressed by temperature

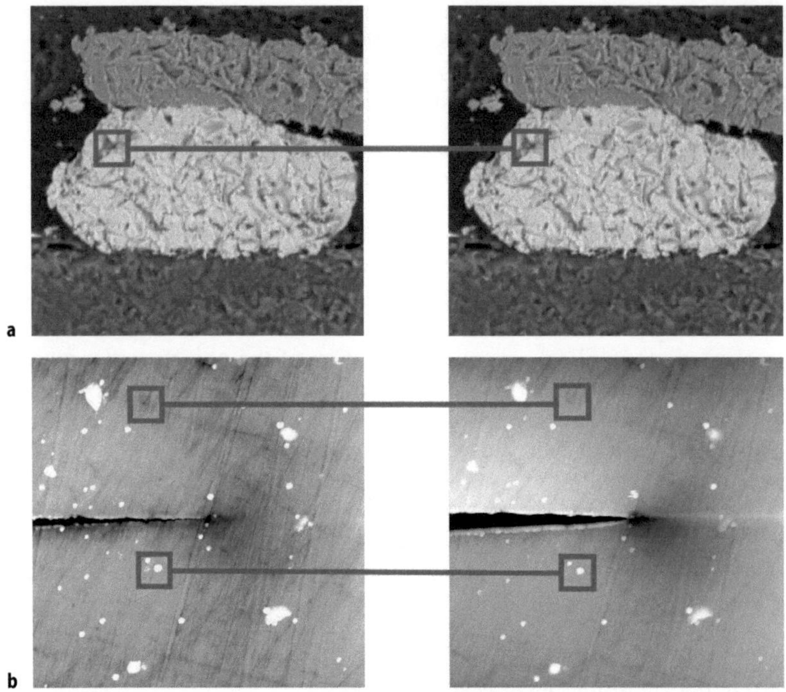

Fig. 8.1a,b. Appearance of local image structures (patterns) during specimen loading. **a** SEM images of flip chip gold bump at room temperature (*left*) and at 125 °C (*right*); **b** SFM topography image of a crack in a thermoset polymer material for different crack opening displacements, scan size 15 × 15 μm

or mechanically. Figure 8.1 shows two examples of images taken by SEM and SFM. Markers indicate typical local pattern of the images. In most cases, these patterns are of stable appearance, even if severe load is applied to the specimens. Even for strong plastic, viscoelastic, or viscoplastic material deformation, local patterns can be recognized after loading, that is, they can function as a local digital marker for the correlation algorithm.

The correlation approach is illustrated by Fig. 8.2. Images of the object are obtained at a reference load state 1 and at a different second load state 2. Both images are compared with each other by using a special cross-correlation algorithm. In the image of load state 1 (reference), rectangular search structures (kernels) are defined around predefined grid nodes (Fig. 8.2, left). These grid nodes represent the coordinates of the center of the kernels. The kernels themselves act as gray scale pattern from load state image 1 that have to be tracked, recognized, and determined by their position in the load state image 2. In the calculation step the kernel window (n × n submatrix) is displaced inside the surrounding search window (search matrix) of the load state image 2 to find the best-match position (Fig. 8.2, right).

This position is determined by the maximum cross-correlation coefficient, which can be obtained for all possible kernel displacements within the search matrix. The computed cross-correlation coefficient K compares gray scale intensity

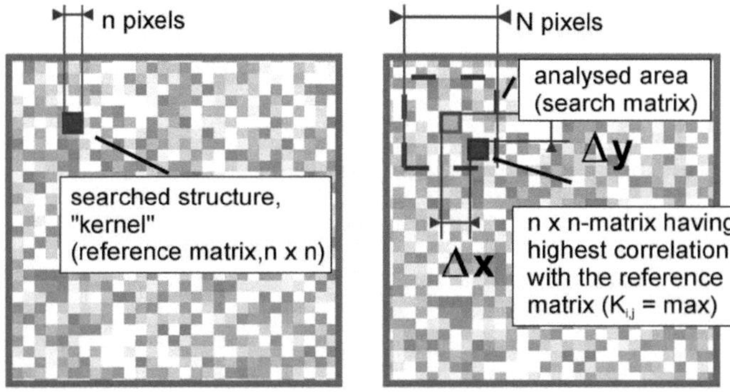

Fig. 8.2. Displacement evaluation by cross-correlation algorithm; reference image at load state 1 (*left*) and image at load state 2 (*right*) used for comparison

pattern of load state images 1 and 2, which have the same size of the kernel. K is equal to:

$$K_{i',j'} = \frac{\sum\limits_{i=i_0}^{i_0+n-1} \sum\limits_{j=j_0}^{j_0+n-1} \left(I_1(i,j) - M_{I_1} \right) \left(I_2\left(i+i', j+j'\right) - M_{I_2} \right)}{\sqrt{\sum\limits_{i=i_0}^{i_0+n-1} \sum\limits_{j=j_0}^{j_0+n-1} \left(I_1(i,j) - M_{I_1} \right)^2 \sum\limits_{i=i_0}^{i_0+n-1} \sum\limits_{j=j_0}^{i_0+n-1} \left(I_2(i+i', j+j') - M_{I_2} \right)^2}} \tag{8.1}$$

$I_{1,2}$ and $M_{I_{1,2}}$ are the intensity gray values of the pixel *(i, j)* in the load state images 1 and 2 and the average gray value over the kernel size, respectively; i' and j' indicate the kernel displacement within the search matrix of load state image 2. Assuming quadrangle kernel and search matrix sizes $K_{i',j'}$ values have to be determined for all displacements given by $-(N-n)/2 \leq i', j' \leq (N-n)/2$.

The described search algorithm leads to a two-dimensional discrete field of correlation coefficients defined at integer pixel coordinates (i', j'). The discrete field maximum is interpreted as the location, where the reference matrix has to be shifted from the first to the second image to find the best matching pattern. Figure 8.3 shows an example of the correlation coefficients inside a predefined search window.

With this calculated location of the best matching submatrix an integer value of the displacement vector is determined.

8.2.2 Subpixel Analysis for Enhanced Resolution

As described in the previous section the calculated displacements by the cross-correlation algorithm are evaluated for integer pixel coordinates. For the calculation of the displacement field with higher accuracy the displacement evaluation has to be improved.

In the reported calculation codes applied by different research groups, the accuracy of the cross-correlation technique is improved in a second calculation step

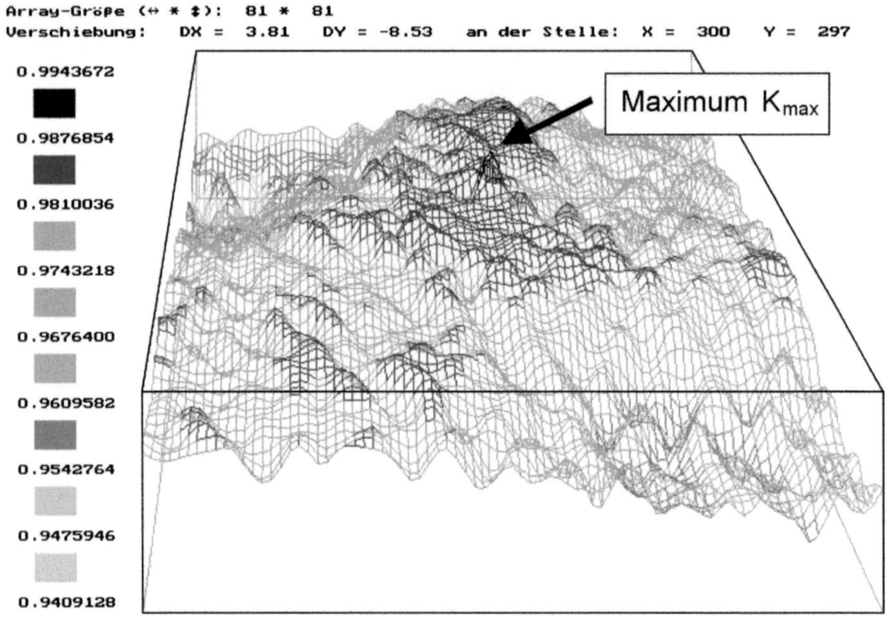

Fig. 8.3. Discrete correlation function $K_{i',j'}$ defined at integer i', j' coordinates; the maximum of the coefficient of correlation is marked by an *arrow*

using a special subpixel algorithm. The presumably simplest and fastest procedure to find a value for the non-integer subpixel part of the displacement is realized in parabolic fitting. The algorithm searches for the maximum of a parabolic approximation of the discrete function of correlation coefficients in the close surrounding of the maximum coefficient $K_{max,discrete}$. The approximation process is illustrated in Fig. 8.4.

The location of the maximum of the parabolic function defines the subpixel part of the displacement. This algorithm implemented quite often allows one to get a

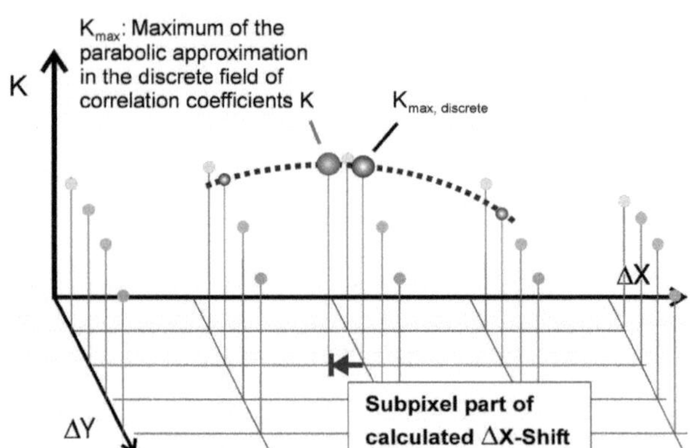

Fig. 8.4. Principle of the parabolic subpixel algorithm

subpixel accuracy of about 0.1 pixel. Even so it must be stated that it can fail considerably and introduce large systematic errors under some circumstances. More advanced algorithms are more accurate and allow one to reach subpixel accuracies up to 0.01–0.02 pixel for common 8-bit depth digitizing, but demand sophisticated analysis and depend on the kind of image sources and of data to be treated.

8.2.3 Results of Digital Image Correlation

The result of the two-dimensional cross-correlation and subpixel analysis in the surroundings of a measuring point primarily gives the two components of the displacement vector. Applied to a set of measuring points (e.g., to a rectangular grid of points with user-defined pitches), this method allows one to extract the complete in-plane displacement field. These results can be displayed in the simplest way as a numerical list which can be post-processed using standard scientific software codes. Commonly, graphical representations such as vector plots, superimposed virtual deformation grids, or color scale coded displacement plots are implemented in commercially available or in-house software packages. Figure 8.5 shows two typical examples of graphical presentations for the results at an SFM image.

Finally, taking numerically derivatives of the obtained displacement fields $u_x(x,y)$ and $u_y(x,y)$ the in-plane strain components ε_{ab} and the local rotation angle ρ_{xy} are determined:

$$\varepsilon_{xx} = \frac{\partial u_x}{\partial x}, \quad \varepsilon_{yy} = \frac{\partial u_y}{\partial y}, \quad \varepsilon_{xy} = 1/2\left(\frac{\partial u_x}{\partial y} + \frac{\partial u_y}{\partial x}\right), \quad \rho_{xy} = 1/2\left(\frac{\partial u_x}{\partial y} - \frac{\partial u_y}{\partial x}\right) \tag{8.2}$$

Derivation is included in some of the available correlation software codes or can be performed subsequently with the help of mathematical software packages.

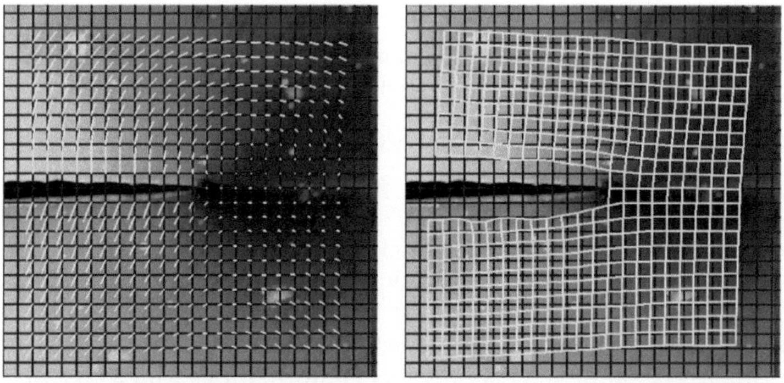

Fig. 8.5. Digital image correlation results derived from SFM images of a crack tip, scan size 15 × 15 μm; *left* image overlaid with user-defined measurement grid and vector plot; *right* image overlaid with user-defined measurement grid and deformed measurement grid; displacement vector and deformed grid presentation are enlarged with regard to the image magnification

8.2.4 Accuracy

The accuracy of the displacement measurement is defined by the object-site field-of-view length $l_{x,y}$, the pixel amount of the micrograph in vertical (y) and horizontal (x) direction m_y and m_x, and the subpixel shift resolution k:

$$\delta u_{x,y} = \left(l_{x,y} / m_{x,y}\right) k \qquad (8.3)$$

The subpixel shift resolution k is a measure of the available displacement resolution and strongly depends on the experimental conditions, the quality of the images obtained in the experiment (e.g., noise, sharpness), and the quality of the software algorithms that are used. The value of k varies in most of the relevant cases between $^1/_4$ to 1/100 pixel. For calculation results obtained by statistic means of multiple point measurement (e.g., rigid body displacement calculated as the mean value of shifts measured at multiple points, linear elastic strain calculated from the slope of a regression line), resolution equivalents to 0.01 pixels and better are reached, although the correlation software itself does not provide this level of accuracy.

Assuming values of a SPM large-scale scan size of $l_{x,}$ l_y = 10 μm, a scan resolution m_x,m_y = 512 and a measurable subpixel shift k = 0.2 (estimated), a measurement resolution of a single one-point displacement value down to 5 nm can be easily obtained with scanning probe microscope applications.

8.3 Displacement and Strain Measurements on SFM images

8.3.1 Digital Image Correlation under SPM Conditions

In comparison to DIC-based measurements treating optical or SEM micrographs, some more essential difficulties have to be overcome for SFM imaging. They correspond to the extreme magnification under SFM conditions. Because SFM image scans are taken over a time interval from one to several minutes, the smallest system drifts can cause significant artificial object deformations. Different drift sources can be distinguished and classified as follows:

- SFM scanner drifts, which are related to time-dependent behavior of the piezo drives
- relative movements between the scanner head and the sample fixture, mainly caused by temperature changes
- drift of sample loading parameters (temperature, forces, load paths, etc.) within testing stages installed at the microscope, and
- incremental object deformations originating from viscous material behavior of test specimens (i.e., time-dependent object deformations which take place even under constant loading parameters).

Substantial concerns regarding stability and reproducibility originate from drifts of the SFM scanner piezo and the thermomechanical loading parameters over time.

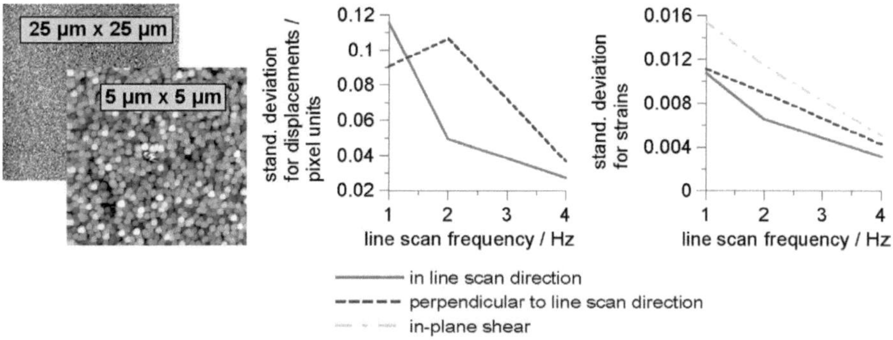

Fig. 8.6. Estimation of measurement errors for nanoDAC by standard deviation of data determined over measurement points of a whole image, subsequent scans from an unloaded test specimen, non-contact scan mode: scan from Si specimens (sample roughness approximately 10–20 nm) (*left*); displacement and strain standard deviation as a function of line scan frequency (*right*)

As a consequence, the accurate selection of SFM equipment and loading stages is a crucial issue. Moreover, the development and implementation of methods of drift control and compensation may be a must for particular applications.

In the following discussion we consider drifts originating from SFM scanner drifts. Figure 8.6 shows a typical result of a respective stability check carried out at an SFM equipment with activated feedback loop of the piezo scanner. For the stability measurement of Fig. 8.6 a series of non-contact topography SFM scans have been picked up from an unloaded and stable mounted object. Topography data was extracted from one scan direction only to suppress artifacts caused by scanner hysteresis. Displacement and strain values were computed for pairs of subsequent scans. The time period between the scans was negligible compared to the interval of scanning. Therefore, determined displacements and strains represent only the arbitrary measurement error of the DIC method superposed by the systematic error due to scanner drift. CMP (chemical mechanical polishing) treated silicon surfaces which exhibit ideal pattern for correlation techniques have been chosen for this analysis. All measurements were carried out after several hours of idle scanning in order to minimize piezo drifts.

By analyzing the data of Fig. 8.6 obtained from an AutoProbe M5 device several conclusions can be made:

1. For suitable SFM choice and installation some of the commercially available equipment exhibits a scanner stability, which does not significantly reduce the possible measurement accuracy as limited by the correlation algorithms (see above). For the SFM micrographs 0.1 pixel (and better) for local displacement values and $4 \cdot 10^{-3}$ for local strain values are feasible. These results relate to topography scans obtained one immediately after the other.
2. Obviously, higher line scan frequencies (i.e., smaller scan time) improve accuracy. Values along the cantilever line scan direction are more accurate, as should be expected from the same stability considerations. The standard deviation for displacements keeps at levels as known from correlation analysis with SEM [10, 24].

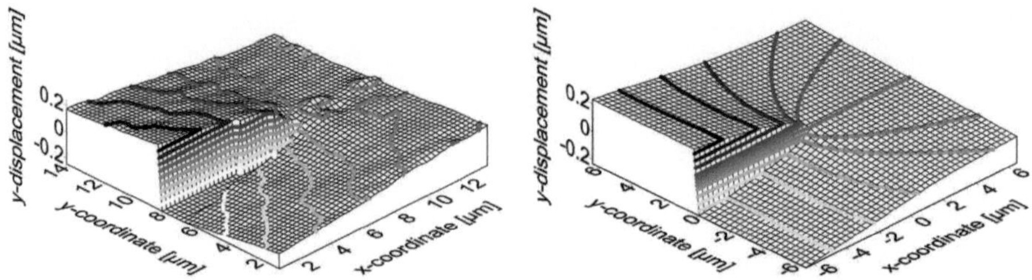

Fig. 8.7. Crack opening displacement field near a crack tip (displacement component perpendicular to the crack boundary): measured by SFM (*left*) and theoretical field (*right*); the incline of the overlaid displacement contour lines indicates a superposition of pseudostrains induced by drift

3. For strain values slightly higher systematic measurement errors are found than under SEM imaging. It is assumed that improvements of strain measurements are possible, because no optimizations of data refinement procedures (smoothing, grid building algorithms) have been included into the analysis to referred.

Drifts introduced by loading stages and the objects under investigation are a separate issue. Already slight drifts of loading forces and applied temperatures as well as material creep can result in large pseudostrains. The cause is a possibly accumulated displacement over the whole sample/stage size, which appears locally at the scan position as a large amount of "rigid body motion". In some cases this "rigid body drift" over the scan time leads to non-negligible values of additional pseudostrains. For example, the accumulation of thermally induced displacements over an aluminum specimen of 1 cm length can give rise to already measurable pseudostrains for temperature drift rates as small as $1 \cdot 10^{-3}$ K min^{-1}. For this estimation an SFM scan size of 1 µm × 1 µm and a scan capture over 5 min has been assumed.

Figure 8.7 illustrates the impact of drift-induced pseudostrains on real nanoDAC measurements. The three-dimensional plot shows the displacement fields nearby a crack trip. The crack was opened by external forces into the direction perpendicular to the crack boundary (Mode I crack opening; also see below). The comparison between the measured and the theoretical crack opening displacement fields reveals slight deviations. In the measurement plots the displacement contour lines of equal y-displacement values are inclined on both sides of the crack boundary in different directions. This behavior is a result of a small, nearly constant pseudostrain in the y-direction superposed to the real crack opening field. Nevertheless, the actual crack opening field dominates and is reproduced in the right way.

8.3.2 Technical Requirements for the Application of the Correlation Technique

There exist two main issues to be discussed if digital image correlation (DIC) is to be applied in SFM: suitable specimen loading must be realized within the equipment and high level scanning reproducibility must be provided.

The placement of thermal and/or mechanical loading stages in the SFM is mainly a question of compact design of loading stages, free access for SFM cantilevers to the specimen surface of interest, and a scanning by cantilever-site actuation. Depending on available equipment and loading stages it can be necessary to install additional spacers in between the cantilever head and the ground plate with x-y-stages for specimen adjustment.

In order to avoid scanning instabilities and different thermomechanical drifts (see above) common tools of environmental isolation offered by equipment suppliers should be installed. This comprises active vibration compensation of equipment tables, acoustic enclosures against ambient sound, and if possible also temperature stabilization. New equipment should be tested for scanner stability and allow reproducibility of subsequent scans of the same scan direction within approximately 0.1-pixel drift difference over the whole scan. Advanced fast line scan equipment can be very helpful to meet this requirement. Because of the possible drift of object loading parameters it might be necessary to develop special tools for drift compensation. These measures can be taken actively by piezo-driven displacement compensation or passively by numerical corrections of measured displacement and/or strain fields.

Generally, every SFM scan mode can be utilized for DIC applications as long as suitable local structures are imaged and the imaged pattern is maintained during the scan. Even so it may be possible that for particular objects some of the modes, like contact scanning, tapping mode, or lateral force microscopy, are not appropriate. The case occurs, for example, if polymers with substantial viscoelastic behavior are processed. Here contact-type scanning can lead to severe decorrelation effects in the scanned pattern resulting in incorrect displacement evaluation.

8.4 Deformation Analysis on Thermally and Mechanically Loaded Objects under the SFM

Most of the published work is aiming at the characterization of materials, either taking into consideration local material structures like grain size or property gradients, for example, refs. [7, 25], or focusing on the determination of material properties on microscopic or nanoscopic structures, for example, refs. [9, 12, 26]. The following two sections present two examples, which give an impression about possible application fields of the DIC under SFM conditions.

8.4.1 Reliability Aspects of Sensors and Micro Electro-Mechanical Systems (MEMS)

Modern sensors and MEMS/NEMS devices consist of extremely fragile functional structures. Because of the desired device functionality, quite different materials in terms of material properties have to be combined with one another. Loading such structures thermally and/or mechanically means implementing severe material mismatch within submicron and nanoscale volumes. Therefore, functional or environmental loading causes local stresses and strains due to different material properties such as coefficient of thermal expansion (CTE), Young's modulus, or time-depended viscoelastic or creep properties. The smallest existing material imperfection or initial micro/nanoscale defects can grow under stress and can finally lead to the failure of the device [10, 27–30].

Thin layers used in sensor and MEMS technology undergo local stresses remote from elastic material behavior, where permanent device alterations are feared after each load cycle. Nowadays, responses of nanomaterials to applied external loads from temperature, vibrations, or chemical agents are not well understood. The same is true for actual failure mechanism and damage behavior. Moreover, quite often those structures escape from simple continuum mechanics and failure modeling and from respective numerical simulation, because material properties on the microscale and nanoscale are not yet known or described in available databases. Furthermore, they may change over exploitation time [17]. Due to these facts efforts have to be made to gain a better understanding of the material responses in submicron and nano regions [9, 25, 31].

The way to achieve this aim is the combination of displacement and strain measurements on the microscale and nanoscale with modeling techniques based on finite element analysis. Parameterized finite element models of MEMS are applied for faster prediction of lifetime and failure modes. The parameterization allows the variation of model geometries and materials in order to accelerate the MEMS design process [32].

8.4.2 Thermally Loaded Gas Sensor under SFM

Sensor applications with local temperature regulation such as the gas sensor shown in Fig. 8.8 are usually thermally loaded with rapid and frequent change in temperature [29]. This thermal cycling and the temperature gradients over the structure imply thermal stresses and may cause failure of the component [33]. In the operation mode of the gas sensors thermal stresses are induced due to the activated microheater.

With in situ SFM measurements on this microsystem the capability of the nanoDAC approach is demonstrated by measuring material deformation resulting from mismatch of material properties.

The gas sensor is designed to tolerate several hundreds of °C thermal loads. The thermal mismatch between the platinum electrodes (CTE = 9 ppm K^{-1}) and the SiO_2 substrate (CTE = 0.65 ppm K^{-1}) leads to high local stresses, if the entire device is heated up. Local displacements resulting from the thermal load have been measured by means of the nanoDAC technique.

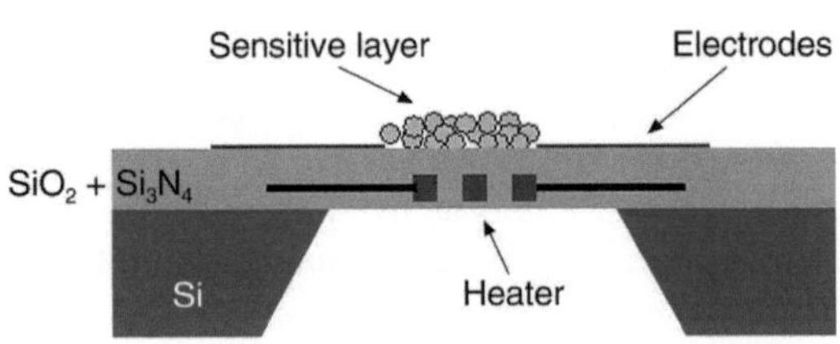

Fig. 8.8. Layout of gas sensor

8.4.2.1
In-Plane Displacements

In situ non-contact SFM scans on top of the gas sensor membrane have been carried out at room temperature and at 100 °C. The area which was observed is illustrated in Fig. 8.9 as location 2. At this area an overlap of the SiO_2 membrane by the platinum electrodes should result in a thermally induced stress/strain field. The temperature was achieved by applying a defined voltage to the microheater of the gas sensor.

The determined thermally induced displacement field shows that the platinum layer with its higher CTE value reveals an inherent expansion towards the edge of the layer (Fig. 8.10). In supplementary tests with heating cycles with maximum temperatures in the range of 450 °C severe delaminations of the platinum layer at the edges to the SiO_2 substrate layer were observed (Fig. 8.11). Details of this testing cycle are described in more detail in ref. [29].

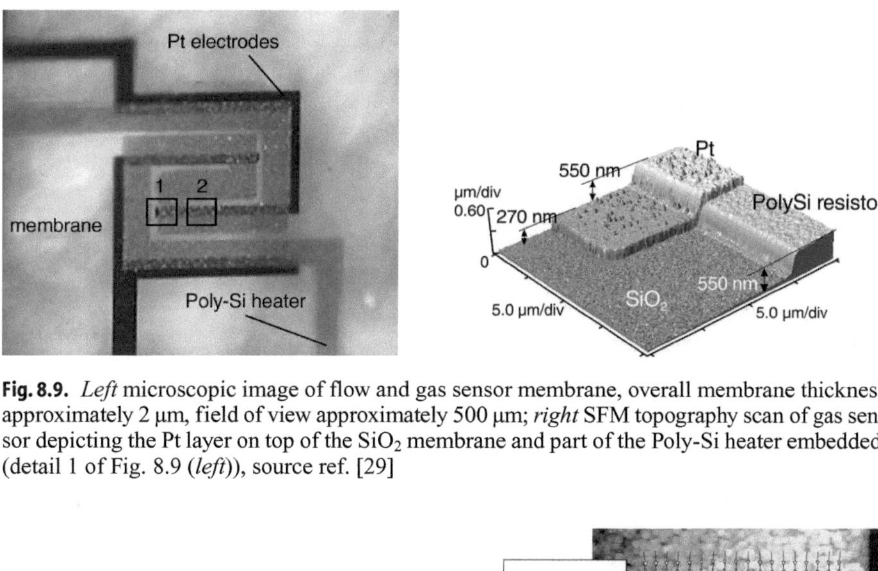

Fig. 8.9. *Left* microscopic image of flow and gas sensor membrane, overall membrane thickness approximately 2 µm, field of view approximately 500 µm; *right* SFM topography scan of gas sensor depicting the Pt layer on top of the SiO_2 membrane and part of the Poly-Si heater embedded, (detail 1 of Fig. 8.9 (*left*)), source ref. [29]

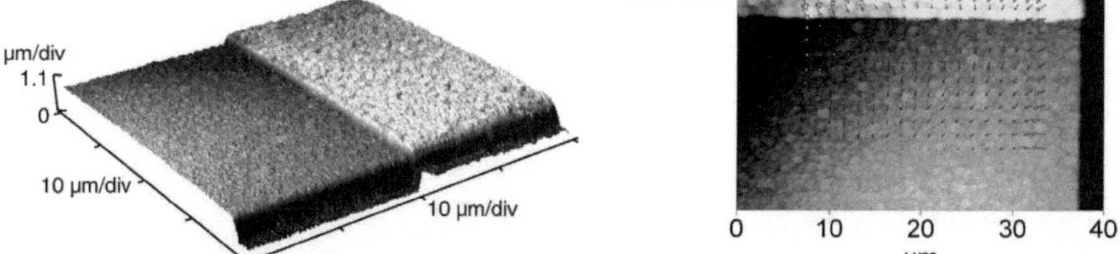

Fig. 8.10. *Left* SFM topography image of platinum and SiO_2 layers; *right* vector plot of displacements *u* measured by nanoDAC

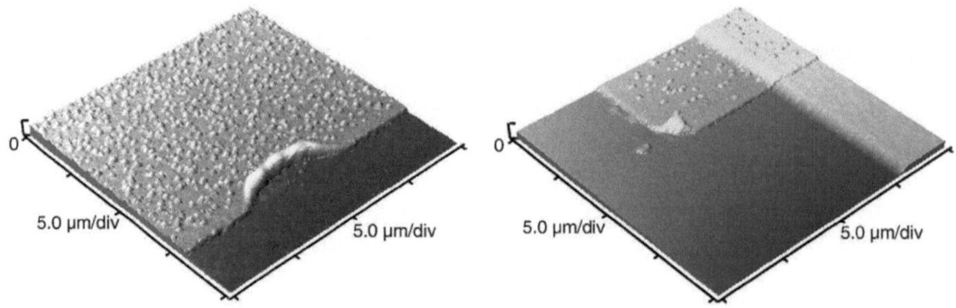

Fig. 8.11. SFM topography scan of membrane layers after tempering at 450 °C, Pt electrode destruction at edge and corners (compare to Figs. 8.9 and 8.10)

8.4.2.2 *Out-of-Plane Displacements*

Besides the information on structural deformation in the *x*-*y* plane the SFM measurement technique allows the determination of the out-of-plane displacement component. The height information of the SFM topography images before and after loading is analyzed for evaluation of movements or deformations in the *z*-direction.

Application of this technique to in situ measurements of thermal deformations by SFM on the top of the sensor membrane has revealed a high value of remaining deformations even after a single heat cycle (25 to 100 °C). Inelastic strains remain after cooling down to room temperature (Fig. 8.12).

8.4.3 Crack Detection and Evaluation by SFM

Tiny defects or cracks in microelectronics components can lead to severe crack propagation and complete failure if electronic devices are stressed. Because of intrinsic stress sources, for example, thermal mismatch, the changes in environmental conditions (temperature, pressure, mechanical vibrations) can initiate crack propagation and cause fatal damage. Experimental crack detection can be a crucial issue bearing in mind original crack sizes of about some micrometers. These cracks

Fig. 8.12. Residual sensor deformation after heat cycle (SFM-based deformation measurement), 3D plot shows part of the membrane layer profile, the *coloring* (gray scale) indicates the remaining vertical deformation after a heat cycle

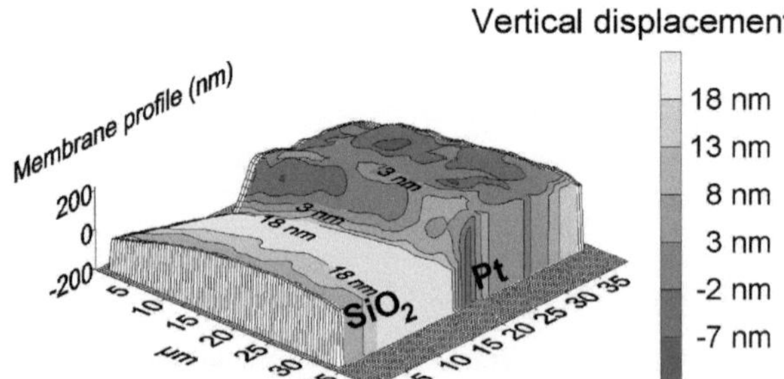

will open only some tens of nanometers or even less under sub-critical load. Their detection, however, is possible by DIC displacement measurements.

In the following section, crack detection and evaluation will be discussed at a cyanate ester resin polymer material. A typical application of this thermoset is the area of microelectronic systems, where it may be used as underfiller between chip and substrate or as matrix material for printed circuit boards. The unmodified resin has a high modulus of elasticity but poor resistance to fracture [3, 34]. Due to the miniaturization of electronic packages and following reliability challenges, micro material testing of such systems will be a key feature for successful design in microelectronic packaging.

8.4.3.1
In Situ Measurement Technique

For the crack detection experiments, a simple specimen configuration is selected to demonstrate the fundamental approach. With a compact tension (CT) crack test specimen as shown in Fig. 8.13, Mode I (opening) loading of the crack tip is enabled. Due to a high accuracy in the machining process of the specimen, in-plane and out-of-plane shear (Mode II and III) components are avoided to a considerable extent.

The CT specimen is loaded by a special tension/compression testing module, which can be utilized for in situ SEM and SPM measurements. Figure 8.13 shows the CT specimen and parts of the loading device under the SPM.

SFM topography scans are taken at different locations of the crack face before and after loading. In the following presentation of measurement results the first location is approximately 50 μm away from the crack tip and the second is directly at the crack tip.

8.4.3.2
Crack Detection

At the first location (crack face) of the CT specimen, the capability of the DIC method for displacement measurements at the nanoscale is demonstrated. SFM

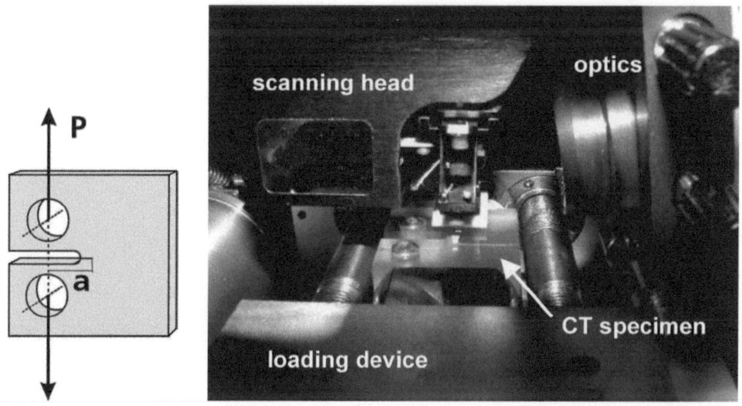

Fig. 8.13. *Left* compact tension (CT) specimen; *right* in situ loading of CT specimen under SPM

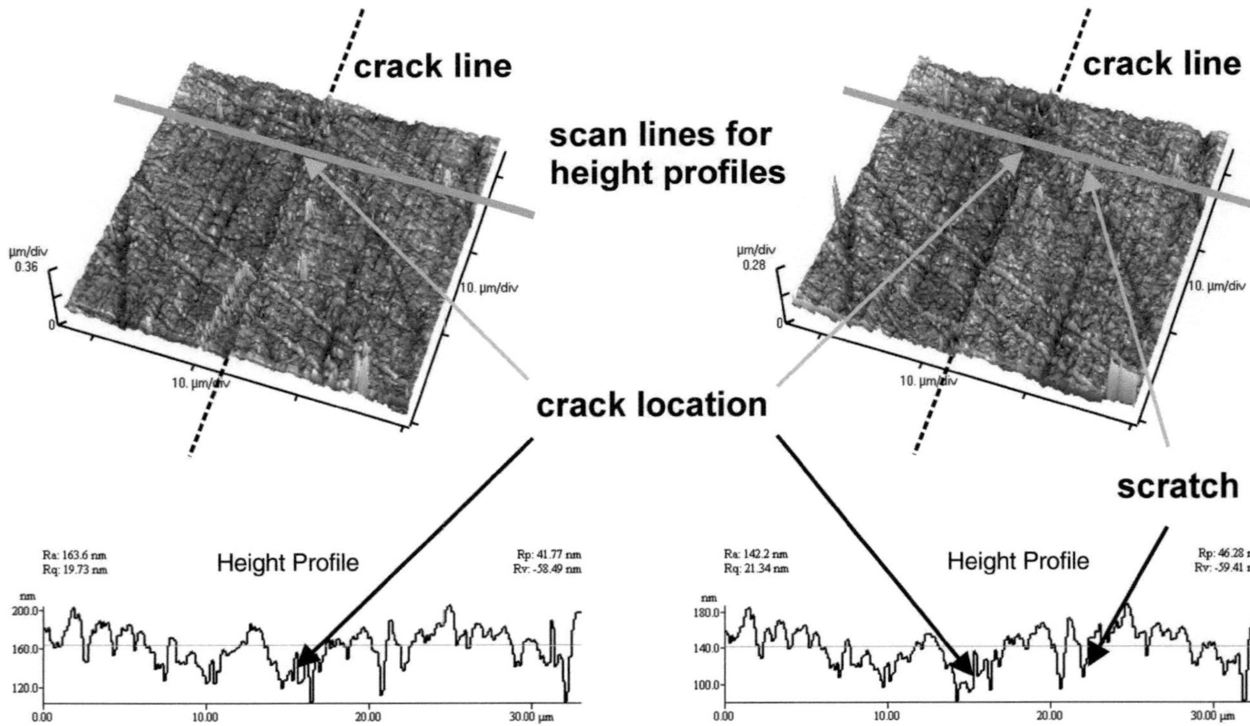

Fig. 8.14. SFM topography images near a crack tip on cyanate ester resin specimen (33 µm × 33 µm image size); SFM scan before crack opening (*left*) and SFM scan after crack opening (*right*)

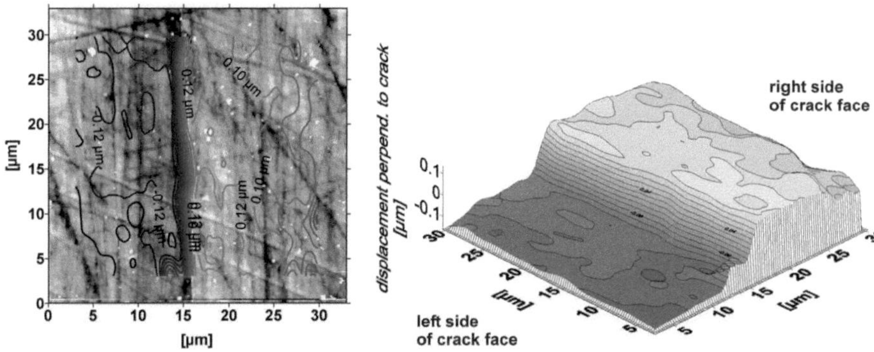

Fig. 8.15. Displacement measurement from SFM images near a crack tip on a cyanate ester resin specimen; SFM topography scan (*left*) with an overlaid contour plot showing the displacements in x-direction, u_x (component perpendicular to the crack boundaries); displacement field u_x as a 3D plot (*right*)

non-contact topography scans are taken at the crack face approximately 50 µm away from the crack tip. The CT specimen is loaded with a force far below the critical fracture load. The SFM images are taken before and after loading with a size of 33 × 33 µm and as 256 × 256 image arrays, that is, the lateral resolution is approximately 130 nm pixel^{-1}.

Figure 8.14 shows the scans before and after loading with height profiles perpendicular to the crack. A comparison of the two topography images and the height profiles of Fig. 8.14 shows that the crack opening is not clearly recognizable due to the low load and the relatively coarse resolution of the scan. There are also scratches on the image which could be identified as cracks.

However, if the digital image correlation algorithm is applied to the images, the crack opening can easily be detected. Figure 8.15 shows the result obtained by nanoDAC analysis. As illustrated, the crack opening due to loading of the CT specimen is about 200 nm. Obviously the crack cannot be identified from the SFM images themselves, because the crack opening is only in the order of 1 image pixel. This fact is illustrated by the height profiles from the topography plot (Fig. 8.14), where scratches and the crack do not clearly differ from each other.

As a conclusion from these measurements, it can be stated that even for larger scan sizes (10–100 µm) displacements in the range of 10 nm can be measured. Cracks in the micron and submicron range can be detected and evaluated. Therefore, it is possible to apply the DIC technique for future reliability issues of MEMS and NEMS.

8.4.3.3
Crack Evaluation by COD Concept

With successful detection of cracks at the microscale and nanoscale the foundation is laid for a more detailed analysis of material faults and defects. The question if available fracture and damage criteria from macro or micro approaches can be

Table 8.1. Crack opening displacement in LEFM for infinite bulk material and Mode I crack opening

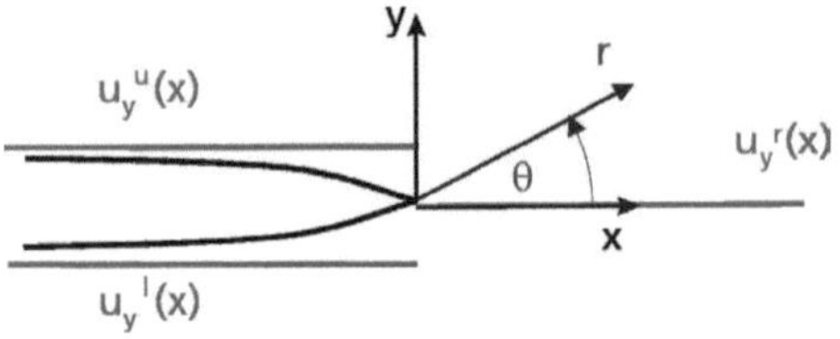

$$u_y^u = \frac{K_I}{2\mu}\sqrt{\frac{x}{2\pi}}(k+1),$$

$$u_y^l = -\frac{K_I}{2\mu}\sqrt{\frac{x}{2\pi}}(k+1) \quad \text{for } x \leq 0 \qquad (8.4)$$

$$u_y^u = u_y^l = 0 \quad \text{for } x > 0$$

transferred to a nanoscopic level is an important issue for reliability evaluation of MEMS and NEMS.

The classical stress intensity factor K is conventionally determined by means of macroscopic fracture tests at standardized fracture specimen such as the CT specimen. This specimen type which is also used for the crack detection test described earlier is used for the verification of the nanoDAC technique for crack evaluation [11]. A straightforward approach for crack evaluation in the SFM is the technique of crack opening displacement (COD) determination. For the combination of the COD concept and the K concept, the following assumptions have to be made:

- linear elastic fracture mechanics (LEFM) apply within the measurement area and for the applied load,
- the specimen consists of homogeneous material.

In order to determine the Mode I stress intensity factor K_I crack opening displacements u_y^u and u_y^l have been measured along both the upper and lower crack boundaries. If determined by LEFM they must equal the values of Table 8.1.

In Eq. (8.4) μ is the shear modulus and k is a function of Poisson's ratio. For the surface of the specimen where plane stress predominates, k is given by $k = (3 - \nu)/(1 + \nu)$ [35].

By taking the square of the difference of upper and lower displacements, we obtain a linear function of the x-coordinate or 0, depending on which side of the crack tip we are:

$$\left(\frac{u_y^u - u_y^l}{2}\right)^2 = Cx \quad x \leq 0 \qquad (8.5)$$
$$= 0 \quad x > 0$$

The expression of Eq. (8.5) does not change if specimen rotation due to load is included into the considerations. In this case, equal rotational terms on both sides of the crack boundary are subtracted from each other. For the equation above, the crack tip is set at location $x = 0$. The crack tip location on the real specimen can be found at the interception of a linear fit of the curve Cx with the x-coordinate

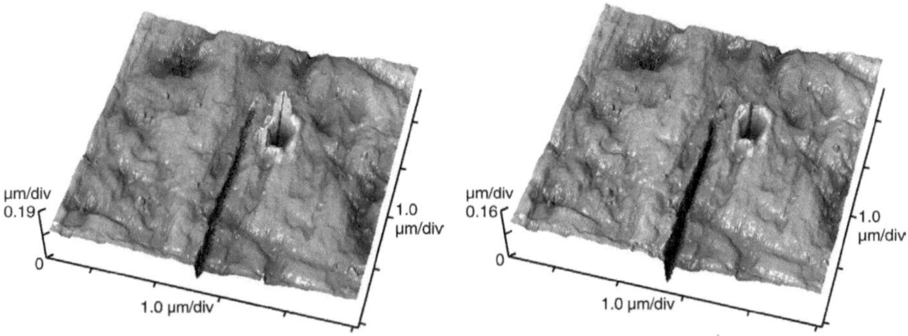

Fig. 8.16. SFM image of crack tip (size 4.6 × 4.6 µm) for the evaluation of stress intensity factors K_I by measuring crack opening displacement around the crack tip; load state 1 (*left*) and load state 2 (*right*); (the *indentation* near the crack tip is a indentation caused by a cantilever approach)

axis. The slope C allows one to estimate the stress intensity factor K_I, which is a measure of the crack tip load and is given by:

$$K_I = \frac{E}{1+\nu}\frac{1}{k+1}\sqrt{2\pi C} \tag{8.6}$$

where E is the Young's modulus.

An example for this procedure will be presented. The crack tip of a cyanate ester resin CT specimen is mapped by the SFM equipment at two different load states. Figure 8.16 shows the corresponding images.

To demonstrate the downsizing capabilities of the DIC approach the scan size of these images have been chosen to be smaller than that of the examples described in the previous section. For a scan size of 4.6 × 4.6 µm the crack opening is already recognizable in the SFM images. Figure 8.17 illustrates the displacement results in the y-direction; u_y is calculated by digital image correlation along with two lines

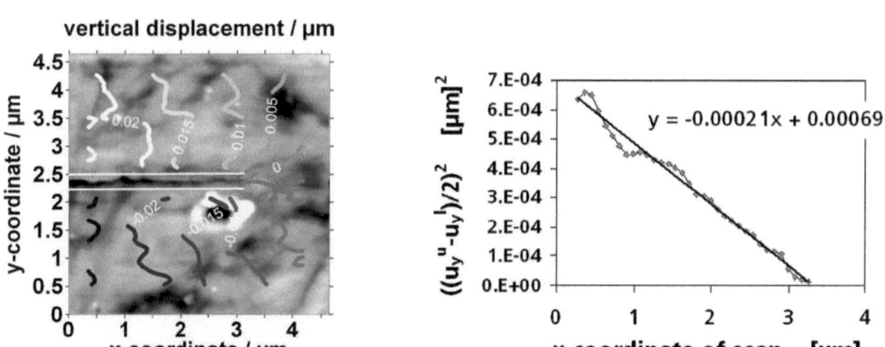

Fig. 8.17. *Left* SFM image of crack tip area (size 4.6 × 4.6 µm) with overlaid displacement results in the y-direction, u_y, lines for the upper and lower crack face are included (*right*) evaluation of slope C for the calculation of the stress intensity factor K_I

marking the upper and lower crack face. The displacement results at these lines are u_y^l and u_y^u used for the determination of the slope C (Fig. 8.17).

The determined value for K_I equals 0.033 MPam$^{1/2}$ which is a value of about 1/20 of the critical fracture toughness K_{IC} for this type of cyanate ester thermoset.

8.4.3.4
Comparison of Experimental Results to Finite Element Analysis

The determined displacement field at the crack tip is compared to a linear elastic plane stress solution by finite element analysis. The measured displacements at the boundary of the deformation field are defined as boundary conditions for the FE model. Figure 8.18 shows the results of the experimental approach and the FE analysis.

The comparison of the experiment to the FEA shows good agreement. The measured displacements in the scanned area are similar to the solution obtained by a linear elastic field approach. Nevertheless, the good agreement between experiment and FEA has to be validated in future research, for example, the influence of viscoelasticity on the measured displacement fields have to be analyzed.

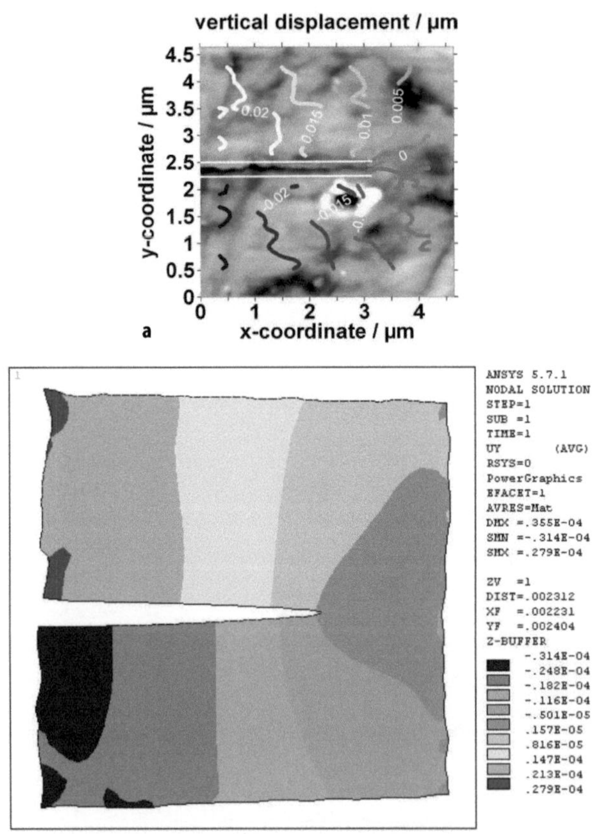

Fig. 8.18. Displacement field u_y determined by nanoDAC **a** and displacement field u_y calculated by finite element analysis **b**

8.5 Determination of Kinematical Properties from SFM Images

One issue in the handling of nano-objects is displacement control. For example haptic devices need to be designed and established to manipulate nano-objects. They require a quantified feedback about any actions performed. Correlation algorithms as discussed above possess potentialities to track the motion of single nano-objects or groups of them. SFM scans comprising the tracked object as well as its vicinity allow one to determine their relative displacements.

Figures 8.19 and 8.20 illustrate how the discussed correlation algorithms can be adopted to measure the motion of particles relative to their neighborhood. Figure 8.19 shows an AFM scan image of a CMP treated Si surface, where a disc-like particle has been added artificially. The displacement between both par-

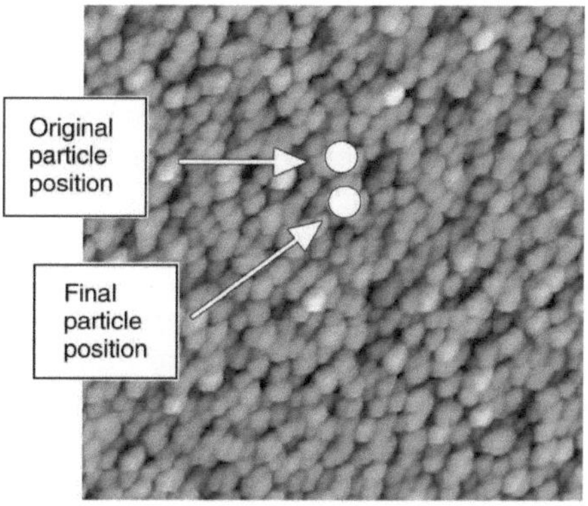

Fig. 8.19. Motion of a disc-like particle over a smooth substrate (simulation based on a SFM topography scan from a Si surface)

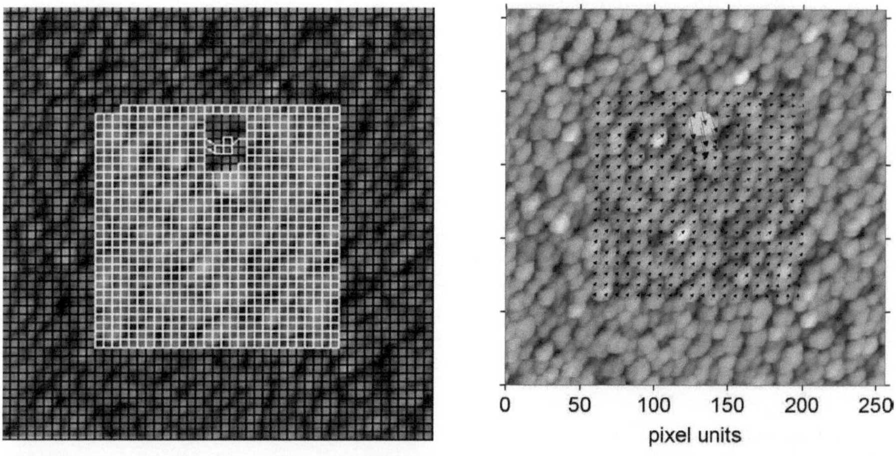

Fig. 8.20. *Left* nanoDAC correlation analysis, part of the grid (measurement node points near the simulated particle) has been shifted downwards; *right* vector displacement plot, particle motion versus neighborhood and slight substrate rotation due to system drift can clearly be distinguished

ticle positions of Fig. 8.19 is calculated automatically by the nanoDAC correlation software.

Figure 8.20 shows a virtual grid presentation of the whole field displacement between the two states. The dark grid indicates the original state, the bright one presents the altered state after particle movement. The relative particle/substrate displacements can be clearly recognized. Figure 8.20 also demonstrates the same motion by the help of a displacement vector plot. The SFM image of the original state (particle/substrate) had been slightly rotated, deformed, and also decorrelated. As seen from the results, such a "disturbance" between two SFM scans is tolerated by the correlation analysis.

8.6 Conclusion and Outlook

It has been shown that SFM images can be utilized to measure displacement and strain fields in very narrow regions of micro- and nano-objects. In that way the versatile information stored in the micrographs can be converted into quantitative data. As a result characterization of materials behavior as a response to external macroscopic loading can be performed aiming at the consideration of the complex influence of the microscale and nanoscale structure. At present digital image correlation methods have been especially attracted to extract displacement and strain data from SFM images. On this occasion it must be mentioned that these tools have been developed over the last few years. One can expect, that capabilities of these tools will be extended within the near future. Stability and reproducibility issues restricting applications today may have less importance, as advanced SFM equipment and measurement approaches will come up. Also the determination of material properties by SFM-based strain measurements is only at the very beginning. More advanced micro material failure analysis will be possible. Probably, fracture criteria like energy release concepts, integral concepts or damage mechanics concepts, such as unified (irreversible) thermodynamic damage formulations (Lemaitre-Chaboche [36], Gurson-Tvergaard [37, 38]) combined with kinetic damage evolution laws as Kachanov creep damage law or the unified damage law suggested by Lemaitre [39] will be applied in the future. Moreover, the better understanding of material behavior in the accessible tiny volumes should have an impact on the development of failure criteria.

Acknowledgements. The authors wish to acknowledge the excellent collaboration with J. Puigcorbé of the Electronic Materials and Engineering (EME) research group, Department of Electronics, University of Barcelona and J.R. Morante of the Centre Nacional de Microelectrònica, Bellaterra, Barcelona. The research was carried out in the Micro Materials Center Berlin (MMCB), which is partially funded by the German Federal Ministry of Research and Education (BMBF) in the MaTech Program under grant 03N6019.

Abbreviations

SFM scanning force microscopy
CMP chemical mechanical polishing
CT compact tension

DAC deformation analysis by correlation algorithms
DIC digital image correlation
LSM laser scanning microscopy
MEMS micro electrical mechanical systems
NEMS nano electrical mechanical systems
SEM scanning electron microscopy
SPM scanning probe microscopy

References

1. Kinoshita T (1998) Stress singularity near the crack-tip in silicon carbide: investigation by atomic force microscopy. Acta Mater 46:3963–3974
2. Komai K, Minoshima K, Inoue S (1998) Fracture and fatigue behavior of single crystal silicon microelements and nanoscopic AFM damage evaluation. Microsys Technol 5:30–37
3. Marieta C, del Rio M, Harismendy I, Mondragon I (2000) Effect of the cure temperature on the morphology of a cyanate ester resin modified with a thermoplastic: characterization by atomic force microscopy. Eur Polym J 36:1445–1454
4. Bobji MS, Bushan B (2001) Atomic force microscopic study of the microcracking of magnetic thin films under tension. Scripta Mater 44:37–42
5. Druffner CJ, Sathish Sh (2002) Improving atomic force microscopy with the adaptation of ultrasonic force microscopy. Proc SPIE 4703:105–113
6. Xie H, Asundi A, Boay CG, Yungguang L, Yu J, Zhaowei Z, Ngoi BKA (2002) High resolution AFM scanning Moiré method and its application to the micro-deformation in the BGA electronic package. Microelectronics Reliability 42:1219–1227
7. Asundi A, Huimin X, Chongxiang L, Boay CG, Eng OK (2001) Micro-Moiré methods – optical and scanning techniques. Proc SPIE 4416:54–57
8. Asundi A, Xie H, Yu J, Zhaowei ZH (2001) Phase shifting AFM Moiré method. Proc SPIE 4448:102–110
9. Chasiotis I, Knauss W (2002) A new microtensile tester for the study of MEMS materials with the aid of atomic force microscopy. Experimental Mech 42:51–57
10. Vogel D, Auersperg J, Michel B (2001) Characterization of electronic packaging materials and components by image correlation methods, advanced photonic sensors and applications II, Nov 27–30, 2001, Singapore. Proc SPIE 4596:237–247
11. Vogel D, Michel B (2001) Microcrack evaluation for electronics components by AFM nanoDAC deformation measurement. Proc IEEE-NANO 2001, Oct 28–30, 2001, Maui, Hawaii, pp 309–312
12. Vogel D, Keller J, Gollhardt A, Michel B (2002) Displacement and strain field measurements for nanotechnology applications. Proc 2002 2nd IEEE conference on nanotechnology, IEEE-NANO 2002, August 26–28, Washington DC
13. Post D, Han B, Ilju P (1994) High sensitivity Moiré. Springer, Berlin Heidelberg New York
14. Sirohi RS, Chau FS (1999) Optical methods of measurement – wholefield techniques. Marcel Dekker, New York
15. Michel B, Kuehnert R (1996) Mikro-Moiré-Methode und MikroDAC-Verfahren anwenden. Zeit Materialprüf 38
16. Vogel D, Auersperg J, Schubert A, Michel B, Reichl H (1997) Deformation analysis on flip chip solder interconnects by microDAC. Proc reliability of solders and solder joints symposium at 126th TMS annual meeting&exhibition, Orlando, pp 429–438
17. Vogel D, Grosser V, Schubert A, Michel B (2001) MicroDAC strain measurement for electronics packaging structures. Opt Lasers Eng 36:195–211
18. Sutton MA, Wolters WJ, Peters WH, Ranson WF, McNeill SR (1983) Determination of displacements using an improved digital correlation method. Image Vision Comput 1:133–139
19. Chao Y-J, Sutton MA (1993) Accurate measurement of two- and three-dimensional surface deformations for fracture specimens by computer vision. In: Epstein JS (ed) Experimental techniques in fracture. VCH, New York, pp 59–93

20. Sutton MA, McNeil SR, Helm JD, Boone ML (2000) Measurement of crack tip opening displacement and full-field deformations during fracture of aerospace materials using 2D and 3D image correlation methods. IUTAM symp on adv optical methods and applications in solid mechanics, pp 571–580
21. Davidson DL (1993) Micromechanics measurement techniques for fracture. In: Epstein JS (ed) Experimental techniques in fracture. VCH, New York, pp 41–57
22. Vogel D, Schubert A, Faust W, Dudek R, Michel B (1996) MicroDAC – A novel approach to measure in-situ deformation fields of microscopic scale. Proc ESREF '96, Enschede, pp 1939–1942
23. Vogel D, Simon J, Schubert A, Michel B (1998) High resolution deformation measurement on CSP and flip chip. Technical digest of the fourth VLSI packaging workshop of Japan, Kyoto, pp 84–86
24. Vogel D, Kaulfersch E, Simon J, Kühnert R, Schubert A, Michel B (2000) Measurement of thermally induced strains on flip chip and chip scale packages. Proc ITherm 2000, Las Vegas, USA, May 23–26, pp 232–239
25. Soppa E, Doumalin P, Binkele P, Wiesendanger T, Bornert B, Schmauder S (2001) Experimental and Numerical Characterization on in-plane deformation in two-phase materials. Comput Mater Sci 21:261–275
26. Cretegny L, Saxena A (2001) AFM characterization of the evolution of surface deformation during fatigue in polycrystalline copper. Acta Mater 49:3755–3765
27. Pinardi K, Lai Z, Vogel D, Kang YL, Liu J, Liu Sh, Haug R, Willander M (2000) Effect of bump height on the strain variation during the thermal cycling test of ACA flip-chip joints. IEEE Trans Comp Pack Techn 23:447–451
28. Schubert A, Dudek R, Walter H, Jung E, Gollhardt A, Michel B (2001) Lead-free flip-chip solder interconnects – materials mechanics and reliability issues. Proc APACK: Int Conf on Advances in Packaging, Singapore, Dec 5–7, 2001, pp 274–287
29. Puigcorbé J, Vogel D, Michel B, Vilà A, Sabaté N, Gràcia I, Cané C, Morante JR (2002) AFM analysis of high temperature degradation in MEMS elements. MME'02, The 13th Micromechanics Europe Workshop, October 6–8, 2002, Sinaia, Romania
30. O'Neal CB, Malshe AP, Schmidt WF, Gordon MH, Reynolds RR, Brown WD, Eaton WP, Miller WM (2001) A study of the effects of packaging induced stress on the reliability of the Sandia MEMS Microengine. Proc IPACK2001: The Pacific Rim/ASME international electronic packaging technical conference and exhibition, July 8–13, 2001, Hawaii, USA
31. Vogel D, Chen Jian, de Wolf I (2000) Experimental validation of finite element modeling. In: Zhang GQ (ed) Benefiting from thermal and mechanical simulations in micro-electronics. Kluwer, Boston, pp 113–133
32. Auersperg J, Döring R, Michel B (2002) Gains and challenges of parameterized finite element modeling of microelectronics packages. In: Michel B (ed) Micromaterials and nanomaterials, no 1, Fraunhofer IZM, Berlin, pp 26–29
33. Puigcorbé J, Vilà A, Cerdà J, Cirera A, Gràcia I, Cané C, Morante JR (2002) Thermo-mechanical analysis of micro-drop coated gas sensors. Sens Actuators A 97:379–385
34. Hamerton I (1994) Chemistry and technology of cyanate ester resins. Blackie Academic and Professional, Glasgow
35. Anderson TL (1995) Fracture mechanics. CRC Press LLC, Boca Raton
36. Lemaitre J, Chaboche J-L (1990) Mechanics of solid materials. Cambridge University Press
37. Santaoja K (1997) Thermomechanics of solid materials with application to the Gurson-Tvergaard material model. VTT Publications
38. Gurson AL Continuum theory of ductile rupture by void nucleation and growth: Part I – yield criteria and flow rules for porous ductile media, Trans ASME. J Eng Mat Tech 99:2–13
39. Lemaitre J (1996) A course on damage mechanics. Springer

9 AFM Characterization of Semiconductor Line Edge Roughness

Ndubuisi G. Orji · Martha I. Sanchez · Jay Raja · Theodore V. Vorburger

9.1 Introduction

Atomic force microscopy (AFM) is an important technique for measurement of the surface roughness and surface features of high technology surfaces such as semiconductor chips and micro-optics. One of the key measurands is the linewidth of semiconductor features [1]. A class of AFM instruments often called CD-AFM has been developed for the purpose of measuring linewidth accurately (Fig. 9.1). The smallest linewidth or hole diameter on a semiconductor circuit is called the critical dimension or CD. Recently, a requirement has arisen in the semiconductor industry for control, specification, and measurement of line edge roughness (LER) for functional semiconductor features, such as processor gates [2–5]. As feature dimensions become steadily smaller, the LER of a single gate is becoming a significant fraction of the gate length itself. Hence, the LER is expected to have a significant effect on properties of the gate such as leakage current. The International Technology Roadmap for Semiconductors (ITRS) [6] specifies a physical gate length for 2002 of 75 nm and a maximum LER of 3.9 nm. The effect of LER on the function of an electronic gate has been modeled by several studies and these models have been verified experimentally. This work has led in part to a succinct

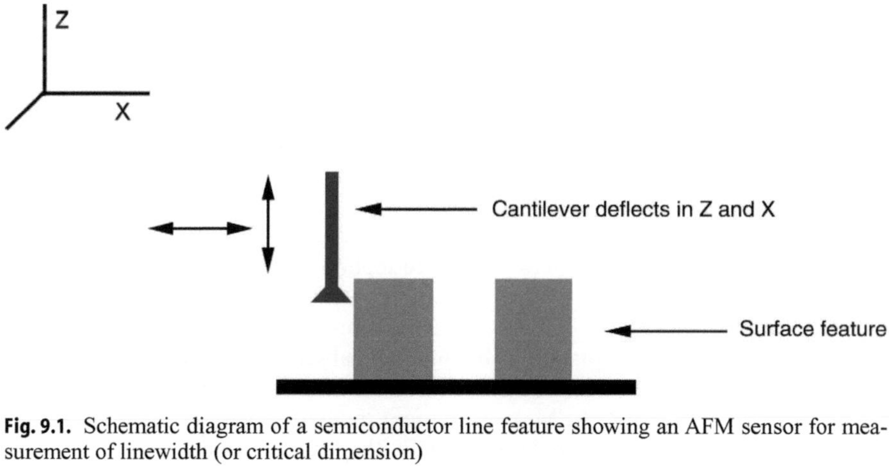

Fig. 9.1. Schematic diagram of a semiconductor line feature showing an AFM sensor for measurement of linewidth (or critical dimension)

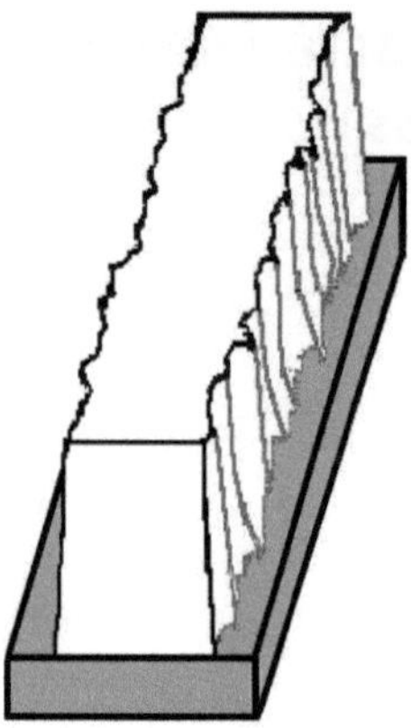

Fig. 9.2. Schematic diagram of a semiconductor line structure showing uneven line edges and sidewall

specification of LER requirements for current and future generations of semiconductor circuits. This specification has been inserted into the International Technology Roadmap for Semiconductors (ITRS) [6].

However, a number of research issues remain. The edge of a semiconductor line has a complex three-dimensional topography (Fig. 9.2). It is not yet clear how the actual topography affects the function of circuits beyond the simple models that have been developed so far, which use simplified characterizations of the actual topography.

The actual topography deviates from the models and this difference may change the functional effects. Therefore, determination of the parameters of the line edge topography that best correlate with functional effects is an open issue. Also, the instruments currently used to perform LER measurements have limitations. SEM [7] is fast and is the tool of choice for many types of dimensional and other measurements of microscopic features. However, its spatial resolution may not be sufficient for measurements of LER specified at 2.7 nm for the year 2004 node in the ITRS and below that for later years. In addition, the common top-down mode of operation for SEM is not capable of profiling sidewall topography. AFM has spatial resolution approaching the atomic scale for measurements of smooth surfaces [8], because the range of interaction in contact mode can be extremely small. However, the finite size of probe tips, which can range from 40 nm to about 1 nm [9], mitigates this advantage. Conventional top-down AFM configurations are also not optimized for profiling the topography of the sidewall. In addition AFM is a much slower technique than SEM because it involves scanning of a mechanical probe instead of a rapidly sweeping electron beam.

Overall, the control, specification, and measurement of LER is a new field with many research issues in play with large technological and economic implications.

In this chapter, we review the AFM measurements that have been done for LER. For an overview of SPM applications in dimensional semiconductor metrology see Marchman et al. [10]. We describe procedures and algorithms that have been promulgated to describe LER, AFM measurement methods, analysis methods, comparisons with SEM measurements, and finally the future outlook.

9.2 Procedures and Algorithms

The specification and measurement of surface roughness is a complex branch of dimensional metrology mainly because surface metrology is also a branch of time series analysis, necessarily involving the statistical analysis of large numbers of data points. Because each data point is effectively an independent parameter, the complete characterization of a series of N data points could involve the specification of as many N parameters. The simplification of the analysis to a few parameters requires statistics, empirical correlation with functional measurements, and judgment. For a field of application like optics, involving the interaction of light with a surface, a relatively small number of parameters and functions can characterize the surface satisfactorily in order to describe surface roughness effects on optical scattering phenomcna [11–13].

For a complex field like the sliding of mechanical parts, the parameters include the properties of bulk materials, films, and a lubricant all interacting under load and in relative motion [14]. A large number of surface parameters and functions have been studied and proposed to explain the effects of surface roughness on tribological processes. So much so that Whitehouse [15] summarized the field with a paper entitled, "The Parameter Rash, Is there a Cure?"

We believe that the phenomena associated with electrical circuits will lend itself more easily to analysis than the tribology of mechanical parts and that the key functional parameters of devices associated with LER may be controllable with a small number of LER related parameters. Although a number of measurable parameters have already been formulated and developed, only one of these parameters so far has resulted from a standardization process (ITRS 2001 [6]). However, the field is likely to change rapidly over the next few years.

In the following, we make a set of observations about LER procedures and algorithms based on existing efforts and previous experience with other branches of surface metrology, where concepts are similar to those of LER measurement.

9.2.1 Observation 1 – A Recommendation

The term LER itself has already been used to refer to many different properties. It would be confusing to confine its meaning to a single defined parameter. LER should be used now as a general term for the irregularities of semiconductor line edges and sidewalls to avoid ambiguities. Other terms, such as linewidth uniformity (LWU) or sidewall roughness (SWR) should be used to describe specific parameters.

9.2.2 Observation 2

LER parameters are primarily calculated from profiles along the long sidewalls of a semiconductor line feature. The profiles may be established by image analysis of SEM or AFM images or by directly profiling the sidewalls.

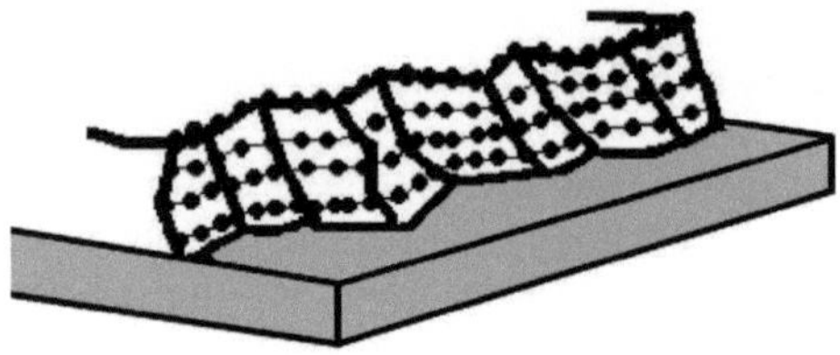

Fig. 9.3. Schematic diagram showing measurement paths along a line edge and sidewall

9.2.3 Observation 3

When measured by AFM, LER is usually measured at or near the line edge, near the bottom (foot), or along some line that represents the middle of the line in height (see Fig. 9.3).

9.2.4 Observation 4

One statistical quantity that should be useful is the rms linewidth uniformity, also called linewidth variation, for which we will use the symbol, *rmsLWU*. It is the rms variation of the linewidth measured about some average mean value or trend (Fig. 9.4).

$$rmsLWU = \left((1/N) \sum_{i=1}^{N} (\Delta w_i)^2 \right)^{1/2} \tag{9.1}$$

where w_i is the width of the line measured at a single place along its length, Δw_i is the deviation of the width from an average value or trend, and N is the total number of these width measurements. The quantity *rmsLWU* is a close analog of the rms surface roughness designated as σ or Rq [18, 19], which is a statistically and functionally important parameter in surface finish metrology. The parameter in the ITRS roadmap is proportional to *rmsLWU*, that is,

$$LWU(\text{ITRS}) = 3\,(rmsLWU) \tag{9.2}$$

Measurands for the *rmsLWU* are definable and measurable with either SEM or AFM.

9.2.5 Observation 5

The sidewall roughness of a single side of a line may provide sufficient correlation with functional properties to be useful. Therefore, the quantity rms sidewall roughness (*rmsSWR*) is likely a useful parameter.

$$rmsSWR = \left((1/N) \sum_{i=1}^{N} x^2(y_i) \right)^{1/2} \tag{9.3}$$

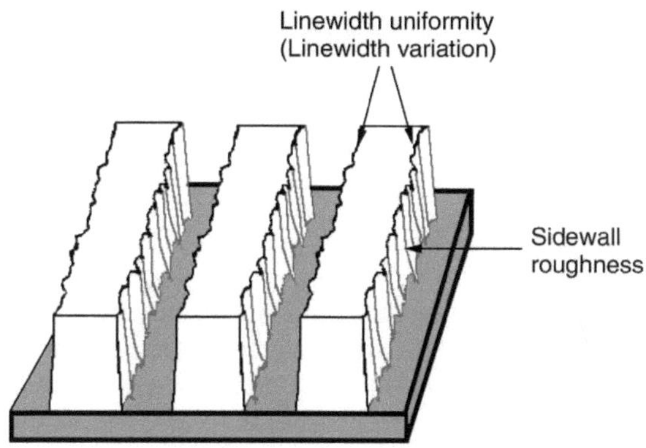

Fig. 9.4. The parameters of *rmsLWU* and *rms SWR*. *rmsLWU* is calculated using the line edge, while *rmsSWR* is calculated using the sidewall

where $x(y_i)$ is a single sidewall profile. Reynolds et al. [16] measured this parameter but didn't use this notation.

9.2.6 Observation 6 – A Recommendation

LER is a wide-band, noise-like property (Fig. 9.5) which necessarily has bandwidth limits when measured. Therefore, the specification or measurement of an amplitude parameter such as *rmsLWU* or *rmsSWR* must include specification of the long and short spatial wavelength limits over which the parameter is calculated. Each of these limits generally depends on one limiting property of the measuring instrument or of the analysis of the measured profile. At the long wavelength

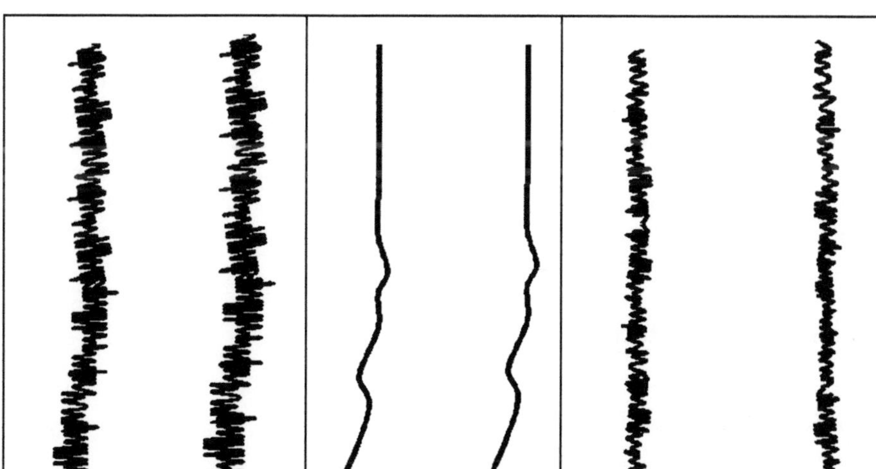

Fig. 9.5. Long and short spatial wavelength aspects of an LER measurement shown schematically. *Left* measured profiles, *middle* long spatial wavelength components, *right* short spatial components

end, the limitation may be the length of the profile or the cutoff length of a high pass filter designed to eliminate longer spatial wavelengths. Currently, the ITRS [6] sets 4 times the technology node as the practical upper cutoff spatial wavelength.

9.2.7 Observation 7 – A Recommendation

The short spatial wavelength cutoff may be determined by one of several factors and likely one of two, the spatial resolution of the measuring instrument or the limit imposed by a numerical smoothing filter based on theoretical or empirical considerations. The spatial resolution of an SEM is likely the interaction volume of the electron beam with the solid surface [17]. The spatial resolution for an AFM is likely determined by the probe size.

It is important that the user understand the spatial wavelength limits of the LER measurement tool being used and that users receive that information from the tool manufacturers stated in a clear credible manner. With that information, an unambiguous statement of an amplitude LER parameter can be stated. It might look like the following:

$$\text{Measured } rmsLWU\ (5\ \text{nm}, 440\ \text{nm}) = 1\ \text{nm} \tag{9.4}$$

Where 5 nm represents the spatial resolution determined by the AFM tip size and 440 nm represents the limit of four times the projected DRAM CD in the ITRS [6] for 2003. Specification of amplitude LER parameters should include both spatial wavelength limits.

9.2.8 Observation 8

Beyond these amplitude parameters researchers have also used the power spectral density function (PSD), which decomposes line profiles into their spatial frequency components. Spatial frequency is the inverse of the spatial wavelength. Once calculated, the PSD can yield rms roughness over any two defined spatial frequency limits, a property that could make the PSD generally useful for statistical correlation with functional properties of the devices. The PSD is described in existing documentary standards of the ASTM [18] and the ASME [19].

9.3 AFM Measurement Methods

9.3.1 Atomic Force Characterization of Patterned Lines

Many of the techniques used to measure LWU and SWR with the AFM were first used to measure line structures and other semiconductor features. To understand the origin and utility of these techniques, we first briefly look at the AFM measurement of line structures. We limit our discussion to those aspects of or im-

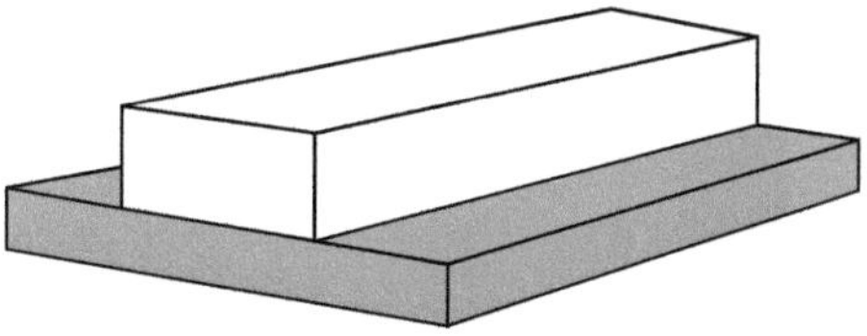

Fig. 9.6. Drawing of an ideal line structure showing straight edges and vertical sidewall

provements to the conventional AFM that enable the imaging of the shape of line structures, and later we will see how they carry over to LER measurements.

The ability of the AFM to image conducting and non-conducting surfaces makes it a natural choice for imaging semiconductor surfaces and features. In this type of measurement, one of the most important measurands is the linewidth. To determine the width of a line structure, one has to determine the edge location. The ideal shape of a line structure is one with a vertical sidewall as shown in Fig. 9.6, but real line structures rarely have smooth sidewalls, they have varying topography and tend to have characteristics like those illustrated in Fig. 9.2.

Knowledge of the shape of the sidewall enables one to determine an average width of the line and also to ascertain the capability of the lithographic process. The top-down tip-sample-force measurement strategy of conventional AFM gives limited information about the sidewall. A measurement procedure or configuration that will provide complete profile or feature information is highly desirable. Some of the methods that have been used in attempts to solve this problem are discussed below. They include the use of special tips, the ability to tilt the sample, and tip characterization techniques.

9.3.1.1
Tip Improvements

In obtaining complete profile or feature information, the tip is perhaps the most important component in force profiling of surfaces. It is the part that actually interacts with the surface. The size, shape, and the tendency of the tip to wear out over a period of time could introduce errors in the image. Not surprisingly the tip is also an area where researchers see the need for improvement. Two such improvements are described below.

The first approach is the fabrication of tips optimized to sense surface proximity in more than one direction. It is possible for the tip to detect forces when brought alongside a vertical sidewall [20]. For common cone-shaped tips, like the one shown schematically in Fig. 9.7, there is only one pointed end and the detection mechanism used is configured to detect forces directly below the tip.

For a two-dimensional tip like the one in Fig. 9.8 with 3 protrusions, forces can be detected in the z and x direction. In this particular implementation by Nyyssonen et al. [21], force detection was based on the attractive force mode proposed by Martin et al. [22]. The tip vibrated in the x and z direction and the amplitudes of the vibration are monitored independently by two heterodyne laser interferometers. The tip did not scan the surface; rather the sample was positioned and brought into contact with the tip in a robot-like manner using a 3-axis piezoelectric driven stage. Another tip shape is that of a flared or boot-shaped tip (shown in Fig. 9.9) proposed

9.7 

9.8 

9.9 

Fig. 9.7. A schematic of a cone-shaped tip

Fig. 9.8. Schematic of a three-point tip proposed by Nyyssonen et al. [21]

Fig. 9.9. Schematic of a flared tip proposed by Martin et al. [1, 23]

by Martin et al. [1, 23]. The detection system was similar to the one described above but used a more sophisticated scanning algorithm to image the sidewall and the surface. Here the vibration frequency of the tip is monitored by interferometers independently used in a feedback loop to servo control the motion of the tip in two directions: in the z direction to control the tip sample interaction on the surface, and in the x direction to control the tip sample interaction along the sidewall. By analyzing the force information, the system could tell what surface was normal to the tip, and obtain data along the contour of that surface.

A second approach is to make tips that are sharper and longer than conventional tips, but still stable enough for imaging. Part of the rationale behind smaller tips, provided they are stable, is that the maximum uncertainty it can introduce into an image is the size of the tip radius, so any reduction in the size of the tip leads to a reduced uncertainty. One method of fabricating such tips is by focused ion beam (FIB). This method starts with a conventional tip, which the FIB sharpens to sizes as small as 5–10 nm in radius. Figure 9.10 shows scanning electron micrographs of a tip before and after FIB sharpening [24].

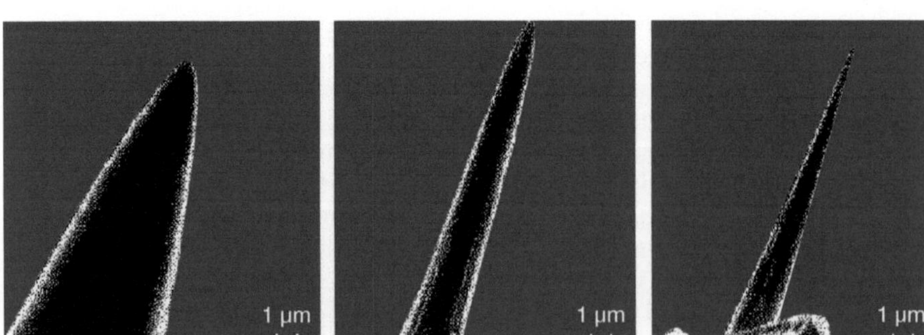

Fig. 9.10. An electron micrograph of a single tungsten tip after each stage of fabrication (Adapted from Vasile et al [24], used with permission)

9.3.1.2 *Sample Tilt*

Tilting the sample is another way to gain increased access to the regions of a feature that conventional AFM cannot adequately image. This enables the tip to scan the sidewall of a feature and obtain information such as sidewall angle and roughness. Generally the sample holder is tilted with respect to the scan coordinate system, usually the x–y plane. The specific method used to tilt the system is not as important as keeping track of the tilt angle, so the effect of the tilt angle can be removed from the image. Fu et al. [25] modeled the error in step height value due to measurement axis and sample axis misalignment and that depending on what algorithm used deviations of up to two percent for a misalignment of only 0.1° are possible.

9.3.2 AFM Characterization of Line Edge Roughness

In this section we will look at how the AFM is used to measure line edge roughness: what measurement strategies are employed, and what modifications are made in order to adequately image this feature. Several groups focused on understanding the origins of LER in resist material and the influence of processing conditions (e.g., Hinsberg et al. [26], Somervell et al. [27], and Patsis et al. [28] among others). We will restrict our discussion to how these measurements are made with the AFM irrespective of the material or processing conditions.

We emphasize here the measurement of LWU and SWR. Figure 9.4 shows sketches of both parameters. As mentioned earlier, we consider LER to be a non-specific term. The LWU is the rms deviation of the average width of a line structure, and the sidewall roughness is the rms roughness of the sidewall. In cases where the measurement configuration cannot allow measurement of the two edges, information from one edge is used. The AFM techniques used to image LER can be categorized as follows: top-down imaging, sample tilting, 90° rotation, and CD-AFM.

9.3.2.1 *Top-Down AFM Imaging*

In this method the feature is measured in a conventional top-down fashion. The sample or the probe can be scanned depending on the design of the instrument. Figure 9.11 shows an example of a tip scanning across a feature. This is the conventional AFM operating mode. The evaluation length can vary depending on the range of the instrument. The main advantage of this method is the imaging of both edges of the line structure. The major limitation is that only LWU information is obtained; little or no sidewall data is available. In spite of these limitations, top-down AFM imaging has been used extensively. By using this method Yoshimura et al. [29] obtained results that suggest that LER in chemically amplified resist (CAR) structures was caused by polymer structures of the base resins used. Yamaguchi et al. [30] also came to similar conclusion that linewidth variations are

Fig. 9.11. Schematic of a cone-shaped tip scanning over a line structure

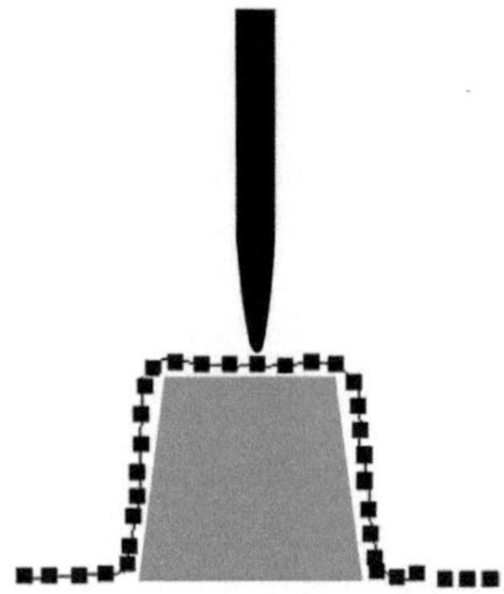

caused by the polymer aggregates in resist films. This paper provided more detailed information on how they measured LER. They defined LER as "the standard deviation of the linewidth from the AFM data at a depth of 5 nm and (an evaluation length) of 800 nm". Yamaguchi et al. used the term *sampling length*, but the term *evaluation length* is more consistent with common practice [19]. He et al. [31] in their study of the process dependence of roughness in a positive tone CAR found that the spatial frequency of the roughness depends on the degree of deprotection, while the amplitude depends on the fabrication process. Kant et al. [32] and Nelson et al. [7] among others have also used this top-down method.

9.3.2.2
Sample Tilt

In this method the sample is tilted to enable the tip to gain access to the sidewall. This is achieved either by using a specially made sample holder or by using a wedge to tilt the sample. In each case, careful attention is paid to the tilt angle. Additionally, the included angle of the tip must be sufficiently small that the image acquired is that of the sidewall, and not the tip. In order to get images of the sidewall, the angle between the sidewall and the z-axis has to be more than the half cone angle of the tip. As the sample is tilted the tip will gain more access to the sidewall. If the angle between the z-axis and the sidewall is less than the half cone angle of the tip, then the acquired image is the same as one obtained using a top-down technique described above. Figure 9.12 shows a tip and line structure at different tilt angles.

The benefit of using this technique is increased access to the sidewall. It is a popular method especially when an image of the lower part of a line structure is desired. Nagase et al. [33] used a sample holder tilted at an angle of 40° to image the bottom part of a 30-nm resist pattern and observed that the roughness of the lower portion is almost the same as that of the top portion. Reynolds et al. [34] explored the effect of different tilt angle on AFM results and found that multiplying their data by $\cos^2\theta$ (where θ is the tilt angle) reduced the results to a constant value. Meyyappan et al. [35] used this method to image the bottom corner of a line structure and pointed out that in some cases, where the tip is at a grazing angle to the sidewall, there may be an averaging effect on the surface contact resulting in a

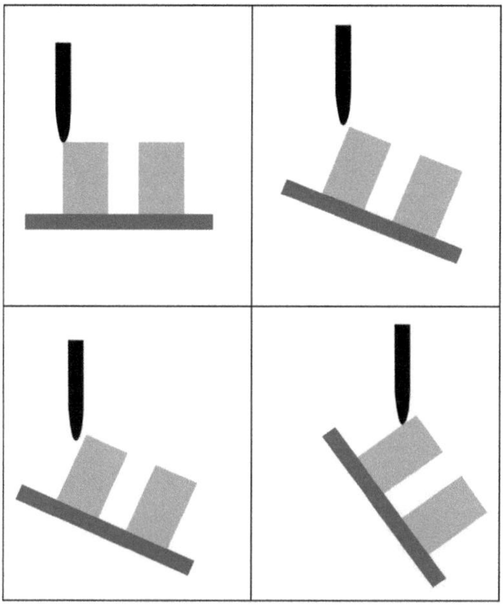

Fig. 9.12. Scanning over a line structure at different sample tilt angles

biased lower value for the roughness. Walch et al. [36] used this technique in conjunction with a flared tip to measure the sidewall. This represents a combination of two methods based on our outline. In this case, the imaging was achieved by using a corner of the tip. Another example of this method is by Nelson et al. [7]. Figure 9.13 shows a schematic of the tilting sample stage used. Some limitations of the tilting sample/tilting sample-stage technique are that it requires thin-film resists, sparse features, or use of long AFM tips to allow imaging near the substrate.

9.3.2.3
90° Rotation

In this technique the sample is cleaved very close to the feature of interest and rotated 90° as shown in Fig. 9.14. This configuration gives the tip full access to the sidewall, and the sample is imaged in a top-down manner. The limitations of this method are apparent. The image acquired is only that of the sidewall (it does not give LWU information). Also cleaving the sample is not only destructive but impractical if the feature of interest is on a large wafer. The major benefit is that it gives the best representation of the sidewall roughness and avoids confusion and errors which may arise when the sample is tilted and the probe contacts the sidewall obliquely. An example of the 90° rotation method can found in Reynolds et al. [16]. Figure 9.15 shows a resist sidewall imaged by Ragson et al. [37], and a schematic of how the data was acquired.

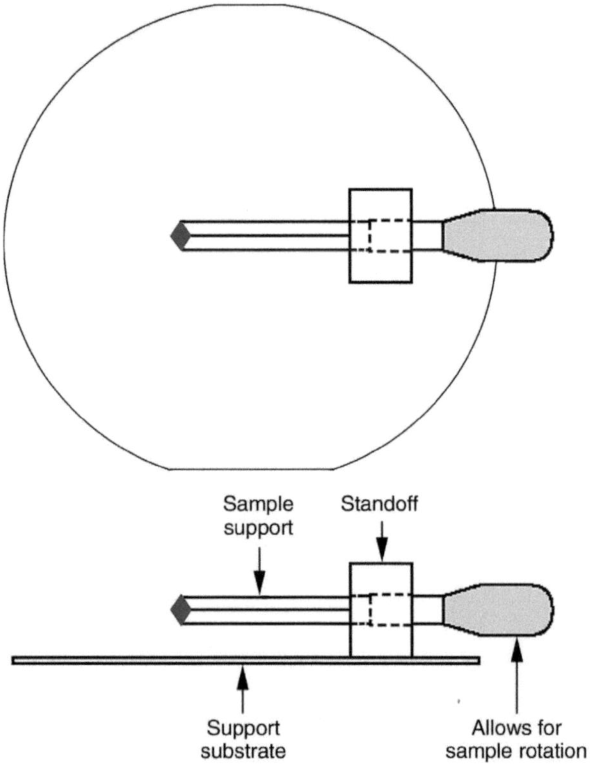

Fig. 9.13. Schematic of a tilting sample holder. Allows the alignment and rotation of the sample. (Adapted from Nelson et al. [7], used with permission)

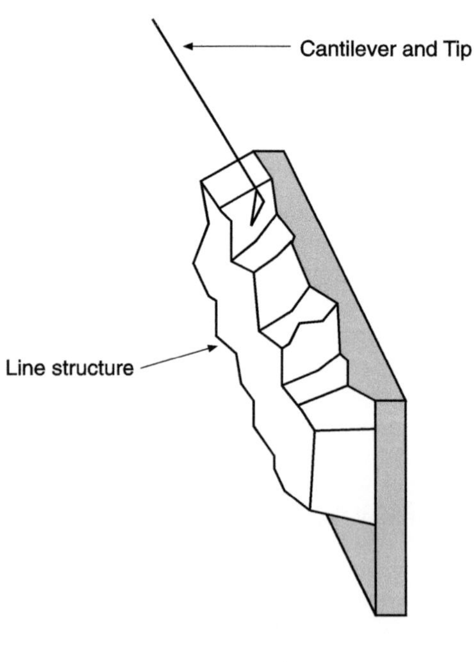

Fig. 9.14. A line structure rotated 90° for full access to the sidewall

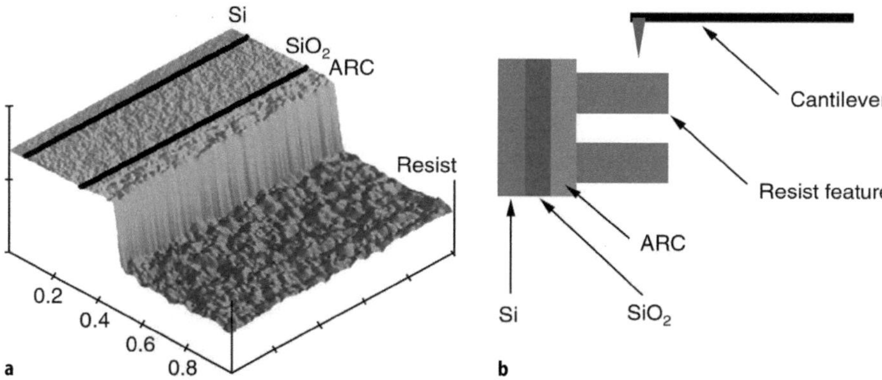

Fig. 9.15. **a** Image of a resist structure acquired using 90° rotation technique. (Image taken by Ragson et al. [37] used with permission), **b** schematic of the profiling direction

9.3.2.4
Critical Dimension (CD) AFM

The dual servo, flared tip AFM described above (also called CD-AFM) may be used to image the sidewall roughness. The dual servo operation mode of this design allows the acquisition of real sidewall data using the end of the flared tip close to the sidewall. The schematic in Fig. 9.1 shows a flared tip in contact with a line structure. Researchers who have used this method include Nelson et al. [7], Kant et al. [32], and Meyyappan et al. [35] among others. Nelson et al. [7] pointed out that one of the limitations of this method was the size of the flared tip, which is too big to image high-frequency roughness. The main advantage is that the sidewall data acquired in the dual servo mode is real x data and does not need to be derived from $z(x)$ data. Table 9.1 summarizes some of the advantages and limitations of the different measurement configurations.

Table 9.1. Advantages and limitations of different LER measurement configurations

	Top down	Tilted stage	Critical dimension	90° Rotation
Parameter	LWU.	Line edge. Sidewall.	Sidewall. LWU.	Sidewall.
Advantages	No modification to the sample stage.	Increased access to the sidewall.	Access to the sidewall. Real x position data is acquired rather than $z(x)$.	Full access to the sidewall.
Limitations	No sidewall information.	Only one side is imaged. Further analysis is needed to determine and remove the influence of the tip.	Tip is too large to resolve smaller features.	Destructive. Only sidewall information is obtained. Difficult to use on wafers.

9.3.3 New Directions

One of the techniques with the potential of measuring LER with improved resolution is the use of carbon nanotube tips. The high aspect ratio of carbon nanotube tips and their small diameter allow them to reach the bottom of a line structure and between line structures, respectively. Shin et al. [38] reported using carbon nanotube tips to reach the bottom of the sidewall without cleaving the sample. Nguyen et al. [39] have shown that this kind of tip could be reliably used to measure DUV and 193-nm resist patterns. It is also possible that carbon nanotubes could be used in conjunction with the sample tilt technique described above. Figure 9.16 shows patterned lines imaged by Nguyen et al. [39], and Fig. 9.17 shows a carbon nanotube attached at an angle that could be used to image sidewalls. This carbon nanotube was attached to a conventional tip using a novel technique described in Nguyen et al. [40]. A carbon nanotube attached at an angle to the probe, if well characterized, can be very useful for LER measurements. The SEM micrograph shown in Fig. 9.17 [41] shows such a configuration.

Another promising technique is the use of two cantilevers in a caliper-like configuration described by Mancevski et al. [42]. The two cantilevers have the ability to sense each other and can provide a zero-width reference before simultaneously approaching both sides of a line structure. Figure 9.18 shows a schematic of a prototype of this instrument. Figure 9.19 shows one of the probes tilted at an angle for contact with the sidewall.

The above descriptions show some of the methods used for measuring LER with the AFM. Some of these methods derived from methods used to image whole line structures, but modified to allow access to the sidewall. In almost all the methods, the use of sharp tips will increase the resolution of the data acquired. The use of carbon nanotube tips will greatly increase the resolution; however, more work is required to determine how high aspect ratio probes interact with the sidewall surface. The dual cantilever approach also holds promise, and could conceivably be fitted with carbon nanotubes for increased resolution.

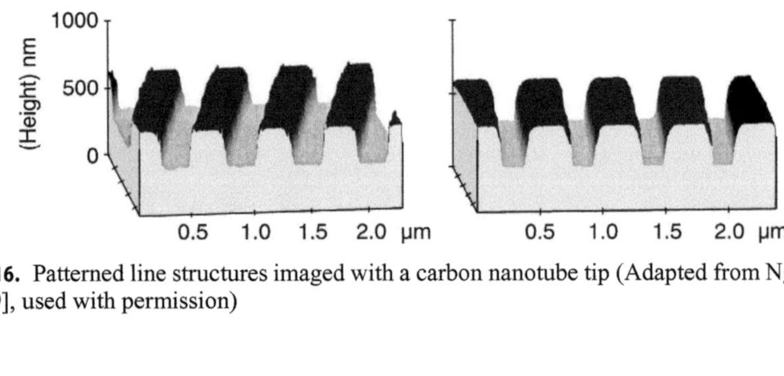

Fig. 9.16. Patterned line structures imaged with a carbon nanotube tip (Adapted from Nguyen et al. [39], used with permission)

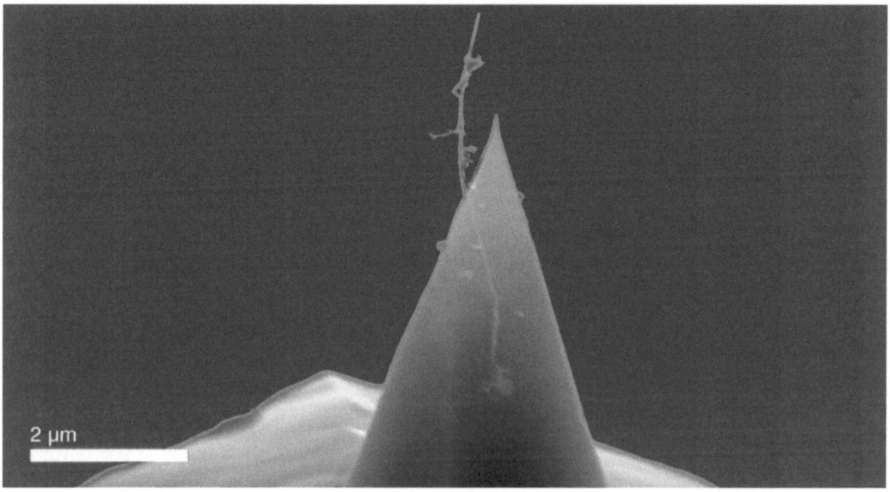

Fig. 9.17. A carbon nanotube tip attached at an angle on a conventional tip (Image by Nguyen at al. [41], used with permission)

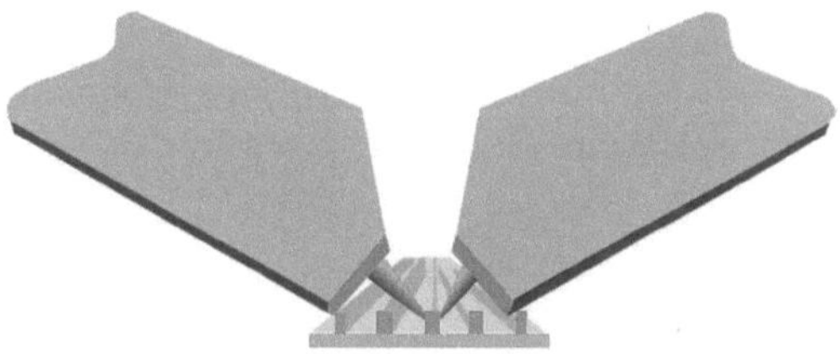

Fig. 9.18. An dual probe atomic force microscope (Adapted from Mancevski et al. [42], used with permission)

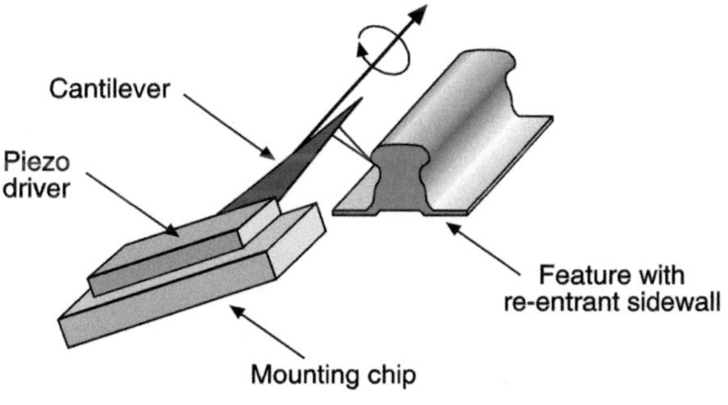

Fig. 9.19. A probe tilted at an angle for contact with the sidewall (Adapted from Mancevski et al. [42], used with permission)

9.4 Analysis Methods

The analysis of AFM LER data relies on methods developed for analyzing optical and engineering surfaces. The objective is to quantify the irregularities or deviations from the mean at the edge of the line and at the sidewall. The schematic in Fig. 9.4 shows a line structure with sidewall roughness and two line edges used to determine linewidth uniformity. The analysis methods may be classified into two broad categories: methods that give a measure of LER amplitude, and methods that give a measure of LER spatial wavelengths. As in most AFM analyses, the data is pre-processed to remove instrument artifacts such as drift and slope. Additional pre-processing is required if the image was acquired by tilting the sample stage as described earlier. Another necessary step before calculating the LER is to determine the location of the edge point of the line structure at each cross section by some criterion. This is important because the line edge roughness will be calculated at this point for the constructed profile of edge points. We now describe in slightly more detail analysis methods described briefly earlier.

9.4.1 RMS Roughness

A fundamental statistical analog of rms LER quantities is the rms surface roughness (also called R_q) defined in Eq. (9.5). Essentially the rms roughness calculates the standard deviation of the surface irregularities with respect to some mean line or curve. It is commonly used to characterize optical components, and in general the lower the rms value the less light is scattered by an optical surface, and hence the better the surface quality [43]. This parameter is regularly used when the distribution of surface irregularities is random. It should be noted that the rms value is not an intrinsic property of a surface, rather it is dependent on the sampling length. For surface profiles, it is given by

$$rms = R_q = \left(\frac{1}{N} \sum_{i}^{N} (z_i)^2 \right)^{1/2} \tag{9.5}$$

Figure 9.20 show a digitized surface roughness profile and the quantities in Eq. (9.5). For LER rms quantities may be specified in several ways. For example Eqs. (9.5) and (9.6) can be used in the following ways:

(1) Specified for one edge of the line structure by reporting it separately for each edge.
(2) Specified for the line structure as a whole by averaging values from both edges.

Which procedure is used may depend on how the image was acquired. If the acquisition method allows the imaging of only one edge, the user has no choice but to use the values separately. The rms values of the scans from each side may be determined and averaged. The rms method is the method of choice used by almost all the researchers encountered in the literature including Nelson et al. [7],

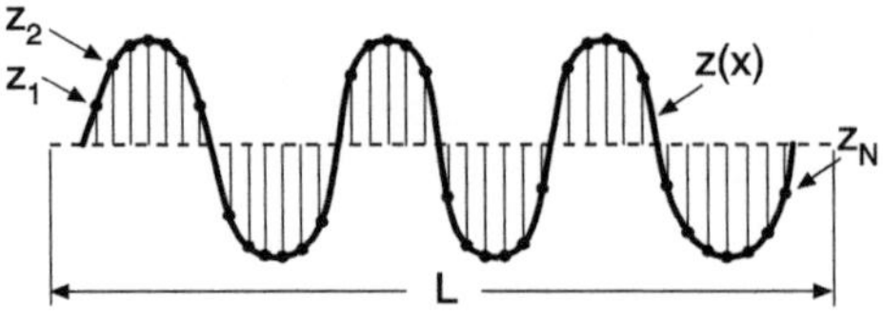

Fig. 9.20. Schematic diagram of a surface roughness profile $z(x)$ and its digitized approximation $z_1, \ldots, z_i, \ldots, z_N$

Reynolds et al. [16, 34], Kant et al. [32], Shin et al. [38] and Winkelmeier et al. [44] among others. But as pointed out by Fang et al. [45] when analyzing the roughness of silicon features, rms is a single statistical value and "two images with exactly the same rms value may have different surface morphology". More parameters may be required to specify LER adequately.

There is a lack of standards on how LER should be evaluated. Although existing standards give choices for describing surface roughness, for example ASTM [18] and ASME/ANSI B46.1 [19], how semiconductor LER should be analyzed is not yet clearly specified. As mentioned above the very definition of LER is still a subject of debate because it is still unclear what roughness parameter most closely correlates with device performance. At the time of this writing there are indications that the ITRS terminology will likely change from linewidth edge roughness to linewidth roughness.

9.4.2 Linewidth Uniformity

Another parameter used to specify LER is the linewidth uniformity, also called linewidth variation and linewidth fluctuation. The ITRS [6] has defined LER in terms of linewidth variation (described below) over an evaluation length of four times the technology node, using all frequency components. This means that if the width of the line structure of interest is 110 nm, the evaluation length for the LER should be 440 nm, and all measured frequency components should be used to calculate the LER. At the time of writing, none of the research literature encountered used this definition. However, it was arrived at by an industrial standardization body. Figure 9.21 shows a line structure with a varying width, indicating linewidth uniformity or linewidth variation.

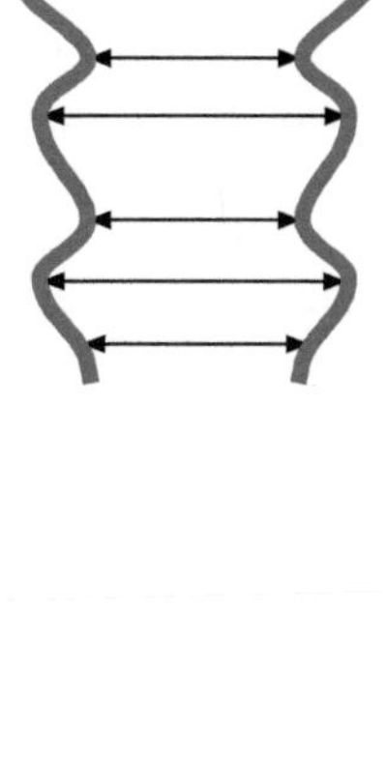

Fig. 9.21. Top-down drawing of a line structure showing width variation

9.4.3 Spatial Wavelength Determination

In traditional surface characterization, a host of statistical functions are used to obtain information beyond what the height parameters provide. Some of these include the power spectral density (PSD), amplitude density, auto-correlation function, and the autocovariance function. So far, only the PSD seems to have gained usage in LER analysis. The PSD decomposes the surface profile into its various Fourier components and may be defined by Eq. (9.6) in analytical form and Eq. (9.7) in digitized form.

$$PSD(F) = \lim_{L\to\infty}\left(\frac{1}{L}\right)\left|\int_{-L/2}^{L/2} z(x)\cdot e^{-2i\pi Fx}dx\right|^2 \tag{9.6}$$

$$PSD(F) = \frac{\Delta}{N}\left|\sum_{j=1}^{N} z(j)\cdot e^{-2\pi F(j-1)\Delta}\right|^2 \tag{9.7}$$

where F represents spatial frequency, Δ is the sampling interval, j is an index, and $i=\sqrt{-1}$. The wide usage of the PSD has to do with the origins of LER, which are believed to be material and process dependent. The various spatial wavelength components of LER are believed to result from specific processes and/or statistical variation in the lithographic method and from the materials used. Figure 9.22 shows plots of a sinusoidal profile, a ground random surface profile, and their corresponding PSD plots. The PSD of the sine wave has one peak located at the spatial wavelength of the sine wave, while the more random profile from the ground surface has a PSD comprising lots of wavelengths.

So using the PSD, one could determine the dominant spatial wavelengths of the profile or image. It is interesting to see how wavelength information is used by different researchers. Kant et al. [32] separated their analyses into two parts, low frequency line edge roughness (LFLER) and high frequency line edge roughness (HFLER) as shown in Fig. 9.5 above. They treated LFLER data as linewidth uniformity, and treated the HFLER as sidewall roughness. Ishida et al. [46] performed an investigation of LER in calixarene fine patterns and found a strong wavelength dependence on the number of monomers forming the cyclic molecule. They were also able to use spectral analysis to distinguish LER caused by mechanical and electrical noise. This would allow exclusion of roughness contributed by the resolution limit of the imaging tool. Nelson et al. [7] integrated the power spectral density of their data and used it to correlate information acquired from different instruments for the same wafer. Eytan et al. [47] used Fourier analysis to verify the detection of specific LER spatial wavelengths by a novel instrument they evaluated.

Overall, the lower the LER the better the quality of the line structure. Some researchers have gone further by using LER to evaluate quantitatively the capabilities of some lithographic processes. Yoshizawa et al. [48] developed an LER method to evaluate the factors affecting the resolution limit in electron beam lithography. They found that the LER of a resist pattern is inversely proportional to the

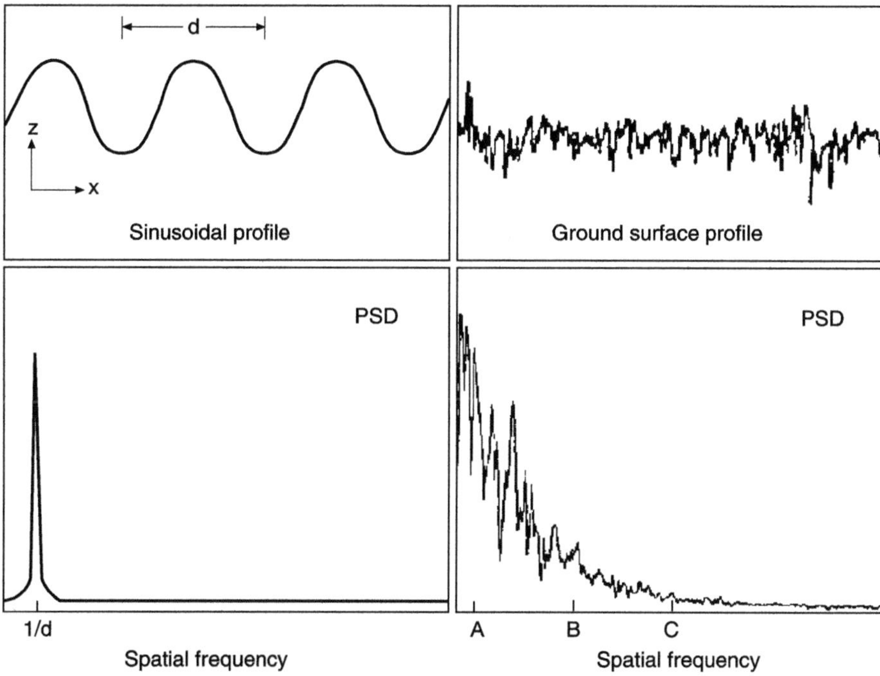

Fig. 9.22. Plots of a sinusoidal profile and a ground (random) surface profile with their corresponding power spectral density

slope of a Gaussian distributed quasi-beam profile (QBP) of an electron beam lithographic system. Since the QPB includes all the factors that affect line pattern formation, such as, the development process, aperture degradation, and electron-optical beam blur among others, the LER of a line structure became a measure of quality for lithography the system as a whole. They further used this method to compare three electron beam lithographic systems [49].

The above summary outlines some of the ways to analyze LER once the AFM information is acquired. An interesting observation from the above literature and others is that there are no precise explanations on how the frequency components were separated (different from how they were detected). None of the profile filters seen in traditional surface metrology (for example, see Raja et al. [50]) were used or even mentioned. There seems to be an ad hoc quality to some of the techniques used. This could be due to the infancy of the field, and a lack of standards for LER characterization. Therefore, there is a need for standardized procedures for analyzing LER. As the field matures, the use of advanced analysis tools such as found in Chen et al. [51] may also be required. The definition in the 2001 edition of the ITRS is a good start; however more research is needed to determine which frequencies are relevant for which node with respect to performance.

9.5 Comparison with SEM Measurements

Since the SEM is currently the imaging tool of choice in the semiconductor industry, we now look at how SEM LER data compare with AFM data. This is not meant to be an overview of the differences between the two instruments. We will focus specifically on AFM/SEM LER comparisons available in the literature. For an overview of how an SEM works and applications in semiconductor metrology see Postek [17] and Postek et al. [52].

In SEM imaging, one has the option of using either low or high acceleration voltages. One of the main differences is in sample preparation. Usually when imaging with high acceleration voltages, the sample needs to be coated to prevent charging due to the high voltage. Coating the sample enhances contrast; however, it may affect the value of the LER depending on the thickness [7] and type of coating used. Low accelerating voltages minimize sample charging, thus allowing the imaging of uncoated samples while providing good surface detail. Nonetheless, low-voltage imaging can result in low image contrast making it difficult for image analysis software to pick out the line edges.

Several researchers have compared SEM and AFM measurements. The general consensus is that SEM and AFM are able to pick out the same roughness spatial wavelengths. Nelson et al. [7] reported that both the AFM and the SEM were able to follow real changes in the roughness as the processing conditions changed. However, based on throughput they recommended the continued use of SEM for the types of measurements they made. Figure 9.23 shows their roughness results from AFM and SEM measurements. The results are highly correlated but differ by a factor of three. Possible reasons for the magnitude difference between AFM and SEM results are discussed in Nelson et al [7]. Eytan et al. [47] compared SEM and AFM data of a specially fabricated sample with predefined amplitude, and spatial wavelength. By using Fourier analysis, they showed that both methods were able to detect the main spatial wavelength component. Figure 9.24 shows a plot of the PSD of data acquired from the AFM and SEM, both instruments show similar spatial resolution, though the amplitude values were different. Other researchers who have performed AFM/SEM comparisons include Reynolds et al. [34] who compared AFM sidewall roughness to top-down SEM LER on chemically amplified resists.

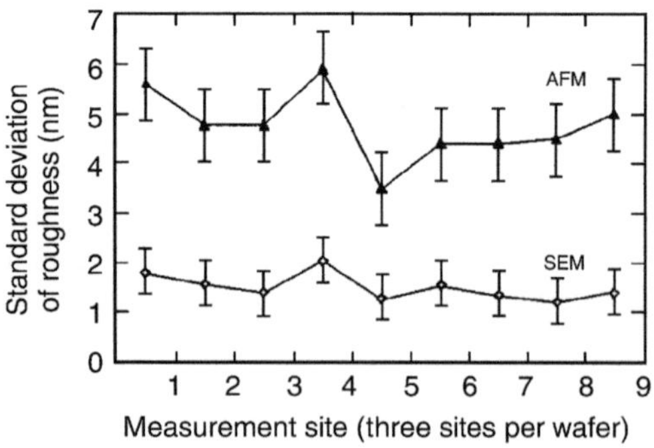

Fig. 9.23. AFM and SEM results from different sites on a wafer. (Adapted from Nelson et al. [7], used with permission)

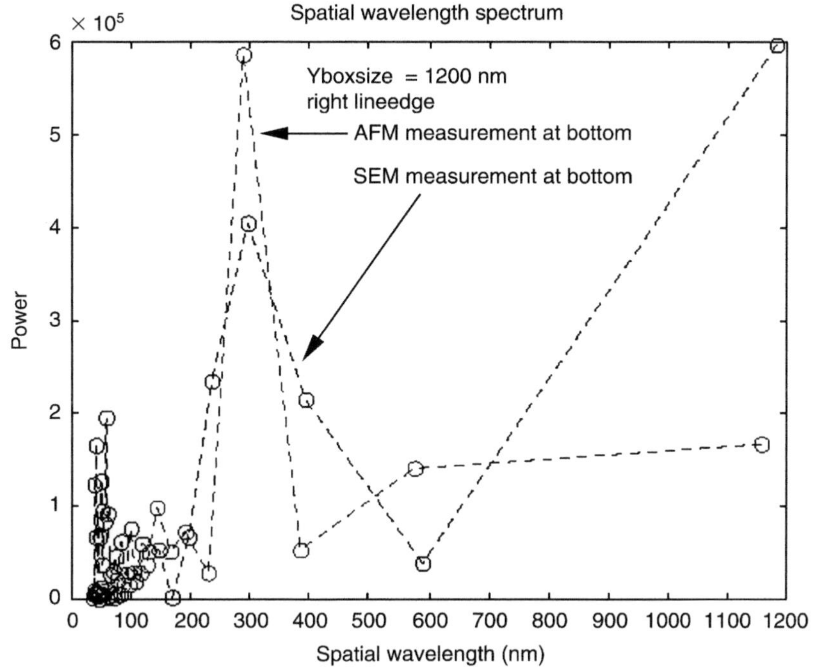

Fig. 9.24. PSD plot of AFM and SEM data. Both methods were able to detect the fundamental spatial wavelength at approximately 300 nm. The reason for the high point at a spatial wavelength of 1200 nm is not clear. (Adapted from Eytan et al. [47] used with permission)

They reported a depth dependence of the sidewall roughness values obtained using the AFM. The roughness values decreased as one went from the top to the bottom of the line structure. Since the SEM is a top-down imaging system, they argued that this may have more of an effect on the magnitude of measurements made with the SEM.

The above comparisons between the SEM and AFM are encouraging insofar as they show strong correlation between AFM and SEM results. However, more work is needed in this area because AFM and SEM not only have different operating principles but may also have to play complimentary roles as the sizes of features being measured approach regimes that greatly challenge their resolution.

Researchers are working on other techniques for measuring LER. One of them is the use of small-angle neutron scattering (SANS). Wu et al. [53] used this technique without any additional sample preparation to nondestructively measure average linewidth of several periodic samples. The scattering intensities from the samples were analyzed and differences in the quality of the samples were visible. This technique, which requires a neutron source, could play a complementary role to established methods as an offline diagnostic tool, especially as node sizes decrease.

9.6 Future Outlook

9.6.1 Physical Standards

The descriptions outlined above underscore the need for standardization of measurement procedures and analysis techniques. The use of physical standards will go a long way in helping alleviate this problem. Fabricating physical standards will involve introducing LER features with a wide range of amplitudes and spatial frequencies ranging from large frequencies to the smallest frequency the lithographic process can support. This will allow users to determine the wavelengths our instruments can measure. This type of effort is being pursued by several researchers. Eytan et al. [47] produced line structures with spatial wavelength deviations of 100 and 200 nm using an e-beam resist, and used it to study the performance of a measurement tool. Similarly, Winkelmeier et al. [44] also produced artificial LER with a 200-nm period using electron beam lithography. Their work focused on methodology for LER quantification. A more comprehensive effort is underway at the International SEMATECH (a semiconductor industry consortium). They plan to induce line edge deviations with spatial frequencies ranging from 50 nm to 16,000 nm, and use these in a study of LER measurability. Other possible studies could also include correlating LER from photomask to final etched features. Such a study could form the basis for commercial LER standards. These efforts will advance as researchers learn more about how LER correlates with performance.

The specification and measurement of line edge roughness is a young field in the semiconductor industry, but an important one. Excessive line edge roughness could be a showstopper in the production of future generation semiconductor chips. This issue has been raised by resist technologists (Hinsberg et al. [26]; Somervell et al. [27]; Patsis et al. [28]; Constantoudis et al. [54]) who are concerned that for resist films operating at the small nodes (feature size), line edge roughness in the developed resist pattern could be a significant fraction of the linewidth itself and could be transferred through to the final circuit element. Line edge roughness in the circuit element has been shown to lead to leakage current in the element. Therefore, LER needs to be controlled and measured.

For measurement of LER, the preferred technique is currently SEM, primarily because of ease and sample throughput. AFM has the potential of higher resolution. However, neither technique has geometry optimized to measure LER directly. For SEM, this could be improved by further development of SEM stereoscopy. AFM could be improved by development of AFM probes with scanning geometry and sensing direction compatible with direct profiling of lines and sidewalls on both sides of a circuit element.

For analysis of LER profiles and topographic maps, we believe that semiconductor metrologists will follow procedures similar to those of conventional roughness metrologists. The parameter of rms surface roughness recast as rmsLWU or rmsSWR will remain as a useful parameter because of its significance for statistical calculations and correlations of many kinds.

The spatial wavelength limits of LER measurements will have to be understood and specified both by the instrument manufacturers and by the instrument users.

Users must understand the manufacturers' spatial limits as well as how their measurement protocols and analyses can change the spatial limits. The parameter of linewidth uniformity (linewidth variation) defined by ITRS 2001 is a good start for standardization. Power spectral density will continue to be used to better understand how LER affects the functioning of electrical circuits. However, broadband amplitude parameters are likely to be more widely specified if we use conventional roughness specifications in optical and mechanical industries as an indicator.

Abbreviations

AFM Atomic Force Microscope
LER Line Edge Roughness
LWU Line Width Uniformity
LWV Line Width Variation
SEM Scanning Electron Microscope
SWR Sidewall Roughness
rms root mean square
SANS Small Angle Neutron Scattering
ITRS International Technology Roadmap for Semiconductors

References

1. Martin Y, Wickramasinghe HK (1995) J Vac Sci Technol B 13:2335
2. Díaz CH, Tao H-J, Ku Y-C, Yen A, Young K (2001) IEEE Electron Dev Lett 22:287
3. Oldiges P, Lin Q, Petrillo K, Sanchez M, Ieong M, Hargrove M (2000) IEEE Electron Dev Lett 21:131
4. Patterson K, Sturtevant JL, Alvis J, Benavides N, Bonser D, Cave N, Nelson-Thomas C, Taylor B, Turnquest K (2001) Experimental determination of the impact of polysilicon LER on sub-100-nm transistor performance. In: Sullivan NT (ed) Proc SPIE metrology, inspection, and process control for microlithography XV v4344. SPIE Press, Bellingham WA, p 809
5. Xiong S, Bokor J, Xiang Q, Fisher P, Dudley I, Rao P (2002) Study of gate line edge roughness effects in 50 nm bulk MOSFET devices. In: Herr DJC (ed) Proc SPIE metrology, inspection, and process control for microlithography XVI v4689. SPIE Press, Bellingham WA, p 733
6. International Technology Roadmap for Semiconductors (ITRS) (2001) Semiconductor Industry Association. San Jose CA
7. Nelson C, Palmateer SC, Forte AR, Cann SC, Deneault S, Lyszczarz TM (1999) Metrology methods for quantifying edge roughness II. In: Singh B (ed) Proc SPIE metrology, inspection, and process control for microlithography v3677. SPIE Press, Bellingham; Nelson C, Palmateer SC, Forte AR, Lyszczarz TM (1999) J Vac Sci Technol B 17:2488
8. Rugar D, Hansma P (1990) Phys Today 43–10:23
9. Griffith JE, Grigg DA (1993) Dimensional metrology with scanning probe microscopes. J Appl Phys 74:R83
10. Marchman HM, Griffith JE (2001) Scanned probe microscope dimensional metrology. In: Diebold AC (ed) Handbook of silicon semiconductor metrology. Marcel Dekker, New York NY, chap 15
11. Church EL, Jenkinson HA, Zavada JM (1979) Optical Eng 18:125
12. Bennett JM, Mattsson L (1989) Introduction to surface roughness and scattering. Optical Society of America, Washington DC, pp 27–31

13. Vorburger TV, Marx E, Lettieri TR (1993) Applied Optics 32:3401
14. Bhushan B (1999) Introduction – measurement techniques and applications. In: Bhushan B (ed) Handbook of micro/nanotribology. CRC Press, Boca Raton FL, chap 1
15. Whitehouse DJ (1982) Wear 83:75
16. Reynolds GW, Taylor JW (1999) J Vac Sci Technol B 17:334
17. Postek MT (1994) Scanning electron microscope metrology. In: Monahan KM (ed) Handbook of critical dimension metrology and process control. SPIE Optical Engineering Press, Bellingham WA, p 46
18. ASTM F 1811-97 (1997) Standard practice for estimating the power spectral density function and related finish parameters from surface profile data. American Society of Testing and Materials, Philadelphia PA
19. ASME/ANSI B46.1 (1995) Surface texture (surface roughness, waviness, and lay). American Society of Mechanical Engineers, New York NY
20. Denk W, Pohl DW (1991) Appl Phys Lett 59:2171
21. Nyyssonen D, Landstein L, Coombs E (1991) J Vac Sci Technol B 9:3612
22. Martin Y, Williams CC, Wickramasinghe HK (1987) J Appl Phys 61:4723
23. Martin Y, Wickramasinghe HK (1994) Appl Phys Lett 64:2498
24. Vasile MJ, Grigg D, Griffith JE, Fitzgerald E, Russell PE (1991) J Vac Sci Technol B 6:3569
25. Fu J, Tsai VW, Koning R, Dixson RG, Vorburger TV (1999) Nanotechnology 10:428
26. Hinsberg W, Houle FA, Hoffnagle J, Sanchez M, Wallraff G, Morrison M, Frank S (1998) J Vac Sci Technol B 16:3689
27. Somervell MH, Fryer DS, Osborn B, Patterson K, Byers J, Willson CG (2000) J Vac Sci Technol B 18:2551
28. Patsis GP, Tserepi A, Raptis I, Glezos N, Gogolides E, Valamontes ES (2000) J Vac Sci Technol B 18:3292
29. Yoshimura, T, Shiraishi, H, Yamamoto J, Okazaki S (1993) Appl Phys Lett 63:764
30. Yamaguchi T, Namatsu H, Nagase M, Yamazaki K, Kurihara K (1997) Appl Phys Lett 71: 2388
31. He D, Cerrina F (1998) J Vac Sci Technol B 16:3748
32. Kant A, Talor G, Samarakone N (1999) Quantitative line edge roughness characterization for the sub-0.25 DUV lithography. In: Singh B (ed) Proc SPIE metrology, inspection, and process control for microlithography v3677. SPIE Press, Bellingham WA, p 35
33. Nagase M, Namatsu K, Kurihara K, Iwadate K, Murase K, Makino T (1996) Microelectr Eng 30:419
34. Reynolds GW, Taylor JW (1999) J Vac Sci Technol B 17:2723
35. Meyyappan A, Klos M, Muckenhirn S (2001) Foot (bottom corner) measurement of a structure with SPM, In: Sullivan NT (ed) Proc SPIE metrology, inspection, and process control for microlithography XV v4344. SPIE Press, Bellingham WA, p 733
36. Walch K, Meyyappan A, Muckenhirn S, Margail J (2001) Measurement of sidewall, line and line-edge roughness with scanning probe microscopy. In: Sullivan NT (ed) Proc SPIE metrology, inspection, and process control for microlithography XV v4344. SPIE Press, Bellingham WA, p 726
37. Rasgon S, Goldfarb D, Mahorowala A (2002) Personal communication. "The authors also gratefully acknowledge DARPA for financial support under contract No. N66001-00-C-8803"
38. Shin J, Han G, Ma Y, Moloni K, Cerrina F (2001) J Vac Sci Technol B 19:2890
39. Nguyen CV, Stevens RMD, Barber J, Han J, Meyyappan M, Sanchez MI, Larson C, Hinsberg WD (2002) Appl Phys Lett 81:901
40. Nguyen CV, Chao KJ, Stevens RMD, Delzeit L, Cassell A, Han J, Meyyappan M (2001) Nanotechnology 12:363
41. Nguyen CV, Stevens RMD, Meyyappan M (2001) Personal communications
42. Mancevski V, McClure PF (2002) Development of a dual-probe Caliper TM CD-AFM for near model-independent nanometrology. In: Herr DJC (ed) Proc SPIE metrology, inspection, and process control for microlithography XVI v4689. SPIE Press, Bellingham WA, p 83

43. Vorburger TV, Dagata JA, Wilkening G, Iizuka K (1998) Characterization of surface topography. In: Czanderna AW, Madey TE, Powell CJ (eds) Methods of surface characterization, vol 5. Plenum Press, New York NY
44. Winkelmeier S, Sarstedt M, Ereken M, Goethals, M, Ronse, K (2001) Microelectron Eng 57–8:665
45. Fang SJ, Haplepete S, Chen W, Helms CR, Edwards H, (1997) J Appl Phys 82:5891
46. Ishida M, Kobayashi K, Fujita J, Ochiai Y, Yamamoto H, Tono S (2002) Jpn J Appl Phys 41:4228
47. Eytan G, Dror O, Ithier L, Florin B, Lamouchi Z, Martin N (2002) Amplitude and spatial frequency characterization of line edge roughness using CD-SEM. In: Herr DJC (ed) Proc SPIE metrology, inspection, and process control for microlithography XVI v4689. SPIE Press, Bellingham WA, p 347
48. Yoshizawa M, Moriya S (2000) J Vac Sci Technol B 18:3105
49. Yoshizawa M, Moriya S (2001) J Vac Sci Technol B 19:2488
50. Raja J, Muralikrishnan B, Fu S (2002) Precision Eng 26:222
51. Chen X, Raja J, Simanapalli S (1995) Int J Machine Tools Manuf 35:231
52. Postek MT, Vladár AE (2001) Critical-dimension metrology and the scanning electron microscope. In: Diebold AC (ed) Handbook of silicon semiconductor metrology. Marcel Dekker, New York NY, chap 14
53. Wu W-l, Lin EK, Lin Q, Angelopolous M (2000) Small angle neutron scattering measurements of nanoscale lithographic features. J Appl Phys 88:7298
54. Constantoudis V, Gogolides E, Patsis GP, Tserepi A, Valamontes ES (2001) J Vac Sci Technol B 19:2694

10 Mechanical Properties of Self-Assembled Organic Monolayers: Experimental Techniques and Modeling Approaches

Redhouane Henda

10.1 Introduction

10.1.1 Self-Assembled Systems

Molecular self-assembly is a process through which molecules organize themselves in an autonomous way, without external intervention. The concept of self-assembly is abundant in nature, and living organisms are typical examples. It is the precise organization of molecules that makes possible the many functions carried out in living cells. For instance, self-assembly is responsible for the formation of liposomes, which are spherical molecular capsules used in clinical trials as drug-delivery systems within the body. They are made of long chain organic molecules with one end attracted to water (i.e., hydrophilic) and the other end repelled by it (i.e., hydrophobic). When placed in an aqueous solution, the molecules form a double layer with hydrophilic ends pointing toward water and hydrophobic ends facing each other. If there are enough molecules in the solution, the double layer will grow into a capsule as depicted in Fig. 10.1. The resulting spherical assembly can hold a drug in its interior and protect it from possible degradation by enzymes. After injection in the body the drug is released via, for instance, rupture of the capsule. The ability to build large functional assemblies from small, often non-functional, entities and molecules makes molecular self-assembly one of the few practical strategies for making ensembles of nanostructures. This "bottom-up" approach is in contrast to the "top-down" approach of miniaturization in silicon-based computer chip technology.

A self-assembled monolayer (SAM) consists of organic molecules, essentially alkane chains, which form a two-dimensional crystal on a solid surface. The resulting dense molecular assembly is formed by a spontaneous chemisorption of a

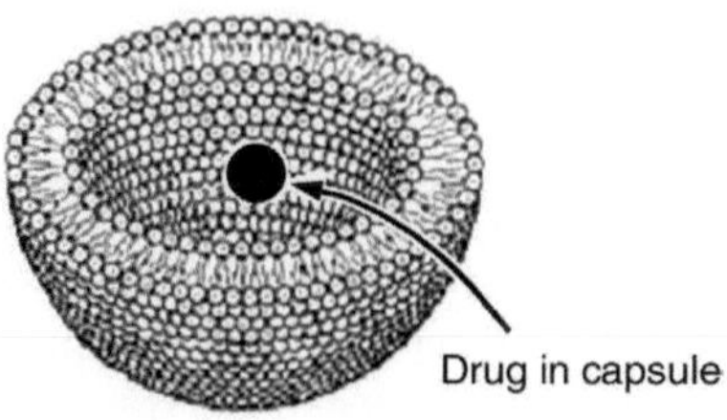

Fig. 10.1. Schematic diagram of a liposome carrying a drug (the lower half of the capsule is shown)

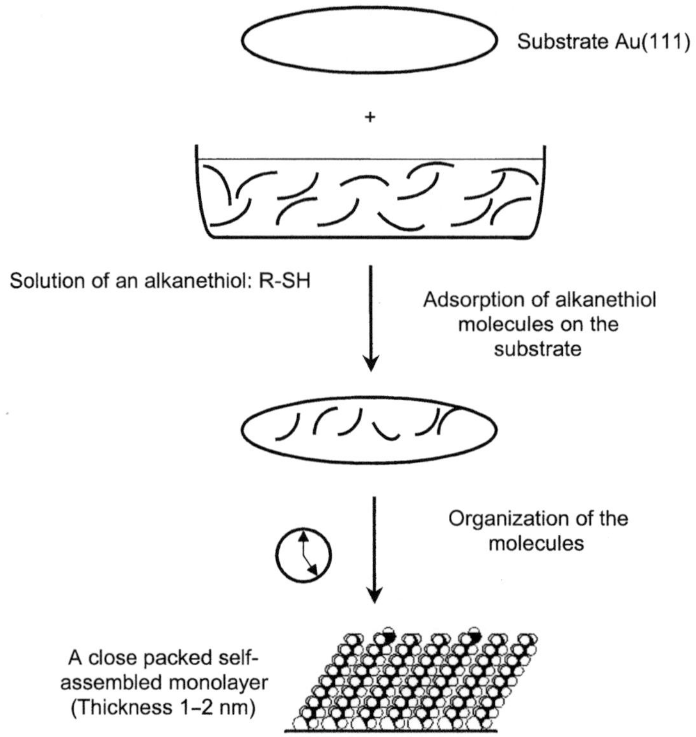

Fig. 10.2. The process of formation of a SAM of alkanethiols on gold. *R* represents the alkane chain with $-CH_3$ as the tail group, and SH (sulfur-hydrogen) is the head group

functional group at one end (called head group) of the long alkane chain onto the substrate as illustrated in Fig. 10.2 for the case of alkanethiols on gold. At the other end (tail group) a variety of functional groups can be affixed, thus altering dramatically the properties of the resulting surface of a SAM. For instance, the wettability of a surface can be controlled with good precision by adsorbing either a hydrophobic or a hydrophilic group onto the "free" surface of a SAM. The same strategy (i.e., by attaching different terminal groups to a SAM), can be used to control other properties, such as adhesion, lubrication, friction, and corrosion, of the surface. Due to their dense and stable structure and the ability to tailor both head and tail groups, SAMs offer unique opportunities to gain fundamental understanding of phenomena affected by competing intermolecular and molecular-substrate interactions. Moreover, the understanding of the structure-property relationships in these systems (i.e., SAMs) is crucial for any progress in the development of new materials through the molecular design approach and self-assembly. SAMs have been intensively studied in the past few years because of their wide range of potential applications in biochemistry, microelectronics, corrosion prevention, microlithography, and other areas [1–3]. Also, SAMs have recently attracted considerable interest as to their potential use as lubricants for such applications as micro-electromechanical systems (MEMS) [4–6].

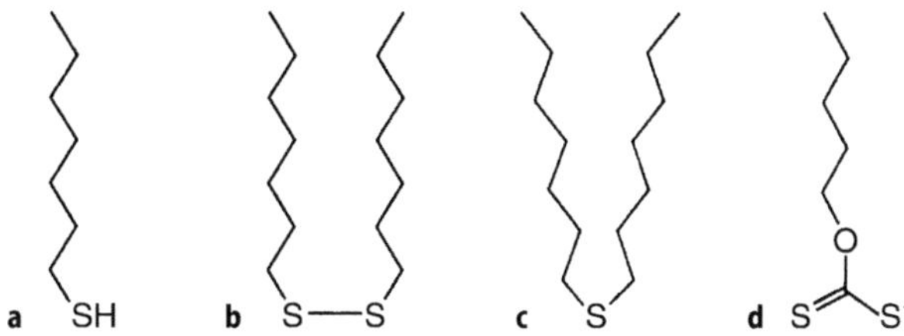

Fig. 10.3a–d. Various surface-active organosulfur alkyl chains: **a** alkanethiol, **b** dialkyl disulfide, **c** dialkyl sulfide, and **d** alkyl xanthate

There are several types of SAMs as shown in Fig. 10.3. These include alkanethiols (Fig. 10.3a) on gold, on silver, and on copper, dialkyl disulfides (Fig. 10.3b) on gold, dialkyl sulfides (Fig. 10.3c) on gold, xanthates (Fig. 10.3d) on gold [1, 2 and the references therein], alkanethiols on germanium [7], and alkanethiols on silicon (e.g., [8, 9]). However, the most extensively studied system of SAMs is that of alkanethiols with a sulfur atom at one end of the hydrocarbon chain attached to a gold surface (Fig. 10.3a). Alkanethiol SAMs on Au(111) surface are most commonly used as a model system for both experimental and theoretical studies. The main reason for opting for gold as the preferred adsorbent is that it does not have a stable oxide [10], at least under ambient conditions.

10.1.2 Origin of Stiction and Mechanical Properties in Small Dimensions

Like many other properties of materials, the mechanical properties begin to deviate from bulk scaling laws when characteristic dimensions, namely, the ratio surface area over volume, become small. As an illustration, attraction causing small particles of dust – situated in the immediate vicinity of a black board in a classroom – to stick to the blackboard does not relate to gravitational or electromagnetic forces, which act from the center of bodies, obey the inverse square law, and can be detected at much greater separations. For instance, the force of attraction be-

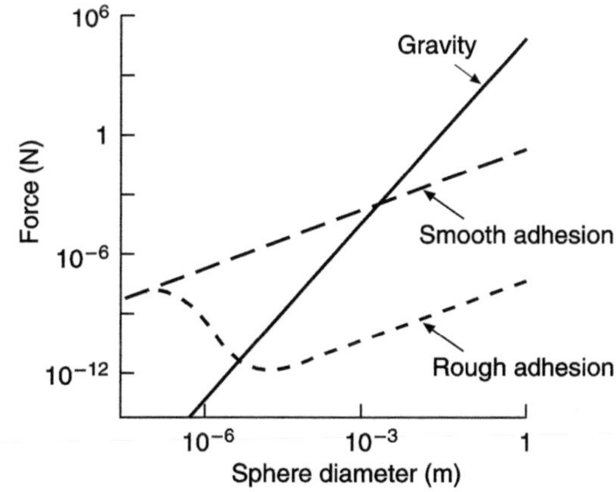

Fig. 10.4. Comparison of the adhesive force and gravity as a function of sphere diameter. Both smooth and rough spheres are represented (reprinted with permission from ref. [11])

tween a particle of 1 μm in diameter and a surrounding surface is more than a million times larger than the force of gravity it experiences as shown in Fig. 10.4 [11]. Therefore, at the submicronic level, namely, below 1 μm, the behavior of materials is more akin to short-range surface forces. In fact, because of the predominance of surface forces over body forces in nanoparticles, the latter tend to stick together without any external assistance. This is in contrast with macroscopic scale engineering materials. This phenomenon is known as "stiction" and is a major limitation in the development of reliable MEMS as these systems shrink towards the nanoscale. One of the possible ways for the reduction of stiction involves the application of a molecular film, such as a SAM, to modify the properties of the surfaces of micromachines [12].

All forces between atoms and molecules are essentially electrostatic in origin, as established from the theory of quantum mechanics and through the elucidation of the electronic structure of atoms and molecules [13]. Intermolecular forces consist of two components: attractive forces to hold molecules together and repulsive forces at short intermolecular distances. The attractive force can be caused by charge transfer interactions, for example, in the case of chemical bonds where the separation of charges causes electrostatic attraction between particles of opposite charges. The repulsive force arises from a quantum mechanical effect known as Pauli exclusion principle. This principle states that no interacting electrons can have the same four quantum numbers. As two or more atoms get very close some interacting electrons, with similar quantum numbers, must occupy higher energy levels so that no two electrons may have the same quantum numbers. The repulsive force develops to prevent electrons from overlapping in order to minimize the increase in the energy of the system of interacting atoms. A typical total force-separation distance curve is illustrated in Fig. 10.5. If we consider the mechanical response of a system of atoms, such as a SAM, to an external load, F, the elastic constant of the system is related to the slope, a, of the curve at the equilibrium separation, x_0, that is,

$$\left.\frac{\partial F}{\partial x}\right|_{x=x_0} = a \tag{10.1}$$

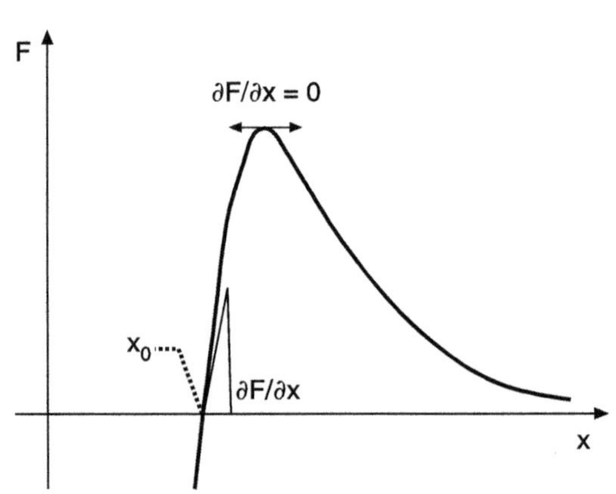

Fig. 10.5. Force-separation distance curve of interacting atoms

At equilibrium the total force (i.e., attractive + repulsive) between atoms is zero. Since the Young's modulus, E, is expressed as the applied stress (force per unit area, A, normal to applied load, F) over the resulting strain, $\varepsilon = (x - x_0)/x_0$, then

$$E = \frac{x_0 a}{A} \tag{10.2}$$

Since the slope given by Eq. (10.1) is a measure of the force required to displace atoms from their equilibrium position, the Young's modulus (or stiffness), E, expresses the resistance of the material to atomic separation. It is also the linear mechanical response of the material. The tensile or breaking strength of the material is the stress that corresponds to the point in the curve (Fig. 10.5) at which the derivative of F vanishes. The toughness, which is a measure of the work needed to break the material, is related to the area under the curve up to the point of failure. Toughness is a combination of both linear and nonlinear mechanical response (plastic, buckling, and failure) of a material. Further mechanical properties can also be extracted from the force-separation distance curve.

In order to calculate the values of the mechanical properties of a material (e.g., E, breaking strength, hardness, toughness, etc.), the total force, F, must be estimated beforehand. This can be done through specialized experimental techniques, such as the ones described below. Alternatively, the total energy, U, of the system needs to be computed, as F is the gradient of U. In this case, the calculation of U can generally be performed through molecular simulation, as discussed later in this chapter.

10.1.3 Goals and Motivation

The mechanical properties of the technologically important substrate-supported organic monolayers, such as self-assembled monolayers, are strongly related to the tribological properties, such as lubrication, friction, and adhesion, of these systems. On the other hand, the study of the response of SAMs to externally applied loads can provide valuable hints as to their reliability and performance in service. This chapter focuses on the response of SAMs to compression and mechanical relaxation, and the quantitative estimation of the Young's modulus from load-displacement curves. Both experimental techniques and molecular simulation approaches are presented. It is assumed that the reader is familiar with first-year college physics and chemistry. Some basic knowledge in statistical analysis will be useful to grasp the concepts of the Monte Carlo method.

10.2 Experimental Techniques

10.2.1 Introduction

Mechanical properties, such as the Young's modulus, hardness, toughness, etc., of materials can be obtained from load-displacement curves. A plot of the force between two interacting surfaces, namely, a probe mounted on a cantilever beam and the sample flat surface, as a function of the probe-sample distance constitutes a load-displacement curve. A variety of available experimental techniques can be used to generate useful data, in the form of a load-displacement curve, for the study of the mechanical properties of materials with sizes in the nanometer range. The advent of the scanning tunneling microscope (STM), the atomic force microscope (AFM), and the interfacial force microscope (IFM) has opened ways for studies of the mechanical properties of materials at the atomic level (these techniques are globally known as scanning probe microscopes). The ideal techniques for organic molecules adsorbed onto a solid substrate, such as a self-assembled monolayer, are AFM and IFM. The label microscope is appropriate insofar as it refers to their capability of measuring the topography of a sample with high spatial resolution. Typically, these microscopes employ a sharp probe that is brought very close to the sample surface and scanned over a specified image area. Through a variety of imaging mechanisms, the probe measures the topography of the sample with very high spatial resolution. On the other hand, mechanical compression and relaxation of a SAM can also be carried out by moving the probe perpendicularly into the SAM surface, and recording the force exerted by the probe on the sample and the vertical displacement of the probe.

In addition to scanning probe microscopes, namely, AFM and IFM, a variety of depth-sensing indentation devices can be used to measure materials mechanical properties. These methods measure the amount of penetration of an indenter tip into a material. However, due to low nanometer scale spatial and load resolutions, current depth-sensing indentation methods have limited capabilities for studying the mechanical behavior of organic thin films on solid substrates. For SAMs, obtaining property information with nanometer spatial resolution is often desired. Further details on depth-sensing indentation methods are given in recent reviews by McHargue [14] and VanLandingham et al. [15].

10.2.2 Atomic Force Microscopy

The atomic force microscope was invented in 1986 by Binnig et al. [16] to remedy the severe limitation of the STM, which can image essentially conducting materials. The AFM operates by measuring attractive or repulsive forces, whose origin is discussed above and is illustrated in Fig. 10.5, between a probe and the sample (Fig. 10.6). In constant-force mode the probe is raster scanned across the surface with feedback mechanisms that enable the piezo-electric scanners to maintain the probe at a constant deflection. A bend in the cantilever corresponds to a displacement, d, of the probe tip relative to an undeflected cantilever and the applied normal force given by $F = -kd$, where k is the cantilever spring constant. To maintain constant the probe deflection, d, the z-scanner moves the position of the probe relatively to the sample surface. It is through this mode that the topography of the sample is mapped during scanning by assuming that the motion of the scanner in the direction normal to the sample corresponds to the sample topography. In the force mode the AFM can image sample elasticity by pressing the probe into the sample. In this case the probe is first lowered into contact with the sample surface, then indented into the surface (loading step),

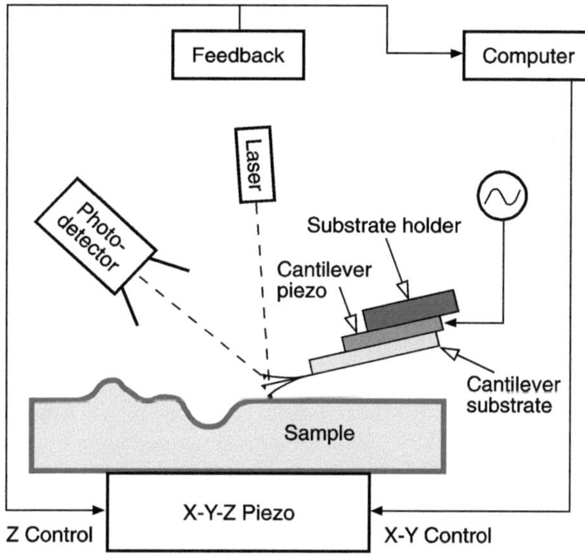

Fig. 10.6. Schematic of an atomic force microscope (material used with permission from Bharat Bhushan (1999) Micro/Nanotribology. In: Principles and applications of tribology. Wiley)

and finally lifted off the sample surface (unloading step) to complete the loading-unloading cycle. Concurrently, a measurement of the probe tip deflection is produced through an optical lever detection system in which a laser beam is reflected off the top of the probe and onto a segmented photodiode (Fig. 10.6). Once the cantilever deflection is measured, the force acting on the probe is obtained using the formula $F = -kd$. That is, a plot of the recorded probe deflection signal as a function of the vertical displacement of the scanner yields the force-displacement curve from which the mechanical properties of the material can be extracted.

10.2.3 Interfacial Force Microscopy

The interfacial force microscope, developed by Joyce and Houston in the early 1990s [17, 18], is a modified version of the AFM but is distinguished by its use of an electrostatically driven, force-feedback sensor (Fig. 10.7). The latter eliminates the so-called jump-to-contact mechanical instability in AFM, and which occurs when the force gradient between the probe and sample equals the spring constant, k, of the cantilever. The mechanical instability gives rise to undesirable discontinuities in the measured load-displacement curves using AFM. Therefore, IFM has superior capabilities for quantitative and stable force measurements over the entire range of interfacial interactions. An electrochemically sharpened tungsten tip is displaced relative to the sample at a given displacement rate by piezoelectric scanners. The tip is then displaced toward the surface at a constant rate until the repulsive interaction between the tip and sample reaches a predetermined force (retract force). At this point the motion is reversed until the original tip position is reached. Throughout an indentation experiment the measured force is recorded as a func-

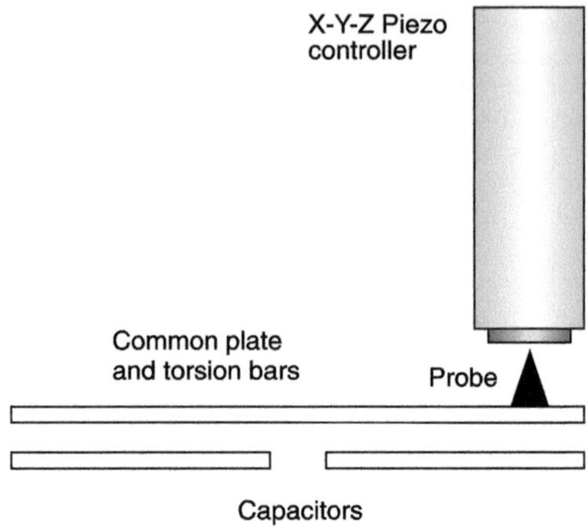

Fig. 10.7. Schematic of the differential-capacitor IFM force-feedback sensor

tion of displacement, and the obtained data allow the quantitative determination of values of the mechanical properties such as the elastic modulus and plastic threshold stress at the nanometer level.

10.2.4 Analysis of Load-Displacement Curves

While the acquisition of experimental load-displacement curves using atomic and interfacial force microscopes is relatively straightforward, the interpretation of the data is rather a challenging task and needs careful examination. This is mainly due to the many phenomena involved (e.g., contact, attraction, repulsion, jump-to-contact, etc.) during probe-sample interactions, and to the fact that the generated experimental data describe the properties of the composite probe-sample system, that is, not the material under study only. Many theories have been developed to describe the mechanics of probe-sample interactions in order to extract the individual properties of the sample. For a detailed and recent review on issues related to the interpretation of experimental data and theoretical contact-mechanics models to successfully determine materials properties from force-displacement curves see Burnham and Kulik [19] and the references therein.

Figure 10.8 shows a typical load-displacement curve for an elastic-plastic material, exhibiting hysteresis between loading and unloading. This is the case of most engineering surfaces including organic films on solids. The material depicted in Fig. 10.8 undergoes both elastic and plastic deformation during probe loading (compression) followed by an essentially elastic deformation during probe unloading (relaxation) [20]. The Young's modulus can be inferred from the initial slope of the unloading curve and is given by

$$\frac{1}{E^*} = \frac{1-\nu_s^2}{E_s} + \frac{1-\nu_p^2}{E_p} \tag{10.3}$$

where E^* is called the reduced modulus of the sample-probe system. E_s, E_p are the Young's moduli and ν_s, ν_p are Poisson's ratios of the sample and probe, respectively.

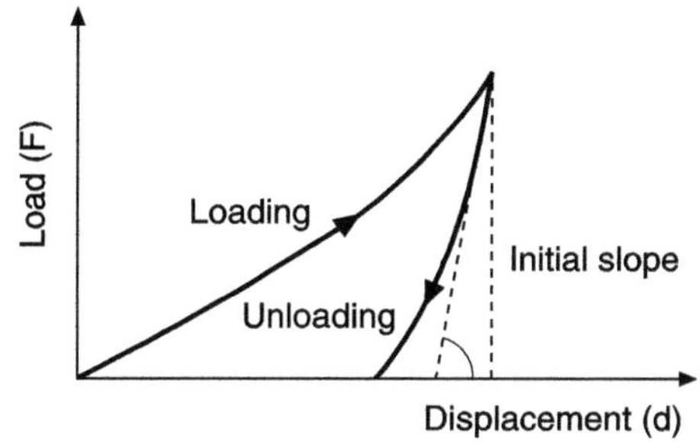

Fig. 10.8. Typical load-displacement curve during an indentation experiment

The reduced modulus is defined as the applied force over projected probe tip-sample contact area over resulting strain. In order to determine the value of E^* the contact area between the probe and sample needs to be calculated. The latter depends on the geometry (e.g., spherical, pyramidal, cylindrical, etc.) and penetration depth of the probe tip. Once the value of E^* is determined, the Young's modulus of the sample, E_s, can be obtained provided that the values of the quantities E_p, ν_s, and ν_p are known.

10.3 Molecular Modeling

10.3.1 Introduction

The understanding of the structure-property relationships in SAMs is a central problem in the quest toward building assemblies for useful applications. As mentioned above, chemisorption on a solid surface is the most important process in the formation of a SAM. The structure of a SAM is determined by the strong interactions between the molecular chains and the crystalline surface, resulting in an adsorption of the head group to specific sites on the underlying surface. Once the molecular head groups occupy all possible sites on the surface, packing and ordering of the molecular chains over the surface proceeds through non-covalent intermolecular interactions (e.g., van der Waals forces [21]). Therefore, packing and ordering in a SAM is the result of intramolecular (bonded) and intermolecular (non-bonded) interactions combined to strong bonds between the head group and the surface (Fig. 10.9). It is this interplay between the various interactions, which is responsible for the two-dimensional order in a SAM. Molecular simulations help in understanding the observable macroscopic properties of a SAM based on information about phenomena at the molecular level. Molecular simulations, thus, can provide insights into the design of organic assemblies at the molecular level, guide the interpretation of experimental data, and enable the prediction of properties not accessible to current laboratory experiments. The properties of a material at the atomic/molecular level can be inferred from the changes in the energy of the system of atoms/molecules due to an external perturbation. Molecular simulations can yield the mechanical properties of a SAM by probing their re-

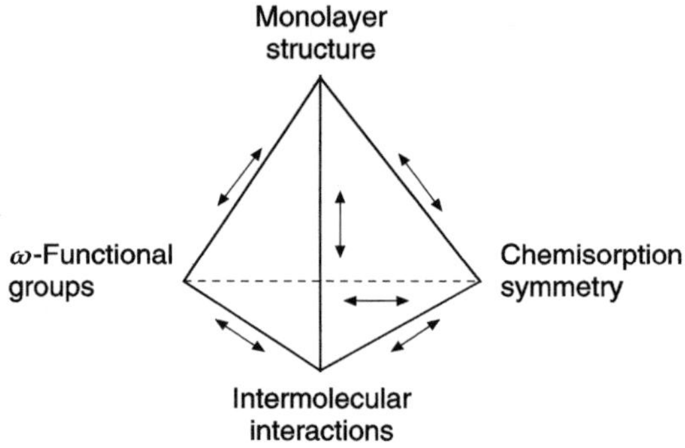

Fig. 10.9. The interplay between chemisorption, intermolecular interactions, and functional groups, is responsible for the structure of a SAM (reprinted with permission from Abraham Ulman (1995) Surface absorption of monolayers. MRS Bulletin 20, no 6)

sponse to compressive loads and mechanical relaxation (the perturbation, in this case). Therefore, of primary importance in any molecular simulation endeavor is the availability of accurate "chemical information" describing all interactions between atoms in the molecular chains in a SAM. This information is known as a *force field* in computational chemistry (e.g., see ref. [22]). There are two approaches to obtain a force field. From quantum mechanical calculations, this is done by solving the Schrödinger equation to obtain the *potential energy surface* either from first principles (through ab initio methods) or via some approximations aided with experimental measurements (case of semi-empirical methods). Although these methods give accurate results in many circumstances, they are computationally expensive, even for molecules with as low as three atoms. The other cheaper approach to obtain a force field is empirical and consists of choosing an analytic form for the function which is to represent the force field, then the parameters of the function are fitted so that experimental data or accurate results of ab initio calculations can be reproduced. The obtained force field can then be used to compute various properties of other molecules composed of units structurally similar.

Since a SAM consists of large functional long-chain hydrocarbon molecules (i.e., with a large number of atoms), methods based on quantum mechanical calculations to compute the potential energy surface are out of reach. Therefore, empirical methods, essentially based on force fields fitted to experimental data, are used. These methods, which have proved to be very successful in describing organic molecules [23], treat the interactions between atoms of a molecule and those of different molecules in the assembly in terms of pairwise potentials. Common methods that provide quantitative information on organic assemblies include energy minimization (also known as molecular mechanics), molecular dynamics, and Monte Carlo. The accuracy of these methods goes hand in hand with the availability of accurate force fields describing the various interactions in a SAM.

The potential energy, U, of the molecular system in question (e.g., a SAM), depends upon the positions in space, denoted here as X_n, of all atoms,

$$U(X_n) = \sum_{i=1}^{n} (U_{str} + U_{bend} + U_{tors} + U_{oop} + U_{vdw} + U_{el} + \ldots) , \tag{10.4}$$

Table 10.1. Typical force field energy functions: b (b_o) is the actual (equilibrium) bond length, θ (θ_o) is the actual (equilibrium) bond angle, and τ and χ are the dihedral and out-of-plane angles, respectively; r_{ij} is the distance between atoms i and j, q_i and q_j are the charges on atoms i and j, respectively; n is an integer number, and k, k_b, k_θ, k_χ, V_n, A_{ij}, B_{ij}, and C_{ij}, are suitable constants

Type of interaction	Force field energy
Bonded interactions	
Bond stretching (U_{str}): between adjacent atoms i and j	$\sum_{bonds} \frac{1}{2} k_b (b - b_o)^2$
Angle bending (U_{bend}): between successive atoms i, j, and k	$\sum_{angles} \frac{1}{2} k_\theta (\theta - \theta_o)^2$
Torsional angles (U_{tors}): between successive atoms i, j, k, and l	$\sum_{dihedrals} \frac{1}{2} V_n (1 + \cos(n\tau))$
Out-of-plane bends (U_{oop}): between atoms i, j, k, and l. Atoms j, k, and l are bonded to atom i	$\sum_{\chi} \frac{1}{2} k_\chi \chi^2$
Non-bonded Interactions	
van der Waals (U_{vdw}): between atoms i and j	$\sum_{ij\,pairs} \frac{A_{ij}}{r_{ij}^{12}} - \frac{B_{ij}}{r_{ij}^{6}}$ (Lennard-Jones)
	$\sum_{ij\,pairs} \frac{A_{ij} e^{-B_{ij} r_{ij}}}{r_{ij}^{12}} - \frac{C_{ij}}{r_{ij}^{6}}$ (Buckingham)
Electrostatic (U_{el}): between atoms i and j	$\frac{1}{k} \sum_{ij\,pairs} \frac{q_i q_j}{r_{ij}}$

where n is the total number of atoms. The mathematical form, Eq. (10.3), of the potential energy function consists of two terms, namely, bonded and non-bonded. The bonded terms describe contributions from atoms which are covalently bound, and are energy functions for bond stretching (U_{str}), angle bending (U_{bend}), torsional angles (U_{tors}), and out-of-plane bends (U_{oop}), respectively. The non-bonded terms represent contributions to potential energy coming from interactions between atoms that are not covalently bound, and include van der Waals (U_{vdw}) and electrostatic (U_{el}) interactions. All energy terms depend upon the Cartesian and internal coordinates of the atoms. The internal coordinates can easily be calculated from the Cartesian coordinates of the atoms of the actual molecule. The different terms are summarized in Table 10.1. A practical introduction to force fields, and data on the various parameters relevant to organic molecules and SAMs are given elsewhere [24–27].

10.3.2 Definition of an Alkanethiol SAM

In order to carry out any sort of modeling of a molecular system, three ingredients are necessary.

I. A description of the structure: the types of atoms present in the molecules are defined, and the way they are covalently connected is described along with non-bonded interactions. This information determines the functional form of the potential energy, *U*.
II. Initial coordinates: in order to search for the new structure of the molecules subjected to an external stimulus (e.g., a mechanical load), the potential energy, *U*, of the system is minimized. This requires a start guess of the geometry, that is, the initial coordinates of all atoms in the molecules should be known.
III. An empirical energy function: given a defined molecular system and its initial atomic coordinates, the function, *U*, describing the potential energy of the system for any configuration, needs to be set up. This step is performed by choosing appropriate force fields among those available.

Alkanethiol SAMs on Au(111) consist of alkyl chains formed by $-CH_2-$ groups connected along the chain backbone. The molecular chain is terminated by a sulfur atom pinned to a gold(111) substrate, and a methyl group, $-CH_3$, at the free surface of a SAM (Fig. 10.10). The structure of these systems has been investigated by IR spectroscopy, NEXAFS, scanning tunneling microscopy (STM), and helium-beam, electron, and X-ray diffraction [28–35]. What is known with confidence is that the monolayers consist of all-*trans* chains packed into a nearly hexagonal lattice with an S···S spacing of 4.97 Å, and that the chains are rotationally ordered and uniformly tilted from the surface normal. The molecular axis is tilted by about 30° from the surfacc normal and has been explained as a result of close-packing requirement to establish optimum van der Waals contacts. This explanation is based on the mismatch between the molecular diameter of 4.5 Å and S···S spacing of 4.97 Å. The simplest alkanethiol on Au(111) system represents a commensurate $\sqrt{3}\, x \sqrt{3}\; R\; 30°$ overlayer structure with one alkanethiol chain in the two-dimensional unit cell, as illustrated in Fig. 10.11. The cell can also contain two chains as observed through the splitting in the methylene asymmetric scissor vibration modes [34], or four different chains as revealed through "extra" diffraction reflections from helium-beam, electron, and diffraction studies [29, 30, 32]. The latter finding could only be interpreted in terms of a $c(4 \times 2)$ superlattice with respect to the simple hexagonal $\sqrt{3}\, x \sqrt{3}\; R\; 30°$ structure. Alkanethiols have two binding modes at the Au(111) surface with either an *sp*-hybridized sulfur (Au–S–C angle = 180°) or an sp^3-hybridized sulfur (Au–S–C angle = 104°). The latter mode is more stable than the former [36].

Finally, given a structural description, atomic coordinates, and the potential energy function, *U*, how does the system compress or relax subject to external loads (Fig. 10.12)? To answer this question on a computer, one needs to employ a few techniques that manipulate the structure given the potential energy of the assembly.

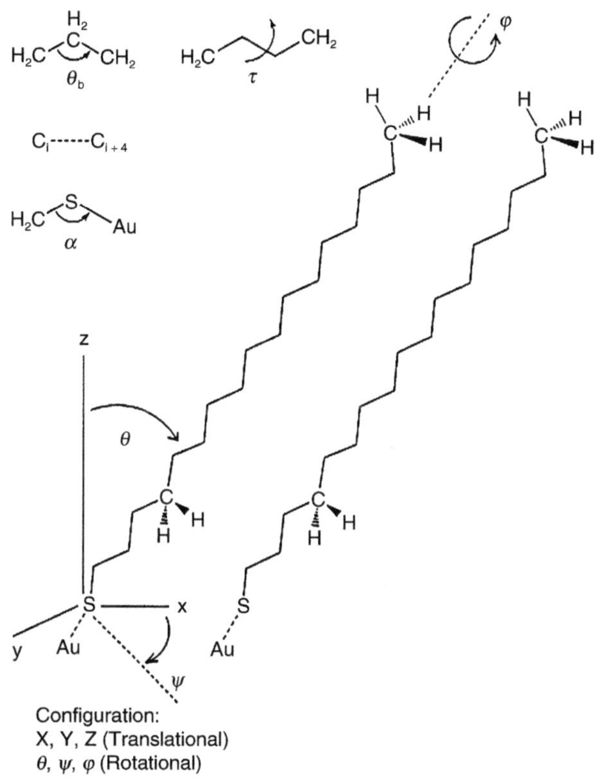

Fig. 10.10. Definition of an alkanethiol SAM in terms of Cartesian (x, y, z), rotational (θ, ψ, φ) and internal coordinates (θ_b, τ, α)

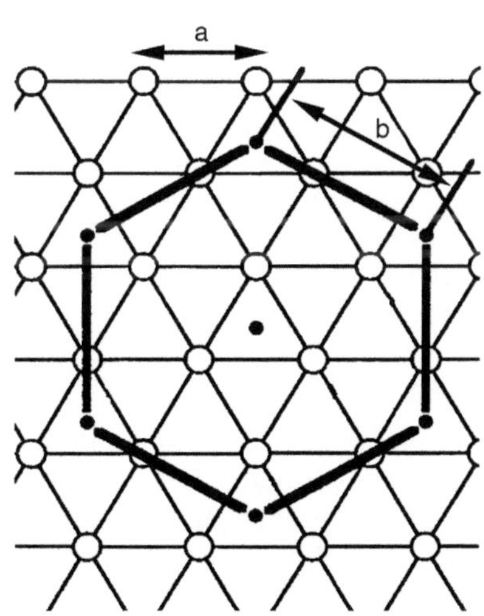

Fig. 10.11. Hexagonal coverage scheme for alkanethiols on Au(111). *Open circles* are gold atoms, and *filled circles* are sulfur atoms. a = Au–Au spacing, b = S···S spacing, $b = a\sqrt{3}$, and the monolayer lattice is oriented by 30° relative to the gold lattice

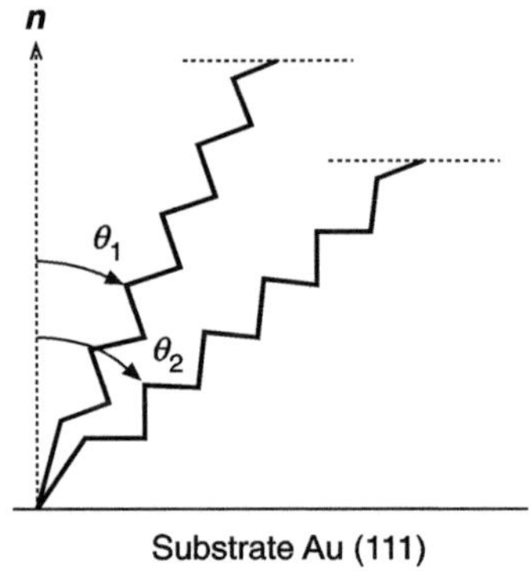

Fig. 10.12. An alkanethiol SAM under compressive forces (two different tilt angles are shown). For simplicity only one molecule per unit cell is depicted. The loaded SAM undergoes configurational changes. Vector $\vec{n}$ is the surface normal

10.3.3 Energy Minimization

The minimization of the potential energy of a molecular assembly involves a search for the minimum of the function U. Generally, to be efficient, the search for the minimum requires calculations of derivatives of the function versus independent variables (the coordinates, in this case). The dimension of the configurational space in a SAM unit cell is equal to $6N$ [$= 3N$ Cartesian $(x, y, z) + 3N$ rotational coordinates (θ, ψ, φ)], where N is the number of symmetrically distinct molecules in the unit cell (see Fig. 10.10). In general, the energy "landscape" of large molecular systems, such as a SAM, possesses a large number of minima, and finding the global minimum is a challenge. Therefore, the minimization methods are iterative, that is, they require on input some initial estimate for the position of the minimum, and provide a better estimate for the minimum as a result. Such methods include the steepest descent and conjugate gradient (e.g., see ref. [37]), which are based on the estimation of the gradient of the function to move for going downhill. If the vicinity of the minimum has the shape of a long, narrow valley, the minimum is reached in far fewer steps using the conjugate gradient method than would be the case using the method of steepest descent. The former method uses conjugate directions instead of the local gradient to move in the direction of the minimum. Each iteration results in a solution that corresponds to a lower value for the energy function, according to the scheme

$$U(X^{start}) > U(X^1) > U(X^2) > \ldots > U(X^{\min}) \quad (10.5)$$

The process of iteration is continued until there is no significant improvement in the position of the minimum. It should be mentioned here that although energy minimization gives the configuration of the molecular assembly at equilibrium (i.e., corresponding to 0 K), it is appropriate for the calculation of the mechanical properties of an alkanethiol SAM. By analogy with organic crystals [21], it is plausible to expect that the static contribution to the mechanical response of a SAM is predominant. This is also corroborated by experimental data discussed in Section 10.4 below.

10.3.4 Molecular Dynamics

The molecular dynamics (MD) method uses Newton's classical mechanics equations of motion and the potential energy function, U, to follow the displacement of atoms in a molecule over a certain period of time, at a certain temperature, and a certain pressure. The total force on an atom, i, is calculated as the sum of the forces due to the remaining atoms, j, in the assembly,

$$\vec{F}_i = \sum_{j \neq i} \vec{F}_{j\,on\,i} \tag{10.6}$$

and is related to the potential energy function by the gradient

$$F_i = -\frac{\partial U}{\partial X_i} \tag{10.7}$$

This force results in the acceleration, a_i, of atom i according to Newton's equation of motion,

$$\vec{F}_i = m_i \vec{a}_i \tag{10.8}$$

where m_i is the mass of atom i. Calculating the trajectories of all atoms involves the solution of a set of second-order differential equations. From atom positions, velocities, and accelerations at any moment in time, one can calculate atom positions and velocities at the next time step. Integrating these infinitesimal steps yields the trajectory of the system for any desired time range. There are efficient methods for integrating these elementary steps, with Verlet and leapfrog algorithms being the most commonly used (e.g., see ref. [38] and the references therein). Generally, one starts a MD simulation with an initial set of atom positions appropriate to 0 K. The first step in the calculation consists of bringing the system to some low temperature, and initial velocities are assigned to each atom so as to reflect equilibrium at the desired temperature according to the Maxwell-Boltzmann distribution. In addition to simulating the time evolution of the molecular system, the calculated trajectory also contains dynamical properties such as time correlation functions, and transport coefficients, such as diffusion constants. The properties of the molecular system can be calculated via the rules of statistical mechanics.

10.3.5 Monte Carlo Method

The stochastic approach, called Monte Carlo (MC) method, is based on exploring the energy surface by randomly probing the configuration of the molecular system. A new configuration is generated by randomly "kicking" one or more atoms of the assembly. Therefore, the system jumps abruptly from configuration l to configuration m, rather than evolving smoothly through time as it does in a molecular dynamics simulation. That is, the MC method takes "snapshots" of the system at different times. This generates a large number of configurations whose respective potential energies are then calculated. Some of the accumulated configurations may also be selected for further minimization. An efficient scheme for the realization of the MC method is called the Metropolis algorithm [39] and consists of the following steps,

1. Specify the initial atom coordinates (e.g., from molecular mechanics minimization), and calculate its potential energy.
2. Select an atom *i* randomly and move it by random displacement.
3. Calculate the potential energy of the new configuration.
4. Calculate the change of potential energy corresponding to this displacement, that is, $\Delta U = U(new) - U(old)$. If $\Delta U < 0$ accept the new configuration and go to step 2. Otherwise, select a random number in the range [0,1]. If the number is less than exp $[-\Delta U/(kT)]$, accept the new configuration and go to step 2. Otherwise reject it and keep the original configuration and go to step 2.

Unlike the MD technique, the MC method cannot be used to study time-dependent properties, but it is appropriate for the calculation of the thermodynamic properties of the system. In fact, organic molecular systems are approached more efficiently with the MC method.

10.4 Results

Experimental studies on the mechanical properties of SAMs have been carried out using AFM [40] and IFM [18, 41]. Joyce et al. [18] have studied the response of SAMs of *n*-alkanethiols ($CH_3(CH_2)_mSH$) on gold (111) using an approximately 500-nm-radius tungsten tip and forces in the μN range which produced pressures of a few GPa. They observed that long-chain thiols respond anelastically to mechanical deformation (Fig. 10.13) while short chains ($m < 10$) do not present any hysteresis in the strain recovery throughout the loading-unloading cycle. They attributed the results to differences in relaxation times between the long-chain and short-chain thiols. According to the authors, during compression the alkane backbones become entangled with one another, and upon unloading of the monolayer long chains have slower recovery times (≈0.08 s) than short chains. An estimate of the reduced Young's modulus for the case of a SAM of *n*-hexadecanethiol gives a value of approximately 30 GPa. This is a gross value since the contact area diameter (≈10 nm) was estimated by simple Hertzian elastic analysis [18]. Recently, Houston and Kim [41] determined the contact radius (area) from a contact-mechanics analysis, namely, using the longstanding Johnson-Kendall-Roberts model of the load-displacement data obtained from IFM measurements for a SAM of *n*-hexadecanethiol. They found a reduced Young's modulus value of approximately 5 GPa, which is six times smaller than the value of the reduced modulus estimated by Joyce et al. [18]. The upper part of Fig. 10.14 shows the load-displacement curve of a SAM of $CH_3(CH_2)_{21}SH$ on Au(111) obtained from AFM measurements [40]. The AFM is based on light reflection from a microfabricated Si_3N_4 cantilever with an approximately 100-nm-radius PtRh tip oscillating at a frequency slightly smaller than its mechanical resonance. Initially, the sample is far away from the tip and the cantilever is in its rest position (see Fig. 10.14, force magnitude = 0). As the sample is moved in the direction of the cantilever tip, the cantilever bends towards the

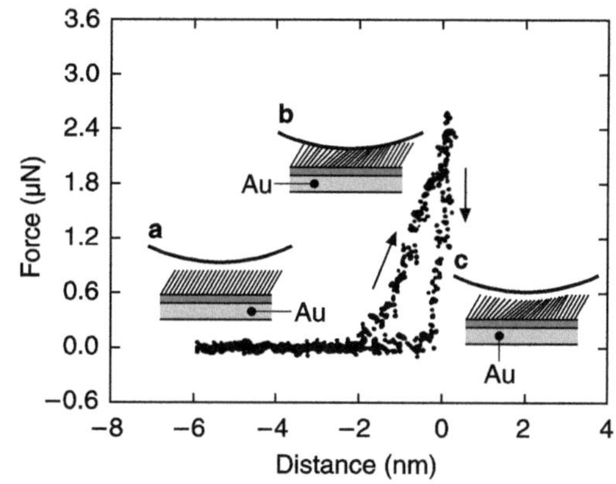

Fig. 10.13. Load-displacement curve of a SAM of *n*-hexadecanethiol ($CH_3(CH_2)_{15}SH$) on Au(111) obtained from IFM measurements. The displacement is expressed as the distance between the tip and the gold surface. Illustrations in the *inset* indicate the proposed film response during various stages of the loading-unloading cycle (reprinted with permission from ref. [18])

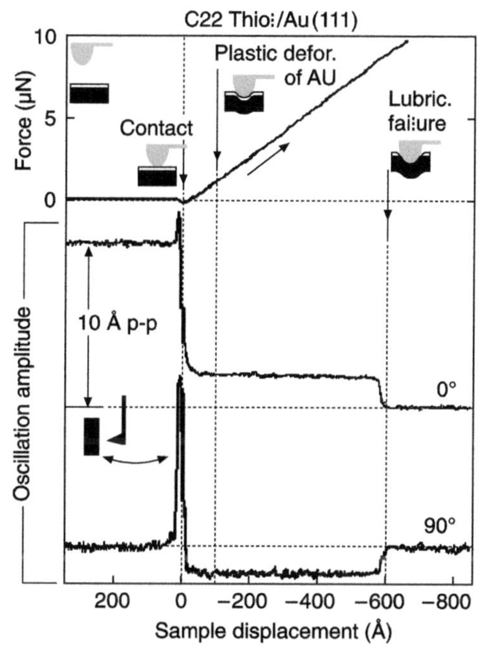

Fig. 10.14. Load-displacement curve (*upper part*) of a SAM of $CH_3(CH_2)_{21}SH$ on Au(111) obtained from AFM measurements. The jump-to-contact locus is illustrated by the dramatic change in oscillation amplitudes (*lower part*) and corresponds to the zero of the displacement axis (reprinted with permission from ref. [40])

sample due to attractive forces (expressed in negative values) until the jump-to-contact occurs. The latter corresponds to the minimum in the load-displacement curve and is accompanied by a dramatic change in the oscillation amplitude of the PtRh tip (Fig. 10.14, lower part). As the sample is moved forward, it pushes the cantilever back through its original rest position until a predetermined sample position is achieved (maximum applied load). Up to this point the SAM has undergone the first half of the cycle (loading half-cycle). At this point, the sample direction is reversed away from the cantilever, and the unloading half-cycle starts. The cantilever movement also changes direction until it separates and returns to its rest position (far away from the sample). The oscillation amplitude does not vanish after con-

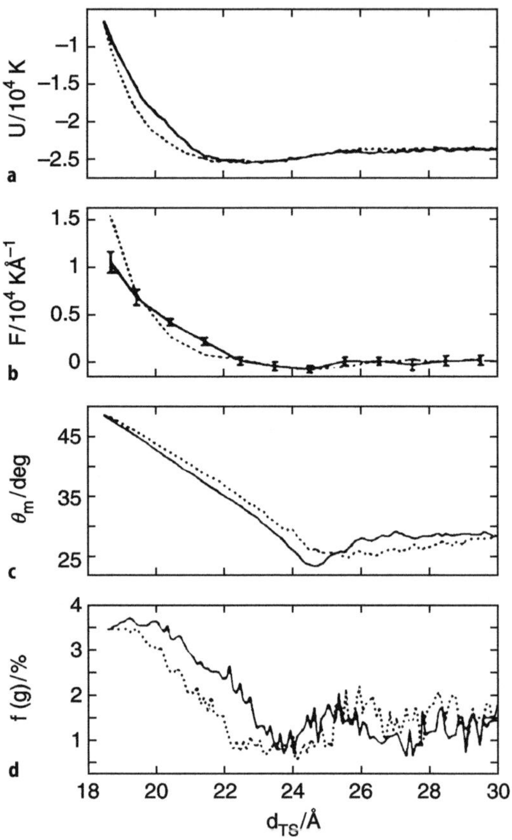

Fig. 10.15a–d. Simulated load-displacement curve **b** of a SAM of *n*-hexadecanethiol on Au(111) during loading (*solid lines*) and unloading (*dashed lines*). The displacement is expressed as the tip-surface separation (d_{TS}). *Error bars* are shown for the loading half-cycle only, but are valid for the unloading half-cycle. The other parts of the figure include **a** potential energy per chain, **c** chain tilt angle, and **d** gauche defects (reprinted with permission from ref. [42])

tact (lower part of Fig. 10.14, displacement range 0 to –600). The authors attributed the residual oscillation to a "cushioning" effect by the SAM as a result of its viscoelastic deformation, in agreement with the results obtained by Joyce et al. [18], and discussed above.

Several groups have carried out molecular simulations to probe the response of SAMs during the loading-unloading cycle. Monte Carlo calculations used by Siepman and McDonald [42] show a nearly elastic (with little hysteresis) response of a SAM of *n*-hexadecanethiol on a gold substrate (Fig. 10.15). The loading-unloading cycle consists of five regions: three regions in the loading half-cycle and two regions in the unloading half-cycle. Prior to tip-sample contact there are no structural changes in the SAM ($d_{TS} > 26$ Å). The jump-to-contact occurs in the displacement range between 26 Å and 24 Å through an abrupt decrease of approximately 5° in the tilt angle (Fig. 10.15c). At this stage, the magnitude of the attractive force (negative) is at its highest absolute value. After the jump-to-contact, a repulsive regime (24 Å $> d_{TS} >$ 18.5 Å) is entered and repulsion stiffens as the tip moves closer to the surface. This is especially obvious from the potential energy of the SAM as depicted in Fig. 10.15a. Throughout this stage the molecular tilt increases almost linearly with decreasing displacement (Fig. 10.15c), and the per-

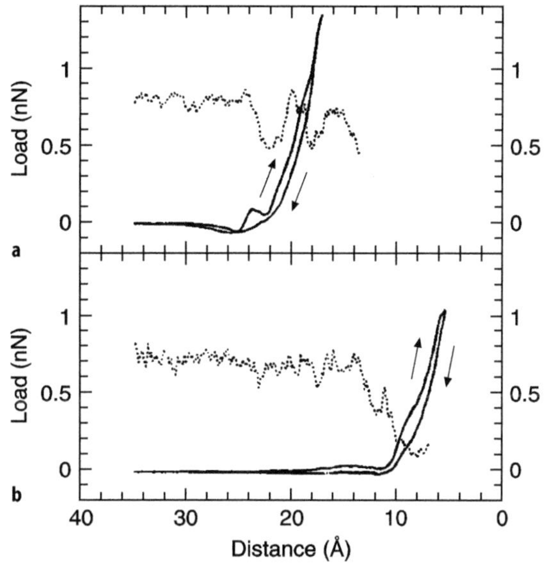

Fig. 10.16a,b. Simulated load-displacement curve for compression of a SAM of *n*-hexadecanethiol on Au(111) with **a** a flat gold (111) surface and **b** a surface containing an asperity. The *arrows* indicate the direction of loading and unloading. The *dashed lines* represent the sum of the peaks in the structure factors. The distance is defined as the separation between the leading edge of the flat surface or asperity and the substrate (reprinted with permission from ref. [43])

centage of gauche defects increases followed by a plateau at the end of the loading half-cycle (Fig. 10.15d). The SAM presents an increase in the number of gauche defects of up to 4% of the total number of dihedral angles, during the loading-unloading cycle. The initial stage (18.5 Å $< d_{TS} <$ 24 Å) of the unloading half-cycle is similar to the third stage (repulsive) of the loading half-cycle, with the exception of little hysteresis in the properties. Finally, the fifth stage ($d_{TS} >$ 24 Å) shows evidence of prolonged contact between the tip and sample. This can be inferred from the slightly smaller values of the tilt angle during this stage than the values of the tilt angle corresponding to the same range of d_{TS} during the loading half-cycle. Tupper et al. [43, 44] performed molecular dynamic simulations of a SAM of *n*-hexadecanethiol compressed under a flat surface and under an atomically sharp asperity (Fig. 10.16). The calculations predict different behaviors of the SAM depending on the contact probe geometry. An increase in the tilt angle and slight decrease in chain length (≈5%), primarily due to bond-length compression, with few gauche defects has been predicted for the flat geometry. On the other hand, considerable decrease in chain length (≈15%), due a decrease in both bond length and bond angle, and formation of a large number of gauche defects has been predicted for the sharp asperity case. The large number of gauche defects allows the SAM to accommodate the asperity. The findings show that the sharp asperity is able to get closer to the substrate, within tunneling range, relative to the flat surface at comparable loads without disrupting the SAM. This can be appreciated from the sum of peaks in the structure factors, which reflect the extent of order in the SAM, and illustrated in Fig. 10.16. Henda et al. [45, 46] have simulated the mechanical response of SAMs of *n*-alkanethiols on Au(111) using the method of energy minimization combined with stochastic global search (a version of the method of molecular mechanics). The response of a SAM has been calculated using different force fields, unit cells, and monolayer configurations. The calculated stress-strain

Table 10.2. Calculated Young's modulus (in GPa) as a function of the force field and number of chains per unit cell. The stress-induced structural changes as a function of the number of chains per unit cell are given for forcefield I[a]

Number of chains/ unit cell	Young's modulus Force field		Structural parameters (force field I)	
	I	II	Configuration x, y, z, ψ, φ	Conformation θ_b, τ
1	12.0	10.5	x, y, z, ψ: NS; φ: max. 15°	θ_b: NS; τ: max. 8° (terminal groups)
2	16.0	9.9	NS[b]	θ_b: NS; τ: max. 8° (terminal groups)
4	18.3	12.7	NS	θ_b: NS; τ: max. 23° (head groups)

[a] The unit cell area = 21.7 Å^2.
[b] *NS* not significant.

relationships for the various SAMs depend on the assumed configuration of the monolayer in the unit cell (viz., 1, 2, and 4 chains per unit cell), the force fields used (viz., all-atom and united-atom approaches), and the chain length (i.e., the number of methylene groups on the chain backbone, m). The differences in the values of the Young's modulus obtained for a SAM of *n*-octadecanethiol ($CH_3(CH_2)_{17}SH$) using all-atom (I) and united-atom (II) approaches are illustrated in Table 10.2. The three experimentally observed structures are depicted. The differences in the values of the Young's modulus are attributed to differences in the equilibrium SAM structures obtained from the two force fields. The calculated values of the Young's modulus are in good agreement with the experimentally determined ones by Joyce et al. [18], and Houston and Kim [41]. The loading-unloading cycle did not produce significant structural changes as summarized in Table 10.2. Except from the conformational changes observed in the head and tail groups, the chains retained their planarity during the loading-unloading cycle for the three unit cell configurations considered as indicated by the order parameter (defined by ref. [47]) in Fig. 10.17. All investigated SAMs recovered from compression after relaxation and no hysteresis has been observed within the conditions of the calculations. Finally, the mechanical response of the SAMs depends on the length of the alkanethiol chain as shown in Fig. 10.18, with long chains ($m \geq 8$) being stiffer (i.e., having a higher Young's modulus) than short chains ($m < 8$). As an example the value of the Young's modulus for SAM with $m = 18$ is 15 GPa compared to 2 GPa for a SAM with $m = 5$. This is, at least, due to the increased contribution of the substrate-chain interaction to close packing in the SAM relative to chain-chain interactions in a SAM with short chains. The dependence of the mechanical properties on chain length correlate well with the chain length dependence of the tribological properties of SAMs of alkanethiols [48, 49].

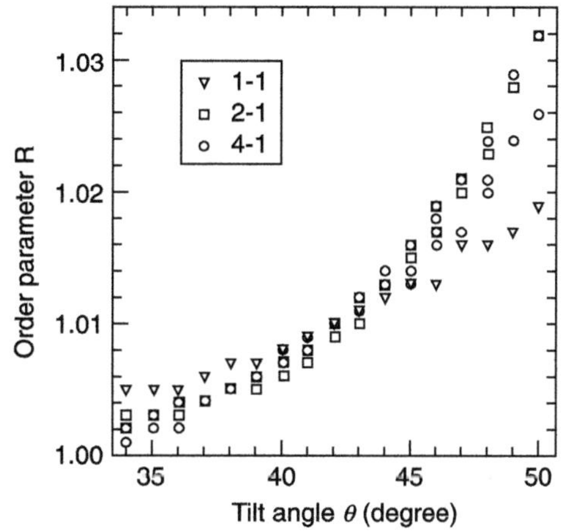

Fig. 10.17. Order parameter, R, of the molecules in the lowest energy 1, 2, and 4 chain(s)/unit cell SAM models as a function of molecular tilt. $R = 1$ for a planar chain. All molecules are represented (reproduced with permission from ref. [46])

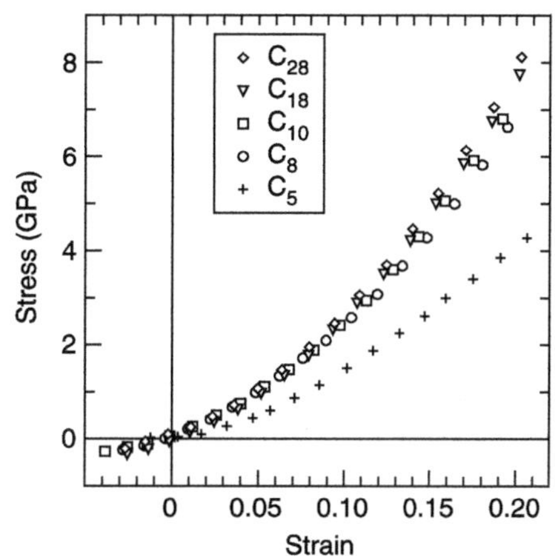

Fig. 10.18. Dependence of the stress-strain curves on the number of monomer units (C_m) in a molecular chain. The *solid lines* serve as guides for the eye (reproduced with permission from ref. [46])

10.5 Conclusions

Self-assembled monolayers are densely packed and ordered molecular systems with a wide range of properties. This is due to the unlimited possibilities of tailoring the head and tail groups of the organic molecular chains, and of modifying the chain length and types of bonds along the chain. The molecular design approach readily possible in SAMs gives rise to a large number of attractive tribological and mechanical properties with potential technological applications, including use in molecular and microdevices. This chapter provides an account of the present understanding of the mechanical properties of nanoscale SAMs and how these properties relate to the molecular features in these systems. The mechanical properties of SAMs are of great interest as they are critical in determining the tribological behavior and performance of these systems.

Innovative experimental techniques are used to study the response of SAMs to external mechanical loads. These techniques include AFM and IFM to probe the mechanical properties of SAMs through load-displacement measurements. While the AFM sensors suffer from the jump-to-contact instability inducing undesirable uncertainties in the obtained results and some difficulties in the interpretation of the data, the IFM provides quantitative and stable force measurements over the entire loading-unloading cycle. For both techniques the lack of the direct knowledge of the contact area between the sample and the tip results in a gross quantitative estimation of the mechanical properties of SAMs. This means that more reliable contact-mechanics models need to be developed. Also, a more systematic investigation of the mechanical properties of SAMs and the many possible combinations through the molecular design approach remain to be explored. Molecular modeling has proved to be very helpful in understanding the structure of molecules making up a SAM and how it relates to the mechanical properties of the assembly as they materialize at the macroscopic level, thus, bridging the gap between the nanoscale and macroscale in SAMs. Computer simulations are also valuable to aid in the interpretation of data obtained through experimental measurements. There still is need for the development of more accurate force fields and faster algorithms, especially if molecular modeling is to be used as a reliable predictive tool.

Finally, SAMs are at the junction of many disciplines including physics, chemistry, chemical engineering, and materials science. Therefore, concerted and joint efforts by researchers in those areas will be decisive to speed the transition of SAMs to large-scale production of molecular and microdevices.

Acknowledgments. The author thanks Prof Douglas Goldsack with the Department of Chemistry, Laurentian University, for reviewing and commenting on the manuscript.

References

1. Ulman A (1991) An introduction to ultrathin organic films: from Langmuir-Blodgett to self-assembly. Academic Press, Boston
2. Ulman A (1995) Surface absorption of monolayers. MRS Bulletin 10:46–51
3. Service RF (1994) Self-assembly comes together. Science 265:316–318
4. Bhushan B (2001) Self-assembled monolayers for controlling hydrophobicity and/or friction and wear. In: Bhushan B (ed) Modern tribology handbook. CRC Press, Boca Raton, Florida, pp 909–929
5. de Boer MP, Mayer TM (2001) Tribology of MEMS. MRS Bulletin 26:302–304
6. Maboudian R (1998) Surface processes in MEMS technology. Surf Sci Rep 30:207–270
7. Han SM, Ashurst WR, Carraro C, Maboudian R (2001) Formation of alkanethiol monolayer on Ge(111). J Am Chem Soc 123:2242–2245
8. Bergerson WF, Mulder JA, Hsung RP, Zhu X-Y (1999) Assembly of organic molecules on silicon surfaces via the Si-N linkage. J Am Chem Soc 121:454–455
9. Zhu X-Y, Boiadjiev V, Mulder JA, Hsung RP, Major RC (2000) Molecular assemblies on silicon surfaces via Si-O linkages. Langmuir 16:6766–6772
10. Somorjai GA (1981) Chemistry in two dimensions: surfaces. Cornell University Press
11. Kendall K (1994) Adhesion: molecules and mechanics. Science 263:1720–1725
12. Maboudian R, Ashurst WR, Carraro C (2002) Tribological challenges in MEMS. Tribol Lett 12:95–100
13. Israelachvili J (1991) Intermolecular and surface forces. Academic Press, San Diego

14. McHargue CJ (1997) Mechanical property determination using nanoindentation techniques. In: Bhushan B (ed) Tribology issues and opportunities in MEMS. Kluwer Academic, Boston, pp 487–508
15. VanLandingham MR, Villarrubia JS, Guthrie WF, Meyers GF (2001) Nanoindentation of polymers: an overview. In: Tsukruk VV, Spencer ND (eds) Macromolecular symposia. Wiley-VCH, New York, pp 15–43
16. Binnig G, Quate CF, Geber CH (1986) Atomic force microscope. Phys Rev Lett 56:930–933
17. Joyce SA, Houston JE (1991) A new force sensor incorporating force-feedback control for interfacial force microscopy. Rev Sci Instrum 62:710–715
18. Joyce SA, Thomas RC, Houston JE, Michalske TA, Crooks RM (1992) Mechanical relaxation of organic monolayer films measured by force microscopy. Phys Rev Lett 68:2790–2793
19. Burnham NA, Kulik AJ (1999) Surface forces and adhesion. In: Bhushan B (ed) Handbook of micro/nano tribology. CRC Press, Boca Raton, Florida, pp 247–271
20. Stilwell NA, Tabor D (1961) Elastic recovery of conical indentations. Phys Proc Soc 2:169–180
21. Kitaigorodsky AI (1973) Molecular crystals and molecules. Academic Press, New York
22. Jensen F (1999) Introduction to computational chemistry. Wiley, New York
23. Pertsin AJ, Kitaigorodsky AI (1987) The atom-atom potential method. Springer, Berlin Heidelberg New York
24. Burkert U, Allinger NL (1982) Molecular mechanics. ACS Monograph 177, Washington, DC
25. Field MJ (1999) A practical introduction to the simulation of molecular systems. Cambridge University Press, Cambridge
26. Rinaldi RP, Pawley GS (1976) An investigation of the intermolecular modes in orthorhombic sulphur. J Phys C 8:599–616
27. Williams DE (1967) Nonbonded potential parameters derived from crystalline hydrocarbons. J Chem Phys 47:4680–4684
28. Anselmetti D, Baratoff A, Guentherodt H-J, Delamarche E, Michel B, Gerber CH, Kang H, Wolf H, Ringsdorf H (1994) Domain and molecular superlattice structure of dodecanethiol self-assembled on Au(111). Europhys Lett 27:365–370
29. Camilone N III, Chidsey CED, Eisenberger P, Fenter P, Li J, Liang KS, Liu G-Y, Scoles G (1993) Structural defects in self-assembled organic monolayers via combined atomic beam and X-ray diffraction. J Chem Phys 99:744–747
30. Camilone N III, Chidsey CED, Liu G-Y, Scoles G (1993) Superlattice structure at the surface of a monolayer of octadecanethiol self-assembled on Au(111). J Chem Phys 98:3503–3511
31. Dubois LH, Zegarski BR, Nuzzo RG (1993) The molecular ordering of organosulfur compounds on Au(111) and Au(100): adsorption from solution and in ultrahigh vacuum. J Chem Phys 98:678–688
32. Fenter P, Eisenberger P, Liang KS (1993) Chain-length dependence of the structure and phases of $CH_3(CH_2)_{n-1}$-SH self-assembled on Au(111). Phys Rev Lett 70:2447–2450
33. Hähner G, Kinzler M, Thümmler CH, Wöll CH, Grunze M (1992) The structure of self-organizing organic films: a near edge X-ray absorption fine structure investigation of thiol layers. J Vac Sci Technol A 10:2758–2763
34. Nuzzo RG, Korenic EM, Dubois LH (1990) Studies of the temperature-dependent phase behavior of long chain n-alkyl thiol monolayers on gold. J Chem Phys 93:767–773
35. Strong L, Whitesides GM (1988) Structures of self-assembled monolayer films of organosulfur compounds adsorbed on gold single-crystals-electron-diffraction studies. Langmuir 4: 546–558
36. Sellers H, Ulman A, Shnidman Y, Eilers JE (1993) Structure and binding of alkyl thiolates on gold and silver surfaces: implications for self-assembly. J Am Chem Soc 115:9389–9401
37. Press WH, Flannery BP, Teukolsky SA, Vetterling WT (1992) Numerical Recipes in FORTRAN: the art of scientific computing, 2nd edn. Cambridge University Press, Cambridge, pp 413–417

38. Rapaport DC (1998) The art of molecular dynamic simulation. Cambridge University Press, Cambridge
39. Metropolis N, Rosenbluth AW, Rosenbluth MN, Teller AH, Teller E (1953) Equation of state calculations by fast computing machines. J Chem Phys 21:1087–1092
40. Salmeron M, Liu G-Y, Ogletree DF (1995) Molecular arrangement and mechanical stability of self-assembled monolayers on Au(111) under applied load. In: Günterodt et al (eds) Forces in scanning probe methods. Kluwer Academic, Boston, pp 593–598
41. Houston JE, Kim HI (2002) Adhesion, friction, and mechanical properties of functionalized alkanethiol self-assembled monolayers. Acc Chem Res 35:547–553
42. Siepmann JI, McDonald IR (1993) Monte Carlo simulation of the mechanical relaxation of a self-assembled monolayer. Phys Rev Lett 70:453–456
43. Tupper KJ, Colton RJ, Brenner DW (1994) Simulations of self-assembled monolayers under compression: effect of surface asperities. Langmuir 10:2041–2043
44. Tupper KJ, Brenner DW (1994) Compression-induced structural transition in a self-assembled monolayer. Langmuir 10:2335–2338
45. Henda R, Grunze M, Pertsin A (1998) Static energy calculations of stress-strain behavior of self-assembled monolayers. Tribol Lett 5:191–195
46. Henda R (2000) Simulation of self-assembled monolayers under normal stress. AIChE J 46:1275–1279
47. Hautman J, Klein ML (1989) Simulation of a monolayer of alkyl thiol chains. J Chem Phys 91:4994–5001
48. Lio A, Charych DH, Salmeron M (1997) Comparative atomic force microscopy study of the chain length dependence of frictional properties of alkanethiols on gold and alkylsilanes on mica. J Phys Chem B 101:3800–3805
49. Xiao X, Hu JH, Charych DH, Salmeron M (1998) Chain length dependence of the frictional properties of alkylsilane molecules self-assembled on mica studied by atomic force microscopy. Langmuir 12:235–237

11 Micro-Nano Scale Thermal Imaging Using Scanning Probe Microscopy

Li Shi · Arun Majumdar

11.1 Introduction

Thermal transport in nanometer scale devices and structures has become an area of active research. A representative example of the nano devices and structures is the metal-oxide field effect transistors (MOSFETs), which have been the driving force of the semiconductor industry for the past two decades. The gate length of the MOSFET has been continuously reduced in order to achieve higher switching speed and lower manufacturing cost. This critical length has been shrunk to 85–90 nm by 2002 and will approach 20–22 nm in 2013 [1]. These lengths are comparable to the scattering mean free paths of electrons and phonons. As a result, nanotransistors exhibit unique electron and phonon transport phenomena that have not been observed in macroscopic devices. Furthermore, as nanotransistors are miniaturized, the power density is increased, leading to localized self-heating and high operating temperatures that can reduce device speed and time to failure. Therefore, it is of both scientific and technological importance to study thermal transport in the nanoscale.

Besides nanotransistors, a variety of other nanoscale devices and structures are being actively developed for electronic, optoelectronic, thermoelectric, electromechanical, and sensing applications. Examples include nanoelectronic devices based on carbon nanotubes [2], nanolasers based on ZnO_2 nanowire arrays [3], and thermoelectric coolers based on superlattices [4] and nanowires [5]. The characteristic size of these low dimensional structures is in the range of 1–100 nm. The short length scale gives rise to unique thermal transport properties. In most cases, these nanoscale thermal transport phenomena cannot be observed using conventional experimental methods developed for macroscopic materials.

Scanning probe microscopy (SPM) techniques have enabled direct observation of physical phenomena with high spatial resolution. A variety of novel SPM-based measurement techniques have been developed to investigate electronic, optical, thermal, mechanical, chemical, and acoustic properties of nanoscale devices and structures. One of such techniques is the scanning thermal microscopy (SThM). The SThM can measure temperature and thermophysical properties with a micro-nano scale spatial resolution. The ability of thermally probing micro-nano scale phenomena has made the SThM a powerful tool for studying fundamental thermophysics and for characterizing micro-nano scale devices and materials. The SThM has demonstrated applications in materials, microelectronic, and pharmaceutical research and development.

A variety of SThM methods have been explored since the invention of the scanning tunneling microscope (STM) and the atomic force microscope (AFM). A popular method relies on thermometry using a built-in thermal sensor on a STM or AFM probe. The spatial resolution of these SThM methods is largely determined by the mechanisms of heat transfer between the thermal probe and the sample. Other methods have been developed without the need of a built-in thermal sensor on the probe. These include scanning tunneling thermometry, thermal expansion thermometry, and near-filed scanning optical thermometry.

This chapter introduces various SThM methods, and discusses heat transfer mechanisms involved in SThM. The structure of this chapter is organized as below. Following this introductory section, we discuss micro-nano scale temperature mapping using probes with built-in thermal sensors, mechanisms of heat transfer between a thermal probe and a sample, microscale imaging of thermal properties, and other SThM methods that often do not require a built-in thermal sensor at the probe, respectively. For a comprehensive review on SThM, readers are referred to ref. [6].

11.2 Micro-Nano Scale Temperature Mapping Using Probes with Built-in Thermal Sensors

This section introduces eight different SThM methods utilizing a thermal probe with a thermal sensor fabricated on the probe for mapping temperature distribution on a sample surface. These eight methods include (i) STM probes with a built-in thermocouple, (ii) wire thermocouple AFM probes, (iii) individually-made thin-film thermocouple AFM probes, (iv) thermocouple AFM probes made by nano-lithography, (v) thermally designed and batch-fabricated thermocouple AFM probes, (vi) AFM probes with a built-in Schottky diode, (vii) AFM probes with a built-in resistance thermometer, and (viii) active thermal feedback AFM probes.

11.2.1 STM Probes with a Built-in Thermocouple

In 1986, Williams and Wickramasinghe pioneered a so-called scanning thermal profiler technique employing a STM probe with a thermocouple fabricated at the probe tip [7]. The schematic diagram of this scheme is shown in Fig. 11.1. A thermocouple sensor was produced at the STM probe tip end by the junction of two dissimilar inner and outer conductors. An insulator separated these conductors in all areas remote from the tip. The size of the thermocouple junction could be made to have dimensions of the order of 100 nm. When a temperature difference exists between the thermocouple junction and the end of the lead wires, a voltage could be measured at the end of the lead wires with a minimum detectable tip temperature change less than 1 mK.

The main purpose of the scanning thermal profiler was not for mapping temperature distribution of a surface, but for regulating the tip-sample distance. The thermocouple STM probe was mounted on a piezoelectric actuator and a heater. The heater raised the temperature of the probe tip. When the tip approached a surface, the tip was cooled by heat transfer through the tip-sample air gap. It was found

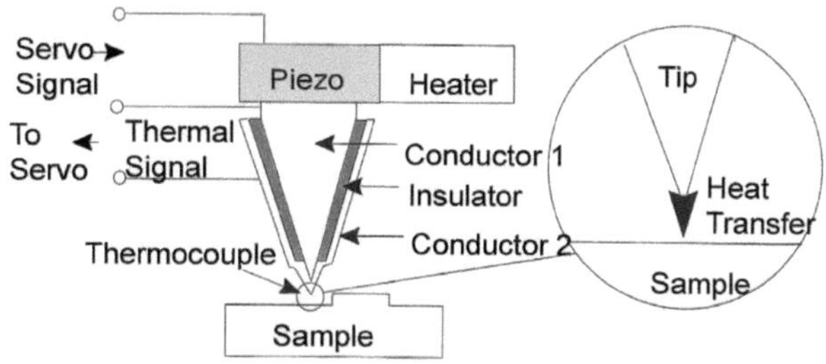

Fig. 11.1. Schematic diagram of the scanning thermal profiler [7]. Reprinted with permission

that the tip temperature was sensitive to the tip-sample distance. In principle, this can allow control of the tip-sample spacing with a feedback loop to move the tip vertically using the piezoelectric actuator for maintaining a constant user-specified tip temperature. However, the dc temperature of the tip may drift with the environment temperature independent of the tip-sample gap. To avoid the problems caused by the dc temperature drifting, the tip to sample spacing was modulated at a frequency near 1 kHz, and the resultant ac temperature modulation was detected and rectified before being sent to a servo loop. The proximity control was therefore provided by maintaining a constant gradient of tip temperature versus tip to sample spacing, rather than the temperature itself. The vertical and lateral resolutions of the scanning thermal profiler were found to be about 3 nm and 100 nm, respectively. Furthermore, unlike a STM, the tip-sample gap control mechanism did not depend on a tip to sample tunneling current, thus the sample surface needs not to be a conducting surface. This allows the scanning thermal profiler for mapping the topography of an insulating surface.

11.2.2 Wire Thermocouple AFM Probes

Although the scanning thermal profiler was not intended for mapping temperature distribution of a surface, it stimulated intense efforts to develop a SPM technique for thermal microscopy. Because the STM requires the sample surface to be conducting, a SThM method developed in the configuration of a STM cannot be applied readily to a dielectric sample. In many applications such as temperature mapping of a microelectronic device, the sample surface is often covered by a dielectric film. Therefore, it is desirable to develop a general SThM method that can be applied to both conducting and non-conducting samples. In 1993, Majumdar et al. [8] introduced a wire thermocouple AFM probe. In an AFM, the tip-sample spacing is regulated by the force acting on the probe tip. As such the AFM can map the topography of both conducting and non-conducting samples. While a thermocouple is fabricated on the probe, heat transfer between the tip and sample changes the temperature of the probe tip during tip scanning. This allows for simultaneous mapping of topography and temperature.

A schematic diagram of the wire thermocouple probe is shown in Fig. 11.2. The probe was fabricated using two thermocouple wires that provided the elastic spring constant necessary for force sensing. The two wires were bonded to form a thermocouple junction as well as an AFM tip. The thermocouple probes were typically

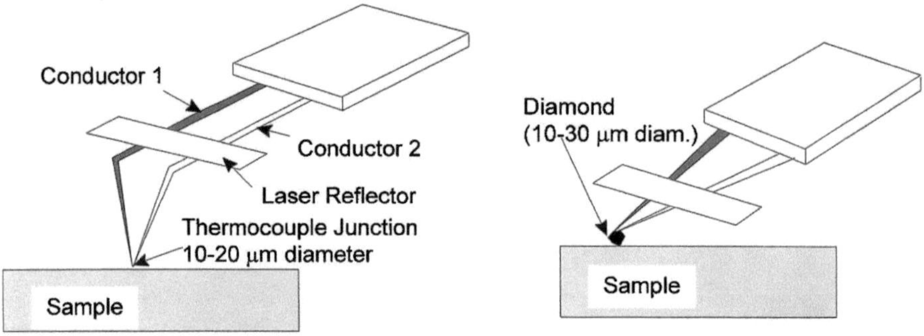

Fig. 11.2. Experimental apparatus and details of the wire thermocouple probe [8]. Reprinted with permission

made of chromel and alumel because these materials had sufficiently low melting points for the bond to be formed during capacitive discharge. However, this also posed a problem because the tip would deform under contact. This problem was solved by attaching a small piece of diamond (about 20 µm in diameter) on the thermocouple junction by using epoxy. Since the bending of the AFM probe is usually monitored by measuring a reflected laser light from the cantilever, it is necessary to attach a laser reflector on the wire probe. The laser reflector was usually an aluminum foil or a piece of metal-coated glass. When the wire thermocouple probe was scanned across the sample, the topographic image was obtained by the standard AFM operation. During scanning, tip-sample heat transfer changed the temperature of the thermocouple, producing a voltage that can be measured at the end of the two thermocouple wires. Therefore, both topography and temperature of the surface can be mapped out simultaneously. Figure 11.3 shows 25 × 25-µm topographical and thermal images of a MOSFET obtained by a wire thermocouple

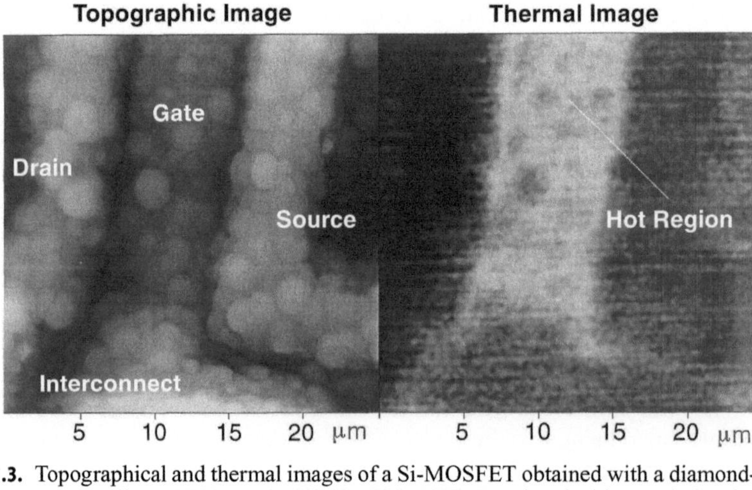

Fig. 11.3. Topographical and thermal images of a Si-MOSFET obtained with a diamond-tip wire thermocouple cantilever SThM probe [9] (© 1995 IEEE). Reprinted with permission from IEEE

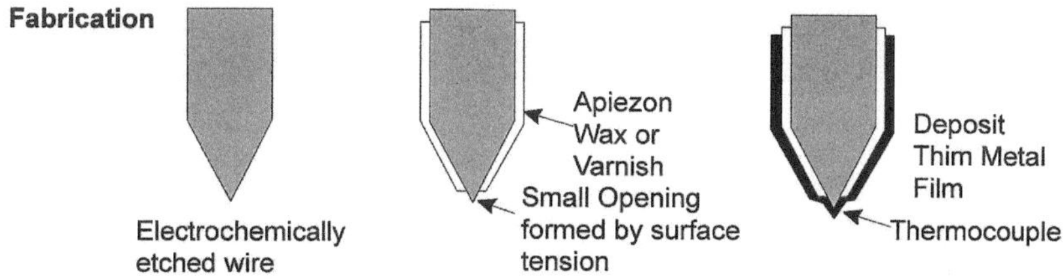

Fig. 11.4. Process for fabricating single-wire thermocouple AFM probes. Reprinted from ref. 10 with permission from Taylor & Francis Ltd. (http://www.tandf.co.uk/journals/tf/10893954.html)

probe with a small piece of diamond attached at the probe end [9]. The thermal image clearly shows that most of the heating is under the gate.

After double-wire thermocouple probes were fabricated and used for SThM, single-wire thermocouple AFM probes were also made. Figure 11.4 shows the fabrication process developed by Nakabeppu et al. [10] for making such wire probes, which involved the deposition of a second metal film on a partially insulated metal wire probe. The size of the thermocouple junction can be controlled by the viscosity and surface tension of the wax or varnish layer to be in the range of 0.3–0.5 μm. Despite improvements in tip sharpness and thermal time constant [11] for these wire probes, the spatial resolution of the thermal images were found to be about 0.3–0.5 μm, although the topographical images showed a resolution better than 0.3 μm. To understand the image contrast mechanism, Nakabeppu et al. measured the temperature change and cantilever deflection as a function of tip-sample distance, as shown in Fig. 11.5. It can be clearly observed that as the tip was

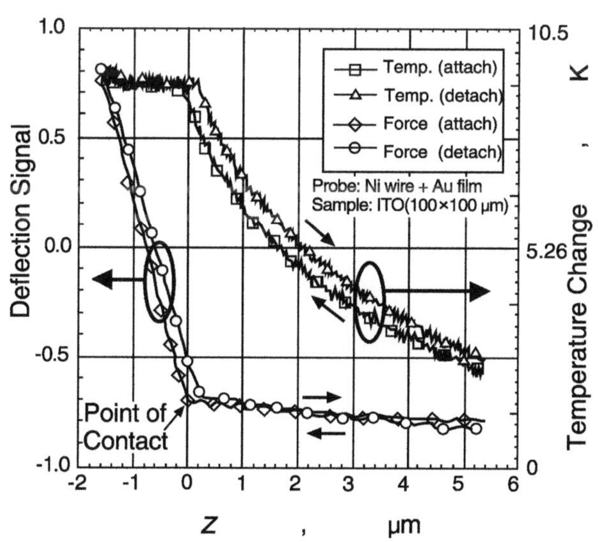

Fig. 11.5. Plot of measured temperature by the wire thermocouple probe as a function of distance from the sample surface. Reprinted from ref. 10 with permission from Taylor & Francis Ltd. (http://www.tandf.co.uk/journals/tf/10893954.html)

brought closer to the sample, the temperature increased because of heating of the surrounding gas. When the tip touched the sample, there was no temperature jump. Instead, the temperature remained constant as the tip was pushed into the sample. The absence of temperature jump indicated that heat conduction through the solid contact was negligible, and tip-sample heat transfer was dominated by gas conductance. As a result, the contrast mechanism in the thermal image was determined by the gas conduction mechanism. Since the average distance that gas molecules travel before they transfer their energy is the mean free path, the spatial resolution of the wire thermocouple cannot be better than the men free path, which is about 0.1–0.3 μm at ambient condition.

11.2.3 Individually Made Thin-Film Thermocouple Probes

Gas conduction dominates tip-sample heat transfer for the wire probes partly because the thermocouple junction was quite large. By reducing the size of the junction, the gas conduction between the sample and the junction can be reduced in principle. However, since the junction is thermally connected with the rest of the probe, gas conduction between the sample and the entire probe can also affect the junction temperature and reduce the spatial resolution. For this reason, it is necessary to thermally isolate the junction from the rest of the probe by choosing low-thermal conductivity materials for the cantilever.

To reduce the influence of gas conduction, Luo et al. [12, 13] developed a technique of fabricating submicron thin-film thermocouple junction on commercially available SiN_x cantilever probes with a much lower thermal conductivity than the metal wire thermocouple probes. Figure 11.6 shows the fabrication process. By using a shadow mask, a thin film of metal 1 was deposited on the cantilever and patterned on the cantilever substrate to form a contact pad. Then a layer of SiO_2 was deposited on the cantilever and substrate to provide the electrical insulation. A film of metal 2 was then deposited on the cantilever and patterned on the substrate to form its contact pad. Next, the tip was brought into contact with a metallic sample by using the AFM. A single or a series of voltage pulses in the range of 110 V was applied between metal 2 and the sample. Because the tip radius was in the range of 20 to 100 nm, the electric field generated at the tip was sufficiently high to evaporate metal 2 and expose the underlying SiO_2 film. Figure 11.7 shows electron micrographs of 100- to 300-nm holes created at the tip by using 3- and 4-V pulses, respectively. Once the hole was formed, the oxide layer was plasma etched to expose metal 1 only at the tip. Then a thin capping layer of metal 2 was deposited on the tip so that a thermocouple between metal 1 and metal 2 was formed.

Thermocouples of several metal combinations (Au-Pd, Au-Pt, and Au-Ni) were fabricated using this technique. The calibrated thermopower of the junction was always lower than the bulk value. It is likely that the evaporated metal films had different Seebeck coefficients than the bulk values.

Figure 11.8 shows the thermal image of a magnetoresistive reading head made of a 30-nm-wide NiFe film and surrounded on both sides by layers of alumina [13, 14]. The head was polished using a lapping process so that the 30-nm-wide NiFe

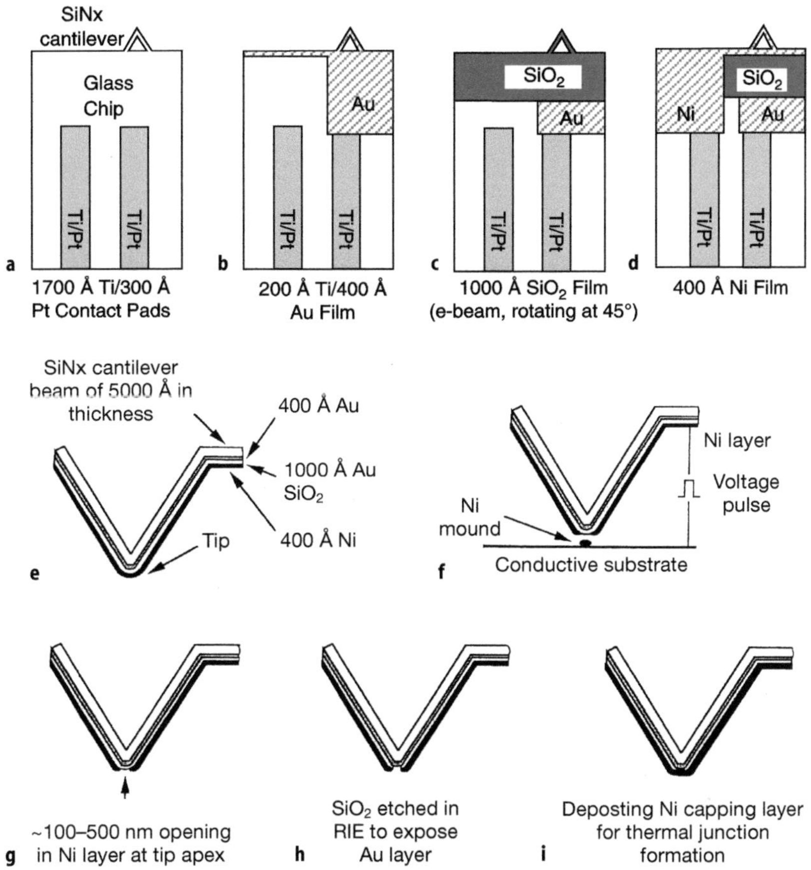

Fig. 11.6a–i. Process for fabricating thermocouple sensors on the tips of commercially available AFM cantilever probes [13]. Thin-film patterns on commercial SiN_x cantilevers and glass chips formed with physical masks. **a** The 1700-Å-thick Ti and 300-Å-thick Pt films as contact pads were deposited with mask A in a thermal evaporator. **b** A 400-Å Au film as metal 1 was deposited with mask B in an e-beam evaporator. The 200-Å Ti film was used as the wetting layer for enhanced adhesion. **c** A 1000-Å SiO_2 layer was deposited with mask C in the e-beam evaporator, with sample rotating at a 45° angle to ensure sidewall coverage. **d** A 400-Å Ni film as metal 2 was deposited with mask B in the e-beam evaporator. **e** Metals 1 (400-Å Au) and 2 (400-Å Ni) were separated by the 1000-Å SiO_2 insulation layer, on the 5000-Å-thick SiN_x surface (pyramidal tip shape). **f** A voltage pulse evaporates the Ni film at the tip end and deposits a nanometer-sized mound on a conductive substrate. **g** The 100- to 500-nm opening formed at the tip end can be used for further sensor fabrication. **h** SiO_2 film within the opening can be etched by reactive ion etching to expose underlying Au layer. **i** Deposition of a Ni capping layer forms a thermocouple junction

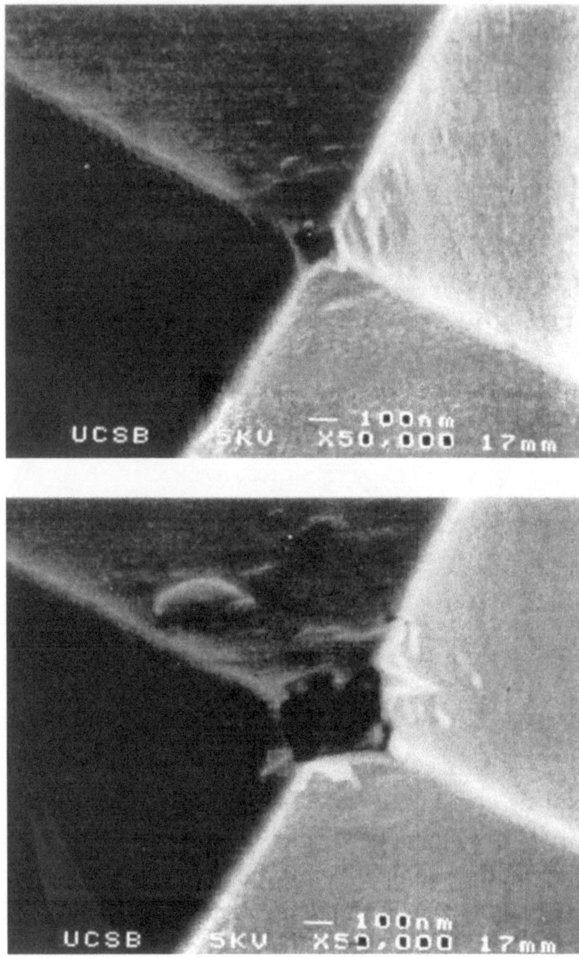

a

b

Fig. 11.7a,b. Electron micrographs of holes created on an Au film covering the AFM tip by a 3-V pulse (**a**) and a 4-V pulse (**b**), respectively [12, 13]

film was exposed on its side. The metal film cannot be observed in the topographical image, although polishing marks can be seen. In the thermal image obtained by the thin-film thermocouple probe, the heated region with a full-width-half-maximum (FWHM) of 0.25 μm around the NiFe heater can be clearly seen. Furthermore, features as small as 25 nm were observed in the thermal image. However, these features correlate with valleys or hills in the topographic image. It is believed that these features were artifacts induced by topographic features. As the probe scanned on the valleys and hills, the temperature of the thermocouple changed due to variation in heat transfer from the tip to the sample through the air and the contact point. This can occur even when no temperature gradient on the sample surface is present, because the thermocouple probe was heated by a laser beam for sensing the cantilever deflection, and the heat loss from the heated probe to the sample changed as the probe scanned on hills and valleys.

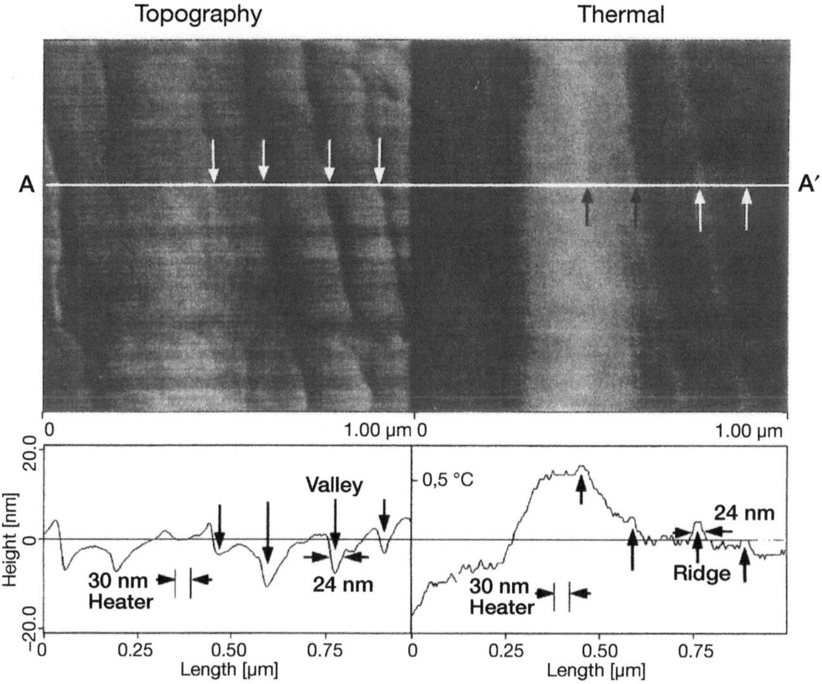

Fig. 11.8. Topographical and thermal (1 μm × 1 μm scan size) images of a magnetoresistive reading head that contained a 30-nm-wide NiFe magnetoresistive element surrounding by alumina coatings [13]

11.2.4 Thermocouple Probes Fabricated Using Nanolithography

Although the individually made thin-film thermocouple probes resulted in significant improvement over wire thermocouple probes in terms of spatial resolution, the probes were made one at a time, and the fabrication was therefore very time consuming. In 1998, Weaver and coworkers [15, 16] developed a technique of nanofabricating thin-film thermocouple probes, as illustrated in Fig. 11.9. A silicon substrate was micromachined to form pyramidal tips. The substrate was then coated with silicon nitride and patterned using electron beam lithography to form cantilevers and tips with a radius of 30 nm. Gold and palladium wires were then patterned using electron beam lithography to form a thermocouple junction at the tip. Finally, the silicon substrate was etched away, leaving the silicon nitride cantilever with the Au-Pd thermocouple on the tip. Thermocouple junctions of a size of about 0.25 μm were fabricated by this method.

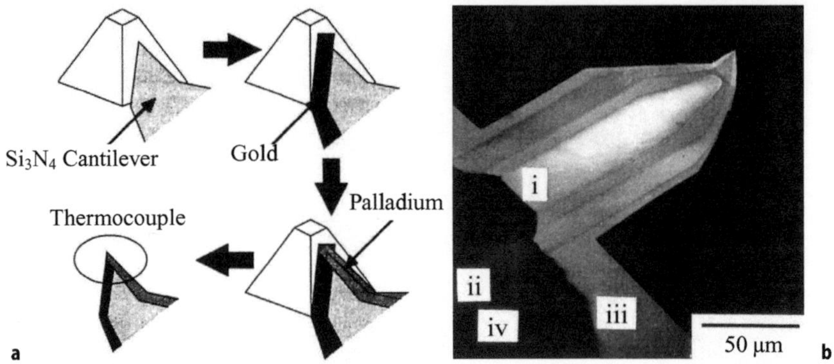

Fig. 11.9. **a** Process of batch fabricating thin-film thermocouple probes on SiN_x cantilevers. **b** Thin-film thermocouple cantilever probe showing *i* mirror for force detection optics, *ii* thermocouple wire, *iii* 0.5-μm-thick SiN_x membrane, and *iv* Si probe base [15, 16]. Reprinted with permission

11.2.5 Thermally Designed Batch-Fabricated Thermocouple Probes

The throughput of electron beam lithography is low. This makes it inefficient for batch fabricating thermocouple probes using the nanolithography method. Furthermore, little attention was paid to the thermal design of the aforementioned thermocouple probes. To address these issues, Shi et al. examined the tip-sample heat transfer mechanisms and developed an all photolithography process for wafer-stage fabrication of thermocouple probes with an improved thermal design [17, 18].

Figure 11.10 shows various heat transfer mechanisms between a thermocouple probe and a sample surface, including solid-solid conduction at the contact, conduction through the air gap between the probe and the sample, thermal radiation, and heat conduction through a water meniscus formed at the tip-sample junction. The water meniscus is formed from water molecules and/or contaminations adsorbed on the sample and tip surfaces. The thermal resistance network in Fig. 11.10

Fig. 11.10. Schematic diagram of a thermal probe in contact with a sample. Also shown are the tip-sample heat transfer mechanisms and a thermal circuit diagram of the system

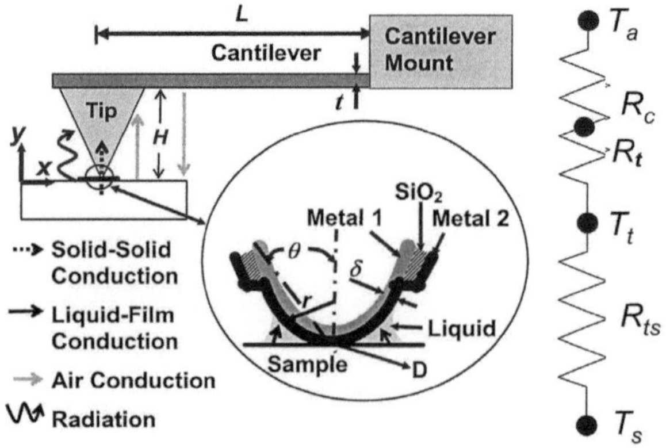

suggests that for a given ambient temperature, T_a, and sample temperature, T_s, the tip temperature, T_t, can be written as

$$T_t = T_s + (T_a - T_s)/(1 + \phi) \tag{11.1}$$

where $\phi = (R_t + R_c)/R_{ts}$; R_t and R_c are the thermal resistance of the tip and the cantilever, respectively, and R_{ts} is the thermal resistance for heat transfer from the tip to the sample. Hence, the change in sample temperature can be related to the change in the tip temperature as $\Delta T_t/\Delta T_s = \phi/(1 + \phi)$. This relationship suggests that the accuracy and sensitivity of sample temperature measurement by the tip depends on ϕ which must be large for better SThM performance. The spatial resolution, Δx, of SThM measurements can be expressed as $\Delta x = \Delta T_n/(dT_t/dx)$ where ΔT_n is the noise in the temperature measurement and dT_t/dx is the temperature gradient on the sample surface. Because the tip and sample temperatures are related through ϕ, the spatial resolution can be expressed as

$$\Delta x = \frac{\Delta T_n}{\left(dT_s/dx\right)}\left(\frac{1+\phi}{\phi}\right) \tag{11.2}$$

This suggests that small values of ϕ lead to poor spatial resolution of SThM.

The aforementioned thermocouple probes were made of high-thermal conductivity materials. Based on a one-dimensional heat transfer model, Shi et al. chose low-thermal conductivity silicon dioxide (SiO_2) and silicon nitride (SiN_x) as the tip and cantilever materials, respectively, so as to increase R_t and R_c as much as possible and to isolate T_t from T_a. It was shown by the modeling results that compared to silicon probes with similar geometry parameters, this design could improve the thermal isolation of the sensor from ambient. In addition, the thermocouple materials were made of Pt-Cr, Cr-Ni, or Cr-Ir to achieve low thermal conductivity and large thermopower. Furthermore, they attempted to minimize the cantilever width, metal line width and thickness, and junction size, and increased the tip height, for increasing thermal resistance of the cantilever and thermally isolating the junction.

The fabrication process of these probes consisted of only wafer-stage processing steps, as shown in Fig. 11.11, with more than 300 probes fabricated on one single wafer. First, a 0.5- to 1-µm-thick low-pressure chemical vapor deposited (LPCVD) SiN_x film was grown on both sides of 100-mm diameter double-side polished silicon wafers with (100) orientation, followed by the growth of a 8-µm-thick LPCVD silicon dioxide or low temperature oxide (LTO) film. The LTO film on the backside was stripped in 5:1 buffered hydrofluoric acid (5:1 BHF) (Fig. 11.11a). The LTO on the front side of the wafers was annealed at 1000 °C for 1 h. The backside SiN_x was then patterned for use as a mask in a subsequent bulk micromachining step.

The probe tips were fabricated out of the 8-µm-thick LTO film by reactive-ion-etching (RIE) and wet etching. A chrome film was sputtered on the LTO, and patterned by photoresist into squares (Fig. 11.11b). Masked by the photoresist and chrome squares, the top 5-µm-thick LTO was etched in CF_4 and CHF_3 plasma. The photoresist plus chrome masks were undercut and the precursor of a sharp-angle tip shape was defined during this etching step. After the RIE etching, the remaining LTO film was etched in BHF 5:1 until the masks were just etched free

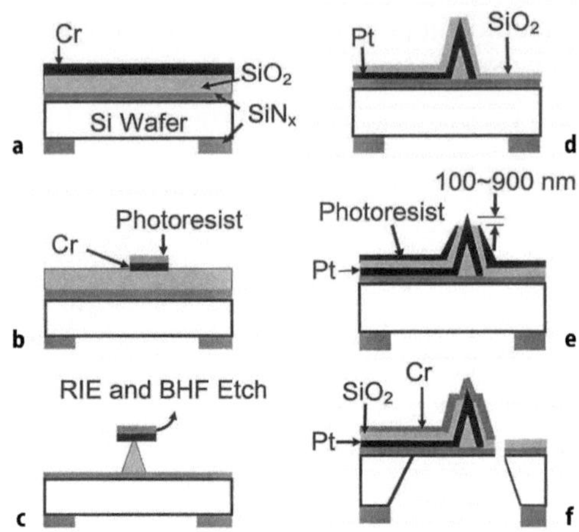

Fig. 11.11a–f. Fabrication process of thermally designed thermocouple AFM probes [17]. **a** Deposit SiN_x and then SiO_2 by LPCVD, strip SiO_2 and pattern SiN_x at the backside of the wafer, sputter Cr; **b** pattern Cr into 10 µm by 10 µm squares; **c** etch SiO_2 by RIE and BHF until the Cr and photoresist mask was etched free, resulting in a SiO_2 tip; **d** sputter and pattern Pt on the wafer, grow SiO_2 on Pt by LPCVD; **e** spin photoresist on the tip, leaving the tip end uncovered, etch SiO_2 from the tip end; and **f** sputter and pattern Cr on the wafer, pattern SiN_x on the front side of the cantilever, release the cantilever in TMAH

(Fig. 11.11c). Before that, the 8-µm-thick LTO film at the unprotected region was etched away. The RIE plus wet etching process could reproducibly yield oxide tips with tip radii about 20 nm and half angles of 10–20°.

After the oxide tips were fabricated, a 50- to 100-nm-thick Pt film (or Cr film) was sputter deposited and patterned on the front sides of the wafers (Fig. 11.11d). A 250- to 300-nm-thick LTO was then deposited to cover the tip. Photoresist with appropriate viscosity was spun on the wafer. At a certain spinning speed, surface tension prevents a sharp tip from being covered by a spin-coated photoresist, leaving the very ends of the tips uncovered (Fig. 11.11e). The exposed LTO film at the tip end was then etched in BHF 5:1. The height of the etched region depends on the photoresist viscosity and coating speed as well as etching time, and could be controlled in the range of 100–500 nm.

After the Pt (or Cr) was exposed at the tip end, the photoresist was stripped and 50- to 100-nm-thick Cr (or Ni) was sputter deposited and patterned to form Pt-Cr (or Cr-Ni) junctions (Fig. 11.11f). After the thermocouples were made, the silicon nitride film was patterned into cantilever shapes. Then 5% tetramethylammonium hydroxide (TMAH) in water was used to etch grooves from the backside of the wafer until the wafer was etched through and the cantilevers were released (Fig. 11.11f). Figure 11.12 shows the cross section and scanning electron micrographs of a finished thermal probe.

The batch-fabricated thermal probes were used for measuring the temperature distribution of micro/nano-electronic devices. Figure 11.13 shows a topographic and thermal image of a Joule-heated multiwall carbon nanotube (MWCN) [19]. The temperature profile across the nanotube shows a full-width-half-maximum

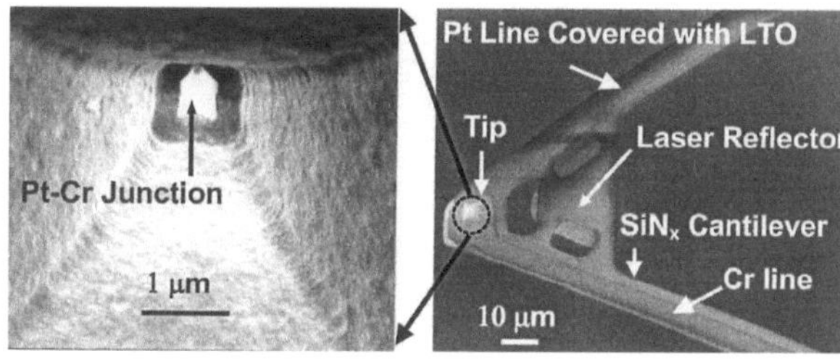

Fig. 11.12. Scanning electron micrographs of a batch-fabricated probe. *Left* a close up of the Pt-Cr junction at the tip end; *right* an overview of the probe [17]

(FWHM) of 50 nm. The thermal images of heated MWCNs were taken at different applied voltages and the thermal profiles were found to increase proportionally with the power input to the nanotube. This suggested that the measured temperature profile was due to temperature gradient in the sample. In addition to carbon nanotube devices, the thermal probes were also used to study localized heating in different very large scale integrated (VLSI) submicrometer W-plug via structures [20] and to map temperature profiles of the cross sectional surface of MOSFETs [21]. The latter study reveals localized heating at a "pinch-off" point near the drain side of the MOSFET.

The thermal time constant of the probe was measured using a sample containing a 350-nm-wide gold line. A sinusoidal current of a frequency f was passed through the gold line, resulting in a modulated temperature of the gold line at a frequency of $2f$. The temperature oscillation led to a $2f$ oscillation in the four-probe electrical resistance of the gold line, which was measured and used to calculate the $2f$ component in the temperature. The amplitude and phase of the $2f$ component of the temperature at the probe tip was measured by a lock-in amplifier while the probe contacted the gold line. The result is shown in Fig. 11.14. The AC tem-

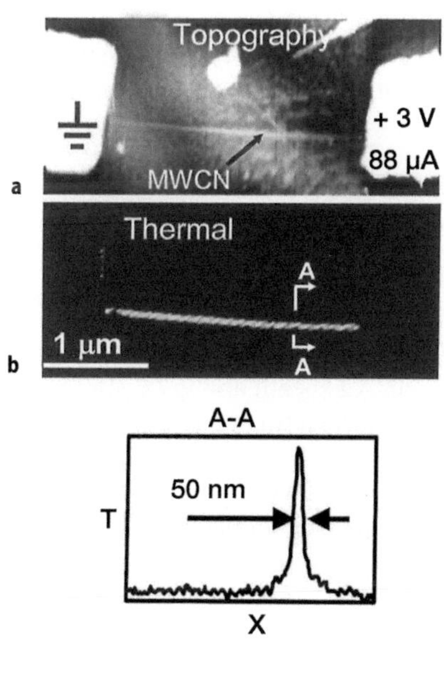

Fig. 11.13a–c. Topographic (**a**), thermal image (**b**), and temperature profile (A-A) across a multi wall carbon nanotube (MWCN) with 88-μA current [19]

Fig. 11.14. Dynamic temperature response of the junction as a function of heating frequency [17]

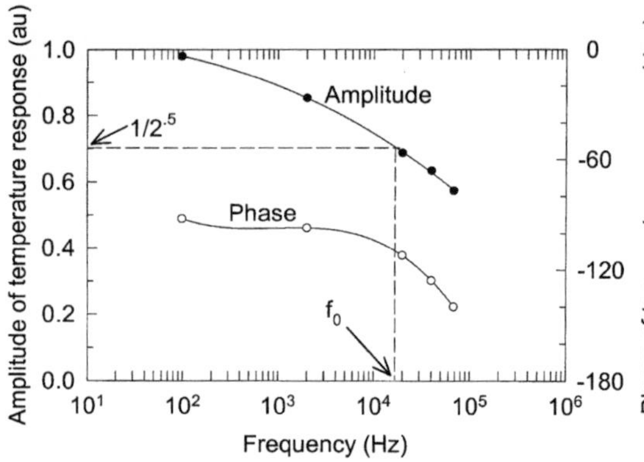

perature showed a typical first-order dynamic response of $T_{ac} = T_{dc} / \sqrt{1 + (2\pi f \tau)^2}$ and decreased to about $1/\sqrt{2}$ at a heating frequency of $f_0 = 18$ kHz. The time constant τ, defined as $1/2\pi f_0$, was calculated to be 8.8 μs.

11.2.6 AFM Probes with a Built-in Schottky Diode

A Schottky diode comprising a metal-semiconductor junction can serve as a sensitive temperature sensor. In 1998, Oesterschulze and coworkers [22] developed a wafer-stage batch-fabrication process for Si cantilever probes with a built-in metal-Si Schottky diode at the end of the Si tip, as shown in Fig. 11.15. The temperature sensitivity of the Schottky diode sensor was about 3 mV K^{-1}, much larger than that of a typical thermocouple probe. In addition, the Schottky diode probe can also be used as a near-field optical sensor for near-field scanning optical microscopy (NSOM). However, because the cantilever and probe were made of Si that has a very high thermal conductivity, the sensor is not thermally isolated from the ambient. As a result, the sensor temperature can be very different from the sample temperature.

11.2.7 AFM Probes with a Built-in Resistance Thermometer

The temperature dependence of electrical resistance has been used to develop SThM probes. Pylkki et al. [23] fabricated a resistance thermometer AFM probe using a Wollaston wire that had a Pt core 5 μm in diameter. The Pt core was encapsulated by an Ag sheath about 75 μm in diameter, as shown in Fig. 11.16. The wire was bent to form a tip and then the Ag was etched away to expose a 200-μm-long section of the Pt wire. The probe was shaped into an AFM-like cantilever probe so that force feedback could be used. The temperature of the probe could be monitored by measuring its resistance by placing the probe resistor in a balanced Wheatstone bridge.

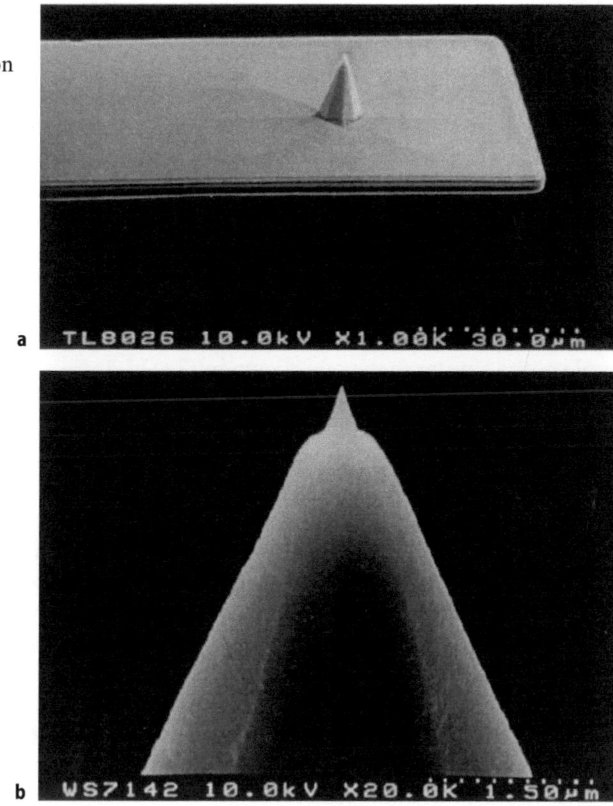

Fig. 11.15. Electron micrographs of an AFM probe made of single-crystal silicon with Schottky diodes on the tips [22]

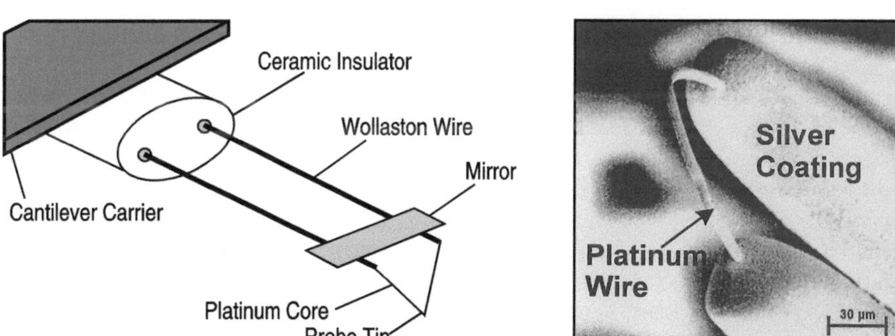

Fig. 11.16. Schematic diagram and electron micrograph of the electrical resistance SThM probe made of a Wollaston wire containing a 5-µm-diameter Pt core surrounded by a 75-µm-diameter Ag wire. Reprinted from ref. 23 with permission

11.2.8 Active Thermal Feedback AFM Probes

The aforementioned SThM probes are passive thermal probes containing a temperature sensor at the tip end. For these passive probes, it was found that a large temperature drop could occur at the tip-sample junction, as discussed in the next section. For this reason, the SThM thermal images can only reveal hot spots, and it is difficult to obtain a quantitative temperature map. For quantitative temperature measurement using SThM, Nakabeppu et al. [24] proposed an active thermal feedback scheme that utilized a thermal feedback loop to keep the tip temperature equal to the sample temperature and thus maintain a zero heat flux between the tip and the sample. The experiment was conducted in vacuum using a wire thermocouple probe similar to the one shown in Fig. 11.4. An additional thermocouple and a heater were built at the probe mount. The temperatures of the two thermocouples were feedback-controlled to be equal by adjusting the heater power. Under this condition, the heat flux from the sample to the tip and that form the tip to the probe mount were set to zero, and thus the tip temperature equaled the sample temperature. Nakabeppu et al. found that the active thermal feedback scheme measured a much larger temperature and temperature gradient of a Joule-heated sample than the same thermal probe without thermal feedback. However, only point measurements were realized because of a large thermal time constant ($\approx$0.5 s) of the wire probe that has a large thermal mass. Nevertheless, the active thermal feedback scheme can potentially be used for quantitative temperature measurement provided that the thermal time constant of the active probes can be reduced to a level suitable for scanning thermal microscopy. This can be achieved by microfabricated cantilever probes that have two built-in thermojunctions and one microheater. The thermal mass of a microcantilever probe is much smaller than that of the wire probe.

11.3 Mechanisms of Heat Transfer between a Thermal Probe and a Sample

The sensitivity and spatial resolution of the aforementioned SThM methods largely depend on the mechanisms of heat transfer between a thermal probe and a sample. As shown in Fig. 11.10, heat transfers between the probe and the sample via solid-solid conduction, conduction through the air gap between the probe and the sample, radiation, and conduction through a liquid meniscus at the tip-sample contact. It can be estimated that radiation contribution is negligible. For the wire thermocouple probes, the experimental result illustrated in Fig. 11.5 suggests that air conduction was the dominant heat transfer mechanism. The experiment did not observe effects of conduction through the solid-solid contact and the liquid meniscus.

The thin-film thermocouple probe, especially the thermally designed and batch-fabricated ones, could minimize the effective area for air conduction by reducing the size of the thermocouple junction and by thermally isolating the junction from the rest of the probe. By reducing the influence of air conduction, the tip-sample heat transfer becomes localized, leading to improved spatial resolution to the sub-100-nm scale.

Various mechanisms of heat transfer between a thin-film thermocouple probe and a sample have been examined. An experiment by Shi and Majumdar [18] has quantified the relative contributions of different mechanisms. The experiment used a thermally designed and batch-fabricated thermal probe and calibration samples containing thin-film metal lines with different width and length. The metal line was Joule-heated during the experiment and its temperature was determined by measuring its temperature-dependant electrical resistance.

The temperature rise of the thermocouple junction was measured when the tip was in contact with the Joule-heated metal line. It was found that the junction temperature rise was about 53%, 46%, and 5% of that of a 50-μm, 3-μm, and 0.3-μm-wide line, respectively, showing a trend of decreasing sensitivity with decreasing sample line width. This suggests that the probe was heated more by air conduction between the tip and the larger hot area of the wider lines. Therefore, air conduction plays an important role in tip-sample heat transfer.

To determine the relative contribution of various tip-sample heat transfer mechanisms, the 350-nm-wide line was Joule-heated to 5.3 K above room temperature. The cantilever deflection and probe temperature response (defined as the ratio of junction temperature rise to the sample temperature rise) was recorded simultaneously when the sample was raised toward and then retracted from the batch-fabricated thermal probe. When the sample approached the tip, the cantilever deflection signal remained approximately constant before the sample contacted the tip, as shown in the deflection curve in Fig. 11.17. In this region, the temperature response was mostly due to air conduction between the probe and the sample, and it can be estimated that radiation contribution is negligible when both the sample and the tip are close to ambient temperature. As the tip-sample distance was reduced, the temperature response due to air conduction increased slowly. Before the sample made a solid-solid contact to the tip, the adsorbed liquid layers on the tip and the sam-

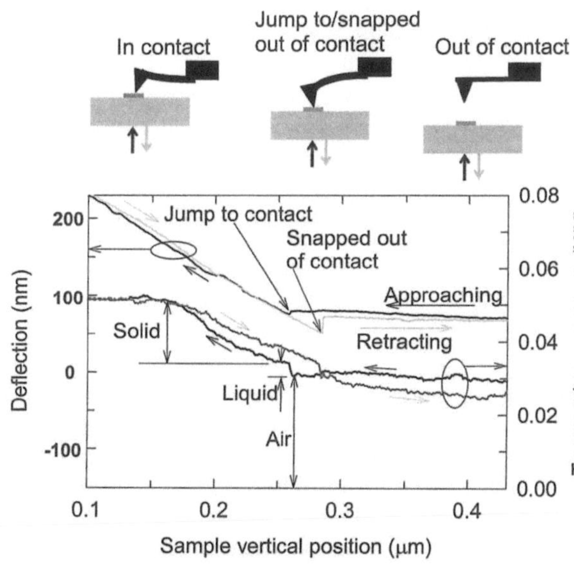

Fig. 11.17. Cantilever deflection and temperature response of the probe as a function of sample vertical position when a 350-nm-wide line was raised toward and then retracted from the tip [18]

ple bridged each other. Initially, this liquid bridge pulled the tip down by van der Waals force, as being seen in the dip labeled as "jump to contact" in the deflection curve. Coincidentally, there was a small jump in the temperature response curve due to conduction through the liquid bridge. As the sample was raised further, both the solid-solid contact force and the tip temperature increased gradually, until the cantilever was deflected more than 100 nm. After this point, the tip temperature remained almost constant as contact force increased.

As the sample was retracted from the tip, the tip temperature again remained almost constant until at a cantilever deflection of 100 nm, the temperature response reduced roughly linearly but at a smaller slope than that found in the approaching cycle. As the sample was lowered further, the tip was pulled down together with the sample by surface tension of the liquid bridge until after a certain point, the restoring spring force of the cantilever exceeded the surface tension and the tip "snapped out of contact" with the sample. Associated with the breaking of the liquid bridge, there was a small drop in temperature response.

The above experiment shows several mechanisms. First, before tip-sample contact, air conduction contributed to a temperature response up to 0.03 K K^{-1}, which was about 60% of the maximum temperature response at large contact forces. Second, conduction through a liquid meniscus was responsible for the sudden jump and drop in temperature response when the tip "jumped to contact" to and "snapped out of contact" from the sample, respectively. Third, solid-solid conduction resulted in the almost linear increase or decrease of temperature response with contact force, which is a well-understood feature for macroscopic solid-solid contacts [25]. Since the tip temperature decreased at a slower slope during unloading (decreasing contact force), there must have been plastic deformation during loading (increasing contact force). For plastic deformation, contact area increased with load [26], resulting in the linear increase of solid-solid contact conductance with contact force. However, since the conductance was still a function of load during unloading, elastic recovery must have been significant.

One question remains as to why the temperature response saturated for cantilever deflection larger than 100 nm. To clarify this question, the contact force corresponding to the 100-nm deflection need to be calculated. The spring constant of the composite cantilever beam was calculated using Roark's formulas [27] to be 0.38 ± 0.11 N m^{-1}. Therefore, the 100-nm deflection corresponds to a contact force $F = 38 \pm 11$ nN.

Assuming plastic deformation, this contact force resulted in a contact spot with a diameter

$$d_c = \sqrt{\frac{4F}{\pi H}} \tag{11.3}$$

where H is the hardness of the tip or sample, whichever is softer. Among the tip and sample materials, the hardness of Au is the lowest and is of the order of 1 GPa. This leads to $d_c \approx 8$ nm. The tip diameter was about 90 nm. However, the details on the tip end are not clearly shown in the SEM image due to the lack of resolution. The linear increase in contact area was probably due to the roughness on the surfaces of the asperity and the sample, since it is well known that in the junction of two ran-

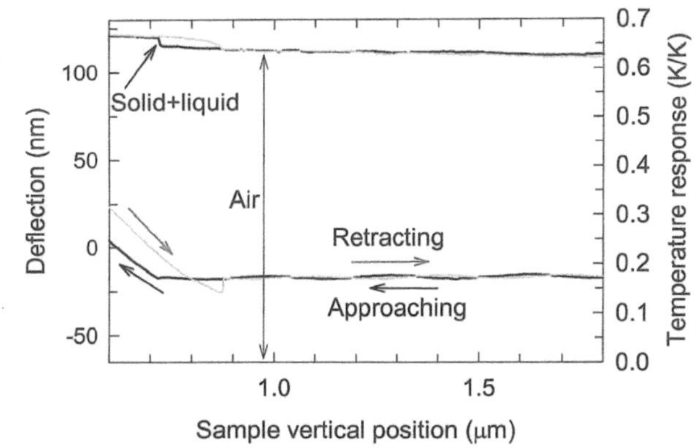

Fig. 11.18. Cantilever deflection and temperature response of the probe as a function of sample vertical position when the 5.8-μm-wide line approached and then retracted from the tip [18]

dom rough surfaces, the contact area increases linearly with contact force [26]. As the contact force increased to about 38 nN, the contact size approached the diameter of the asperity. At this point, the contact area could not increase further with contact force, until the asperity was completely pressed into the sample by a contact force much larger than those used in the experiment. As a result, the temperature response in Fig. 11.17 remained almost constant for deflection larger than 100 nm.

The point contact experiment was repeated for a 5.8-μm-wide and 2000-μm-long line and the result is shown in Fig. 11.18. While the increase of temperature response due to solid and liquid conduction was similar in magnitude to those in Fig. 11.17, the temperature response due to air conduction was about 0.6 K K^{-1}, one order of magnitude higher than the corresponding one (0.03 K K^{-1}) in Fig. 11.17. Therefore, it is clear that for the 5.8-μm-wide line, air conduction dominated tip-sample conduction and was responsible for the large temperature response.

These experiments reveal that the contribution of air conduction in tip-sample heat transfer depends on the size of the heated region on the sample. For larger heated regions, air conduction dominates tip-sample heat transfer. As the size of the heat source reduces, the contribution of air conduction decreases and solid-solid conduction and liquid conduction become important. For micro/nano-devices with localized submicron heated features, such as carbon nanotube circuits, air conduction contribution can be reduced to a level smaller than that from solid-solid and liquid film conduction.

To estimate the magnitude of solid-solid and liquid film conductance, Shi and Majumdar [18] developed a one-dimensional heat conduction model considering various microscale heat transfer mechanisms. Using the model, it was inferred from the experimental data in Fig. 11.17 that the thermal conductance of the liquid meniscus was $G_{lf} = 6.7 \pm 1.5$ nW K^{-1}, and solid-solid conduction varies linearly with the contact force with a contact conductance of 0.76 ± 0.38 W/K-N. For contact forces larger than 38 ± 11 nN, solid-solid conductance saturated at $G_{ss} \approx$ 30 nW K^{-1}, because the contact size between the sample and an asperity on the tip end approached the approximate 10-nm diameter of the asperity. At this solid-solid contact with an effective contact radius $b \approx 5$ nm, it is shown in the following section that

$$G_{ss} = \pi k_{ef} b; \quad \frac{2}{k_{eff}} = \frac{1}{k_s} + \frac{1}{k_{tip}(\tan\theta)^2} \tag{11.4}$$

where k_s and k_{tip} are the thermal conductivity of the sample and the tip, respectively, and θ is the half angle of the tip. With $\theta \approx 18°$, $k_s \approx 200$ W mK^{-1} for the thin-film gold metal line, $k_{tip} \approx 15$ W mK^{-1} for the Pt-Cr film at the tip end [18], and $b \approx 5$ nm, it can be estimated that $G_{ss} \approx 49$ nW K^{-1}. This is within the vicinity of the saturation value of 30 nW K^{-1} estimated from the thermal model.

Due to the small contact thermal conductance at the nanoscale tip-sample contact, the temperature response of the thin-film thermocouple probe was below 0.1 K K^{-1} when micro/nano-devices with submicron localized heat source were imaged. In addition, due to the influence of air conduction, the measured temperature profile by the SThM was found to deviate from the true one. This makes it necessary to correct the SThM measurement results by careful thermal modeling. Alternatively, the influence of air conduction can be eliminated by performing the SThM in vacuum with active thermal feedback.

11.4 Imaging of Thermal Properties

The aforementioned applications of SThM have focused on mapping surface temperature distribution of electronic devices. The SThM can also be used to map thermal properties of a sample, including thermal conductivity, diffusivity, heat capacity, and thermal expansion coefficient. This capability has been developed using the resistive probe shown in Fig. 11.16. These techniques are often collectively referred as micro thermal analysis (µTA), and are finding increasing applications for characterizing polymer samples and drugs.

In contrast to thermocouple probes, the resistive probes can be used in two modes – passive and active. In the passive mode, the temperature is estimated by measuring the voltage across the Wheatstone bridge when a constant current is passed through the resistor. The voltage is recorded while the probe is scanned across the sample. The current used to measure the resistance is kept low so that Joule heating in the probe resistor is negligible compared with the heat flow between the tip and the sample. In the active mode, larger currents are passed through the probe resistor to induce Joule heating. The heat flow between the tip and the sample is influenced by the sample thermal conductivity as well as the resistance at the tip-sample junction. If this resistance is kept constant, then by scanning the tip in the active mode the thermal properties of the sample can be mapped out. The active mode can be operated in either constant current or constant temperature mode. In the first mode, a constant current passes through the probe, and changes in heat flux between the probe and the sample are detected as changes in probe resistance and temperature. In the constant temperature mode, a feedback loop is used to measure the probe resistance and to change the voltage applied to the probe to maintain a constant resistance.

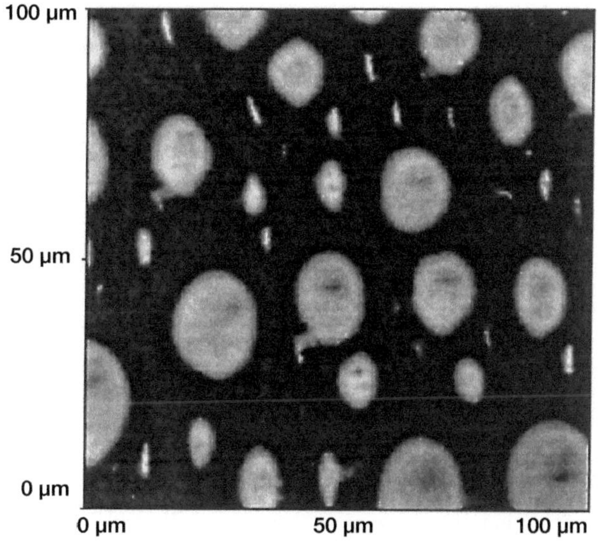

Fig. 11.19. Thermal image of a polymer blend containing poly(vinyl chloride) (PVC)/polybutadiene (PB) obtained using the electrical resistance SThM probe under active mode of dc operation where the probe temperature was maintained at 40 °C [30]. Reprinted with permission

Many groups have used the electrical resistance probes for thermal imaging, property measurement, and calorimetry [23, 29–33]. Hammiche and coworkers [29–31] used this technique to study the thermal behavior of different types of polymer blends. Figure 11.19 shows an image of a sample containing poly(vinylchloride)/polybutadiene blend. The image was obtained under the active dc mode at a constant probe temperature of 40 °C. The image contrast is an indication of thermal conductivity variations within a depth of a few micrometers. Polybutadiene has a higher thermal conductivity (0.24 W mK^{-1}) than PVC (0.14 W mK^{-1}) and appears as the bright spots because it requires more current through the probe to keep the probe temperature to be constant. The ac mode of operation can provide thermal diffusivity contrast, as shown in Fig. 11.20. The left image is that of a blend

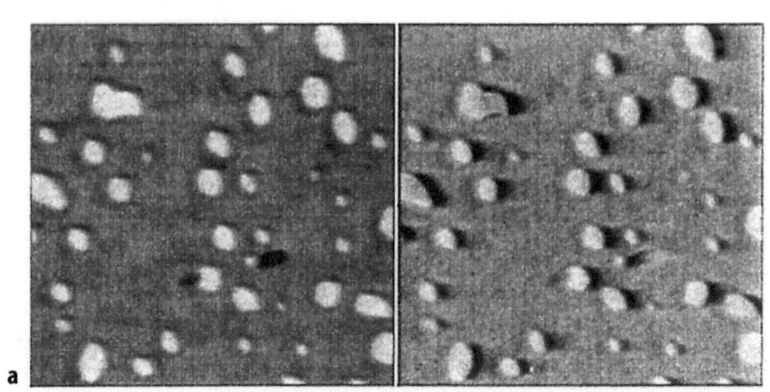

Fig. 11.20. Thermal images of a polymer blend containing poly(methyl methacrylate) PMMA)/chlorinated polyethylene (CPE) obtained under (**a**) dc operation at 40 °C probe temperature and (**b**) ac operation at probe temperature modulation of 5 °C at 10 kHz. The ac image is the phase shift measured by a lock-in amplifier [30]. Reprinted with permision

of poly(methyl methacrylate)/chlorinated polyethylene under dc operation at 40 °C. Here, poly(methyl methacrylate) has higher thermal conductivity (0.193 W mK^{-1}) than that of chlorinated polyethylene (0.144 W mK^{-1}). The right image is the phase shift in the ac voltage across the probe when the probe temperature was modulated by 5 °C at 10 kHz around a dc temperature of 40 °C. At 10 kHz the thermal diffusion length in the sample is typically 2 μm.

It is necessary to understand how the signals under dc and ac modes depend on the thermal properties of the sample. Consider the tip-sample point contact to be such that tip-sample heat flow occurs over a small region of radius b. Assuming that heat flow occurs through the point contact and spreads spherically in the sample, the steady-state heat conduction in the sample is governed by

$$\frac{d^2(\Theta r)}{dr^2} = 0 \tag{11.5}$$

where $\Theta = T(r)–T_\infty$ and T_∞ is the ambient temperature far away. At $r = b$, $\Theta = T_o - T_\infty = \Theta_0$, where T_o is the sample temperature at the contact point. The probe temperature is kept constant at T_p by using feedback control in the active mode of operation. Assuming that the sample is sufficiently thick compared with b, the temperature profile can be found by solving Eq. (11.5) to be $\Theta/\Theta_0 = b/r$ for $r > b$. Conservation of energy at the tip-sample interface suggests that

$$G_p(\Theta_p - \Theta_0) = -2\pi b^2 k_s \left.\frac{d\Theta}{dr}\right|_{r=b} = G_s\Theta_0;\ G_s = 2\pi k_s b \tag{11.6}$$

where k_s is the sample thermal conductivity, G_p and G_s are the contact thermal conductance at the tip and sample side, respectively, and $\Theta_p = T_p - T_\infty$. Using a similar approach, it can be shown that

$$G_p = 2\pi k_{tip} b(1 - \cos\theta) \tag{11.7}$$

where θ is the half angle of the tip. The solid-solid conductance is

$$G_{ss} = \left(G_p^{-1} + G_s^{-1}\right)^{-1} = \pi k_{eff} b;\ \frac{2}{k_{eff}} = \frac{1}{k_s} + \frac{1}{k_{tip}(1 - \cos\theta)} \tag{11.8}$$

Under dc operation the power required to maintain a constant probe temperature should be equal to the heat flow rate from the probe to the sample. This can be calculated to be

$$P_s = \frac{G_s\Theta_p}{1 + \left(G_s/G_p\right)} \tag{11.9}$$

For the power to be sensitive in spatial variation in G_s, G_s must be much smaller than G_p. This further requires that k_s be much smaller than k_{tip}, as illustrated by the experimental result [34] shown in Fig. 11.21, which indicates that the power is less sensitive to k_s for $k_s > k_{tip}$.

Although thermal conductivity contrast has been observed under dc operation [23, 30], accurate measurement of thermal conductivity has not been made. This

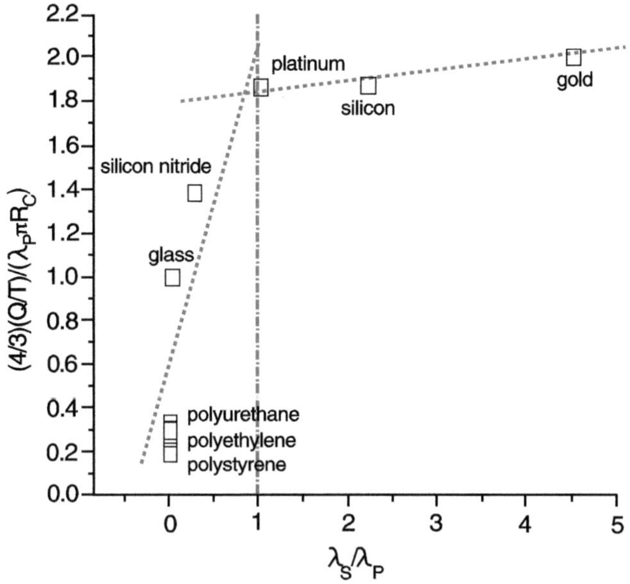

Fig. 11.21. Normalized power dissipation as a function of the ratio of the sample thermal conductivity (λ_s) to the probe thermal conductivity (λ_p). Q power dissipation, R_c effective contact radius. Reprinted from ref. 34 with permission from Elsevier

is partly because a significant fraction of the power was dissipated through the Ag leads to the surrounding gas, which is called P_g. By assuming the Ag wire to act like a fin to the hot Pt wire, the total heat loss through the Ag leads can be estimated from fin theory [35] to be

$$P_g = \sqrt{2\pi^2 d_{Ag}^2 k_{Ag} k_g}\Theta_p \tag{11.10}$$

where d_{Ag} is the diameter of the Ag wire, k_{Ag} is the thermal conductivity of Ag, and k_g is the gas conductivity. It is assumed that most of the heat conduction in the Wollaston wire is through the Ag sheath and not through the Pt core. It is also assumed that the conductance per unit area between the Ag wire and the surrounding gas can be estimated as k_g/d_{Ag}. The ratio of the two power losses can be written as

$$\frac{P_s}{P_g} = \sqrt{2}\left(\frac{b}{d_{Ag}}\right)\frac{k_s}{\sqrt{k_{Ag}k_g}}\left(\frac{1}{1+\left(2\pi k_s b/G_p\right)}\right) \tag{11.11}$$

For contact of the probe with polymer samples whose thermal conductivity is 0.1 W mK^{-1}, one finds that $\dfrac{k_s}{\sqrt{k_{Ag}k_g}} \approx 0.03$. In addition, because the spatial resolution (roughly equal b) observed in the images is approximately 1 μm, whereas $d_{Ag} = 75$ μm, one can estimate $P_s/P_g \approx 6 \times 10^{-4}$ if it is assumed that $G_s/G_p \ll 1$. Hence, most of the heat loss is through the Ag leads and into the surrounding gas.

For ac operation the sample temperature is governed by

$$\frac{\partial(\Theta r)}{\partial t} = \alpha_s \frac{\partial^2(\Theta r)}{\partial r^2}; \Theta = T_s(r) - T_\infty \tag{11.12}$$

where α_s is the thermal diffusivity of the sample. When the temperature is modulated at frequency ω, the oscillating component of the temperature profile in the sample can be derived to be

$$\frac{\Theta(r)}{\Theta_0} = \left(\frac{b}{r}\right) e^{-(i+1)(r-b)\sqrt{\omega/2\alpha}} \tag{11.13}$$

where Θ_o is the temperature oscillation at $r = b$. Hence, the oscillating component of the heat flow rate into the sample is

$$P_s = \frac{2\pi k_s b\left[1+(i+1)\,b\sqrt{\omega/2\alpha_s}\right]\Theta_p}{1+\left(2\pi k_s b/G_p\right)\left(1+(i+1)\,b\sqrt{\omega/2\alpha_s}\right)} \tag{11.14}$$

As opposed to the dc conductance of the sample, which is $2\pi k_s b$, the ac conductance is $2\pi k_s b\,[1 + (i + 1)\,b\sqrt{\omega/2\alpha_s}\,]$. The quantity $\sqrt{2\alpha_s/\omega} = \delta$ is the thermal penetration depth into the sample. Hence, the ac conductance is essentially $2\pi k_s b$ $[1 + (i + 1)b/\delta]$. Under the condition that $b \ll \delta$, the problem reduces to that of the dc case.

The energy balance for the probe under ac power dissipation P_o at frequency ω is

$$mCi\omega\Theta_p = -\frac{2\pi k_s b\Theta_p\left[1+(i+1)\,b\sqrt{\omega/2\alpha_s}\right]}{1+\left(2\pi k_s b/G\right)\left[1+(i+1)\,b\sqrt{\omega/2\alpha_s}\right]} - \frac{\pi d_{Ag}k_{Ag}}{4}\Theta_p\sqrt{\frac{i\omega d_{Ag}^2}{\alpha_{Ag}}+\frac{4k_g}{k_{Ag}}} + P_o \tag{11.15}$$

where mC is the thermal mass of the Pt wire. Here, the first term on the right-hand side is the ac heat loss to the sample, the second term is the ac heat loss through the Ag wire and into the surrounding gas, and the last term is the electrical heating term, which drives the temperature oscillation at frequency ω. For most of the frequencies used in previous studies $\omega d_{Ag}{}^2/\alpha_{Ag} \gg 4k_g/k_{Ag}$ and $b\sqrt{w/2\alpha_s}$ is order unity. Therefore, the amplitude of temperature oscillation for a given oscillating power can be written as

$$\Theta_p = \frac{P_o}{G_r + iG_{im}} \tag{11.16}$$

where G_r and G_{im} are the real and imaginary parts of the complex ac conductance, which can be expressed as

$$
\begin{aligned}
G_r &= 2\pi k_s b\left(1 + b\sqrt{\frac{\omega}{2\alpha_s}}\right) + \frac{\pi d_{Ag}^2 k_{Ag}}{4}\sqrt{\frac{\omega}{2\alpha_{Ag}}} \\
G_{im} &= \omega m C + 2\pi k_s b^2\sqrt{\frac{\omega}{2\alpha_{Ag}}}
\end{aligned}
\tag{11.17}
$$

where it is assumed that $2\pi k_s b/G_p \ll 1$. Note that Θ_p in Eq. (11.12) was measured as change in resistance by the lock-in amplifier, which provided the amplitude image. The phase image was formed by measuring the phase lag φ by the lock-in amplifier. The advantage of measuring the phase lag is that it is independent of the temperature dependence of the electrical resistivity of Pt wire and the power input P_0 to the probe and is simply given as

$$
\tan\phi = -\frac{G_{im}}{G_r} \tag{11.18}
$$

Note that if the ac conductance of the Ag wire is much higher than the ac conductance of the sample, the phase lag would be rather insensitive to the changes in sample thermal properties. If the frequency is about 10 kHz and the sample is a polymer, the ac conductance of the Ag wire is about four orders of magnitude higher than the ac conductance to the sample. In addition, it should be noted that both the amplitude and the phase of the signal are influenced by the penetration depth $\delta=\sqrt{2\alpha_s/\omega}$. Because the change in frequency changes the penetration depth, this method allows for some element of depth profiling and subsurface imaging.

Hammiche and coworkers [29–31] also used SThM with the electrical resistance probe to perform localized ac calorimetry. Here they ramped the temperature of both the sample and the probe at about 15 °C min^{-1} while adding an ac temperature modulation of 1 °C at 10 kHz by the probe. They found that any phase transition in the sample gave rise to a change in the phase signal in ac mode. Figure 11.22 plots the phase lag as a function of temperature for various polymers where the phase transitions result in indicated kinks in the curves. This can be observed more clearly in the first derivative of the phase as a function of temperature, which is plotted in Fig. 11.22. By scanning the sample under this mode, calorimetric analysis can be performed locally.

Although the Pt wire probes provide a simple method for thermal imaging and calorimetry, the 5-μm diameter of the wire inherently limits its spatial resolution to about 1 μm. To improve the spatial resolution of SThM with electrical resistance probes, Wendman and coworkers from Digital Instruments fabricated thin-film sensors on commercially available silicon nitride cantilever probes. Figure 11.23 shows an electron micrograph of such a probe. An Al film is first deposited on the tip side of the cantilever in such a way that it forms two electrodes. These electrodes are then joined by depositing an organometallic resistor on top of the tip by using focused ion beam deposition. The tip resistance is typically 1 KΩ, and the temperature resolution is approximately 1 °C.

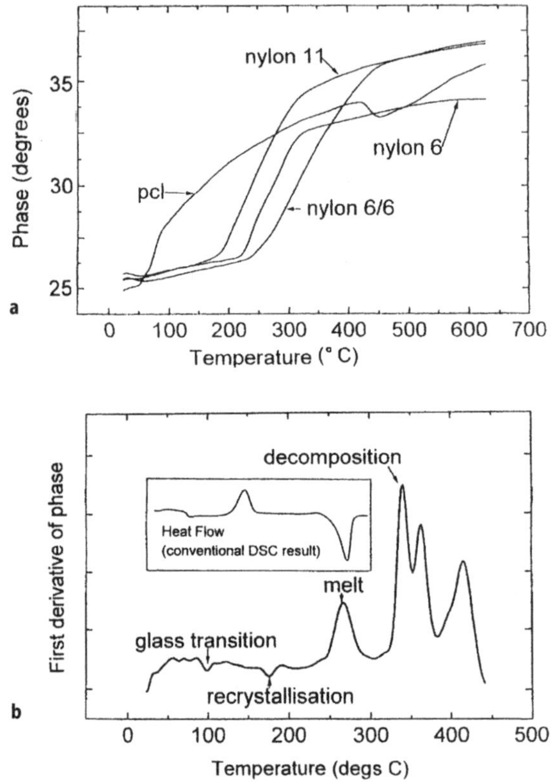

Fig. 11.22. Phase lag of the thermal signal as a function of sample temperature. Phase of the signal undergoes a kink in the curve when the sample undergoes a phase transition. First derivative of the phase lag clearly shows the phase transition temperatures (DSC, differential scanning calorimetry) [30]. Reprinted with permission

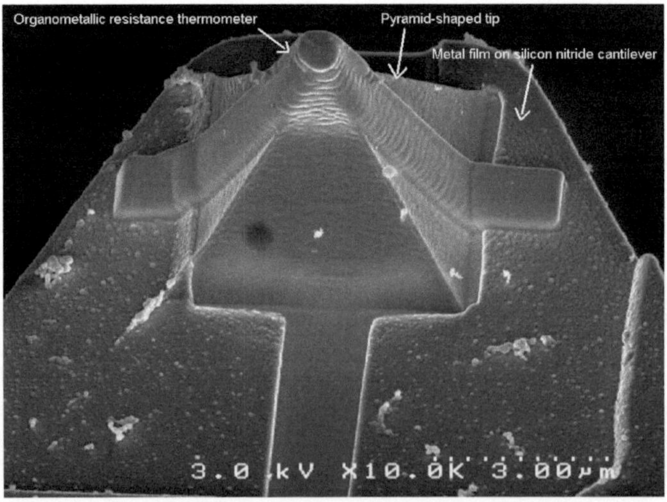

Fig. 11.23. Electron micrograph of an organometallic thin-film resistor deposited on the tip of a SiN_x AFM probe using focused ion beam deposition. Reprinted with permission from M. Wendman

11.5 Other Scanning Thermal Microscopy Methods

The aforementioned SThM methods rely on the fabrication of a thermal sensor at a STM or AFM probe for imaging the variation of temperature or thermal properties on a sample surface. Several other methods have been developed for thermal imaging, including scanning tunneling thermometry, thermal expansion thermometry, and near-field optical thermometry. These SThM methods usually do not require a built-in thermal sensor at the probe for thermal imaging.

11.5.1 Scanning Tunneling Thermometry

The thermal probes with a built-in thermal sensor achieved a spatial resolution in the range of 50–100 nm at best. This resolution is still one or two orders of magnitude lower than that of the STM. In 1989, Weaver et al. [36] developed a scanning tunneling thermometer shown in Fig. 11.24. The technique is based on a "tunneling thermocouple" which, in principle, can measure "temperature" down to the atomic scale. In the experiment, the junction region between an STM and a metal film was illuminated by a chopped laser tuned to a wavelength of spectroscopic interest, resulting in periodic heating of the tunnel junction. Because the tip and sample were made of dissimilar conductors, a potential difference is developed across an external circuit. Variations in the amount of light absorbed by the specimen led to variations in the temperature of the junction and hence to variations in the voltage. Thus the thermoelectric voltage was directly related to the optical absorption with the spatial resolution of the STM.

In 1992, Nonmenmacher and Wickramasinghe [37] implemented the tunneling thermometry method in the configuration of an AFM. A metal film was deposited on the surface of a Si substrate with embedded SiO_2 film. The AFM tip was heated by a laser. When the tip scanned over the metal film, the heat transfer to the sub-

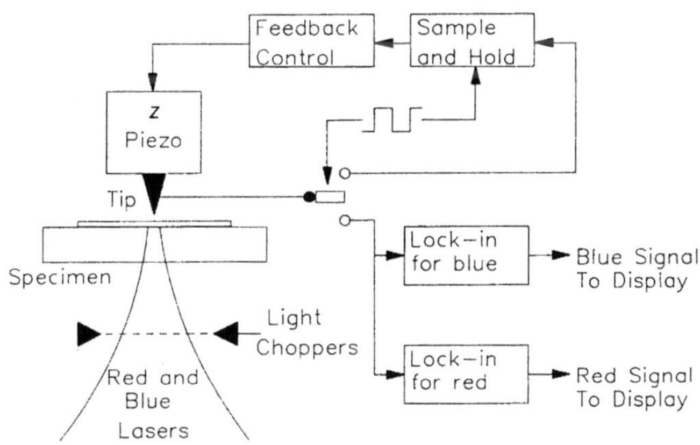

Fig. 11.24. Schematic diagram of the experimental setup used for the tunneling thermometer [36]. Reprinted with permission from Nature Publishing Group (http://www.nature.com)

strate was reduced when the tip was above the region of the embedded SiO_2. This resulted in a change of the tip temperature. The tip temperature was measured by measuring the thermoelectric voltage generated at the tunneling thermocouple and recorded simultaneously with the topographic image. The obtained thermal image thus contained the information on the thermal conductivity of the substrate. In 2002, Zhou et al. [38] employed the tunneling thermocouple approach for mapping temperature distribution of electronic devices. A thin metal film, preferably with a high Seebeck coefficient, was deposited on the sample surface to allow tunneling thermocouple measurement. The spatial resolution is thus limited by the thickness of the metal film, which can be 10–20 nm. The Seebeck coefficient of the tip can be much smaller than that of the metal film. Therefore, the measured thermoelectric voltage is mainly due to that developed in the metal film. In addition, the measured thermoelectric voltage is not affected by the high tip-sample contact thermal resistance, potentially allowing quantitative thermal imaging with a spatial resolution in the range of 10–20 nm.

Although the scanning tunneling thermometry may be able to map temperature variation with a resolution comparable to the STM, it is unlikely that temperature can vary within a distance of atomic spacing. Statistical thermodynamics requires that temperature can be defined only when local thermodynamic equilibrium is established over a length scale larger than the energy relaxation length. The energy relaxation length is usually 1–10 nm for amorphous materials and 10–100 nm for crystalline materials. The relaxation length posts a fundamental limit for any nanoscale thermometry techniques.

11.5.2 Thermal Expansion Thermometry

Besides the scanning tunneling thermometry, another approach that requires very little or sometimes no sensor fabrication is based on thermal expansion measurement. Three schemes have been developed for expansion thermal microscopy, namely a bimaterial cantilever technique, the scanning Joule expansion microscopy, and the scanning thermal expansion microscopy.

11.5.2.1 *Bimaterial Cantilever Technique*

In 1994, Gimzewski et al. [39] showed that if a SiN_x AFM cantilever is coated with an Al film, the cantilever becomes an excellent temperature sensor. A temperature change of the cantilever induces a bending moment due to mismatch in thermal expansion coefficient of the two cantilever materials, which results in a cantilever deflection. The measurement of deflection is, therefore, a measurement of temperature. Because the optical detection scheme in an AFM has a resolution of approximately 0.1 Å, this relates to a temperature resolution of 10^{-5} K, which is far better than that of a thermocouple or an electrical resistance thermometer.

Nakabeppu et al. [40] used this approach for thermal microscopy. An Al film was deposited on a commercially available SiN_x AFM cantilever. The heating in the sample was modulated at a frequency *f*. The temperature modulation caused a can-

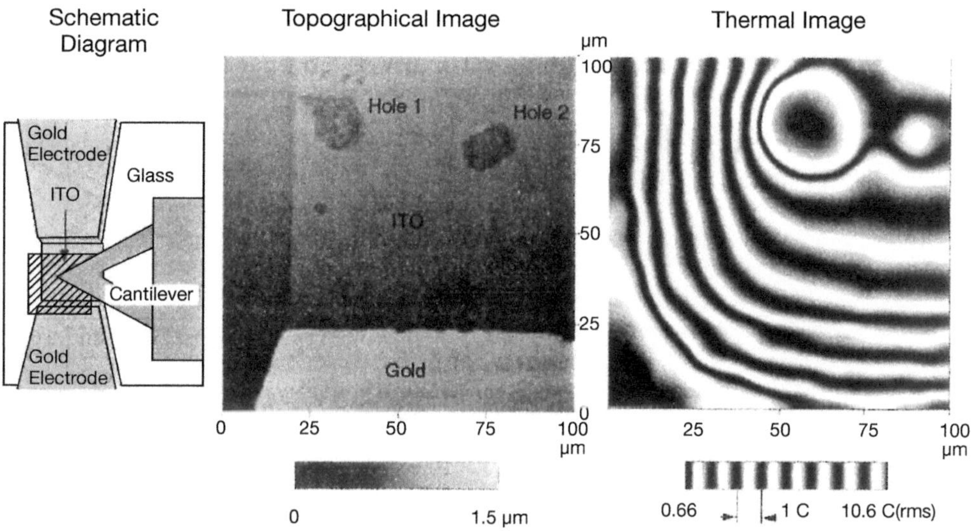

Fig. 11.25. Schematic diagram, topographical and thermal images of an indium-tin-oxide (ITO) resistor that contained two defects, as observed in the topographical image. The thermal image shows two peaks caused by current crowding around the two defects [40]

tilever deflection that produced an oscillating photodiode signal at the same frequency. This signal was measured by a lock-in amplifier. Figure 11.25 shows the topographical and thermal expansion image of an indium-tin-oxide heater containing two defects. The increase in current density between and around the two defects results in two hot spots clearly revealed in the thermal expansion image.

As opposed to thermovoltage measurement, which relies on a sensor fabricated on the probe tip, the bimaterial cantilever technique uses the whole cantilever as a temperature sensor. Under ambient conditions, the conduction through the surrounding gas dominates the heat transfer from the sample to the cantilever. Hence, the experiment must be operated in vacuum, or the probe must be designed such that heat flow through the tip is enhanced.

11.5.2.2
Scanning Joule Expansion Microscopy

In 1998, Varesi and Majumdar developed a technique called scanning Joule expansion microscopy (SJEM) for thermal imaging [41, 42]. Figure 11.26 shows a schematic diagram of the SJEM system. An AFM was used to bring a sharp tip into force-controlled contact with the sample surface and to perform a raster scan. Applied to an electrically conducting sample was a sinusoidal or pulse voltage, which produced sample Joule heating and resulted in sample thermal expansion. The AFM photodiode detects the cantilever deflection due to both sample thermal expansion and sample topography. Because the feedback controller of the AFM had a bandwidth of 5 kHz, the photodiode signal below 5 kHz was processed for feedback control of the z-piezoactuator to image surface topography. The Joule heat-

Fig. 11.26. Schematic diagram of the experimental setup for scanning Joule expansion microscopy [41, 42]

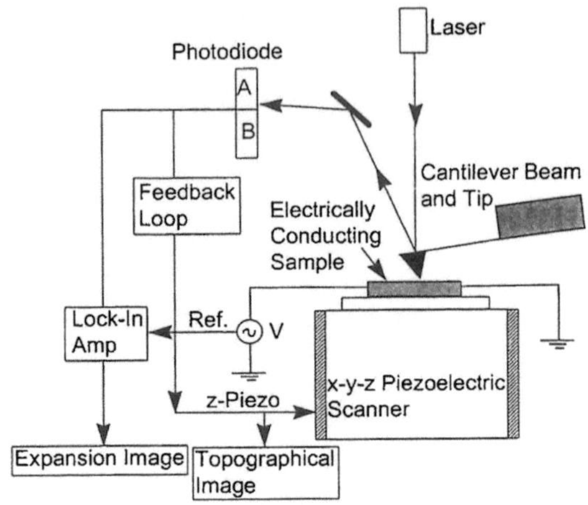

ing frequency was kept higher than 5 kHz to avoid feedback response. The lock-in amplifier was tuned to the Joule heating frequency, which detected only the expansion signal and provided this to an auxiliary AFM channel to form the expansion image.

The advantage of the SJEM is that it eliminates the need of fabricating a sensor on the probe tip, and, therefore, commercially available silicon nitride or silicon probes can be used. Because the SJEM does not rely on tip-sample heat transfer and the tip does not need to draw from or supply any energy to the sample, the performance of the technique improves with tip sharpness.

Figure 11.27 shows images of two parallel gold lines fabricated by electron beam lithography on a SiO_2 substrate. A voltage of 1.4 V at 30 kHz was applied to the sample, and the expansion image was obtained with a bandwidth $\Delta f = 26$ Hz. The lines had full-width-half-maximum (FWHM) values of 630 and 430 nm, respectively, and both were 160-nm thick. The expansion signal was about 24 pm, whereas the noise was about 1.5 pm. Note that, although the narrower line was 430-nm wide, the sharpness of the expansion image suggests the resolution to be sub-100 nm.

The SJEM technique has been used for thermal imaging of interconnect structures in VLSI circuits [43]. A thin polymer film was coated on the sample surface

Fig. 11.27. Topographical and SJEM images of two Au lines on a silicon dioxide substrate. The Au lines were 160-nm thick. A voltage of 1.4 V at 30 kHz was applied to the sample, and the expansion image was obtained with $\Delta f = 26$ Hz [41, 42]

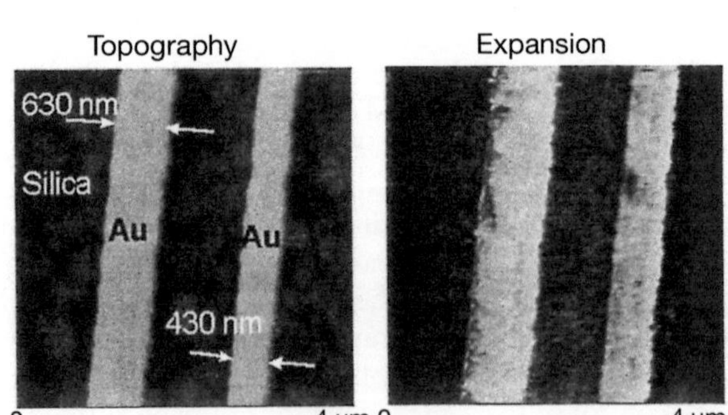

to allow large expansion signal. The spatial resolution was thus limited by the thickness of the polymer film to about 50 nm. A major issue of the SJEM is that it is difficult to obtain the dc temperature rise on the sample from the ac expansion signal.

11.5.2.3
Scanning Thermal Expansion Microscopy

In 2000, Hammiche et al. developed a scanning thermal expansion microscopy (SThEM) method for imaging spatial variation of thermal expansion coefficient of polymer composites [44]. The STHEM differs from the SJEM in that a resistive probe is used as both heater and expansion sensor. The resistive probe could be either the Wollaston wire probe shown in Fig. 11.16 or a probe made by nanolithography as illustrated in Fig. 11.9. An external source was used to apply an alternating current at frequency ω_C to the probe when it was scanned across the surface at a constant contact force. The ac current modulated the tip temperature at $\omega = 2\omega_C$. Heat transfer from the tip to the specimen resulted in thermal expansion of the surface at the same frequency. This caused modulated deflection of the probe height which was detected using the AFM. Both the amplitude and phase of the height modulation were recorded, providing contrast in thermal expansion of different constituents in a polymer sample. Figure 11.28 shows the images of topography, amplitude and phase of expansion of a polyethylene terephthalate (PET)/ resin composite. The resin shows brighter amplitude contrast, confirming that its coefficient of thermal expansion is higher than that of PET.

11.5.3
Near-Field Scanning Optical Thermometry

A variety of optical microscopy methods have been used for temperature measurements. These include laser reflectance method, infrared thermometry, Raman spectroscopy, etc. The spatial resolution of a standard optical microscope is limited by diffraction to a theoretical spatial resolution of $d = \lambda/NA$, where λ is the wavelength of light and NA is the numerical aperture of the lens used. The highest numerical aperture one can achieve is about 1.6, and, hence, the best spatial resolution of optical microscopes is approximately 0.2–0.3 µm although typically a lower spatial resolution of approximately 0.5–1 µm is achieved. To overcome the diffraction limit, near-field scanning optical microscopes (NSOMs) have been developed. In this technique a tapered optical fiber is coated with an opaque metal film leaving a small aperture of diameter d at the fiber tip. This tip is then brought in close proximity (within 1–10 nm) of a sample surface by controlling the shear or the normal force between the tip and the sample surface. The fiber can be used as the light source or a light collector. By scanning the tip laterally, an optical image can be obtained. Because diffraction is a far-field effect, the close tip-sample proximity allows it to overcome the diffraction limit. The spatial resolution of NSOM is limited by the aperture size d, which is usually approximately 50 nm. Efforts to make apertures in the sub-30-nm range have proven to be very difficult, if not impossible. Additionally, the aperture needs to be larger than twice the optical

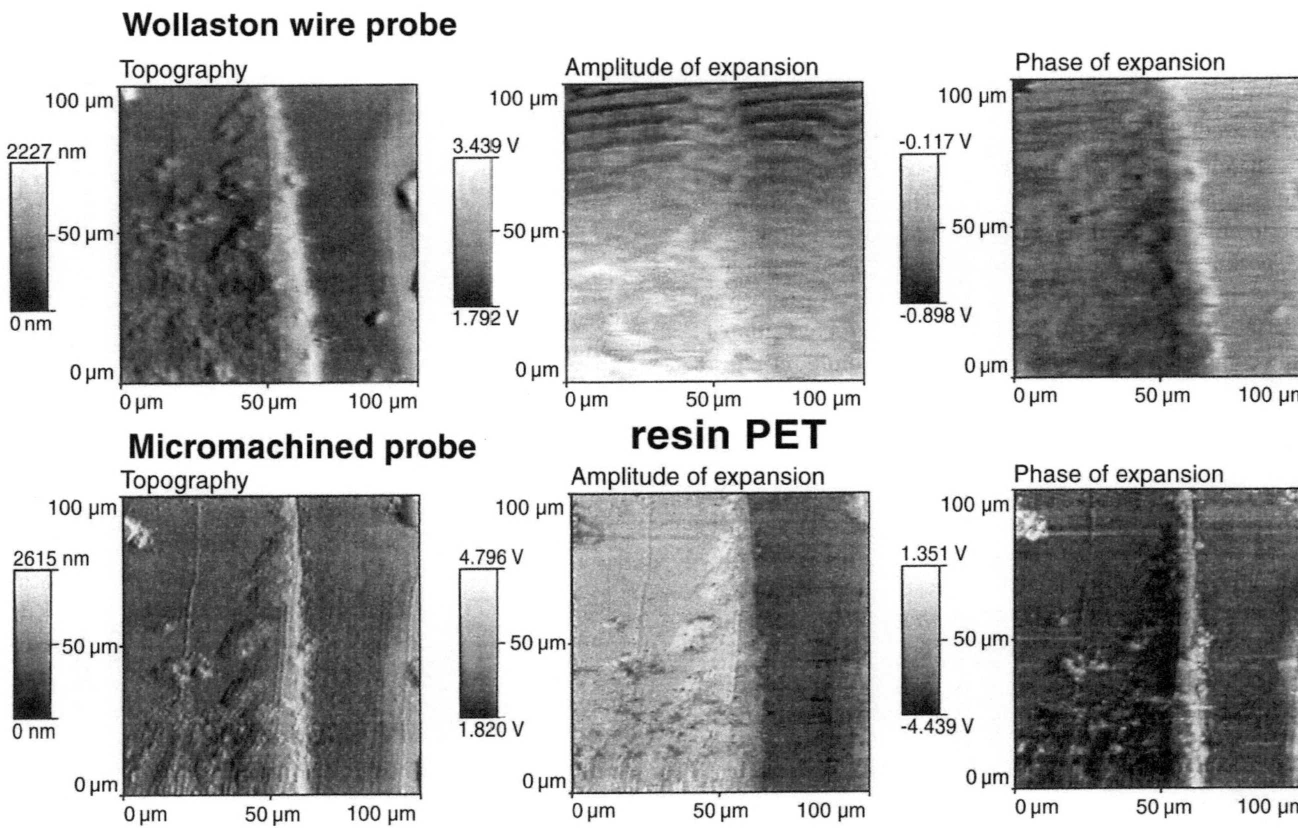

Fig. 11.28. SThEM modulated thermal expansion contrast. Sample PET/resin composite. Probe temperature modulation amplitude 10 °C, frequency $\omega = 350$ Hz. Images showing cross-section of the boundary between resin (*left side*) and PET (*right*). Reprinted from ref. 44 with permission

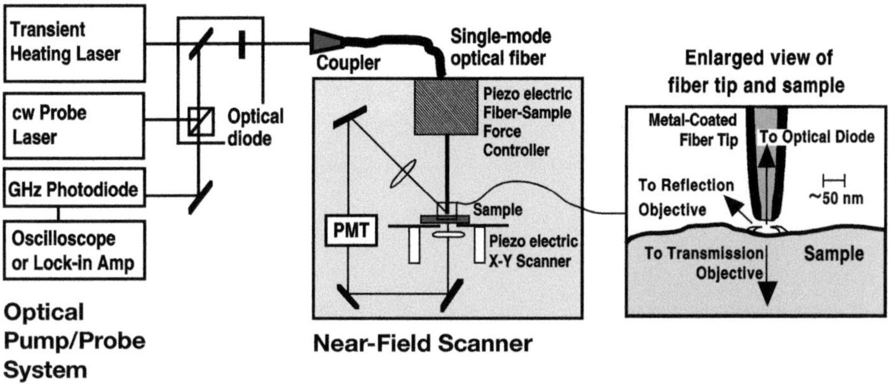

Fig. 11.29. Experimental setup for near-field optical thermometry (NFOT). Reprinted from ref. 45 with permission by Taylor & Francis Ltd. (http://www.tandf.co.uk/journals/tf/10893954.html)

skin depth in the metal coating, which is approximately 24 nm in Al at 633-nm wavelength.

The NSOM can be used for temperature measurement using reflectance thermometry, Raman spectroscopy, or infrared thermometry. Figure 11.29 shows the setup of the near-field optical thermometry experiment developed by Goodson and coworkers [45]. The experiment can be operated in two modes. In the first mode a continuous wave laser is coupled into the single mode fiber and transmitted through the 50-nm-diameter aperture at the end of the tapered fiber. The typical transmission efficiency is approximately 10^{-5}, which depends strongly on the size of the aperture and the geometry of the tapered fiber. Higher transmission efficiencies are now being reported with chemically etched fibers. Radiation can be collected by a photomultiplier tube using one of two far-field optical paths, which capture the reflected or transmitted light. Radiation reflected by the fiber can also be routed back through the fiber and separated from the incident beam by using a polarizing beam splitter and a quarter-wave plate. In principle, thermometry can be performed using the change of the collected radiation power caused by temperature changes of the sample. The collected radiation for each path depends on the optical constants of the sample, which vary with temperature. If the sample temperature is modulated at a frequency f, then the temperature dependence of the collected radiation signal can be obtained using a lock-in amplifier tuned to f. Figure 11.30 shows the thermal image of two adjacent Al interconnects with only the right interconnect heated at 2 kHz.

Whereas the first mode of operation is used for temperature measurement, the second mode of operation can be used for thermal property measurement at high resolution. Here, a pulsed laser is used to excite the sample, and a probe laser is then used to measure the reflectance as a function of time. This is recorded by the high-speed photodiode and an oscilloscope. Because the optical constants of materials are temperature dependent, the temporal behavior of the optical signal provides some idea of the transient temperature of the sample, which is closely coupled to the sample thermal properties. This is widely exploited for

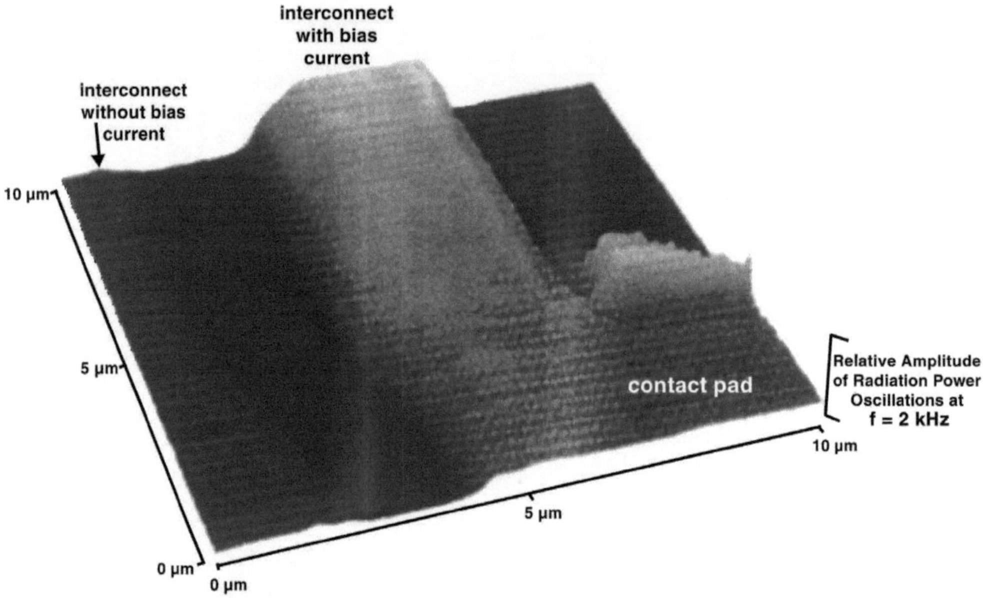

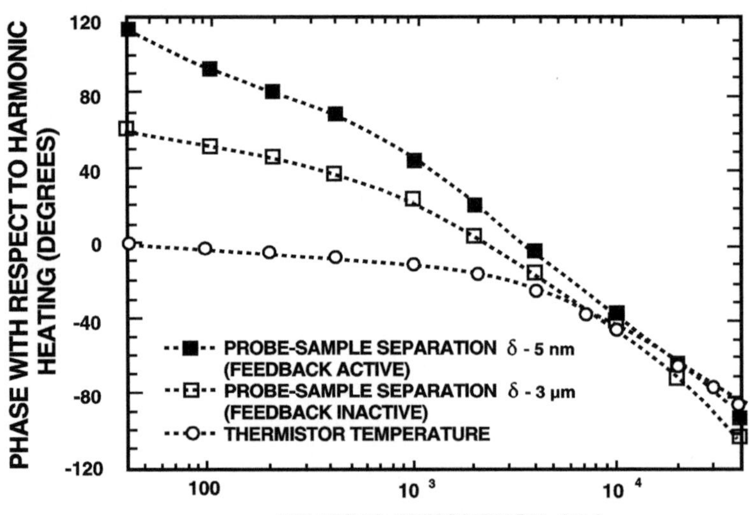

Fig. 11.30. NFOT image of a biased Al interconnect adjacent to an unbiased interconnect. Also shown is the phase difference between the electrical bias and the thermo-optical signal plotted as a function of heating frequency. Reprinted from ref. 45 with permission from Taylor & Francis Ltd. (http://www.tandf.co.uk/journals/tf/10893954.html)

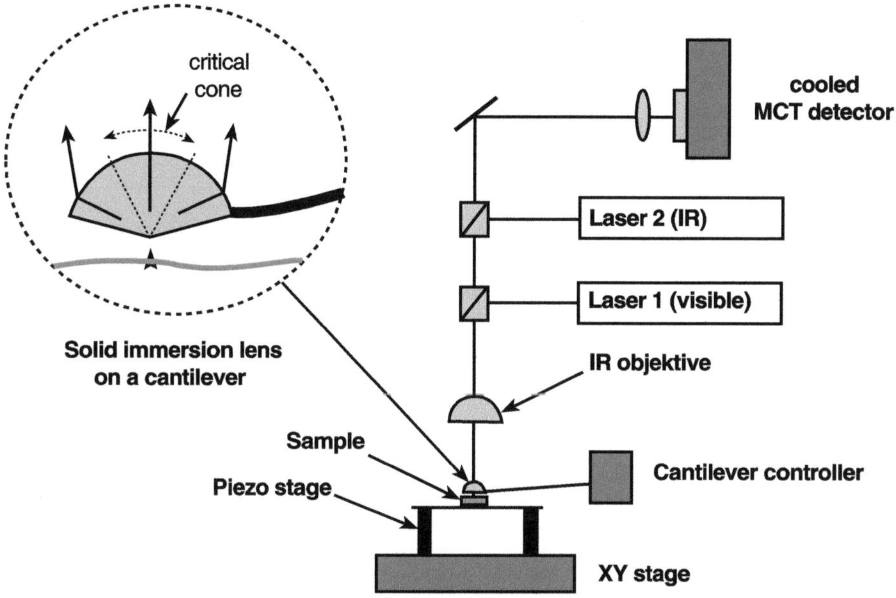

Fig. 11.31. Schematic diagram of an AFM cantilever probe containing a solid-immersion lens used for near-field infrared thermometry [46]

optically measuring thermal properties of samples by far-field techniques. The use of the NSOM allows measurement of thermal properties at sub-wavelength resolution.

In addition to using NSOM for thermometry, Fletcher et al. [46] developed another technique, which is illustrated in Fig. 11.31. Here, a solid immersion lens made of a material that is transparent in the infrared is mounted at the end of a cantilever and scanned across the sample. The solid immersion lens works both as an AFM tip and for collecting infrared emission. Near-field radiation coupling at the tip as well as the reduction of radiation wavelength due to the high refractive index of the lens material allow for sub-wavelength resolution for infrared imaging.

References

1. 2001 International technology roadmap for semiconductors, http://public.itrs.net
2. Dekker C (1999) Physics Today 52:22
3. Huang MH, Mao S, Feick H, Yan H, Wu Y, Kind H, Weber E, Russo R, Yang P (2001) Science 292:1897
4. Hicks LD, Dresselhaus MS (1993) Phys Rev B 47:12727
5. Heremans JP, Thrush CM, Morelli DT, Wu MC (2002) Phys Rev Lett 88:216801
6. Majumdar (1999) Annu Rev Mater Sci 29:505
7. Williams CC, Wickramashinghe HK (1986) Appl Phys Lett 49:1587
8. Majumdar A, Carrejo JP, Lai J (1993) Appl Phys Lett 62:2501
9. Lai J, Chandracood M, Majumdar A, Carrejo JP (1995) IEEE Electron Device Lett 16:312
10. Nakabeppu O, Igeta M, Hijikata K (1997) Microscale Thermophys Eng 1:201

11. Fish G, Bouevitch O, Kokotov S, Lieberman K (1995) Rev Sci Instrum 66:3300
12. Luo K, Shi Z, Lai J, Majumdar A (1996) Appl Phys Lett 68:325
13. Luo K, Shi Z, Varesi J, Majumdar A (1997) J Vac Sci Technol B 15:349
14. Luo K, Lederman M, Majumdar A (1997) Microscale Thermophys Eng 1:333
15. Mills G, Zhou H, Midha A, Donaldson L, Weaver JMR (1998) Appl Phys Lett 72:2900
16. Zhou H, Midha A, Mills G, Thoms S, Murad SK, Weaver JMR (1998) J Vac Sci Technol B 16:54
17. Shi L, Kwon O, Miner A, Majumdar A (2001) J Microelectromechanical Sys 10:370
18. Shi L, Majumdar A (2002) J Heat Transfer 124:329
19. Shi L, Plyasunov S, Bachtold A, McEuen PL, Majumdar A (2000) Appl Phys Lett 77: 4295
20. Shi L, Kwon O, Wu G, Majumdar A (2000) 2000 IEEE international reliability physics symposium, p 394
21. Kwon O (2000) Thermal design, fabrication, and imaging of MEMS and microelectronic structures. PhD dissertation, University of California at Berkeley
22. Leinhos T, Stopka M, Oesterschulze E (1998) Appl Phys A 66:S65
23. Pylkki RJ, Moyer PJ, West PE (1994) Jpn J Appl Phys 33:3785
24. Nakabeppu O, Igeta M, Inoue T (1999) Thermal Sci Eng 7:1
25. Williamson M, Majumdar A (1992) J Heat Transfer 114:802
26. Johnson KL (1985) Contact mechanics. Cambridge University Press, New York
27. Roark RJ, Young WC (1989) Roark's formulas for stress and strain. McGraw-Hill, New York
28. Langer G, Hartmann J, Reichling M (1997) Rev Sci Instrum 68:1510
29. Hammiche A, Pollock HM, Song M, Hourston DJ. Meas. Sci. Technol. 7:142
30. Hammiche A, Hourston DJ, Pollock HM, Reading M, Song M (1996) J Vac Sci Technol B 14:1486
31. Hammiche A, Reading M, Pollock HM, Song M, Hourston DJ (1996) Rev Sci Instrum 67:4268
32. Maywald M, Pylkki RJ, Balk LJ (1994) Scanning Microsc 8:181
33. Maywald M, Pylkki RJ, Reineke FJ, Balk LJ (1996) Prog Nat Sci 6:S103
34. Tsukruk VV, Gorbunov VV, Fuchigami N (2003) Thermochimica Acta 395:151
35. Incropera FP, DeWitt DP (1990) Fundamentals of heat and mass transfer, 3rd edn. Wiley, New York
36. Weaver JMR, Walpita LM, Wickramasinghe HK (1989) Nature 342:783
37. Nonnenmacher M, Wickramasinghe HK (1992) Appl Phys Lett 61:168
38. Zhou J, Yu C, Hao Q, Kim D, Shi L (2002) Proc. 2002 Int Mech Eng Congress Exp HT-32112
39. Gimzewski JK, Gerber C, Meyer E, Schlittler RR (1994) Chem Phys Lett 217:589
40. Nakabeppu O, Chandrachood M, Wu Y, Lai L, Majumdar A (1995) Appl Phys Lett 66:694
41. Varesi J, Majumdar A (1998) Appl Phys Lett 72:37
42. Majumdar A, Varesi J (1998) J Heat Transfer 120:297
43. Igeta M, Banerjee K, Wu G, Hu C, Majumdar A (2000) Electron Device Lett 21:224
44. Hammiche A, Price DM, Dupas E, Mills G, Kulik A, Reading M, Weaver JMR, Pollock HM (2000) J Microscopy 199:180
45. Goodson KE, Asheghi M (1997) Microscale Thermophys Eng 1:225
46. Fletcher DA, Crozier KB, Quate CF, Kino GS, Goodson KE, Simanovskii D, Palanker DV (2000) Appl Phys Lett 77:2109

12 The Science of Beauty on a Small Scale. Nanotechnologies Applied to Cosmetic Science

Gustavo Luengo · Frédéric Leroy*

12.1 Introduction to Cosmetic Science

From the early stages of human development man has had two basic needs, among others, to fulfill: a need to protect oneself and a need to communicate to others. The latter focalized on the use of products to impress or enhance certain feelings, while the former, more advanced, saw cosmetics as a way to protect the original biological surfaces from the external agents [1].

But what is a cosmetic? In general terms it refers to any article to be applied to the various external parts of the human body for cleaning them, perfuming them, changing their appearance and/or correcting body odors and/or keeping them in good condition.[1]

The success of a cosmetic product is based, not only on its properties, but also on the way it interacts with the biological substrate [2] (see Fig. 12.1). This is one of the challenges of the cosmetic scientist – to develop chemicals that eventually will interact with hair or skin in such a way, that the main needs are fulfilled: a durable cosmetic effect, a protective action, and obviously harmlessness. The heterogeneous structure of hair and skin is not helpful in achieving these goals, taking into account its innate variability of morphology due to its biological origin.

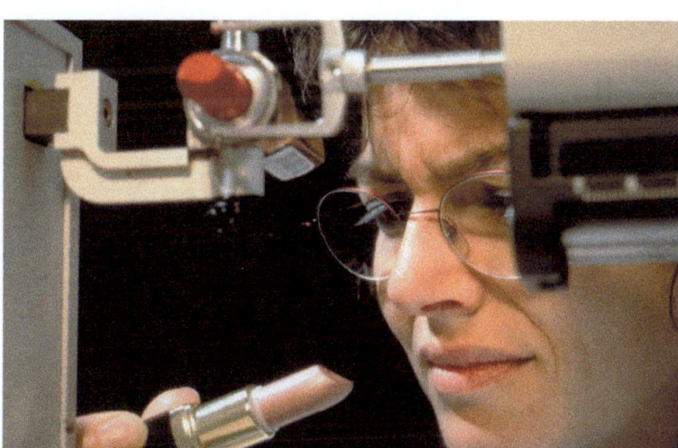

Fig. 12.1. In the development of cosmetics, special care must be taken to understand the final interaction of the product with the human substrate (L'Oréal)

[1] The Cosmetic Directive 76/768 of the Council of European Communities, March 2000.

* To whom correspondence should be addressed.

Consequently, the knowledge of the surface of hair, skin, and nail is essential to develop innovative, proper cosmetic products. For example, the hair surface is known to be hydrophobic but turns to hydrophilic in presence of water. Behind this macroscopic evidence there exists a molecular mechanism that is still poorly understood, the control of which could lead to significant applications.

As we will see throughout the chapter, Surface Science is of great importance for the formulation and development of cosmetics products, since most products will eventually interact with the cosmetic surface. In this respect, and if we add the morphological complexity of hair and skin, scanning probe microscopy (SPM) methods can provide essential information about the structure of the surface, its properties and their evolution after an applied treatment.

12.2 The Cosmetic Substrate: Hair and Skin and Their Modification

The main substrates of interest in cosmetic science are hair and the outermost layer of skin, called the *stratum corneum*. Both are composed of dead cells mainly filled with a specific type of protein called keratin. This protein is characterized by a high content of cystine, an aminoacid that has the capacity to crosslink the protein by its intermolecular disulfide linkages. The structure and assembly of these proteins (and other components like lipids) in the cells together with the cohesion of the cells give hair and stratum corneum their specific properties, mechanical and tribological among others.

12.2.1 Hair

If we want to study a biocomposite material, hair is a good example (see Fig. 12.2). This complex biological fiber is roughly cylindrical (~50–100 μm in diameter). The overall number of hairs covering a normal human head is 120,000–150,000, which leads to an important surface area (typically ~6 m^2 for 20-cm-long hair). This is a large surface to cleanse when shampooing for example.

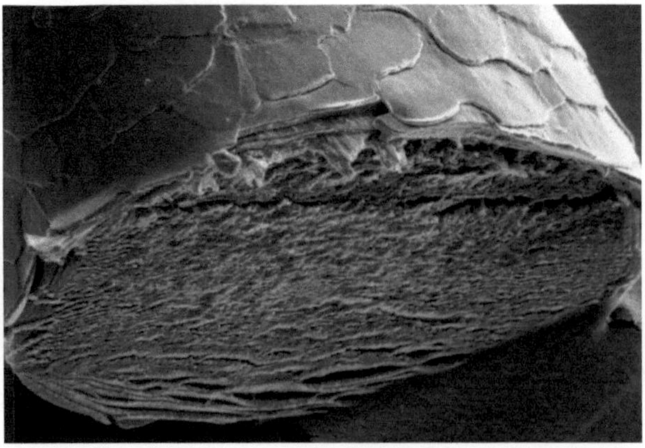

Fig. 12.2. Scanning electron micrograph (SEM) of a section of human hair fiber. The diameter of the fiber is approximately 80 μm. The image of the transversal cut shows the complex structure of this nanocomposite; the cortex and the cuticle are clearly visible. Note the typical scaled structure of the cuticle surface (L'Oréal)

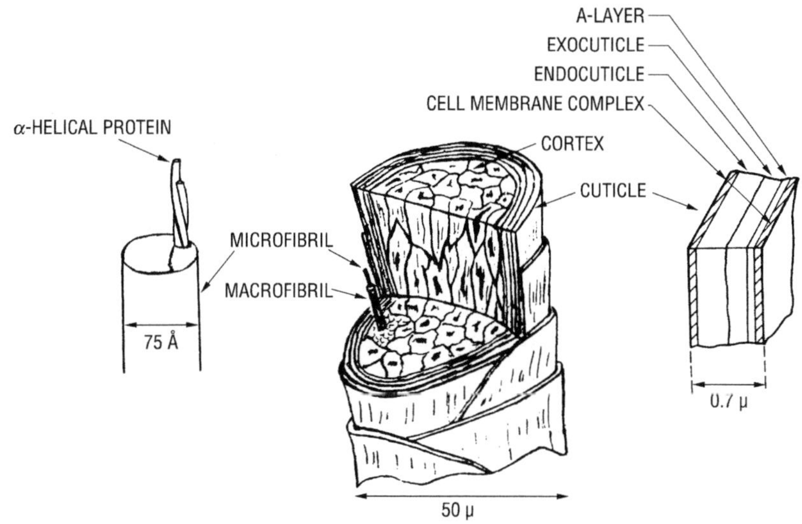

Fig. 12.3. Scheme of the hair fiber structure from the α-helical protein, to the microfibril, the macrofibril, and the cortical cells. The structure of the cuticle is also shown (reprinted with permission Springer)

Hair structure is complex and can be divided from the center into three layers: the medulla, the cortex, and the cuticle (see Fig. 12.3).

12.2.1.1
The Medulla and Cortex

The medulla is in the center and is composed of honeycomb-like cells with empty spaces aligned along the longitudinal axis of the hair.

The cortex is the main part of the hair. It is formed of elongated cortical cells (~100 μm in length and ~1–6 μm in diameter) aligned along the axis of the fiber and filled with partially crystallized keratin. This peculiar structure of the cortex gives the hair its unique mechanical properties.

Keratin has the ability to crystallize in the form of an α-helical structure, basic element of a 4-strand rope arranged in a coiled-coil configuration [3, 4]. Serine, glycine, *n*-acetylserine, and cystine are the main amino acids and the chains are stabilized by many intermolecular and intramolecular interactions such as electrostatic, van der Waals, hydrogen, and covalent (disulfide bonds). These units form hexagonal crystals called *microfibrils* (~7.5 nm in diameter) or intermediate filaments (IFs). Moreover, these IFs are embedded in an amorphous protein (matrix), as a semicrystalline substructure of the cells called *macrofibrils* (~100–400 nm in diameter) [5–7].

These imbricate crystalline structures, strongly anisotropic, convert the hair in a true bionanocomposite, giving it its unique mechanical properties, namely, high rupture stress and elongation, high elastic modulus, and rapid and total recovery even at high deformation [8].

12.2.1.2
The Cuticle

The cuticle forms the outer surface of hair. It has an overlapping scale-like structure protecting the cortex. A normal hair has about 6–8 overlapping cells in close contact. In addition and thanks to earlier electron microscopy studies, we know that the cuticle is divided into three layers: the innermost endocuticle, the exocuticle, and the outermost epicuticle.

The endocuticle, composed of cytoplasmatic and nuclear remnants has a low cystine content (~3%) and is supposed to be the weakest component of the cuticle structure.

The exocuticule is composed of cystine-rich proteins component; the part of the layer closer to the epicuticle is referred to as the A-layer. Its high cystine content (>30%) suggests that is a highly crosslinked material, therefore producing a more compact material which could explain its high mechanical resistance and behavior as a water barrier.

The final structure of the outermost layer of the cuticle is of special interest: the surface properties of hair depend on the physicochemical properties of this layer.

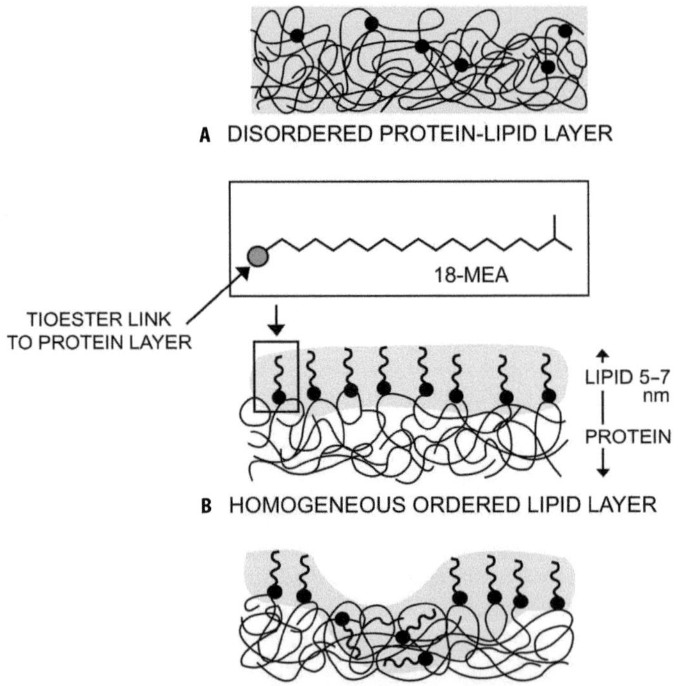

Fig. 12.4A–C. Scheme representing the diverse possible scenarios that could take place at the outermost layer of the hair surface. **A** Covalently linked lipids like the 18-MEA disorderly attached to the protein, **B** the lipids are homogeneously ordered at the surface of the protein, **C** a mixed model is proposed and therefore an heterogeneous coverage can be expected (L'Oréal)

The most recent findings have shown that the epicuticle membrane of hair fibers contains highly crosslinked protein (~75%) and fatty acids (~25%). Among these adsorbed fatty acids, the 18-methyleicosanoic acid (18-MEA) appears to be covalently grafted through thioester linkages to the protein and is considered to play an important role in the physicochemical and tribological properties of hair. Based on these observations a model has emerged to explain the fine molecular structure of the hair surface [9, 10]. Further measurements need to be done to evaluate the accuracy of this representation (see Fig. 12.4), as well as the extent or ordering of this suspected monolayer, the degree of coverage on the scale structure, its dynamics in the presence of surfactants or various treatments, etc.

12.2.2 Skin

Our whole body is covered by skin, whose surface area can reach ~1.6 m^2 and whose main role is protection. The structure of skin includes three compartments: epidermis, dermis, and subcutaneous tissue. The epidermis structure is made up of several cell layers: basal, spinous, granular, and horny. Horny or stratum corneum is the outermost part of the skin. From the basal layer, the cells migrate upwards, progressively flatten, loose their nucleus, and become filled with keratin filaments (*keratinization*) embedded in the cell protein remnants. The cells of the horny layer are referred to as *corneocytes* (see Fig. 12.5), and are roughly hexagonal in shape overlapping at the edges with neighboring cells. The actual arrangement of the keratin is still discussed (α-helix-like hair versus β-pleated sheet), although recent results using X-ray diffraction are more consistent with a model of self-folded keratin β-sheets [11].

The stratum corneum acts as a mechanical, thermal, and chemical barrier against the external environment and is without doubt, the first target of skin-care

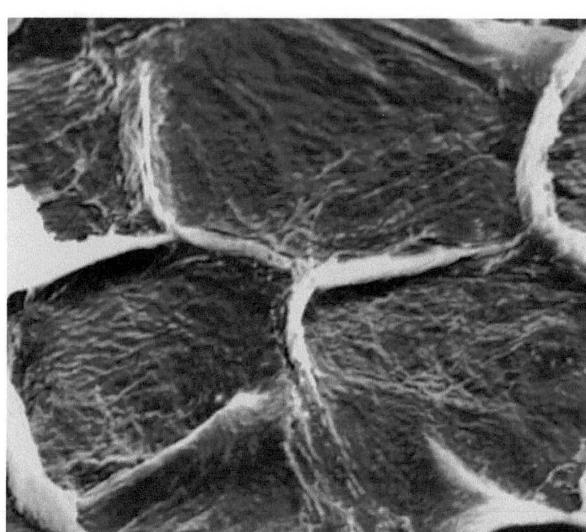

Fig. 12.5. SEM micrograph showing the corneocytes that form the surface of skin and its outermost layer: the *stratum corneum* (L'Oréal)

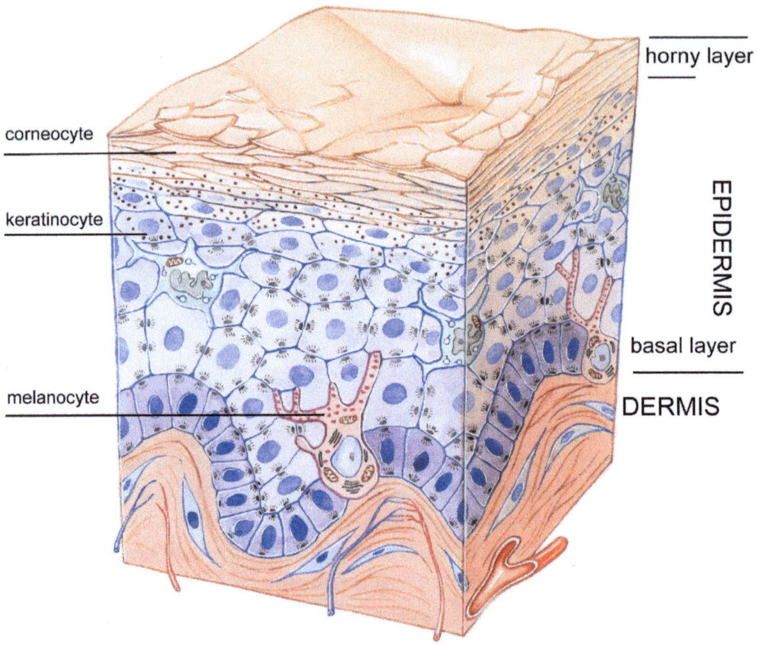

Fig. 12.6. Scheme representing the complex structure of skin with its different layers: dermis, epidermis, and finally horny layer or stratum corneum (L'Oréal)

products. "Indeed the *raison d'être* of the epidermis is to make the stratum corneum; this is its specific biological mission" [12] (see Fig. 12.6).

It is remarkably impressive that the 15-μm-thick layer of stratum corneum is able to prevent the whole human body against water loss and the external physical and chemical aggressions. The barrier effect and the cohesiveness of the stratum corneum are due to the highly structured organization of the corneocytes and the intercellular matrix [13]. In effect, from the simplest structural model widely accepted [14], the matrix is mainly composed of a lamellar phase of various lipids organized in bilayers that act as a glue keeping the cells together and providing a barrier effect.

The lipids are mainly composed of ceramides, cholesterol, fatty acids, and cholesterol esters. It is interesting to note that they do not contain phospholipids which are commonly found other biological membranes. Ceramides, which constitute approximately 40% of all stratum corneum lipids, are a type of surfactants with a polar head (hydroxyl terminal groups) and a double non-polar saturated tail. They would naturally seem to be ideally suited to form highly ordered and impermeable membranes. There are six types of ceramides (I–VI) depending on the molecular structure of the polar head. Ceramide I (see Fig. 12.7) is believed to play a fundamental role in the bilayer and lamellar structure of the lipids in the intercellular matrix as it is covalently bound to the corneocytes and has a long saturated chain that may extend more than one bilayer distance and therefore act as a link to keep two membranes tightly associated.

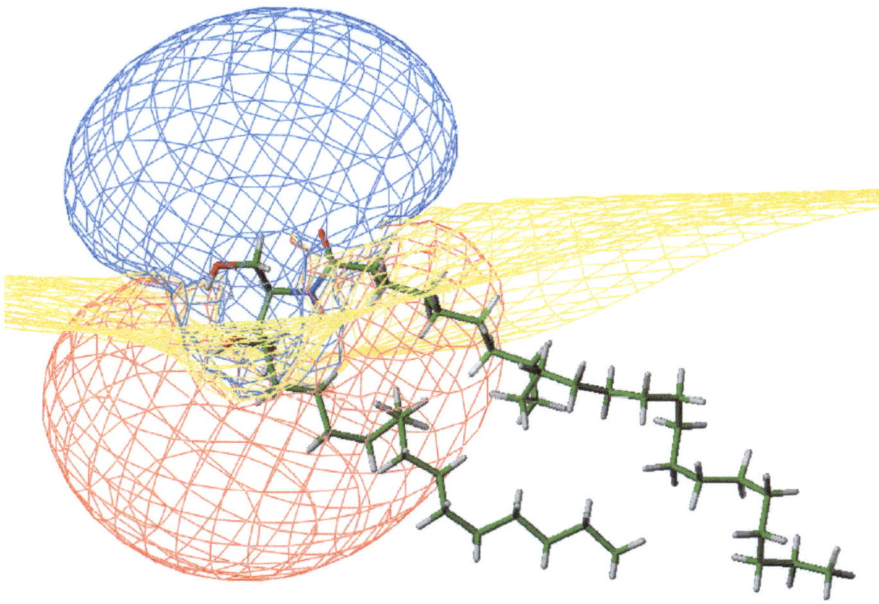

Fig. 12.7. Scheme representing the ceramide I or acylceramide. This molecule consists of a sphingoside base with an amide-linked ω-hydroxyacid and a nonhydroxyacid ester linked to the ω-hydroxyl group. The molecules have the structure of a surfactant with double hydrophobic tail and a polar head

12.3 The APM as a New Tool. From the Microscopic World to the Cosmetic or Sensory Perception

The availability of the scanning tunneling microscope (STM) and the later development of the atomic force microscopy (AFM) [15, 16] enabled a much easier access to the nanometric world. These tools have helped Richard Feymann's vision on nanotechnology: the possibility to create extremely small structures through nanoscale manipulation [17]. But this use is only one of the many applications that have appeared with the use of the AFM and that the cosmetic scientist can benefit from.

Nowadays, these two techniques have evolved and have been expanded by the addition of very new diverse methods using electric, thermal, and even optical probes and are known in general as Scanning Probe Microscopy (SPM) techniques. These are more commonplace in many academic and industrial environments and with the speed of technical advances, nanoscale manipulation is in a continuous state of development and improvement. In the rest of the chapter most of the applications will be based on AFMs and that will be the term used from now on.

AFM is a relatively easy instrument with which we can visualize microstructures and nanostructures. Also, it can be used for measurements in many environmental conditions, unlike the comparable resolution techniques of scanning or transmission electronic microscopy (SEM or TEM). For example, until recently, biologists needed to make use of the latter vacuum techniques to image surface samples, thereby preventing in vivo analysis, since, in general, biological materials cannot withstand prolonged vacuum. In addition, SEM operates on electrically conducting samples and surfaces need to be coated with metal before imaging.

Moreover, it is important to emphasize that other techniques are of interest for the understanding of the physics at the nanoscale. Precisely, the Surface Forces Apparatus (SFA) developed by Israelachvili [18] is probably still the most quantitative method to measure surface forces. The SFA allows one to measure the interaction between two symmetric mica surfaces. The distance is measured by an interferometric method and it also allows one to control the actual surface of contact. The mica surfaces can be coated with many other materials, surfactant monolayers [19–21], polymers [22, 23], microemulsions [24], etc. which offer a great variety of applications. The use of SFA nevertheless remains more oriented towards fundamental research; but we believe it provides a useful complementary tool for the understanding of cosmetic substrates and more precisely the role of skin and hair lipids.

Returning to the AFM, we will give a basic description. The instrument is based on the interaction of a sharp point (tip) attached to the end of a cantilever with the surface of interest (see Fig. 12.8). The interaction is measured by the cantilever bending that is itself monitored by the displacement of a laser beam reflected on the back of the cantilever to an XY detector. The lateral movement of the tip or the surface is accomplished by using a piezo-electric element which provides scanning capabilities.

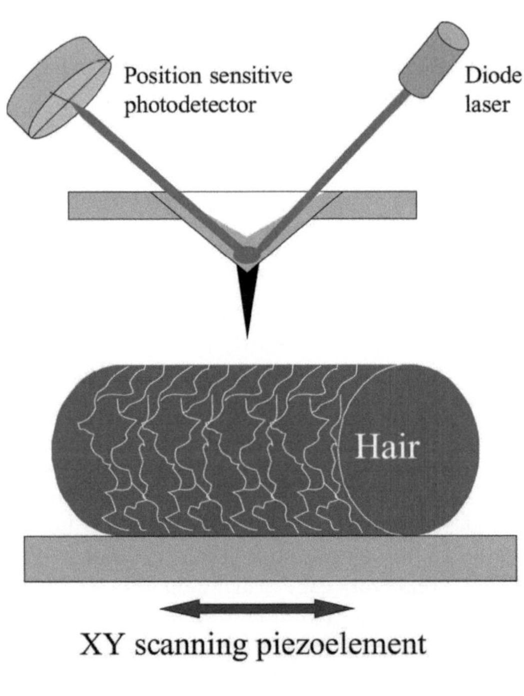

Fig. 12.8. Basic AFM scheme which shows how the tip's small size is perfectly suited to characterize human hair fibers. The possibility of working in a liquid environment is further valued

One of the few drawbacks of this technique is the need of information relative to the surface state of the tips used; as a result, the need for extremely well-characterized tips has produced a boom in the tip-manufacturing and tip-modification business.

But how can these new nanotechnologies based on the SPM be beneficial to cosmetics designed for hair or skin? We come back to the importance of the biological substrate. In order to reach an optimal adecuation of the product to the hair fiber, not only it is important to have in-depth knowledge regarding hair structure, but also its alterations following applied chemical treatments, as currently used on hair. For example, there is still little knowledge of the process whereby subtle chemical modifications of the hair fiber can propagate to create the sensory perception, as reflected by sensorial tests or combing force measurements. The AFM can therefore help to explore basic mechanisms that are underlying these cosmetic effects, thanks in part to its specific design to interact with the submicrometric surface structures.

In addition, AFM is also of maximum interest as an analytical tool to characterize the very different types of new materials (polymers, surfactants, etc.) that form part of new innovative shampoos, creams, and other cosmetic products.

And finally, imagery in general is part of the ethics and concerns in the cosmetics. It is not surprising that AFM provides an attractive imaging technique to visually demonstrate the effects of products or active ingredients.

12.4 Recent Uses of AFM on Cosmetic Substrates

When the AFM instruments became available, they were first applied to research animal fibers, like wool. From the outset, the textile industry soon recognized them as the best-suited tool for close visualization and characterization of fibers. Several institutes, such as the Australian Textile and Fiber Technology Division of the Australia's Commonwealth Scientific and Industrial Research Organisation (CSIRO), and the Griffith University incorporated the AFM techniques in their studies [25, 26–28].

Many studies have then shown the applicability of AFM to diverse cosmetic issues, namely for the visualization of deposits on hair [29, 30], or the effects of shampoos on hair cleaning [31].

Note that most of the applications are hair oriented, although few studies have been included on skin lipids that foresee the potential of AFM in this field for the future.

12.4.1 Morphological Studies

As mentioned in the previous section, the relatively friendly imaging potential of AFM represents a major field of interest in the development of the cosmetic industry. The possibility of obtaining true topographic information without any special constraint (no vacuum, no metallic coating) and the possibility of imaging in water are some characteristics, among others, which are highly appreciated by the

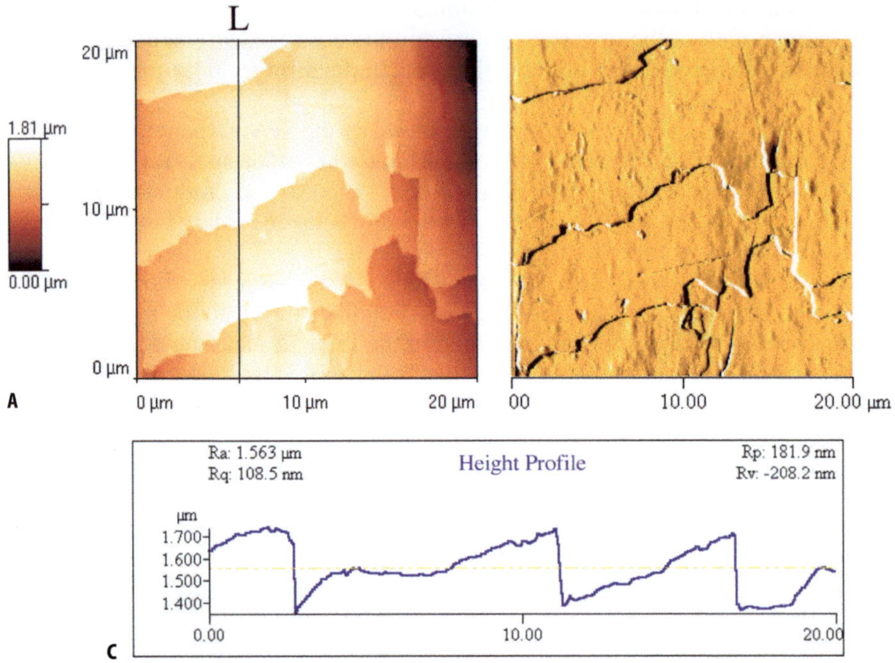

Fig. 12.9A–C. Example of the type of images obtained by AFM on the hair surface: **A** height or topographic image and **B** error signal image that enhances the small details of surface roughness. True height information is only available from the topographic image. In **C** the height profile through *line L* is displayed (L'Oréal)

cosmetic scientist. Figure 12.9 shows a typical AFM image of the hair surface where the fine details of the hair surface as well as a profile through the indicated line can be clearly seen.

As the measurement of the topographic height profile is straightforward and quantitative, experiments based on the quantification of the scale step height and surface roughness give insight into the hair structure and its sensitivity to external parameters (physical or chemical). If we look closer at hair surface structure, the scales of the cuticle overlap determining a saw-tooth profile characterized by a step height. Most studies reflected a wide distribution of cuticle step heights, now easily measured by AFM. Several reports by CSIRO scientists already confirmed these findings using wool fibers [27]. On human hair fibers, O'Connor et al. measured an average value of approximately ~500 nm but no further statistics were given [32]. More precise statistical data have also been reported: 340 ± 90 nm [33]. This variation is not a method-related bias. It shows that we must speak in terms of distribution of cuticle heights [34] which now can be easily obtained (see Fig. 12.10).

One of the most important factors is water or air moisture, which is known to swell hair fibers. This swelling is attributed to water molecules which increase the interstitial spaces between the cuticular cells or within macroprotein structures. This swelling also affects the structure of the cuticle scales, spreading out by ap-

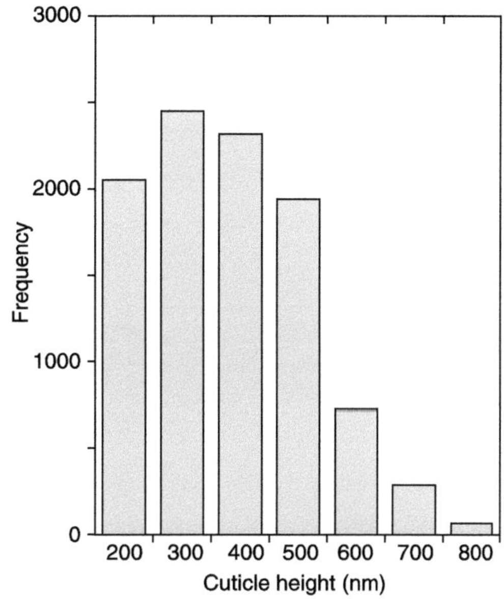

Fig. 12.10. Histograms showing the cuticle height distribution from approximately 10,000 steps measured from 10 AFM images of European brown hair (reprinted with permission from the *Journal of Microscopy*)

proximately 21% in the case of wool fibers [27] and by 50–150% in the case of human hair when hydrated [32].

The effect of pH on hair morphology has also been studied by immersing in phosphate/HCl/NaOH buffer solutions showing an increase in the step height with pH [32]. This finding was attributed to the elimination of residual commercial conditioners (cationic polymers) on hair, although a decrease of the step height has also been reported recently [33]. Despite these discrepancies, it is already known that the effects of alkalis generally are stronger on structural distortion of the cuticle due to easier "attachability" to disulfide bonds presents on the keratin [8].

In a relatively ingenious way, hair expansion due to humidity has also been measured by AFM [35]. The authors correlated the topographic images of the same area at different ambient humidity by computer matching of the different features.

Finally volume expansion has also been observed as a function of temperature in a punctual study [33].

All these applications contribute to a better knowledge of the structure of the hair surface in many close-to-real conditions, and specifically in a very quantitative manner. The use of AFM to investigate the morphology of hair is common practice for the cosmetic scientist.

12.4.2 Tribological Approaches

As we mentioned before, one of the advantages of AFM has been the ability to retrieve more information about the physical, chemical, and mechanical properties of the surface studied. The first interest was to explore the micro and nanotribology (adhesion and friction) of hair surfaces as both provided a way to de-

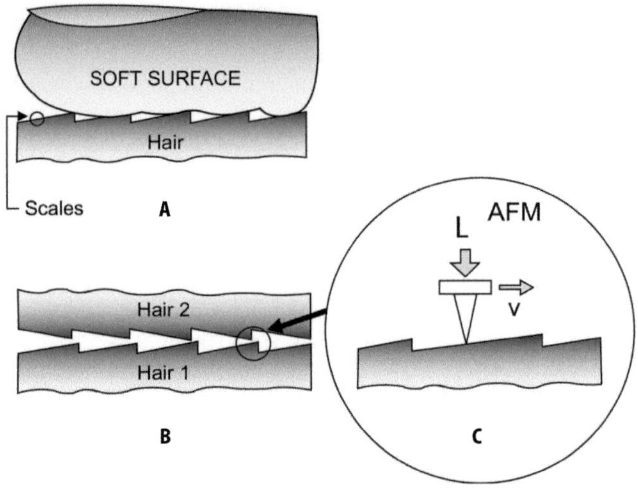

Fig. 12.11A–C. Scheme to illustrate the importance of the nanoscale friction on hair. The well-known asymmetric friction effect is well observed macroscopically, and is mainly due to the specific orientation of the scales from the root to the hair tip. This effect is easily detected by just sliding the fingers through the fiber in both directions (**A**) and is related to the compliance of the sliding object (the finger in this case). In contrast, when we are looking into fiber-fiber interactions for example, the contact is made between two relatively rigid bodies and takes place in contact zones of submicrometer dimension (**B**) that can better studied by sliding the AFM tip at a specific load and velocity (**C**) (L'Oréal)

tect the presence of surface mechanical changes and the influence of active ingredients.

As we see in Fig. 12.11 at the macroscopic level, the friction of hair is asymmetric. This is mainly due to the orientation of scales [36]. These effects are of importance when a macroscopically deformable object like a finger slides over the fiber. In contrast, when looking into fiber-fiber interaction, the role of the submicron structures is believed to be dominant [36–39]. Hair-hair friction has been studied as a macroscopic phenomenon but it was only with the advent of AFM techniques that we have access to the local friction and progress towards an understanding of both the hair surface and the molecular origin of macroscopic friction. The latter is, among others, a key factor when studying the effects produced by hair products such as hair-setting, hair-styling, conditioners and shampoos.

The application of the friction mode of the AFM (sometimes called lateral force microscopy, LFM) is therefore of great use for these applications. The torsion of the AFM tip during the normal scanning of a surface is measured by the displacement of the laser reflection on the detector in the horizontal axis. Friction forces are the biggest contribution to this torsion and can be analyzed in comparison with the simultaneously obtained topographic images.

Although this would be sufficient in the case of completely flat surfaces, more real systems such as the hair surface, are more irregular. In the latter case there is an influence due to topography which can be eliminated to a large extent by recording both the forward and reverse traces and their subsequent subtraction (image subtraction technique). This helps eliminate much of the topographic influences that are of a different sign on each trace and, therefore, can cancel each other out. There are several methods for an absolute calibration of the friction force [40, 41], but generally they are not required when the objective is to compare different surface regions of a sample.

Philips and his colleagues at CSIRO [27] were the pioneers in the use of these techniques on keratin fibers by detecting the effects of a surface oxide treatment on wool through changes in the homogeneity of the friction force distribution on the surface (no image contrasts were observed on untreated fibers).

Recent applications of the lateral force microscopy include three main directions – the improvement of knowledge of hair surface structure, the quantitative measurement of local friction and its relation to the underlying molecular structure – all in order to get a basic understanding of the final sensorial cosmetic effects like touch, texture, softness, etc.

As we discussed before, the outermost layer of hair is composed of specific lipid structures. The nanotribology of thin films and in particular of monolayers made of surfactants or polymers (grafted or not), is nowadays a subject of great interest. It is believed to result from the influence of molecular organization of the film on subtle friction forces [42–44]. Therefore it is expected that the fine structure of the hair surface can be further analyzed by the influence of these thin layers on friction.

Figure 12.12 shows the fine details of friction contrast recently observed [45] by this method on human hair fibers in which the non-covalently linked fatty acids

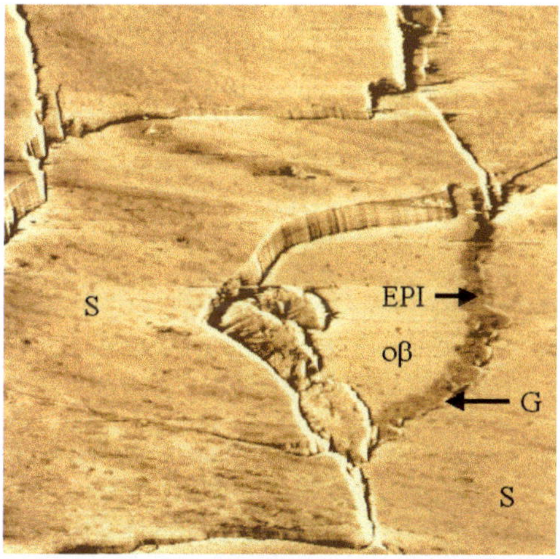

Fig. 12.12. Lateral force microscopy image of a brown European hair surface previously treated with a 70:30 chloroform/methanol solution for 5 h in order to extract the unbound fatty acids. The image shows a cuticle step in which part of the cuticle is missing (the original scale followed the path *G*). The original outer cuticle (*S*) is seen, as well as the surface of a newly exposed cuticle surface. The friction contrast reveals the presence of two layers hardly detected by topography. The epicuticle (EPI) and the supposedly covalently linked layer of 18-MEA (outer β layer, *oβ*). Field of scan 10 μm (reprinted with permission from the *Journal of Microscopy*)

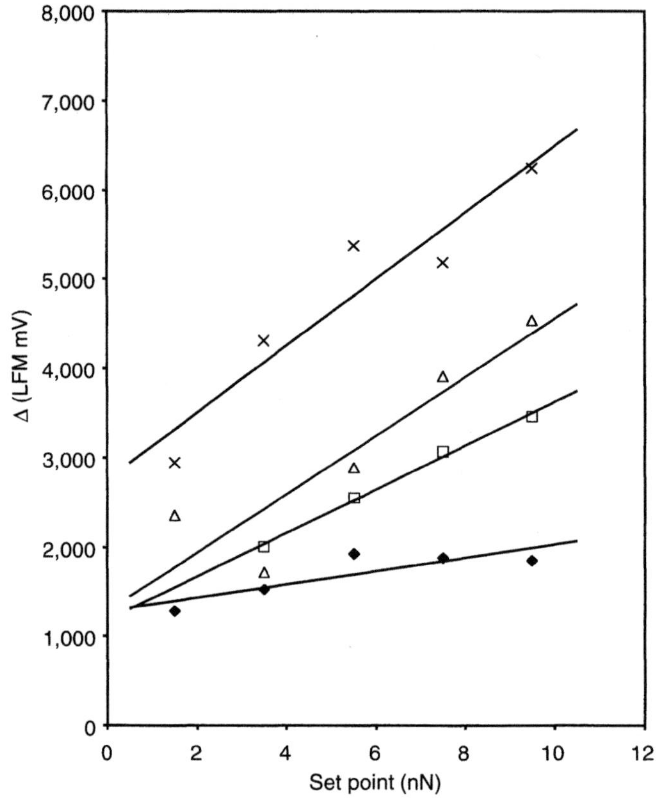

Fig. 12.13. Friction (in arbitrary units) dependence on load (*L*) for different treated hairs. Native (◆), bleached (□), delipidized (△), and copolymer treated (X). From Mc Mullen and Kelty [47]; reprinted with permission from *Scanning*

had been removed. The studied area showed a frictional contrast among three different layers that probably correspond to the δ-layer, the epicuticle, and the outer β-layer of the cuticular cell membrane complex (CCMC). The presence of this last layer is detected because it exhibits a low friction contrast and is therefore ascribed to as the outer β-layer that contains the presumably monolayer of saturated fatty acid 18-MEA.

Use of a variety of cosmetic products, such as shampoos and conditioners, can change the tribological characteristics of hair at the macroscale. To a large extent, sensory perceptions are controlled by the way fibers slide among each other or when a combing element is run through. For example the applied charge or load (L) and speed (V) can determine if hairs will or will not disentangle when combed out.

It is clear that further better quantification of friction forces and their dependence on speed and load become necessary to progress in the understanding of hair fiber tribology, especially when this type of study is aimed at being extended to cosmetics-treated hair. In this way McMullen and Kelty [46, 47] examined the effect of load on the friction behavior of several treated hair samples (see Fig. 12.13). Looking at the different results, the friction seemed to follow a typical Amontown's law ($F = \mu L$). A pseudo-friction coefficient (the friction force was measured in volts) was drawn from the results allowing for a friction classification: the native fiber showed the lowest friction followed by bleached, solvent-extracted, and finally polymer-treated hair respectively.

It is interesting to observe that there was friction force even at zero load. Although the authors attributed this difference to changes in imaging conditions, this is not likely considering the image subtraction method used. A finite friction at zero load is a clear indication of an adhesion-controlled friction mechanism [22, 48]. Due to the small area of contact of an AFM tip this effect could be due either to a special chemical affinity of the tip, tip penetration or capillary forces. By taking into consideration that these experiences were done in air, the latter explanation seems the most plausible.

Although a correlation between nano and macrotribological results is not straightforward, McMullen et al. found a correlation between their AFM friction results and dry combing forces (and work) measured with a miniature tensile tester [46] (method to measure the forces required to comb through a bulk of hair fibers as a function of the tress length).

When looking into the few AFM applications to skin (to date), we foresee a higher interest in the future. Most interest in this study is focused on the morphology of surface cells and the role of lipids as a barrier against the environment. This is in part due to the soft character, dynamic modes that are more adapted to help understand the fine structure of the skin surface (see below).

12.5 Advanced Applications of AFM

12.5.1 Dynamic Methods

The introduction of dynamic modes (intermittent or non-contact modes) are appealing owing to the smaller level of interaction between the AFM tip and the substrate.

In a dynamic experiment the tip vibrates close to the surface, just enough to be sensitive to the van der Waals potential.

The dynamic modes (also known as *tapping* modes) have extended the applications of AFM to more biological and soft materials [49–51], like the visualization of living cells, or even the visualization of liquids and biomolecules [52, 53]. In essence, with almost no loss in resolution, soft matter is also accessible by AFM techniques. Apart from the obvious interest of visualizing the hair or skincare materials (we will discuss this in more detail in the next section), these methods can be helpful in understanding the adsorption of softer complex liquids (like the presence of soft viscous polymers) on the hair [54–56] or the skin surface.

From a morphological point of view, the possibilities of imaging in a liquid environment makes it relatively easy to explore the morphology of the outer skin

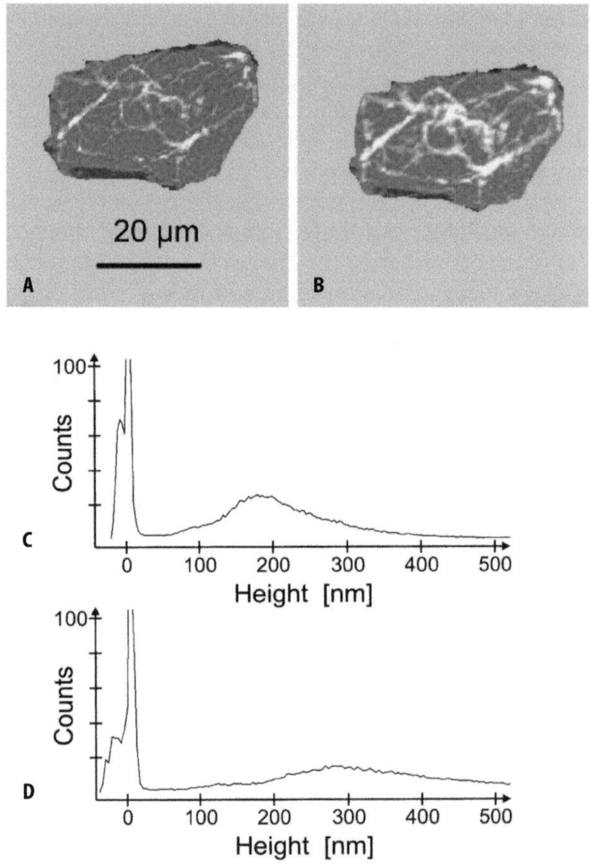

Fig. 12.14A–D. Direct comparison of the topographic images of corneocytes cells in ambient conditions (**A**) and swelled in water (**B**); the histograms of height distribution are able to detect a shift from approximately 200 nm to approximately 300 nm after water exposure (**C** and **D**). (from Richter et al. [57])

surface cells (i.e., the corneocytes). Richter et al. [57] measured the swelling of individual human skin corneocytes in distilled water and compared it to ambient conditions using this method (see Fig. 12.14).

As we mentioned before, the lipids found in the bilayers of stratum corneum (mostly ceramides, cholesterol, and free fatty acids) play a vital role as a barrier for the body. It is not surprising that an important research field at the moment deals with the way the different lipids interact with each other (if they mix or if they phase-separate, their specific or peculiar role, etc.). These studies are usually carried out ex-situ by preparing Langmuir-Blodgett monolayers on Si-wafers [58, 59] using dynamic modes. There is even certain evidence of imaging the molecular architecture of the stratum corneum: a repetitive pattern (distorted orthorhombic) observed on high-resolution scanning size (2.5 nm) images of the surface of water-immersed corneocytes is arguably interpreted as the ordered lipids [60].

The cosmetic scientist normally uses dynamic modes as a visualization tool. But, apart from the improvement of imaging aspects, the capability of measuring the phase lag of the real tip vibration with respect to the imposed one (so-called phase methods), are probably of more potential interest. Actually, changes in the phase lag reflect also changes in the mechanical properties (adhesion and hardness) of the sample surface, accentuating the contrast among mechanically different zones of the sample.

By simultaneously collecting the phase image and the topographic image, researchers have been able to look with more detail to the relative mechanical properties of the substrate. We say "relative" because it is often intricate to extract quantitative values from these measurements and usually too time-consuming from an industrial point of view.

Nevertheless, if there is an inhomogeneous distribution of mechanical properties due, for example, to phase separation or self-organization of a material (like copolymers), the phase image can enhance this difference. Unfortunately, it is difficult to easily separate the topographic influence on these images, especially if the substrate is not flat. Most of the studies of phase imaging are related to the materials used in cosmetics as a characterization tool and may form part of the substantiation of a new product. We foresee further interest in these methods in the study of hair mechanical properties.

12.5.2 Forces at the Hair Surface

In an effort to extract other information, researchers are looking into using other potentials of AFM, such as the ability to measure forces at the surface, even by deforming it (indentation). *Force spectroscopy* allows the continuous measurement of the interaction of the tip with the cosmetic substrate. This force is measured as a function of distance (*force curves*). Figure 12.15 shows the typical parameters taken into

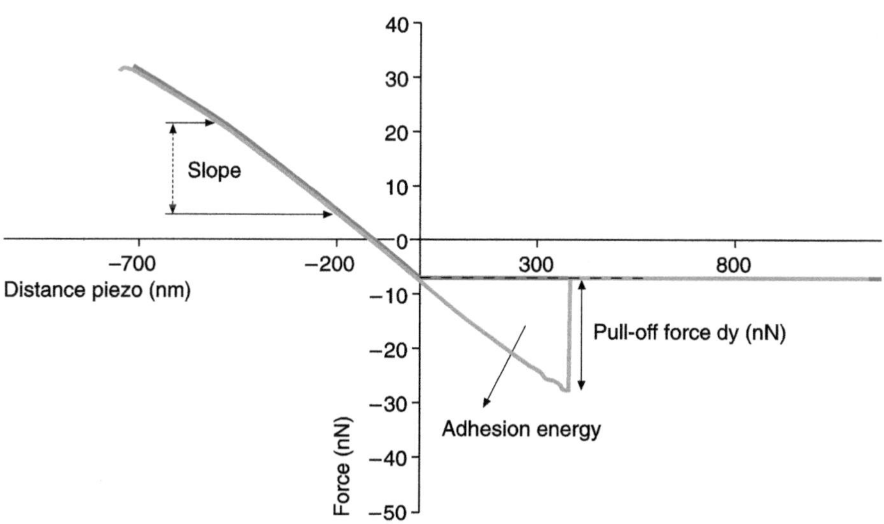

Fig. 12.15. Example of force curve showing the main parameters of interest from a mechanical point of view. The slope can be related to capability of tip penetration on the substrate while the pull-off force is somehow related to the adhesion energy when the geometry of contact is well established

account in these curves. Although there are still many uncertainties about the physics concerning contact mechanics of AFM tips on substrates, many comparative studies have turned out to be very satisfactory, opening a whole new spectrum of research.

In essence, the scanner brings the sample and the tip together while monitoring the force induced on the cantilever that holds the tip. The system records both the approaching and retracting curve as force applied versus distance traveled by the piezo. The retracting or unloading curve is usually the one that holds information regarding the strength of the interaction; a minimum in the retracting curve (pull-off force) is associated to the adhesion.

Depending on the load applied on the surface (loading force) the tip may penetrate the substrate (*indentation*). The degree of penetration can give information about the Young modulus of the material.

Researchers at the Griffith University, in collaboration with CSIRO, decided to exploit these surface force detection capabilities of AFM to help understand the structure and properties of the native lipid layer present on wool surfaces [25, 26, 61]. In concrete, relatively small discontinuities observed on the approach curves were modeled on the basis of a multilayer of different compliant elements helping infer the presence of a lipid layer of 3–8 nm and also estimate the effective spring constant (rigidity) of approximately ~0.12 N/m^{-1}, although this value can change due to water exposure. It is known that hair treated with potassium hydroxide losses this lipid layer, therefore leaving the fiber more unprotected. Actually, pull-off forces and estimated surface stiffness decreased by almost 10 times suggesting an easier penetration and plastification of the fiber [25]. Recent experiments on hair [28], showed very similar patterns: after 1-h aqueous exposure the stiffness and adhesion decreased by 2–3 nN (see Fig. 12.16). The effect of adhesion is probably due to reorganization of the molecular groups at the surface of the fiber, exposing polar groups that transform an initially hydrophobic hair surface into a hydrophilic one, as observed in wettability experiments [62].

A more advanced understanding of the role of these lipids is necessary, and this requires sometimes to work ex situ on more ideal surfaces. For example, Smith et al. [63] looked at the adhesion of 18-MEA grafted on mica surfaces observing that its branched character made it exhibit lower adhesion forces than straight chain analog lipids.

More quantitative results can be obtained by using diamond tips of calibrated shapes (usually pyramidal tips) and performing *nano-indentation* experiments. Although their use is relatively uneasy and time-consuming, scientists from the Neuroscience Research Institute (Santa Barbara, USA) in collaboration with The Wool Research Organization of New Zealand (WRONZ) [64] performed Young modulus measurements on a 1-µm-thick transversal section of a wool fiber. Despite using a rather uncommon method of analyzing the approach curve (Sneddon method) instead of the retracting curve (Oliver and Pharr method [65]), the authors were able to identify the different internal regions (cuticle and cortex) from their different mechanical properties. Actually, they observed a clearly higher value of the modulus for the exocuticle (~20 Gpa) compared to the cortex and endocuticle (~4 Gpa). A sulfur-reduction process eliminated this difference suggesting that an important number of disulfide bonds of the exocuticule was at the origin of its higher rigidity (cf. structure of hair surface above).

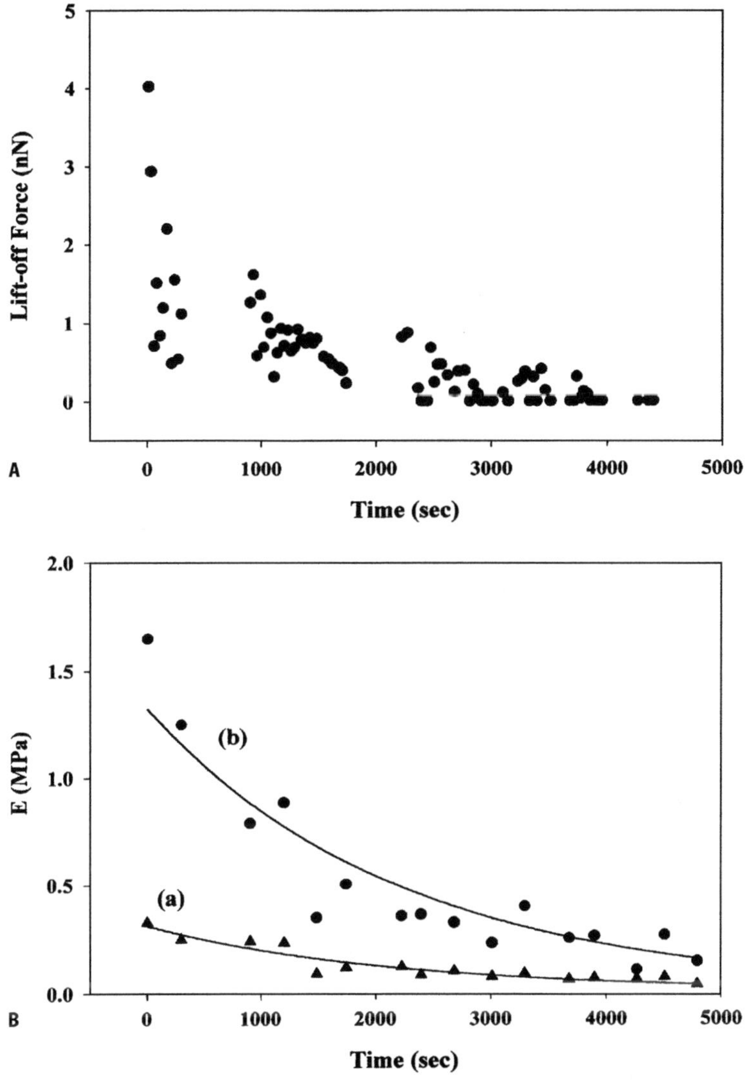

Fig. 12.16. Relative change in pull-off force (**A**) and effective Young modulus (**B**) for a hair surface treated by a commercial shampoo/conditioner with as a function of aqueous exposure. Indentation size approximately 30 nm (**a**) and >100 nm (**b**). Reproduced from Blach et al. [28] with permission from *International Journal of Cosmetic Science*

Due to the similarity of wool to hair, this method could help in the future to localize the effect of cosmetic treatments on hair. Figure 12.17 shows a contact image of one hair section showing the different layers that form the cuticle. Several indentations can be observed that show the high positional capabilities. In recent measurements we obtained relatively lower values for the hair fiber of approximately 1 Gpa (unpublished results).

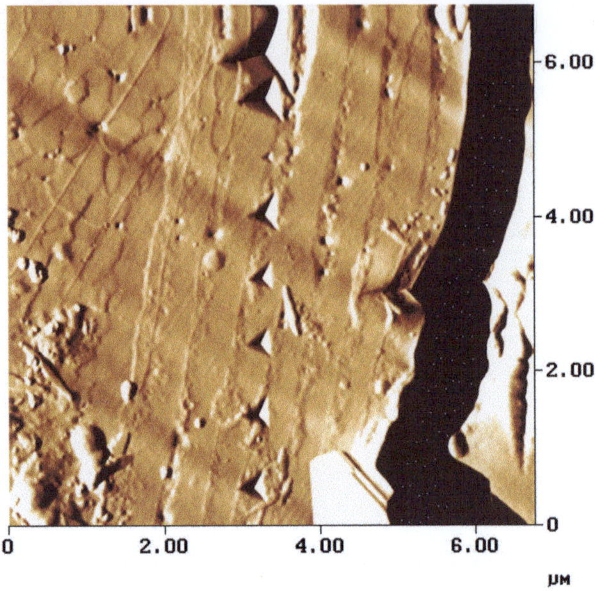

Fig. 12.17. AFM topographic image showing small indentations on the different cuticle layers on a transversal cut of a human hair fiber (L'Oréal, unpublished results)

Although extracting mechanical information like Young modulus or elasticity from AFM is possible, classical nano-indenters are probably better suited in terms of reproducibility or automatism, although lacking the fine position, detection and imaging capabilities of AFM, which is an important technical advantage when working on such complex materials as hair or skin. Hybrid systems have also been marketed in which the AFM tip is substituted by a vertical nano-indenter diamond tip.

12.6 The Influence of Hair and Skincare Treatments

We have seen how the AFM has given the cosmetic scientist a better understanding of the biosubstrate. Above, we emphasized the usefulness of AFM in providing images to illustrate and substantiate the effects of applied treatments. More than ever, imaging techniques are powerful candidates to help understand the properties and behavior of a cosmetic product.

Therefore, AFM imaging methods are currently applied as a complementary tool to other more established ones like confocal microscopy or SEMs. AFM can help to characterize various hair treatments, like bleaching [31], perming [64], use of conditioners or shampooing (i.e., see Fig. 12.18).

For example, cationic polymers have been used widely in the cosmetic industry due to the special affinity they have for the keratin substrate. We can find these polymers in many haircare products from shampoos to conditioners. Because of their importance in the creation of these products, a number of studies have been carried out on cationic polymers using other techniques such as electron spectroscopy for chemical analysis (ESCA). However, these techniques hardly give

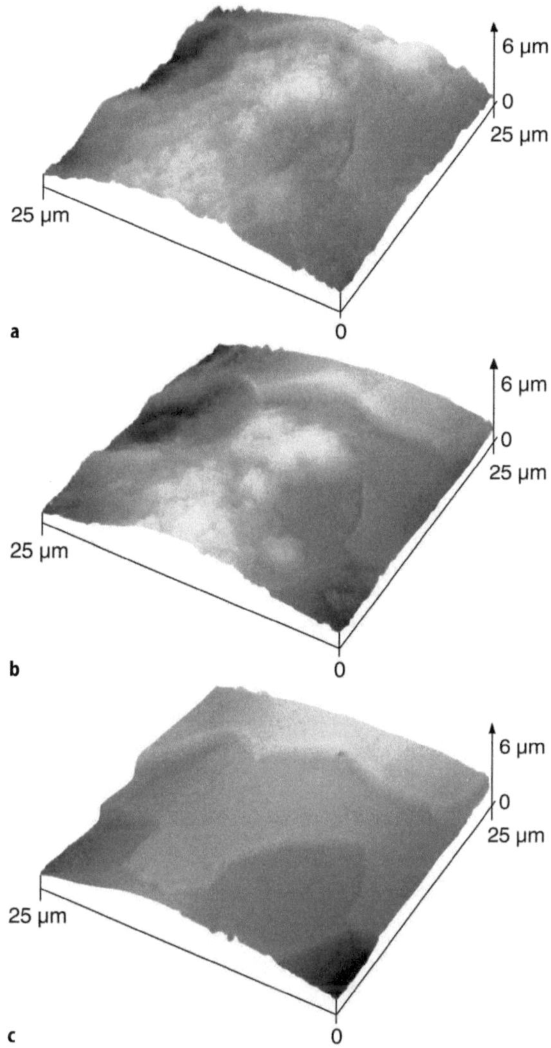

Fig. 12.18a,b. AFM images of human hair in liquids with some natural grease on the surface (**a**), after adding 0.07% shampoo to the water (**b**) and after further addition of shampoo. From Will et al. [31]; reprinted with permission from the *IFSCC*

visual proof of the presence of adsorbed polymers (e.g., dye or fluorescence techniques). This is where AFM has helped progress.

One of the first applications of AFM as a visualizing tool was using mica as a model surface for keratin (both being negatively charged). Goddard and Schmitt [54] immersed a mica surface on a 0.1% polyquaternium-10 solution (Polymer JR400) and observed a complete coverage as reflected by a definite increase in the surface roughness. It is interesting to note that SEM images were unable to detect this presence.

Since conditioning polymers for hair are frequently applied in anionic surfactant-based shampoo, in a subsequent study [66], the authors further made the system more complex by incorporating small amounts of an ionic surfactant. They observed different types of texture as a function of the polymer used or the concentration of the solution. In general the adsorbed polymer layers were less homogeneous than when adsorbed alone suggesting a complexation of the polymer with the anionic surfactant.

Many of these subtle variations are dependent on the right choice of the model surface. This is an issue of great importance because any model surface may resemble hair on certain physicochemical aspects (contact angle, surface potential, etc.) but not all. It is true that the improved flatness of the mica substrate can help us concentrate on the dynamics of a complex mixture of cosmetic interest. There is also the possibility of using polished keratin (horn or hoof) for a model surface, albeit it lacks the fine structure of the hair surface.

This complex interaction between cationic polymers and surfactants has been further and more recently investigated on real hair fibers [30, 55, 56, 67]. The use of non-contact modes and phase images as well as the capability of repositioning and washing in situ helped the authors to attribute the net-like structures seen to the polymer-surfactant complexes [55, 56].

The adsorption of these polymer-surfactant complexes also changes the tribological properties of the hair fiber; for example, recent results show that shampoos and conditioners may affect the adhesion at the nanoscale, and Blach et al. [28] reported a clear decrease in the pull-off force on wet fibers.

Due to the inherent variability of hair, the recognition of these nanometric structures is not always evident. Investigators are forced to increase enormously the number of examined sites. Fortunately the development of repositioning techniques can help in many cases. By repositioning, the same area can be explored before and after a treatment, even in a wet environment, therefore eliminating the variability of the hair surface morphology.

12.7 Conclusions

The knowledge of the surface of cosmetic substrates, for example, hair and skin, and their interactions with applied products has always been a major concern for basic research in cosmetic science. The purpose of this chapter was to illustrate the role and the use of SPM in the development of cosmetic products with optimally controlled properties.

The continuous improvements of the technique and the availability of more sophisticated working modes allowing the tip to measure the electric potential (Kelvin force microscopy), capacitance, or thermal changes on the surface allows us to foresee a successful development of these techniques. Nevertheless, further improvements are still necessary. The immediate goals are to obtain higher data acquisition speeds, automatism, and better control of tip characteristics.

References

1. Mitsui T (1997) New cosmetic science. Elsevier, New York
2. Zviak C (1986) The Science of hair care. Marcel Dekker, New York
3. Pauling L, Corey RB (1953) Nature 171:59
4. Crick FHC (1953) Acta Cryst 6:689
5. Fraser RDB, Macrae TP, Suzuki E (1976) J Mol Biol 108:435
6. Briki F, Busson B, Doucet J (1998) Biochim Biophys Acta 1429:57
7. Parry DAD (1996) Int J Biol Macromol 19:45
8. Robbins CR (2002) Chemical and physical behavior of human hair, 4th edn. Springer, Berlin Heidelberg New York
9. Negri A, Rankin DA, Nelson WG, Rivett DE (1996) Text Res J 66:491
10. Negri AP, Peet DJ, Rivett DE (1996) J Text Inst 87:608
11. Garson JC, Doucet J, LevêQue JL, Tsoucaris G (1991) J Invest Dermatol 96:43
12. Kligman AM (1964) The biology of stratum corneum. In: Montagna W, Lobitz WC (eds) The epidermis. Academic Press, New York, p 387
13. Wertz PW, Downing DT (1989) Stratum corneum: biological and biochemical considerations In: Swarbrick J, Guy RH (eds) Transdermal drug delivery. Marcel Dekker, New York, chap 1, p 1
14. Wertz PW, Madison KC, Downing DT (1989) J Invest Dermatol 92:109
15. Binnig G, Quate CF, Gerber CH (1986) Phys Rev Lett 56:930
16. Binnig G, Gerber CH, Stoll E, Albrecht TR, Quate CF (1987) Europhys Lett 3:1281
17. Feynman RP (1960) Eng Sci 23:22
18. Israelachvili JN (1973) J Colloid Interface Sci 44:259
19. Israelachvili JN, Wennerström H (1992) J Phys Chem 96:520
20. Leckband DE, Helm CA, Israelachvili J (1993) Biochem 32:1127
21. Yoshizawa H, Israelachvili JN (1993) J Phys Chem 97:11300
22. Luengo G, Schmitt F-J, Israelachvili J (1977) Macromolecules 30:2482
23. Schmitt F-J, Yoshizawa H, Schmidt A, Duda G, Knoll W, Wegner G, Israelachvili J (1995) Macromolecules 28:3401
24. Giasson S, Kuhl T, Israelachvili JN (1998) Langmuir 14:891
25. Gibson CT, Myhra S, Watson GS, Huson MG, Pham DK, Turner PS (2001) Text Res J 71:573
26. Crossley JAA, Gibson CT, Mapledoram LD, Huson MG, Myhra S, Pham DK, Sofield CJ, Turner PS, Watson GS (2000) Micron 31:659
27. Phillips TL, Horr TJ, Huson MG, Turner PS, Shanks RA (1995) Text Res J 65:445
28. Blach J, Loughlin W, Watson G, Myhra S (2001) Int J Cosmetic Sci 23:165
29. Hössel P, Schrepp W, Sander R (1995) Proceeding IFSCC, p. 349
30. Hössel P, Sander R, Schrepp W (1996) Cosmet Toilet 111:57
31. Will T, Kelch A, Riedel JH, Hintze U, Wittern KP (1998) Proceedings of the 20th IFSCC Congres. Beiersdorf, p. 165
32. O'Connor SD, Komisarek KL, Baldeschwieler JD (1995) J Invest Dermatol 105:96
33. You H, Yu L (1997) Scanning 19:431
34. Smith JR (1998) J Microscop 191:3
35. Resch R, Ehn R, Tichy H, Friedbacher G (1998) Appl Phys A 66:S607
36. Adams MJ, Briscoe BJ, Wee TK (1990) J Phys D Appl Phys 23:406
37. Schwartz M, Knowles DC (1963) J Soc Cosmet Chem 14:455
38. Johnson KL (1996) Langmuir 12:4510
39. Johnson KL (1997) Proc Royal Soc London Ser A 453:163
40. Ogletree DF, Carpick RW, Salmeron M (1996) Rev Sci Instrum 67:3298
41. Ruan JA, Bhushan B (1994) J Tribol 116:378
42. Gee ML, Israelachvili JN (1990) J Chem Soc Faraday Trans 86:4049
43. Chen YL, Helm CA, Israelachvili JN (1991) Langmuir 7:2694
44. Yoshizawa H, Chen YL, Israelachvili J (1993) J Phys Chem 97:4128
45. Smith JR, Swift JA (2002) J Microscopy (Oxford) 206:3
46. McMullen RL, Jachowicz J, Kelty SP (2000) IFSCC Magazine 3:39

47. McMullen RL, Kelty SP (2001) Scanning 23:337
48. Luengo G, Israelachvili J, Granick S (1996) Wear 200:328
49. Tamayo J, García R (1997) Appl Phys Lett 71:2394
50. Nony L, Boisgard R, Aime JP (2001) Biomacromolecules 2:827
51. Nony L, Cohen-Bouhacina T, Aime JP (2002) Surf Sci 499:152
52. Revenko I, Proksch R (2000) J Appl Phys 87:526
53. Hansma HG (1996) J Vac Sci Technol B 14:1390
54. Goddard ED, Schmitt RL (1994) Cosmet Toilet 109:55
55. Andre V, Norenberg R, Hossel P, Pfau A (1999) Macromol Symp 145:169
56. Hössel P, Dieing R, Nörenberg R, Pfau A, Sander R (2000) Int J Cosmet Sci 22:1
57. Richter T, Muller JH, Schwarz UD, Wepf R, Wiesendanger R (2001) Appl Phys A 72 Suppl: S125
58. Tengrotenhuis E, Demel RA, Ponec M, Boer DR, Vanmiltenburg JC, Bouwstra JA (1996) Biophys J 71:1389
59. Schaffer H, Bakowsky U, Rettig W, Kresse H (1999) Mol Cryst Liq Cryst Sci Technol Sect A 329:2
60. Chen TL, Wiedmann TS (1996) J Invest Dermatol 107:15
61. Gibson CT, Watson GS, Mapledoram LD, Kondo H, Myhra S (1999) Appl Surf Sci 145:618
62. Kamath YK, Dansizer CJ, Weigmann HD (1978) J Appl Polym Sci 22:2295
63. Smith JR, Eaton PJ, Waddington K, Alexander C, Tsibouklis J, Swift JA (2000) Proceedings of the 10th International Wool Textile Research Conference, HH-2, 1–10. Aachen, Germany
64. Parbhu AN, Bryson WG, Lal R (1999) Biochemistry 38:11755
65. Oliver WC, Pharr GM (1992) J Mat Res 7:1564
66. Schmitt RL, Goddard ED (1994) Cosmet Toilet 109:83
67. Pfau A, Hossel P, Vogt S, Sander R, Schrepp W (1998) Macromol Symp 126:241

Part III
Industrial Applications

13 SPM Manipulation and Modifications and Their Storage Applications

Sumio Hosaka

SPM technology has the potential not only to observe and evaluate fine structure on a sample but also to manipulate an individual atom or cluster and modify a sample surface. This means that as previous microscope technology such as optical and electron microscopes has been applied to fabricate fine patterns for VLSI devices and storage devices, an application of the SPM technology to fabricate them with atomic and nanometer size can also be expected. So far, many researchers have introduced the potential or method to make a fine pattern using electric field, electron induced current heating, field oxidation, induced charge, near-field light heating, electron induced chemical reaction, etc. in SPM systems.

In this section (§13), first of all we will introduce some principles of atomic manipulation and surface modification such as field evaporation, electron induced current, nano-indentation, and evanescent light heating using STM, AFM, MFM, and SNOM, and then introduce one of the approaches of AFM-based storage and SNOM-based storage for trillion bit (Tera bit) storage with faster than 10 Mb s^{-1}.

Second in §14, polarization-type SCaM-based storages using ferroelectric material will be introduced. A spin-on VDF-TrFE copolymer thin film and p-type silicon wafers were used as the ferroelectrics and the semiconductor, respectively. It is shown that the ferroelectrics were poled and capacitance modulation caused by the direction of the polarization was detected. It has been dynamically demonstrated that bits with 0.3- to 0.4-μm spatial frequency were recorded and reproduced in a contact mode SCaM, which corresponds to >1 Gbit cm^{-2} in area recording density.

Third in §15, an unique approach of optical near-field to high-density optical storage using super aperture in unique film on a disk will be presented. This technique is one of optical near-field or thermal recordings as well as SNOM-based storage. But, the small aperture is formed in the near-field aperture layer on disk substrate instead of the SNOM probe. The technique is called super-resolution near-field structure (super-RENS). The super aperture is formed by changing optical property at a localized area of special film such as an antimony (Sb) layer and a silver oxide layer by raising the temperature of the localized area when illuminating and focusing a far-field laser beam into the area. According to the aperture, near-field optical probe is formed to record or read out a small bit. It is demonstrated that many bits with around 100-nm size have been written and read out in an ordinary optical recording system.

Finally in §16, a single electron transistor (SET) and a single electron memory fabricated by field oxidation using AFM will be described. The principles of field oxidation, fabrication of SET, and memory will be introduced. The potential to control electrons in future Si devices will be described. A single electron transistor showed the clear Coulomb gap of 200 mV and Coulomb oscillation characteristics with the periods of 2.1 V at room temperature. A single electron memory showed the hysteresis loop by the return trip of the memory bias when starting from 0 V to 10 V and again coming back to 0 V and stored about 25 electrons at room temperature.

13.1 Atomic Manipulation and Surface Modification and Their Storage Application

The area bit density of data storage has increased at a rate of 30 times over the last 2 years. However, because of optical diffraction and extremely weak signals from the small bit area, the ceiling density seems to be several tens Gb in^{-2} in current magnetic recording and optical recording devices.

Many researchers have proposed various SPM-based storage techniques such as STM-based storage [1–3], AFM-based storage [4], MFM-based storage [5, 6], and SNOM-based storage [7, 8]. These techniques are supported by atomic manipulation and nanosurface modification for writing, and atomic- or nanometer-sized probe detection for reading. For data storage, however, it is necessary to write nanobits smaller than 100 nm and to read them out at a rate over 10 MHz.

Here, theory and experiment of atomic manipulation and surface modification are briefly presented, and then some of the advantages of the different SPM-based storage techniques in regard to bit size and readout speed are described.

13.2 Nanometer-Sized Bit Formations

Future data storage requires primary techniques such as nanobit writing and high-speed readout. Figure 13.1 shows some desirable techniques for developing high-density data storage. When replacing induction heads with the SPM heads, we need to develop nanometer-sized domains, high-speed readout, highly responsive gap control, and highly packed tracking.

In these techniques, nanometer-sized domain writing and high-speed readout should be investigated with the following performance:

(1) Writing bit size with less than 100 nm.
(2) Readout speed with over 10 MHz.
(3) The substrate surface should have very smooth structure with a 1-nm depth and a 2-nm grain size.

The SPM technique has the potential to make a nanometer-sized bit structure using a field evaporation, nano-indentation, an electron induced local heating or evanescent light heating. Regarding the readout speed, a regulation gap between the probe and sample can be controlled only in a 10-kHz response range. I considered that a key point is at least whether compensating for the servo error is possible in the readout by using the gap spacing dependence on detected signal intensity or cantilever spring. The need for (3) is very related to the writing method of (1). The

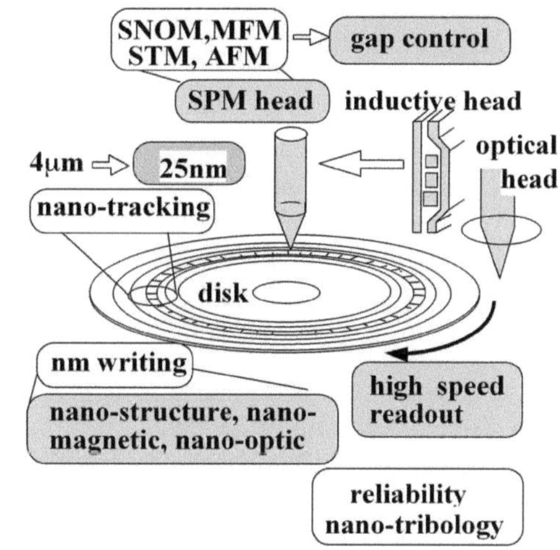

Fig. 13.1. Scheme of some techniques to be developed for a SPM-based data storage

writing mechanisms will be presented first, followed by the possibility of writing a nanometer-sized bit structure and prototype systems for high-speed readout. Here, we describe 3 mechanisms (field evaporation, nano-indentation, and local heating) to form a nanometer-sized bit in a sample surface.

13.2.1 Field Evaporation

Recently, atomic manipulation using field evaporation in a scanning probe microscope has been studied by many researchers. This technique is expected to be instrumental in achieving high-packed storage memory and high-speed high-density electronic devices. The scanning tunneling microscope (STM) [9] has proved useful not only for observing surfaces, but also for manipulating atoms [2, 3, 6, 9–14]. Field evaporation [2, 3], which is used in atomic manipulation, makes it possible to remove and deposit an individual atom or a cluster in the sample surface. This technique is expected as a single atom manipulation as described later.

The field evaporation using STM, however, can be applied only to conductive surfaces, although many material surfaces are oxidized. We have various needs for atomic manipulation for ultrahigh-density recording using both STM and atomic force microscope (AFM) [15–18]; (1) to make nanometer-sized metal dots on both a conductive material and an insulating one, (2) to make dots or pits on a wide, super-flat substrate such as a Si wafer, and (3) to make dots or pits at room temperature in air.

Here, we theoretically and experimentally describe the feasibility of making nanometer-sized dots on a conductive surface and an insulator by field evaporation using an STM and an AFM. Theoretical and experimental studies consider whether it is possible for field evaporation to occur in both STM and AFM using a thin insulating or a conductive substrate.

13.2.1.1 *Theoretical Consideration of Field Evaporation*

Field evaporation can occur in both STM and AFM. In these systems, potentials due to a probe and a sample of these systems overlap with each other. Here, we use a charge exchange (CE) model [19] for the calculations. The theoretical considerations in STM have been presented by Tsong [19, 20]. Furthermore, the consideration in an AFM configuration is similar to the STM configuration, which has been presented by Hosaka et al. [21].

(1) STM configuration

Figure 13.2a shows atomic and ionic potentials in STM with a metal substrate. They are assumed to have a hump of height Q_m at distance r from the probe surface and distance d from the substrate surface. Q_m may be expressed as follows [19, 20]:

$$Q_m = \Lambda + \sum I_i - n\phi - 3.6n^2/r - 3.6n^2/d - nrV/(r+d) \quad [eV] \qquad (13.1)$$

where Λ is the binding energy of the atom of the probe, $\Sigma I_i - n\phi$ is the ionized energy of a probe atom with n^+ charge, ϕ is the work function of the surface, r_0 is the

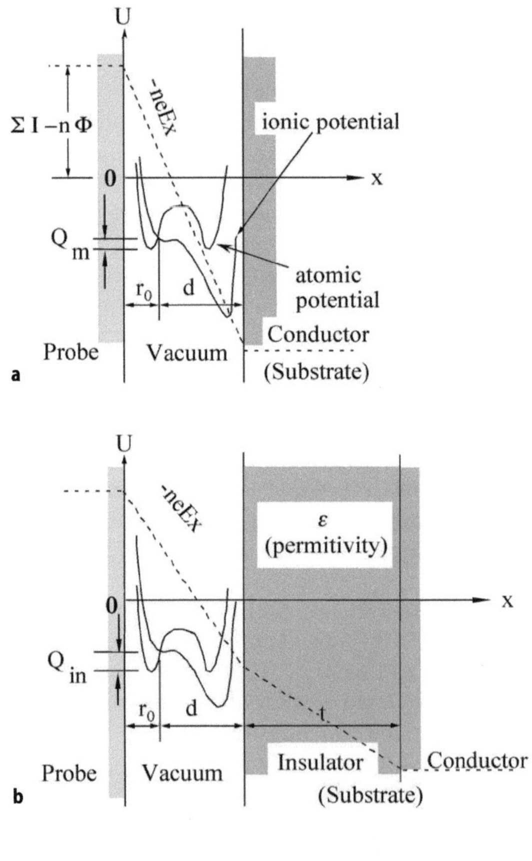

Fig. 13.2a,b. Energy diagrams of **a** the STM (AFM) metal probe/vacuum/conductor configuration and **b** the AFM metal probe/vacuum/insulator/conductor configuration

distance between the surface and the hump position, and V is the voltage applied between the probe and the substrate. The fourth and fifth terms on the right-hand side are the image potentials for the probe and substrate, respectively, and the last term expresses the field effect. With a negative charge, $\Sigma I_i - n\phi$ is replaced by $\phi - E_{aff}$ as described in Tsong's works [19, 20], where E_{aff} is the electron affinity of the atom.

(2) AFM configuration

The AFM configuration using an insulator/conducting substrate can be represented by the potential energy diagram shown in Fig. 13.1b. A hump Q_{in} in the potential is assumed to be at a distance r, as in the above STM configuration. According to Eq. (13.1), assuming $d \ll t$, Q_{in} can also be represented as follows:

$$\begin{aligned} Q_{in} = \Lambda + \sum I_i - n\phi - 3.6n^2/r_0 - \left(3.6n^2/d\right)(\varepsilon - 1)/(\varepsilon + 1) \\ -3.6n^2/(t \cdot \varepsilon) - nr_0 V/(r_0 + d + t/\varepsilon)\,[eV] \end{aligned} \quad (13.2)$$

where ε is the permittivity of the insulator, and t is its thickness. The fifth and sixth terms are the image potentials for the insulator and the conducting substrate, respectively. The last term represents the field weakened by the presence of the insulator.

A change in the hump height ΔQ, induced by inserting the insulator between the vacuum and the conductor, can be estimated by $Q_{in}(V) - Q_m(V)$ as follows:

$$\begin{aligned} \Delta Q = \left(3.6n^2/r_0\right)\left\{[2t\varepsilon - d(\varepsilon + 1)]\big/[dt\varepsilon(\varepsilon + 1)]\right\} \\ + nr_0 V\,(t/\varepsilon)\big/\{d_0\,[d_0 + (t/\varepsilon)]\} \end{aligned} \quad (13.3)$$

where $d_0 (= r_0 + d)$ is the tip-sample separation. The first term on the right side indicates that the barrier height of the image potential induced by the insulator/conducting substrate is larger than that in the metal/vacuum/conductor configuration. The second term indicates that the electric field is weakened by inserting the insulator into the metal/vacuum/conductor configuration.

On the other hand, according to the hump height Q (Q_{in} or Q_m), field evaporation takes place at the rate r, as

$$r = \nu \exp(-Q/(kT)) \quad (\nu\text{: lattice vibration}) \quad (13.4)$$

(3) Calculated results

Figure 13.3 shows the field evaporation voltage V as a function of tip-sample separation d in Au^-, Au^+, Pt^+, and Pt^{2+}, when permittivity of the insulator is about four, thickness of the insulator is 15 Å, r_0 is 2 Å, the evaporation rate is $10^3\ s^{-1}$, and the temperature is 300 K. The calculations are carried out using Eqs. (13.1), (13.2), and (13.4).

From the figure, the field evaporation voltage (absolute value) of Au^- is smaller than that of Au^+, while that of Pt^{2+} is smaller than that of Pt^+ in another calculation. These results agree with the STM configuration results calculated by Tsong. Furthermore, from the comparison of these results with experimental threshold voltage shown in Fig. 13.5, the tip-sample separation is considered to be about 4.5 Å. This is because experimental threshold voltage is –8 V, which is applied to the probe against the sample. In the d_0 region above 4.5 Å, the separation renders field

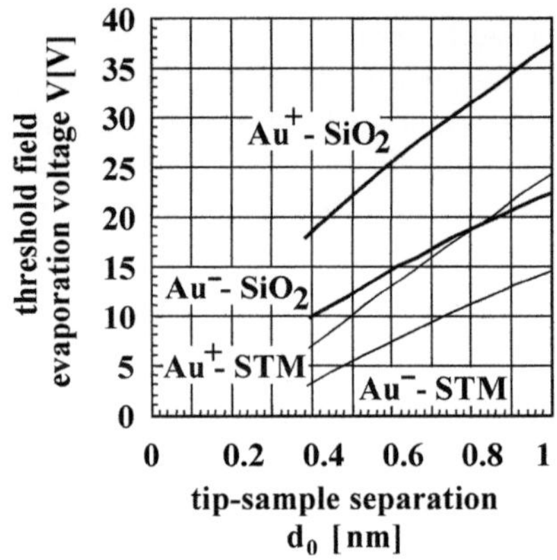

Fig. 13.3. Calculated results of threshold field evaporation voltage for tip-sample separation in both Au ions in STM configuration (Au+ –STM) and in AFM-SiO_2/conductor configuration (Au+ –SiO_2); permittivity = 4, evaporation rate = $10^3\ s^{-11}$ at room temperature, $r_0 = 2$ Å

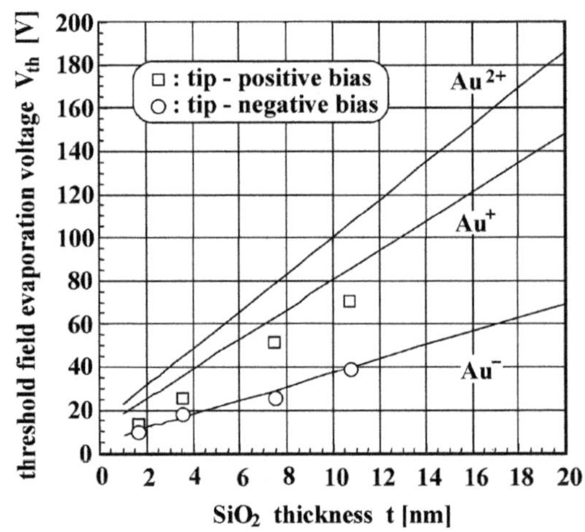

Fig. 13.4. Experimental results and calculated results of relationship between threshold field evaporation voltage and SiO_2 thicknesses in AFM gold probe-SiO_2/conducting substrate configuration; □, ○ experimental results, and *lines* in calculation

evaporation voltage larger than that when the d_0 is 4.5 Å. This means that (1) the tip-sample separation is about 4.5 Å in this AFM operation, and (2) the tip-sample separation larger than 4.5 Å corresponds to the metal removal from the AFM tip. When the tip-sample separation is 12 Å, as caused by xy-scanning, z-controlling, or anomalously large field evaporation, the field evaporation voltage needed for Au^- increases to 25 V. As mentioned above, slightly removing the covered metal from the tip of the AFM probe increases the threshold voltage.

Figure 13.4 shows changes in the field evaporation voltage V for various insulator thicknesses in the cases of Au^-, Au^+, and Au^{2+}. These changes were calculated at an evaporation rate of $10^3\ s^{-1}$ and a tip-sample separation of 4.5 Å at room temperature.

As shown in the figure, field evaporation voltage needs to be large as the thickness increases. In Au^-, the voltage is 40 V when the insulator is thinner than 10-nm thick, while V is less than 80 V in Au^+. These results agree with the experimental results. Moreover, in regards to the permittivity, the voltage decreases as the insulator's ε value increases. When silicon dioxide ($\varepsilon = 3.9$) and silicon nitride ($\varepsilon = 7$) are used as the insulator, the voltages of Au^- and Pt^{2+} at SiO_2 insulator with

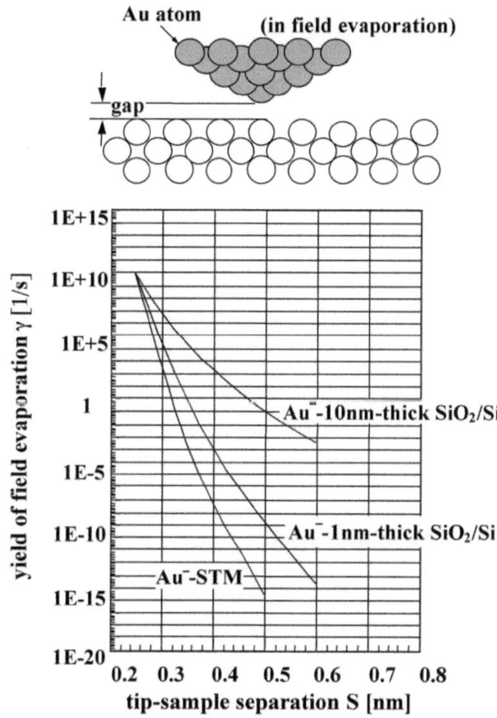

Fig. 13.5. Calculated results of yield of field evaporation as a function of tip-sample separation in the STM configuration and the AFM configuration with 10-Å-thick SiO_2 insulator, permittivity of 3.9, and a tip-sample separation of 4.5 Å at room temperature

the thickness of 10 nm are about 35 V and 85 V; they are about 20 V and 50 V at Si_3N_4 insulator. Consequently, it is easy to make metal dots on an insulator with an insulator thickness of less than 10 nm.

Figure 13.5 shows field evaporation yield against tip-sample separation in both STM and AFMconfiguration.

The evaporation yields increase rapidly as the separation decreases. In the STM case, the gradient is about 10^6 s^{-1} $Å^{-1}$ while it is 10^5–10^3 s^{-1} $Å^{-1}$ in the AFM case in a region of the separation of 4–5 Å. This indicates that the slight addition of voltage to the threshold voltage exponentially increases the field evaporation rate. The result predicts that it is possible to manipulate a single atom at a tip of the probe or a single atom of the sample surface at the opposite side of the tip by field evaporation because the field acts only on the atom following high evaporation rate.

These calculated results also indicate that (1) metal dots can be made on a thin insulator/conducting substrate by field evaporation using the AFM configuration, (2) it is easier to form metal dots on a higher-permittivity insulator, and (3) it is possible to field-evaporate the metal on the insulating layer at high evaporation rates.

13.2.1.2
Experimental Results

The experiments were carried out using a home-made AFM [21] with optical-beam deflection for the atomic force. We used a thin-gold-film-covered AFM probe as

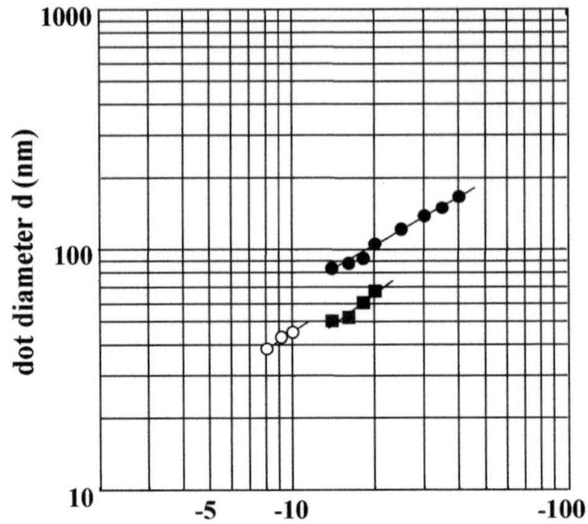

a

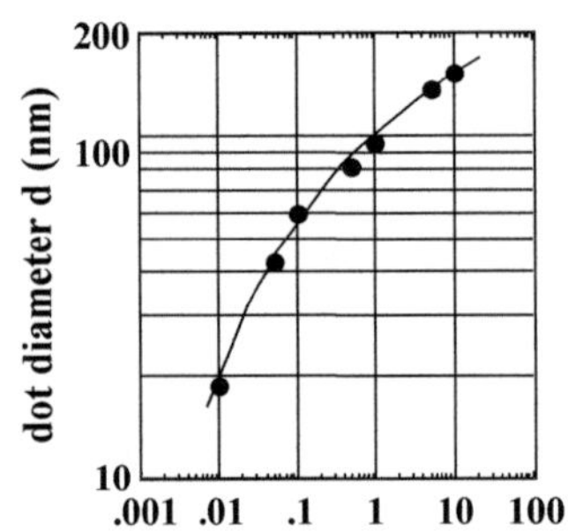

b

Fig. 13.6. Experimental results of dot size dependence on pulsed voltage and pulsed width

the field evaporation source. Gold was selected because it is inert and the required field evaporation voltage is low, as described above. The probe was prepared by vacuum evaporation of thin gold film over a SiO_2 birdbeak-type cantilever [21] fabricated by a Si microprocess. The sample was a piece of Si wafer with a thin insulating layer of natural SiO_2 film on the substrate. The SiO_2 film was about 17-Å thick, as measured with an ellipsometer. The AFM has a wide-range scanner covering about 6 µm-cubic at 100 V. For both AFM observation and metal dot formation, the constant repulsive force control was set to around 5×10^{-9} N.

The evaporated dot size increases with the applied voltage (Fig. 13.6). In the figure, the diameter is defined as that at half the height of the dot.

The figures shows some results using a few probes. Threshold voltages are dispersed from –8 V to –12 V. This is caused by changes in the tip structure of the probe. According to the scanning electron microscope (SEM) observations, the tip structure changes upon removal of the gold film at the tip. The separation makes the threshold voltage large. Figure 13.3 indicates that additional voltage of about 2 V $Å^{-1}$ to the ideal threshold voltage for the separation is needed in the metal/vacuum(4.5 Å)/insulator(15 Å)/conductor system. If the ideal threshold voltage is –8 V, the removing length from the tip in the removal of the metal covered on tip

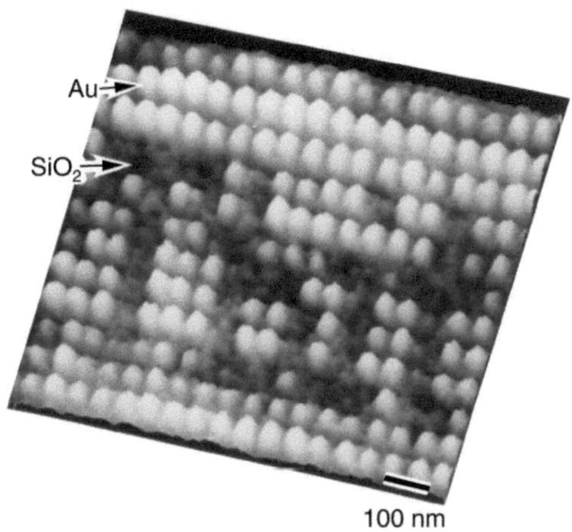

Fig. 13.7. AFM image of nanometer gold dot pattern forming the words 'Tbit' (Trillion bit) on SiO_2/Si surface using field evaporation in concave pattern at applied voltage of –30 V for 5 ms, which was observed with the same gold-film-covered AFM probe

is about 2 Å in the initial evaporation voltage of –12 V (Fig. 13.6). These theoretical and experimental results indicate that it is possible to field-evaporate metal atoms in the AFM metal probe/vacuum/thin insulator/conductor configuration. However, the removal of metal from the tip poses serious problems in the stable formation of metal dots on a thin insulator. It causes the threshold voltage to increase or stops field-evaporation of the metal atoms.

On the other hand, we confirmed that selected probes could make stable nanometer-sized gold dots with the shots of more than 10^6 times. Figure 13.7 shows an AFM image of nanometer-sized Au dots on the insulator of a natural SiO_2/Si substrate after applying the pulsed voltage.

The experiments were carried out by applying a voltage of –30 V with 3-ms pulse width to the probe. The image indicates that field evaporation in the AFM configuration can actually make gold dots on an insulator/conducting substrate. We can obtain ultra-small gold dots less than 15 nm in diameter on the insulator. Furthermore, we confirmed that gold dots can be obtained at a pulse width of 10 μs. This result and Fig. 13.5 clarify that it is possible to make gold dots on thin insulating layer at rates higher than 10^5 Hz.

13.2.1.3
Summary

The feasibility of forming a nanometer-sized bit structure on both insulating or conductive substrate has been studied both theoretically and experimentally. It is clear that in both STM and AFM configurations, field evaporation can occur when voltage of less than 100 V is applied. Furthermore, we have demonstrated that ultra-small metal dots were formed on a natural SiO_2/Si substrate with a threshold voltage of less than 10 V, and that ultra-small gold dots with 15-nm diameter were made on the insulator. This technique should facilitate (1) ultra-high recording with a density of more than 1 Tbit in^{-1} in air, and (2) direct writing of nanometer-sized patterns for use in electronic quantum devices.

13.2.2 Nano-Indentation of a Probe Tip into a Sample Surface

We can use not only a strong electric field but also a fine force of AFM system for surface modification. Using this force, nanometer-sized pit structures can be formed in a sample surface. This is due to nanometer-sized plastic deformation, which is called as nano-indentation. This method is quite simple. The nano-indentation technique using AFM was introduced by T.A. Jung et al. and E. Hamada and R. Kaneko, who demonstrated nano-pit formation with a hole of 50–100 nm in diameter [22, 23]. This technique can be applied to make a nanometer-sized fine pit of less than 25 nm.

13.2.2.1 *Force Modulation of AFM Probe for Nano-Indentation*

The small pit structure is made by cold plastic deformation (nano-indentation) for the modification process. This was done by pushing the AFM tip into a sample surface with the piezo actuator. With this method, the force can be precisely controlled at a resolution of a few nN. Here, a polycarbonate disk is used as an organic material of sample. This is because we need a material which has a small yield stress, so that we can deform the sample surface even at a few tens nN. The fine force makes a small pit structure as a result of using a fine tip of the AFM probe.

13.2.2.2 *Nano-Pit Formation Using Nano Force Modulation [24]*

Figure 13.8 shows a simple model and an equivalent model of the probe and sample. When the cantilever is pushed at a distance of X_0 from the sample, the tip penetrates the surface to a depth of X.

The working force of the tip can be measured from the deflection of the cantilever. The important points are a spring constant and a dashpot of the sample surface in the model. When the spring constant and the dashpot are k^* and η, respectively, we can obtain the following equations.

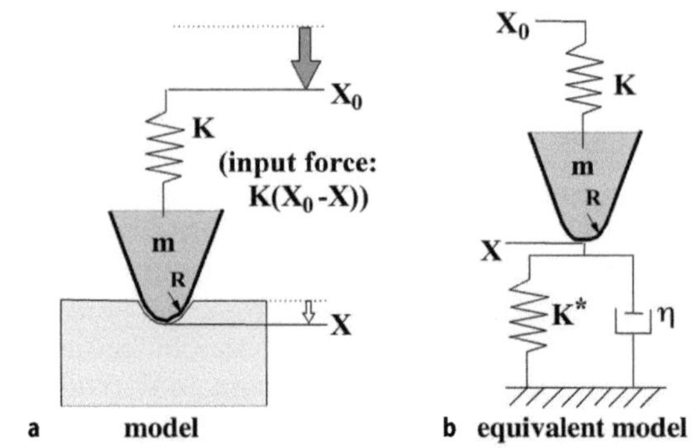

Fig. 13.8. Schematic diagram of pushing cantilever tip by external piezo device and the tip penetration into sample surface; **a** simple model and **b** equivalent model

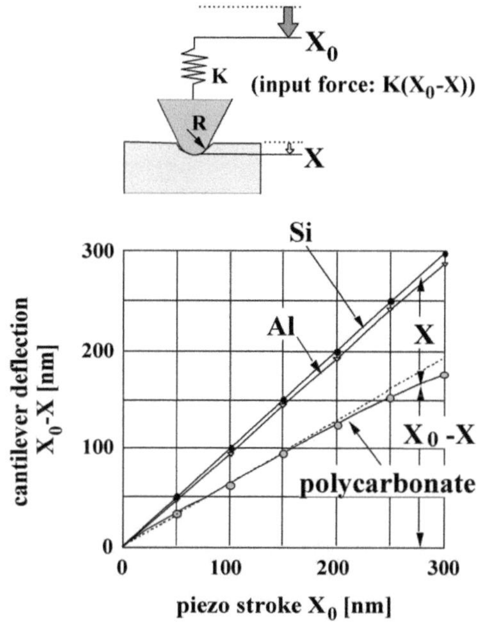

Fig. 13.9. Cantilever deflection due to piezo stroke for tip penetration into sample

$$m\ddot{X} = k\,(X_0 - X) - k^*X - \eta\dot{X} \tag{13.5}$$

$$\frac{X}{X_0} = \frac{k}{\left(k^* + k + \eta S + mS^2\right)} \qquad (S = j\varpi t) \tag{13.6}$$

Assuming that the time transient of the tip movement can be neglected at a frequency of less than 10 MHz ($(k^* + k) \gg (\eta S + mS^2)$), the tip penetration depth can be given as follows.

$$X = \frac{k}{(k^* + k)}X_0, \quad X = \frac{k}{k^*}\,(X_0 - X) \tag{13.7}$$

Equation (13.7) shows that the tip penetration depth is in proportion to the tip working force and that it is in inverse ratio to a sum of spring constants of the tip and the sample surface in a region of elastic deformation.

The relationships between the cantilever deflection $X_0 - X$ and the piezo stroke X_0 are measured as shown in Fig. 13.9.

We performed the experiments in air using silicon(001), 1-μm-thick aluminum on silicon and polycarbonate as the samples. The tip is silicon nitride. The cantilever deflection is the same as the piezo stroke in the case of a silicon surface. On the other hand, the cantilever deflections of aluminum and polycarbonate gradually become less than the piezo stroke. In particular, the cantilever deflection of polycarbonate is very much smaller than those of silicon and aluminum. This means that silicon, or aluminum, has a hard surface but polycarbonate has a soft surface. The tip penetration depths shown in Fig. 13.10 were obtained from these curves.

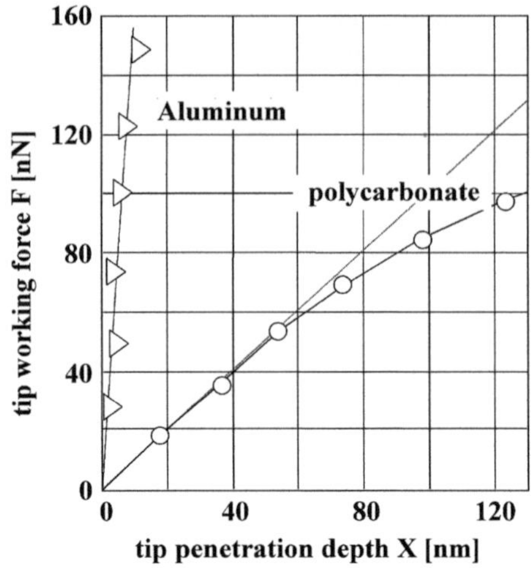

Fig. 13.10. Variations of tip working force for tip penetration into sample surface (tip: silicon nitride)

The depth in the aluminum surface is less than 10 nm when the working force is less than 100 nN. In the polycarbonate surface, however, the tip penetrated into the surface to a depth of 40 nm at a working force of 40 nN. The elastic region may be up to 40 nN in the polycarbonate. The working force is gradually separated down with the tip penetration depth. This means that the critical force for plastic deformation is about 40 nN. In the elastic region, the working force is in proportion to the penetration depth. From this figure, the spring constant of the polycarbonate surface is about 1 N m^{-1}, and the relationship between the tip penetration and the working force follows Hooke's law, but not Hertz theory.

Figure 13.11 shows variations in the pit size on the polycarbonate surface due to a pulsed force for a pulse width of 6 ms. The pit size can be controlled by the pulsed force.

From the figure, the minimum size is less than 10 nm in diameter and the threshold force for plastic deformation is about 40 nN. The threshold force agrees with the critical force in Fig. 13.10. The plastic deformation occurs at a force of larger

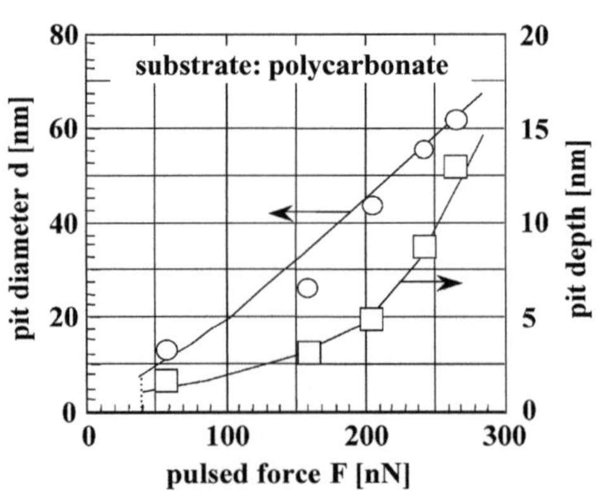

Fig. 13.11. Variations of pit diameter and depth against pulsed force applying to the AFM tip by plastic formation

than 40 nN and it can be used for writing nano-pits to achieve a trillion-bits-ultra-high density recording.

13.2.2.3
Summary

A nano-pit structure formation using nano-indentation (cold plastic deformation) by force modulation of AFM probe is proposed and the possibility to form fine pit structure with a diameter of less than 10 nm in polycarbonate surface is demonstrated. We reached such a conclusion as this technique is expected to be an attractive candidate for future ultrahigh-density recording. And we achieved the following:

(1) A minimum pit size of around 10 nm in diameter was formed by cold plastic deformation of polycarbonate disc surface.
(2) The tip penetration depth into the polycarbonate surface was over 40 nm when the plastic deformed.
(3) The nano-pitting in the polycarbonate surface was formed at a force of over 40 nN.

13.2.3
Electron Induced Local Heating and Evanescent Light Heating

These heating processes using scanning probe microscopes (SPMs) are applied to control a magnetization and a material phase in a sample surface. We can expect these techniques as a candidate of ultrahigh-density rewritable data storage with magnetization switching and phase changing for the stable and reliable memory operations.

Here, the heating process is considered using tunneling current or field emission current and evanescent light, and discussed in terms of temperature rise and a localized area of the sample surface.

13.2.3.1
Theoretical Consideration

The temperature distribution is calculated by the thermal diffusion Eq. (13.8) using the finite element method to find how high the temperature rises in the sample.

$$C \times \rho \times \partial T\,(x, y, z, t)\,/\partial t = div\,(\kappa \times gradT\,(x, y, z, t)) + Q\,, \tag{13.8}$$

where C is the specific heat, ρ is the density of the sample, T is the temperature, κ is the thermal conductivity, and Q is the writing power. In the experiment, the writing pulse width is kept constant at a few µs, which is long enough to saturate the heating procedure. The time differential of the temperature might be negligible under the above condition. Thus the Eq. (13.8) is simplified to the Eq. (13.9)

$$div\,(\kappa \times gradT\,(x, y, z)) + Q = 0\,. \tag{13.9}$$

By varying the amplitude and position of applied writing power Q, the temperature profile of a sample surface is calculated by the finite element method with the differential equation solver language (DEQSOL) [25] program.

13.2.3.2
Electronic Induced Heating

Before the calculation of Eq. (13.9), we have to estimate the input power Q and its area distribution as follows. First, the tunneling current is estimated by integrating the current density calculated by tunneling equation [26]. The applied power is estimated using this current. Second, the applied power distribution is assumed to be Gaussian. And third, the sample surface is heated during the electrons are scattering in the sample; thus, the standard deviation of the applied power distribution can be assumed to be same as the mean free path of the electron. We vary it from 30 nm to 50 nm in the present calculation. Based on this model, the value of Q can be estimated to be varied from 0.68 mW to 3.9 mW by applying pulsed voltages from 1.5 V to 2.3 V.

Figure 13.12 shows the calculation of the temperature rise in the sample. The diameter of the heated area higher than *Tc* becomes smaller with decreasing the amplitude of the pulsed voltage. It has been found that the diameter of the area becomes less than 100 nm for the applied pulsed voltage less than 1.8 V. Besides, the temperature rise of the heated area depends on the standard deviations of the applied power distribution.

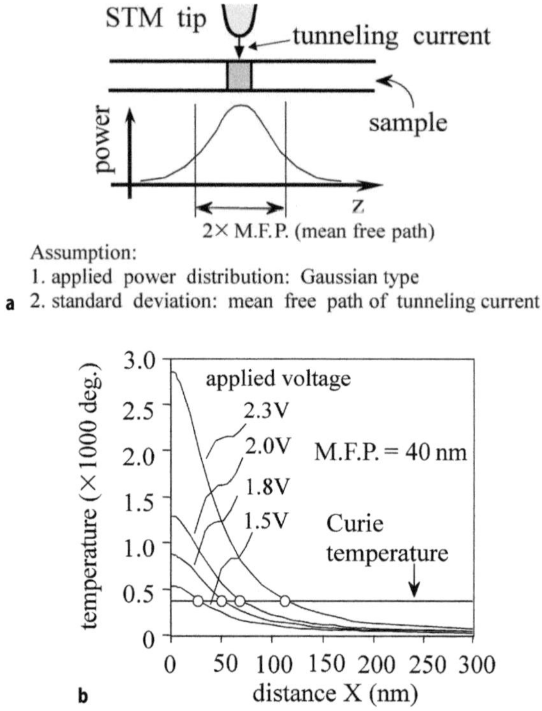

Fig. 13.12. Temperature rise in the sample when applying the current to sample in the STM; **a** model and **b** calculated result

13.2.3.3
Switching a Magnetization [27]

The experiment can be performed by STM, AFM, and MFM for switching the magnetic domains. The Co/Pt multilayered film with perpendicular magnetization was used for the sample. The film is suitable for the present experiment because its surface covered with Pt layer has a good conductivity and the tunneling current can be stably flowed even in air. The characteristics of the sample are as follows: the total thickness of the magnetic film is 15 nm (Pt: 0.3 nm/Co: 0.83 nm, 13 layers), the saturation magnetization is 0.6 Wb m^{-2}, the coercive force is 40 kA m^{-1}, and the Curie temperature (Tc) is 380 °C. The formation of the magnetic domains was carried out as follows: (i) the sample was at first initialized by an external perpendicular magnetic field (15 kOe); (ii) the STM probe was approached to the sample and the spacing gap was kept constant by setting the bias voltage and tunneling current 0.2 V and 0.3 nA, respectively. In case of MFM and AFM, we operate them at a force of 5 nN in contact mode; (iii) with the probe held at constant height, the domain formation was carried out by applying pulsed voltage between the tip and the sample. The pulse width was fixed 1 ms and the amplitude of the pulsed voltage was varied from –5.0 V to 8.0 V. These values depended on the conditions of setup and sample. Finally, we observed the magnetic domains and the surface morphology of the sample using a polarized optical microscope or the MFM and the SPM. All the experiment was done in air.

Figure 13.13a shows the polarized optical microscope images of the magnetic domains formed by applying the pulsed voltage to the STM tip. The amplitude of the

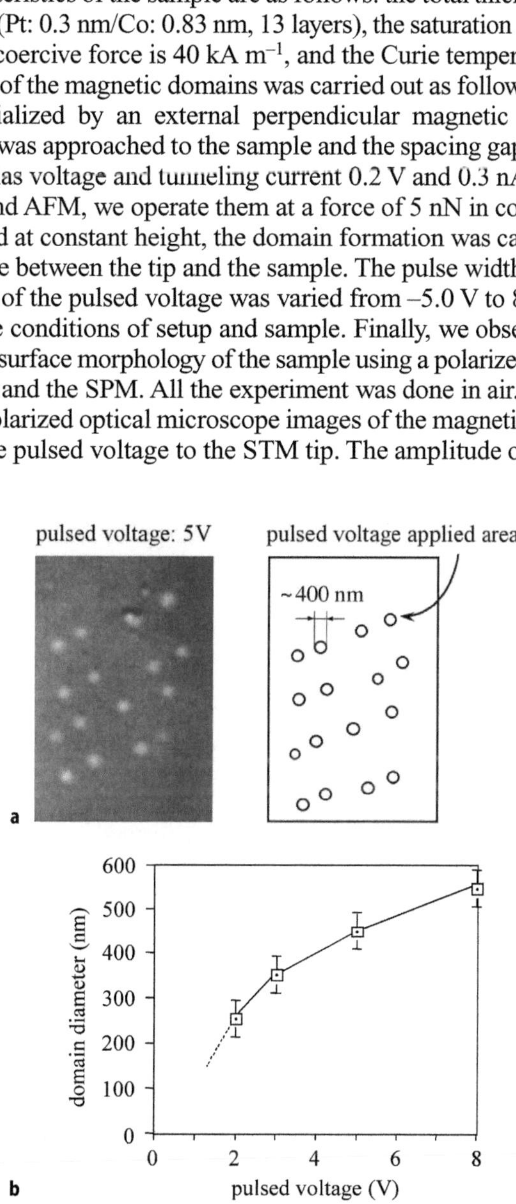

Fig. 13.13a, b. Magnetized domain diameter dependence on pulsed voltage in the STM with MO film (Pt/Co multilayer); **a** polarized optical image and **b** variation of domain diameter (a diameter of less than 0.2 μm cannot be observed by the optical microscope)

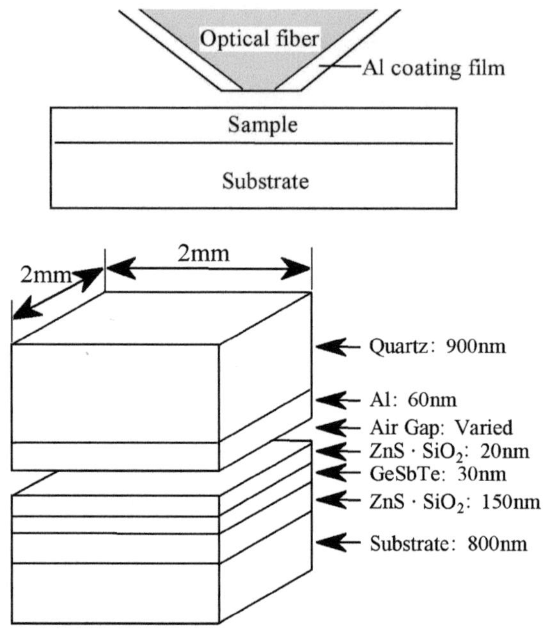

Fig. 13.14. **a** Schematic diagram of optical fiber tip and the sample; **b** simplified model of the tip with detailed structure of the sample

pulsed voltage is 5.0 V. The bright points shown in the figure are magnetic domains because they changed to be dark ones by changing the angle of the polarizer, and the spotted patterns could be all erased by the application of large external perpendicular magnetic field. Figure 13.6 shows the dependence of the domain diameter on the amplitude of the pulsed voltage. The magnetic domains were formed by the pulsed voltages with both polarities. By decreasing the amplitude of the pulsed voltage, the diameter of the magnetic domain becomes smaller and magnetic domains of about 200-nm diameter can be obtained in the case of applying a 2.0 V pulsed voltage. The domains obtained by the present experiment are larger than the heated area higher than Tc estimated by the calculation. However, both experiment and the calculation have good qualitative agreement on the characteristic of the dependence on the amplitude of the pulsed voltage. This indicates that the magnetic domains smaller than 200 nm diameter can be formed by the application of the pulsed voltages less than 2.0 V. We also observed the sample surface by STM just after the application of the writing pulsed voltage. No surface modification could be observed by applying a pulsed voltage of below 3.0 V, and only magnetic domain formations could be observed. Furthermore, we could confirm small magnetic domains using the MFM setup as describe later.

13.2.3.4
Evanescent Light Heating in SNOM [8, 28]

In SNOM, the switching a magnetization or a material phase is performed by point heating of samples using the light energy from the fiber tip. Here some types of heating mechanisms can be considered: evanescent light power, transmission light power of far-field light, thermal conduction from the tip to the sample, and the radiation power from the tip. Among them, evanescent wave power and thermal conduction are considered to be main causes of a sample heating. The roles of the other two mechanisms are negligible. This is because transmission light power is very small under the condition that the aperture at the tip is more than 10 times smaller than the wavelength of the incident light, and the radiation power calculated using the Stefan-Boltzmann law of radiation is also small.

An investigation of the contributions of the two mechanisms, evanescent light power and thermal conduction, is necessary to optimize conditions of the SNOM operation and thermal designs of the sample. We focus on this point and discuss two mechanisms in phase-change materials. For the evaluation of the temperature increase, we also use the thermal diffusion Eqs. (13.8) and (13.9).

For simplicity, the optical fiber tip of SNOM is modeled in the following way. Figure 13.14a shows a schematic diagram of the SNOM tip made of an optical fiber. The tip which is sharpened by pulling the fiber during CO_2 laser heating is considered. It was then coated with thin Al films by angled evaporation, resulting in the formation of a small aperture at the tip. As shown in Fig. 13.14b, tip structure is simplified to a 900-nm-thick quartz layer and a 60-nm-thick Al layer for the benefit of calculation, which represent the optical fiber and the Al film, respectively.

The sample is a phase change recording medium consisting of $ZnS \cdot SiO_2$ (protection layer)/GeSbTe (phase change material)/$ZnS \cdot SiO_2$ (protection layer) thin films. The thicknesses of these films are 20 nm, 30 nm, and 150 nm, respectively. The phase of the GeSbTe layer changes from amorphous to crystalline when heated to higher than 180 °C. The 800-nm-thick polycarbonate is used for the substrate. The lateral size of the calculated area is 2 mm × 2 mm. This sample structure is the same as the one used by Hosaka et al. [8] in their experiment, where 60-nm-sized phase change domains were formed by applying laser light with the power of 7.3 mW and the pulse width of 5 ms. The values for the thermal conductivity of each layer are given in Table 13.1. The aperture is assumed to be 60 nm in diameter.

The evanescent light power is absorbed only by the GeSbTe layer, because the air gap layer and the $ZnS \cdot SiO_2$ layers are transparent and thus no absorption occurs. For the incident laser light with 780 nm, the absorption rate of the GeSbTe layer is calculated to be 0.35 by solving Maxwell's equations based on the electromagnetic boundary conditions of the film [29]. Here we assume that the power distribution is Gaussian and its deviation is a half of the tip aperture (i.e., 30 nm), because, in

Table 13.1. Thermalconductivity of each layer

Layer	Thermal conductivity (W mK^{-1})
Quartz	1.9
Al	240
Air gap	0.024
$ZnS \cdot SiO_2$	0.66
GeSbTe	0.58
Substrate	0.22

our calculation, the distance between the tip and the GeSbTe film varies from 30 nm to 50 nm (thickness of the air gap layer plus the protection layer), which is larger than the radius of the aperture, and the power distribution of the evanescent wave at this distance from the tip is similar to the Gaussian calculated by Kawata using Bethe's theory [30]. On the other hand, in the thermal conduction, the incident light power heats the tip and the sample is heated in turn by thermal conduction from the tip. The incident light power which heats the tip is assumed to be localized in the 60 nm × 60 nm × 60 nm aperture region of the Al layer. By applying these powers separately, temperature increases in the sample are calculated.

Considering heat sources for the phase change, we must estimate the temperature of the heated area in the GeSbTe film. In our experiments, the phase change occurred when the temperature of the recording film was over 180 °C. This is because the 5-ms-pulse width of recorded laser is sufficiently long to change the film phase from amorphous to crystal, since the crystal growing time is estimated to be about 2.3 ns, which is deduced from the temperature dependence of the crystal growing speed (Arrhenius plot).

Figure 13.15 shows the calculated result of the temperature increase at the center of the phase change layer (GeSbTe) by direct evanescent light heating. The gap between the tip and the sample is set at 10 nm. These conditions were the same as those in the experiments [6]. This result indicates that the sample can be heated to only 100 °C, even for the transmission rate of 0.005 which is rather high for evanescent light transmission of the aperture. Therefore, it is difficult to change phases of the sample from amorphous to crystalline by applying only evanescent light power.

On the other hand, we have melted the aluminum at the tip of the probe with incident power larger than 13 mW. We also calculated the temperature rise of the aluminum layer of the tip, assuming that the metal thickness is 60 nm and the light

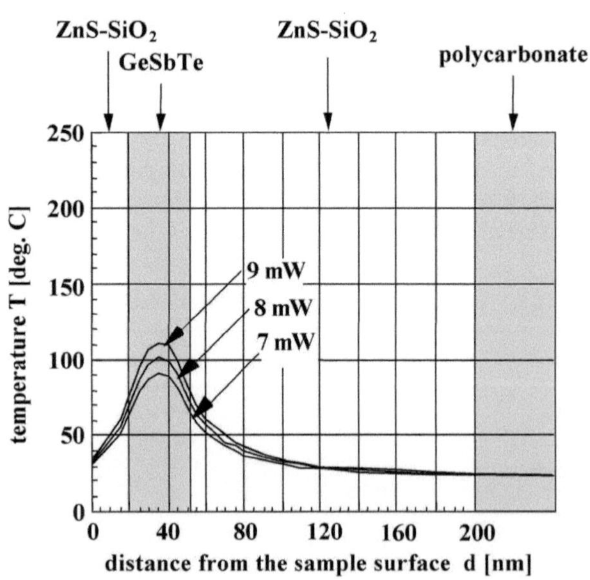

Fig. 13.15. Calculated temperature of the recording media due to evanescent light illumination (optical mode), where indicated power are incident powers into the probe

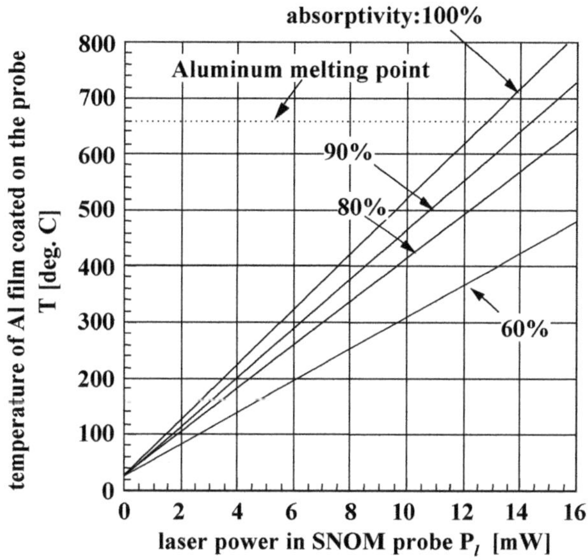

Fig. 13.16. Calculated temperature rise of Al layer of the probe induced by laser illumination into the probe

illuminated spot size is about 60 nm in diameter. Figure 13.16 shows the temperature rise versus laser power for different values of light absorptivity of the tip structure. When the absorptivity is near 100%, aluminum melting occurs at a laser power of about 12.8 mW, which is consistent with the experimental data as described above.

Also, the absorptivity of 100% agrees well with the experimental result that we did not detect the reflected light from the tip through the probe fiber. The calculation suggests that the SNOM tip is heated by the laser illumination, and heat transfer from the tip to the recording film raises the temperature of the recording film. When writing at the power of 7.3 mW, the tip temperature reaches 400 °C. Therefore, the temperature rise for the phase change is supported by heat transfer from the tip to the recording film. At 7.3 mW, temperature increments due to the light illumination (optical mode) and thermal conduction (thermal mode) are estimated to be about 70 °C and 90 °C, respectively.

As mentioned above, the evanescent light heats the sample, and the thermal conduction from the heated tip heats the sample simultaneously.

13.2.3.5
Summary

We have proposed some of the new methods to form the small domains and found that it is possible to make magnetic domains using the tunneling current. The calculation indicated that the heated area could be smaller than 100 nm in diameter. Both calculation and the experimental results have good qualitative agreement on the characteristic of the dependence on the amplitude of the pulsed voltage. This suggests the possibility of forming the magnetic domain smaller than 100 nm in diameter. This method can be one of the candidates for the new data storage technique with ultrahigh density.

And we have calculated the temperature of the local area by illuminating near-field light on the sample. The calculation has proved that the local area is heated by both photon mode and thermal mode so that the phase change material is changed to opposite phase.

13.3 Various SPM Storages [31]

13.3.1 STM-Based Storage

STM-based storage incorporates field evaporation or tunneling current induced heating for writing, and modified surface detection for reading. Field evaporation technique can be used to write ultimately small bits with a single atom as information units, which corresponds to 6 Pb in^{-2} [3]. Figure 13.17 shows the schematic diagram of the STM-based storage using field evaporation and an STM image of the field-evaporated surface of cleaved MoS_2.

In Fig. 13.17a, the STM probe was controlled at 0.3 nm above the surface for field evaporation of S-atom (writing), or 1 nm above the surface for STM observation (reading). Figure 13.17b shows an STM image of a small message formed with vacancies of the S-atoms, which were detached by applying a pulsed voltage

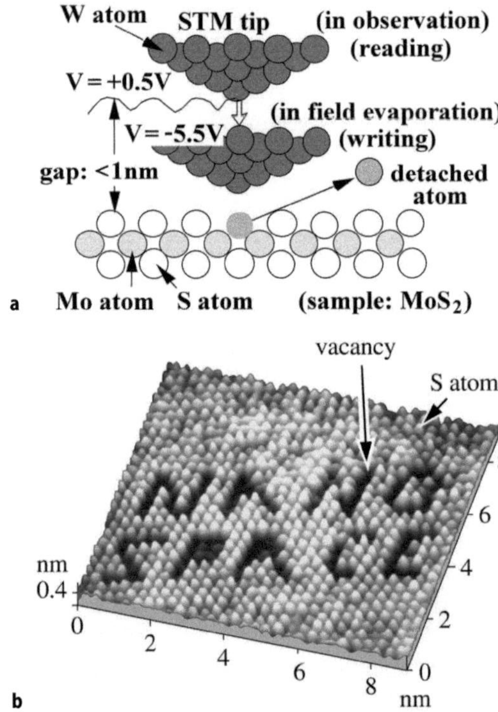

Fig. 13.17. Schematic diagram of STM-based storage; **a** writing and reading, and **b** STM image of cleaved MoS_2 surface after sulfur atom detachment from the surface using STM-field evaporation

of –5.5 V for 70 ms to the probe. The image suggests that it is possible to write 0.3-nm-sized bits by using a vacancy as an information unit.

Other methods using electron induced heating have been proposed for writing nano-holes on graphite [1] and nano-sized highly conductive domains.

13.3.2 AFM-Based Storage

13.3.2.1 *Field Evaporation-Type AFM Recording*

AFM-based storage incorporates field evaporation or crystal modification for writing, and modified surface detection for reading. The technique has 2 major features to modify the insulating surface and to compensate for the servo error with an AFM cantilever spring. The AFM-based storage proposed is shown in Figs. 13.18a and b. Figure 13.18c shows an AFM image of field evaporated Au dots with diameters of 20–40 nm. In the experiment, the smallest gold bit obtained on a thin SiO_2/Si substrate in air has a diameter of 15 nm, which corresponds to 6 Tb in^{-2} [4].

13.3.2.2 *Nano-Indentation-Type AFM Recording*

This method uses plastic deformation of a sample surface when applying a small force to the surface by the AFM tip as shown in Fig. 13.19a. So far, previous work has reported that this method has a potential to write a small pit with a diameter of around 50 nm. The diameter is insufficient to achieve an ultrahigh-density recording. It can be considered that the sample to be deformed was metal surface, so that it has a problem not to make a fine pit structure in the surface. In such an experiment, a force of more than 1 μN was needed for plastic deformation of the metal. In the force region, a contact area becomes larger than the tip radius based on Hertz theory. Figure 13.19b shows small pit array with a diameter of 20 nm, which corresponds to more than 1 Tb/in^{-2}. This method has an ad-

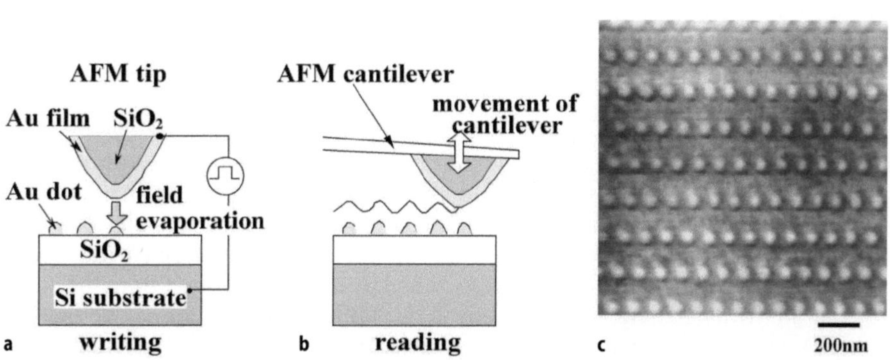

Fig. 13.18a–c. Scheme of AFM-based storage using field evaporation; **a** writing (field evaporation), **b** reading (AFM observation), and **c** AFM image of Au dots formation on a thin SiO_2/Si

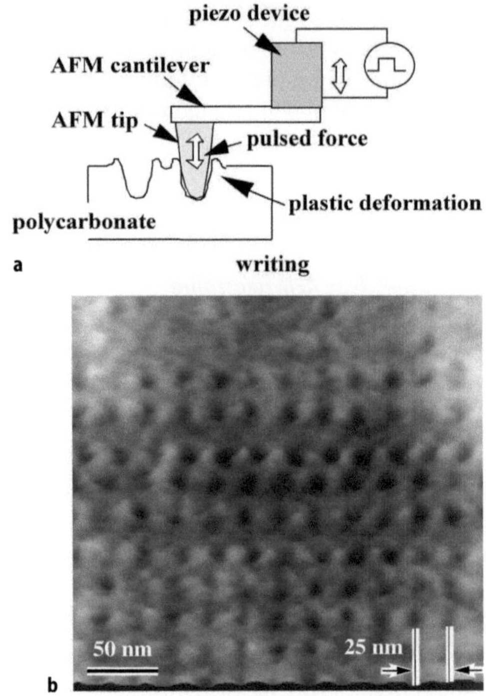

Fig. 13.19. **a** Scheme of force modulation AFM pit recording and **b** AFM image of the indented pit array on polycarbonate

vantage in stability and reliability for writing compared with other SPM-based storage [24].

13.3.3 MFM-Based Storage

MFM-based storage incorporates nano-magnetic domain formation with a bias magnetic field [5] or a tunneling current induced heating [6] for writing, and detection of magnetic force gradient induced from the formed domain for reading. This technique controls magnetization in the recorded film, and can be used for rewritable recording. The smallest domain size obtained in the bias magnetic field was 200 nm, but the domain in the heating was 60 nm × 240 nm. These sizes are equivalent to 15 Gb in^{-2} and 40 Gb in^{-2}, respectively.

The method proposed is shown in Fig. 13.20. This method uses the current induced heating for writing and the MFM observation for reading. In writing, the MFM is operated as contact-type AFM to induce the current to flow from the probe to the sample. The MFM probe used consisted of SiO_2 base, a PtCoCr-alloyed film used as a magnetic probe, and a carbon film to make the probe conductive as shown in Fig. 13.20a. The magnetization was switched by demagnetization from the current heating with magnetic field bias due to the MFM probe. Figure 13.20c shows an MFM image of the switched magnetic domains of a Pt/Co multilayered film. The size (60 nm × 240 nm) can be estimated by MFM image interpretation. These dimensions indicate that forming a 60 nm-sized domain (170 Gb in^{-2}) is possible.

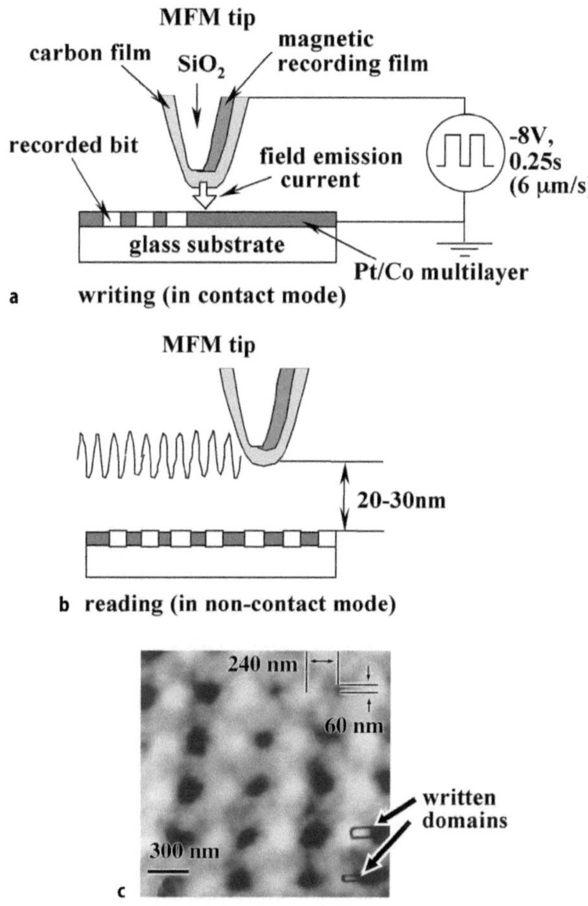

Fig. 13.20a–c. Schematic diagram of MFM based storage; **a** electron induced local heating (writing), **b** MFM observation (reading), and **c** MFM image of written domains

13.3.4 SNOM-Based Storage

SNOM-based storage incorporates magnetic or optical modifications for writing, and polarized and reflected light detection for reading. Since it uses a nanometer-sized optical probe, ordinary optical recording mechanisms such as magneto-optical (MO) and phase change (PC) recordings can be applied to this technique. The smallest domain size obtained in both recordings is 60 nm in diameter, which corresponds to 170 Gb in^{-2} [7, 8]. In Fig. 13.21a, the proposed SNOM-based storage using PC recording is shown. A small localized probe with an evanescent light through the SNOM aperture heats the nanometer area in a 30 nm-thick PC film of GeSbTe. The rise in temperature due to light irradiation changes the film phase from amorphous to crystalline such that the reflectivity of the irradiated area increases. A reflection SNOM image of the phase-changed nano-optical bit array is shown in Fig. 13.21b. Reflectivity varied at a pulsed light (wavelength of 785 nm) of 7.3 mW (in the probe) for 5 ms. This means that the smallest domain size is

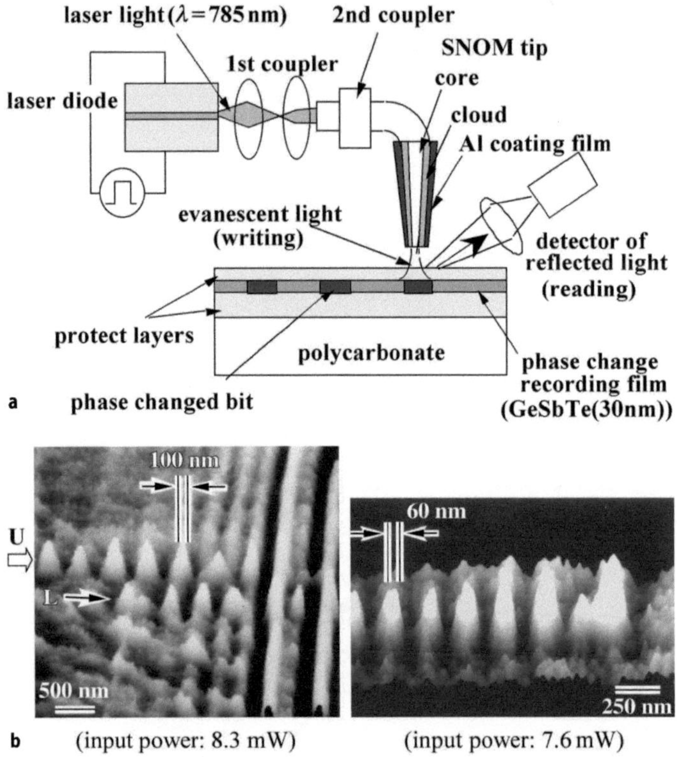

Fig. 13.21a,b. Schematic diagram of SNOM-based storage; **a** reading/writing, and **b** reflection SNOM image of phase-changed domain array

60 nm in diameter. I believe that a smaller domain can be obtained by developing an optical aperture smaller than 50 nm.

13.3.5 Other SPM-Based Storages

In addition of above SPM-based storages, SPM-based storages using a charge and a polarization have been proposed. These storages use a scanning capacitance microscopy (SCaM). The writing is done by applying a pulsed voltage between the probe and a sample. The charge is kept by boundary condition of permittivities of the dielectric layers. In the writing, the SCaM is worked as contact mode AFM to induce charges into the sample, while it is worked as a detector of capacitance. Storage information is determined by change of capacitance caused by the existence of charge in the dielectric material. In experimental results, bit size, which consists of charges, is under 100 nm in diameter. The charge storage is considered to have the potential to achieve a high-density recording with several 10s Gbit in^{-2}. So far, many companies have researched this type of SPM storage in Japan. The polarization-type SPM storage is almost the same as the charge storage, but it is only different from writing in charge type. A change of polarization of ferroelectric medium is done by electric field when applying a pulsed voltage between a probe and a sample. This method will be introduced in next section.

13.4 High-Speed Readout [32]

In order to achieve a high-speed readout of over 1 Mb s^{-1}, we have to study the readout speed in two aspects: detection response and gap control between the tip and the disc surface. The former is an unknown parameter because detection speed will be improved in the future. The latter is very important because gap variation causes the readout signal to deteriorate. Here, we only discuss the gap controls.

The gap control methods are mainly of 2 types; noncontact and contact (Fig. 13.22). The gap cannot be perfectly controlled due to low resonant frequency of the actuator. For example, the resonant frequency of a piezo device with a stroke of larger than 1 µm appears at less than 10 kHz, so that the frequency response band of the servo control might be less than 10 kHz. Therefore, the compensation for servo error is very important. Here, we can consider two compensation methods. The first method compensates for the error by using redundancy in the detected signal gap properties in noncontact recordings such as STM, noncontact-AFM, MFM, and SNOM-based recordings. The second method uses a flexible probe such as an AFM cantilever in contact AFM recording.

13.4.1 Noncontact Recording

The SNOM recording technique is the most suitable for high readout speed in the noncontact recordings as shown in Fig. 13.23. This is because the detected signal gap property has a flat region in a proximity region above the disc surface within a radius of the SNOM aperture (Fig. 13.23c). When the aperture is 60 nm in diameter, the flat region is about 30 nm, which permits control error within 15 nm. The value is sufficient to protect against readout signal error caused by surface roughness when a silicon substrate is used. This is because the roughness of the Si surface is around 1 nm.

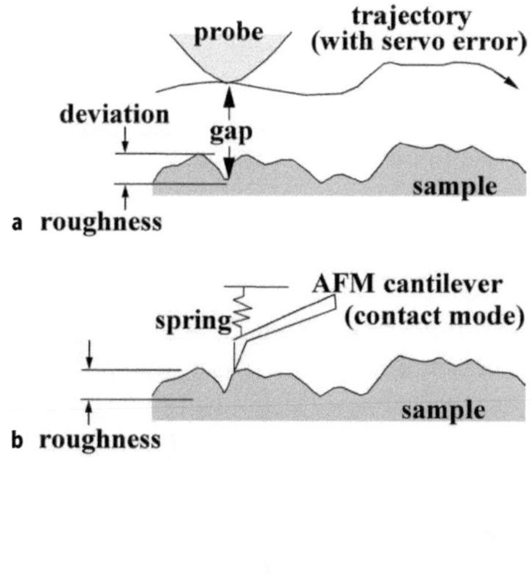

Fig. 13.22a, b. Schematic diagram of probe and sample in high-speed scanning; **a** noncontact and **b** contact mode

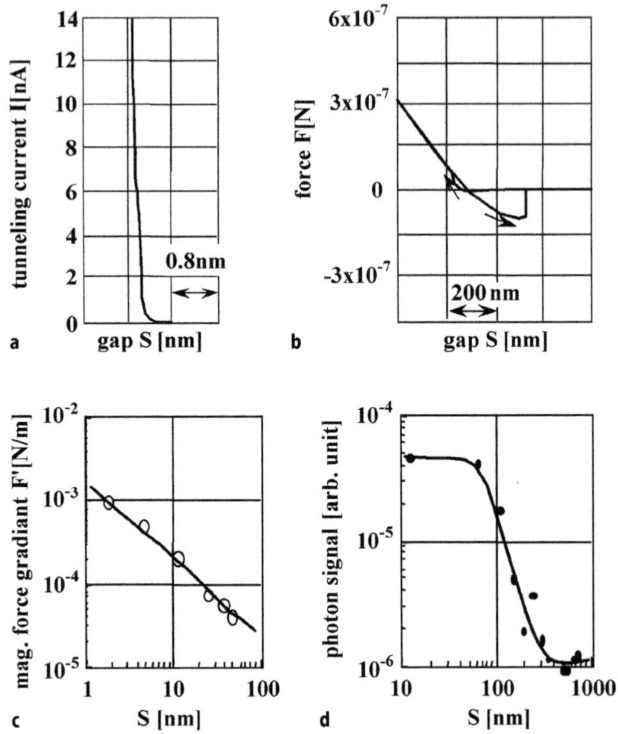

Fig. 13.23a–d. Variations of detected signal intensity for gap between the tip and the sample surface; **a** in STM, **b** AFM, **c** MFM, and **d** SNOM

The STM, noncontact-AFM, and MFM recording techniques have a small margin of error in the servo control within 0.1 nm or a few nm (Figs. 13.23a and 13.23b). Their responses of gap controls are within 10 kHz when using a piezo device so that the readout speeds are basically restricted within 10 kHz.

13.4.2 Contact Recording

The contact AFM can compensate for servo error by working its cantilever spring as a suspension device. The resonant frequency of the AFM cantilever determines readout speed. A cantilever with a resonant frequency of around 10 MHz and a spring constant of less than 1 N m^{-1} is considered possible.

As described above, we consider that both SNOM- and AFM-based data storages are close to being available for practical use.

13.5 Possibility of SNOM-Based Storage [4]

SNOM recording can use magneto-optical (MO) change [5] or phase change (PC) [4] of the recording film for writing, and polarized or reflected light detection for reading. We studied SNOM-PC recording because it detects more photons emitted from the recorded bit than SNOM-MO reading.

13.5.1 Expectable Readout Speed

The readout speed is determined by signal detection rather than the gap control in SNOM recording as described in the previous section. The intensity of the detected light (photon number per bit) is the dominant factor affecting high-speed readout. In the experiment using reflection SNOM, the efficiency of evanescent light propagation through the probe aperture was 5×10^{-3}. We can estimate the number of photons reaching the PMT cathode surface from the efficiency (Fig. 13.24). If assuming a critical number of 4×10^3 photons bit^{-1} for high SNR (signal noise ratio), the maximum readout speed is 10 MHz in the reflection SNOM-PC reading with an illumination laser power of 2 mW. This technique is expected to be used for high-speed readout. On the other hand, the MO recording is hardly expected because its efficiency is 2 orders smaller than that in PC recording due to its small Faraday or Kerr rotation.

This technique also compensates for the servo error by using a redundancy of the detected signal gap property as shown in Fig. 13.23d. The error margin is about 100 nm when it is defined as half of the intensity at contact. The readout speed is determined by the signal detector rather than the control ability. The intensity of detected light (photon number per bit) is dominant for high-speed readout.

13.5.2 Rotation-Type SNOM Recording System

For an optical storage, high-speed mechanical scanning is needed to get a high-speed readout. So, the rotation-type SNOM recording system has been prototyped (Fig. 13.25).

The system consists of an air-spindle motor, a SNOM slider head, an illuminating optics, and a photodetector optics for transmitted light to use both the SNOM aperture monitor and the readout signal detection. The sample used here

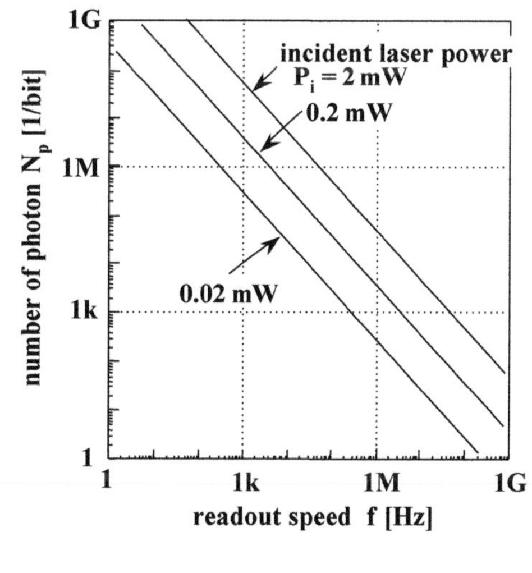

Fig. 13.24. Estimation of number of photon detected from 60-nm-diameter bit for readout speed derived from the experimental setup

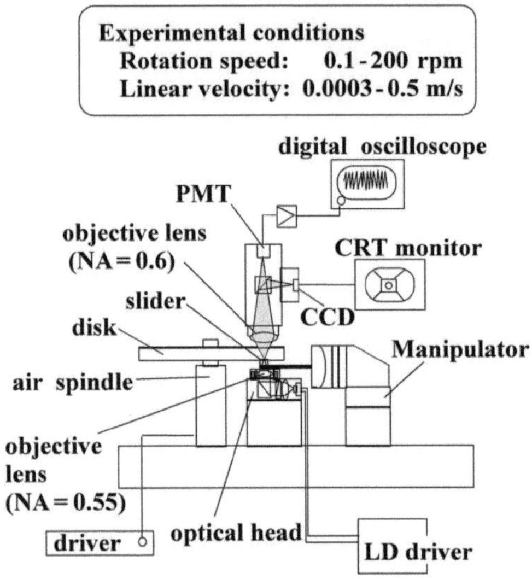

Fig. 13.25. Experimental setup using slider type SNOM head

is tungsten line and space patterns on quartz disk. The photodetector optics consist of objective lens (NA: 0.8) for a collection of the transmitted light, a splitter, a CCD (charge coupled device) camera, a PMT (photomultiamplifier tube) and a digital oscilloscope for the signal detection. The mechanical scanning in microscopy has a speed of around 10 μm s^{-1}. This speed is extremely slower than that in storage, which requires a readout speed of higher than 0.1 m s^{-1}. The microscopy keeps a gap between the probe and the sample constant by using shear-force detection. Here, the slider is used as a gap controller the same as that in magnetic recording. Thus, the high-speed mechanical scanning is operated. The air-spindle can control a rotation speed of 0.1–200 rpm within an accuracy of 0.1%. The CCD camera monitors whether the head is floated or not. When the head is floated at a gap of around 10 nm, an image of the head is stopped in the monitor, but when the head is scanned at contact operation, it is monitored that the head moves at random due to a stick slip mechanism. The PMT detects an optical signal through the line and space pattern on the quartz disk. The line pattern has widths of 1 μm and 100 nm.

Two type slider heads are fabricated. One is with a connection of the flat quartz slider and an optical fiber, which is called flat slider (Fig. 13.26a), and the other consists of a slider head with the tip and the illumination optics, which is called pyramidal tip slider (Fig. 13.26b). The flat slider has an aperture on the flat edge of the fiber. The pyramidal tip slider has an aperture on a top of the tip. The apertures are fabricated by FIB (focused ion beam) drilling or slicing, respectively. The flat slider has a throughput of 10^{-4}, which is not enough to get high-speed readout. Its operation is unstable in stick slip motion.

The pyramidal tip slider has the tip on rectangular quartz slider, and an aperture of 50–200 nm in diameter. The throughput of the head is 10^{-2} to 10^{-3}, which is expected to get a high-speed readout signal. For smooth sliding, the pyramidal slider has three pads on the slider. One of them has a pair of 4 small pads and a pyrami-

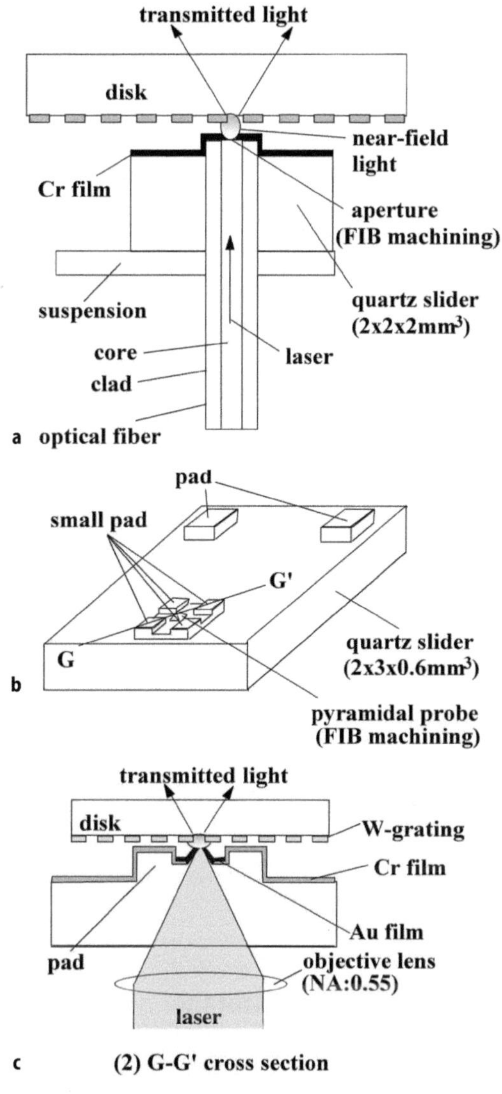

Fig. 13.26a–c. Schematic diagrams of prototyped SNOM slider heads; **a** flat type SNOM probe on slider, **b** pyramidal SNOM probe on slider, and **c** enlarged SNOM probe of **b**

dal tip at a center of the pair. On the other hand, the flat slider has only one pad, for which the fiber edge surface is used. In the experiment, the laser beam which is emitted from a laser diode is focused onto the aperture by objective lens (NA: 0.55). This illumination optics uses a part of DVD (digital versatile disc) head. In the readout experiment, the laser beam with a power of 10 mW and a wavelength of 650 nm is illuminated and focused onto the SNOM aperture through the head substrate. The focused laser beam on the aperture is about 0.5 μm in diameter. The focusing is adjusted by aligning the illumination optical axis on the aperture so as to obtain the maximum light power at the PMT. The readout experiment is performed at a linear velocity of 0.38 m s^{-1} and a rotation speed of 150 rpm. As shown

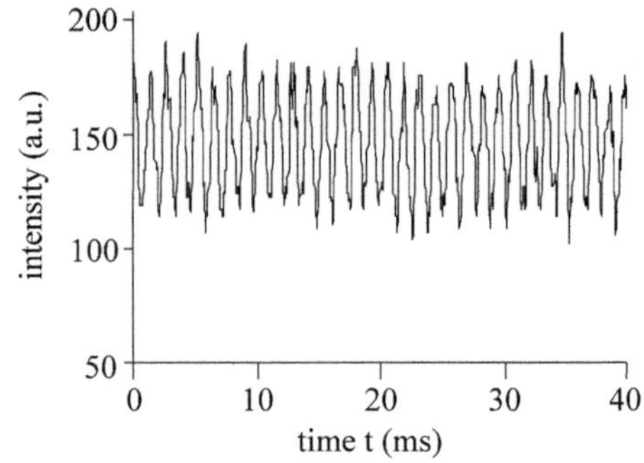

Fig. 13.27. Readout signal of 250 nm-wide line and space pattern at rotation speed of 150 rpm (the signal corresponds to readout speed of about 2.2 Mb s^{-1} when using EFM)

in Fig. 13.27, the readout signal of 0.6-μs period was obtained using 0.25-μm-wide line and space pattern. The frequency of the signal is about 780 kHz, which corresponds to a rate of 2.2 Mb s^{-1} when using EFM (eight fifteen modulation). The CNR (carrier noise ratio) is higher than 40 dB. From this experimental result, ultrahigh-density SNOM-based storage can be expected in the future [33].

13.5.3 Future Issues

The experimental results indicated a possibility to achieve an ultrahigh-density recording with a density of more than 100 Gb in^{-2} and a readout speed of faster than 2 Mb s^{-1}. We expect that this technology can be applied to an ultrahigh-density recording beyond 100 Gb in^{-2}. However, some technical subjects still remain unsolved. For example, these are (1) how to fabricate a nanometer-sized optical aperture having high yield, (2) how to control the gap between the tip and the sample precisely, and (3) how to transfer the heat to a nano-area of the recording film. By solving these problems, the next step forward in developing a SNOM recording system can be reached.

13.6 Possibility of AFM-Based Storage

13.6.1 Expectable Readout Speed

The resonant frequency of an AFM cantilever can determine readout speed because the servo error is compensated by working the cantilever spring as a suspension. AFM is operated at around 10^{-8} N, at which the deflection of cantilever responses on the roughness of sample surface are as shown in Fig. 13.22b. The ordinary resonant frequency of the cantilever is around 40 kHz in force-free conditions. By applying a repulsive force of 10^{-8} N to the cantilever (a spring constant of 1 N m^{-1}), the frequency increases 5 times to 200 kHz. To achieve a readout speed of 10 MHz or more, we must develop a more advanced cantilever with a resonant frequency of higher than 2 MHz and a spring constant of less than 5 N m^{-1}, which is probably possible.

The signal-noise ratio (SNR) is determined by surface roughness and pit depth or dot height. When the dot height is 2 nm, the roughness should be within 0.2 nm to obtain an SNR of 20 dB. If a permitted diameter of the roughness structure as a noise is less than 2 nm, an SNR of 40 dB can be obtained when the diameter of pit or dot is about 20 nm. Thus, a roughness of the sample surface is very important to get high-quality signal in a contact-type recording. When we make such a surface of a disk, the SNR can be about 60 dB, which is enough for practical use.

13.6.2 Prototyping of Rotation-Type AFM Recording System

We prototyped a rotation-type AFM read/write experimental system. The system consists of an AFM head system including a laser beam deflection detection system, a z-axis fine and coarse actuator for approaching the tip to the sample surface, a controller, and an air-spindle motor (Fig. 13.28). The revolution of the air-spindle was controlled from 0.01 to 4,000 rpm within a deviation in speed of 10%.

13.6.2.1 *Writing and Reading [24]*

In the experiments, nano-pit arrays with a 20-nm pitch in line and a 40-nm pitch in track were demonstrated (Fig. 13.29). The nano-pit arrays were written at a pulsed force of 60 nN for 1 ms (Fig. 13.29b). The array pattern means 1.2 Tb in^{-2} in the 1/2(2,7) code recording technique. Furthermore, a fine bit array with less than a 10-nm diameter bit and a 25-nm pitch on the polycarbonate disc was formed (Fig. 13.29c). These were made by applying a pulsed force of about 40 nN for 60 µs to the tip. The nano-pit arrays were recorded at a yield of 100%. We confirmed that the cantilever probe was able to write more than 10 million nano-pits.

After the writing process, we read a part of the pit array at a readout speed of 1.25 Mb s^{-1} without tracking (Fig. 13.30). The rotation speed was 100 rpm, and the constant force was set at about 10 nN. The minimum pit size was about 25 nm in

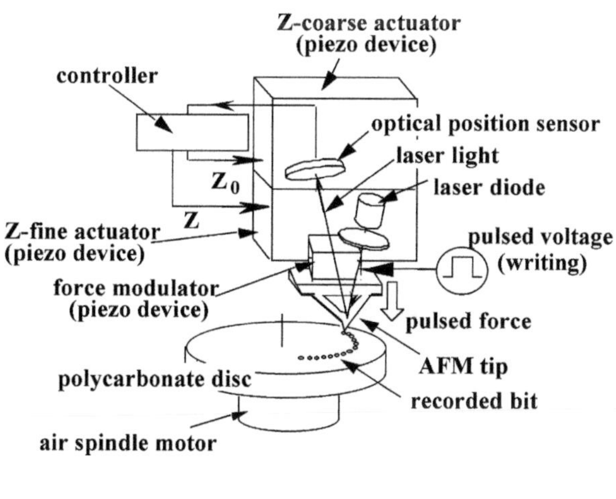

Fig. 13.28. Schematic diagram of rotation type AFM recording system using force modulation AFM pit recording

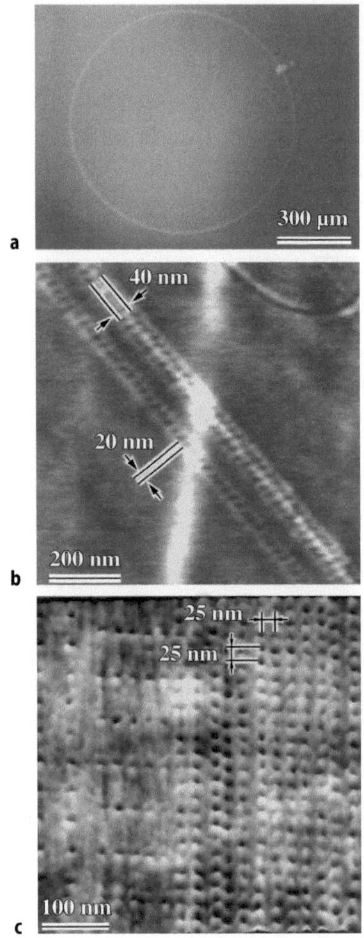

Fig. 13.29a–c. Recorded bit arrays; **a** overview of the pit arrays (SEM image), **b** AFM image of nano-pit arrays which corresponds to 1.2 Tb in^{-2} in 1/2(2,7) code recording, and **c** AFM image of high packed pit array (1.5 Tb in^{-2}) with 25-nm pitch

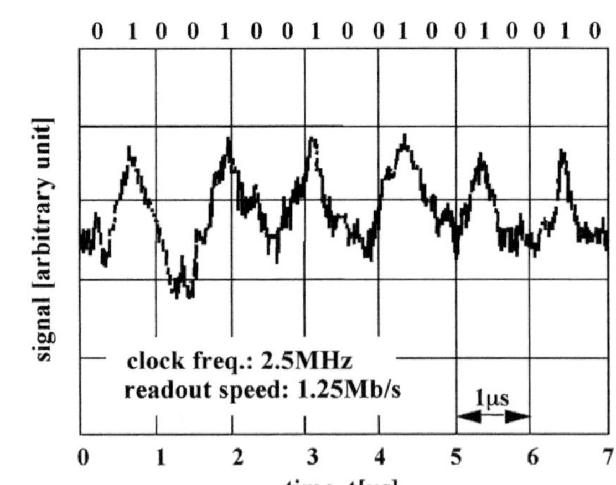

Fig. 13.30. Readout signal of a part of the nano-pit array with a pit diameter of 25 nm at a spindle revolution of 100 rpm, which corresponds to 1.25 Mb s^{-1} in 1/2(2,7) code recording (preset force: 10 nN)

diameter. As described above, AFM pit recording has the potential to achieve ultrahigh-density recording with a density of over 1 Tb in^{-2} and a readout speed of over 1 Mb s^{-1}. Furthermore, a readout speed of over 10 Mb s^{-1} is considered possible by developing a cantilever with a higher resonant frequency.

13.6.2.2
5 Mb s^{-1} Readout [34, 35]

In high-speed readout using AFM, one of the key points is the resonant frequency of the cantilever. A new small cantilever has been developed as follows. High-speed readout was then done and is discussed below.

(1) New small cantilever

(i) Design of high-resonant frequency AFM cantilever

Our target is to develop a cantilever with a resonant frequency of >2 MHz, a spring constant of <4 N m^{-1}, and a tip. Figure 13.31 shows a contour map of both resonant frequency and spring constant due to length and thickness of the triangular cantilever.

The figure shows that smaller length increases the resonant frequency and smaller thickness decreases the spring constant. This result indicates that our tar-

Fig. 13.31. Resonant frequency and spring constant of AFM cantilever due to the AFM cantilever length l and thickness t

$$\mathbf{k} = \frac{\mathbf{E\ t^3}}{3\sqrt{3}\ l^2}$$

$$\mathbf{f} = \frac{\mathbf{1.789\ t}}{\pi\sqrt{3}\ l^2}\sqrt{\frac{\mathbf{E}}{\rho}}$$

(in case of triangle cantilever)

(E; Young modulus (1.9x10^{11} N/m), ρ; density (Si; 2.33 g/cm^3)

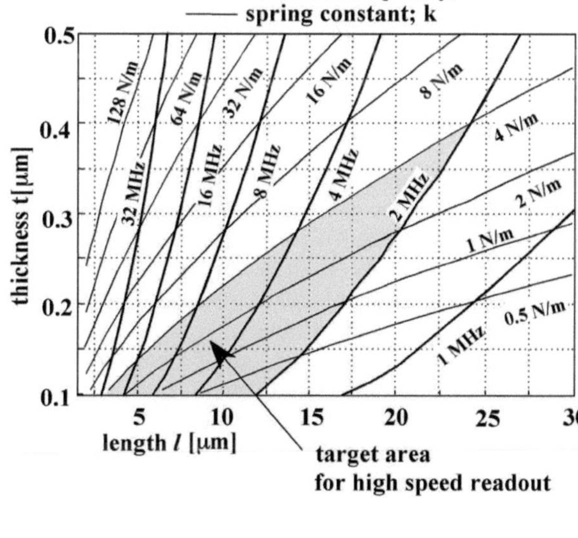

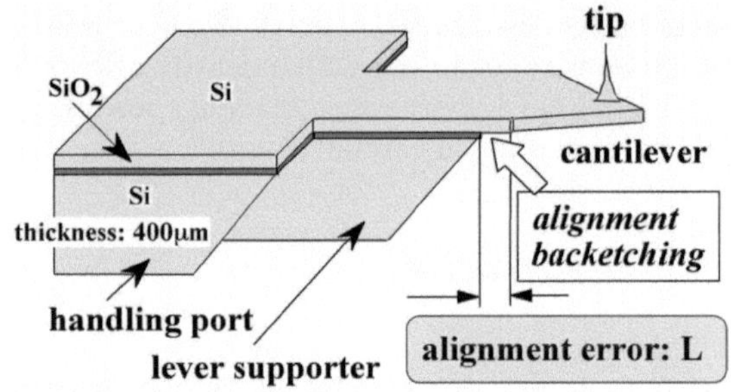

a

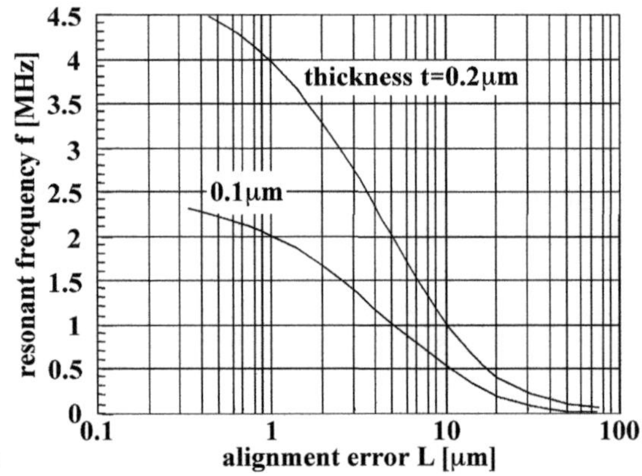

b

Fig. 13.32a, b. Lowering resonant frequency by misalignment (triangle cantilever); **a** scheme of triangle cantilever, and **b** calculated result of the frequency

get is to develop a cantilever with a length of less than 20 μm and a thickness of less than 0.3 μm in order to achieve the desired mechanical properties as described above. Also, we have to consider an easy use and operation of the cantilever. For example, we easily attach the handling port of the cantilever to the cantilever holder in AFM, and the tip always contact the sample surface while the cantilever holder does not contact it. As a design, we adopt the cantilever structure as shown in Fig. 13.32a. The cantilever is held on a lever supporter, which is further held on a handling port.

(ii) Controlling the lever length

In this structure there is a significant issue of how to control the length of cantilever. Even when we use a Si microprocess with a dual side aligner, a small misalignment between front and backside may change the length of the cantilever which in turn increases or decreases the resonant frequency. Figure 13.32b shows that the decrease of the resonant frequency is about 20% when the misalignment is only 1 μm, and it is about 60% when the misalignment is 5 μm. In order to eliminate the change, we use a concept of a buffer step as shown in Fig. 13.33a. This step is formed at a boundary of the designed cantilever. When the step is about 0.5 μm, the influence on the frequency is very small (Fig. 13.33b). Consequently, the margin of alignment error of 5 μm is allowed by the buffer step so that we can use the aligner, which has an alignment error within 3 μm. Furthermore, lowering

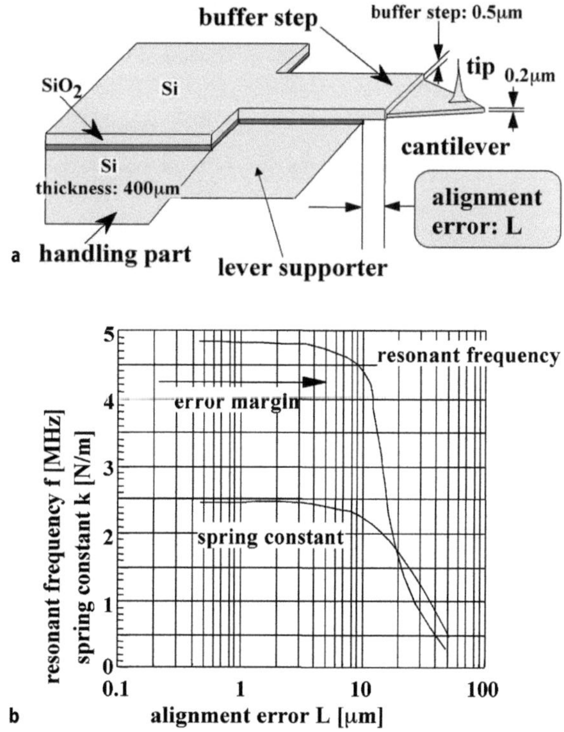

Fig. 13.33. **a** Scheme of new cantilever structure and **b** resonant frequency and spring constant of the cantilever with buffer step due to alignment error (calculated result)

the spring constant does not occur within an alignment error of 5 μm. This step is so useful that we can make an accurate mechanical property of the cantilever.

(iii) Controlling the lever thickness

This Si microprocess with SOI wafer consists of some etchings. The etching process has an issue of how to control the thickness because we can only confirm the lever thickness after the etching. There are some changes of etching speed and active layer thickness of SOI wafer. These cause the change of the thickness. In order to protect the change, we use the pre-step formed on the active layer. This means that the pre-step is kept during SF_6 reactive ion etching. Finally, the depth of the pre-step corresponds to thickness of the lever, when we can precisely control a stop of the etching. In practice, we can detect the end by observing whether an area of transparency appears or not in the wafer. Therefore, in this method, it is very important to form the pre-step with accurate etching depth. In our experiments, machining accuracy of the thickness is less than 0.1 μm because the depth of pre-step can be formed within an error of 0.1 μm.

(iv) Procedure for the cantilever

Our procedure for fabricating this small Si cantilever, using a Si microprocess with the SOI wafer, is shown in Fig. 13.34. The procedure was carried out using dual-

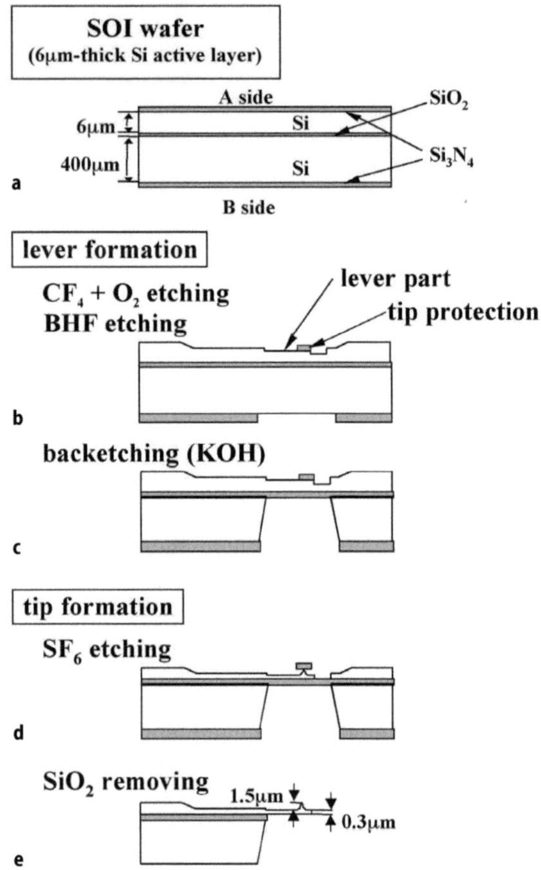

Fig. 13.34. Process flow for fabrication of the AFM probe with a high resonant frequency

side aligner with contact printing, anisotropic Si-etching, and dry etching. These were employed to achieve precise machining of the cantilever length within an alignment error of 5 μm. The process techniques used here can offer a protection manner, buffer step (Fig. 13.33), against aligner's error. In addition, we designed a support plate which supports the cantilever with a thickness of 0.7 μm. This design allows the misalignment (Fig. 13.33a) and allowable error is 5 μm. The process consists basically of two stages: the first stage is preformation of the cantilever. The buffer step with a depth of 0.5 μm and pre-step with a depth of 0.2 μm were fabricated by $CH_4 + O_2$ etching. Then, the tip mask pattern was made from SiO_2 by BHF etching and a 400-μm-thick Si substrate was back-etched by KOH etching (Figs. 13.34b, 13.34c). The second stage is for fabrication of both the cantilever and the tip (Figs. 13.34d, 13.34e). The cantilever is triangular in shape. The shape of the lever was already defined by $CF_4 + O_2$ pre-etching in the Si active layer (Fig. 13.34b). Then, the tip and the lever were fabricated at the same time using SF_6 RI (reactive ion) etching. Finally, we removed the SiO_2 middle layer of SOI wafer. Figure 13.35 shows a scanning electron microscope (SEM) image of the prototyped

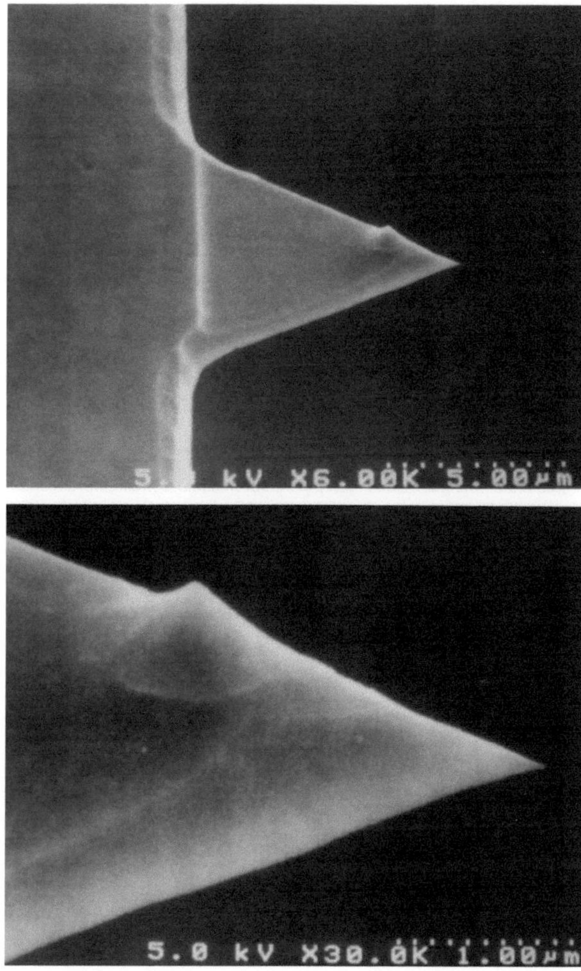

a

b

Fig. 13.35a, b. SEM images of prototype of high resonant frequency AFM Si cantilever, **a** overview and **b** enlarged tip

cantilever. This cantilever has a length of 7 µm and a thickness of 0.2 µm. The tip radius was about 30 nm. The spring constant was estimated to be about 4 N m^{-1}. We fabricated some cantilevers with a length of <20 µm and a thickness of <0.3 µm.

13.6.2.3
Resonant Frequency Cantilever and High-Speed Readout

We measured its frequency spectrum using newly developed laser beam deflection detection systems [33]. In particular, we used new laser beam deflection detection with a high numerical aperture (NA) objective lens for small cantilever with a length of less than 15 µm [4]. As a result, a maximum resonant frequency of about 6.6 MHz was measured as shown in Fig. 13.36. Using this cantilever, a high-speed readout at higher than 10 Mb s^{-1} in AFM-based recording can be expected because

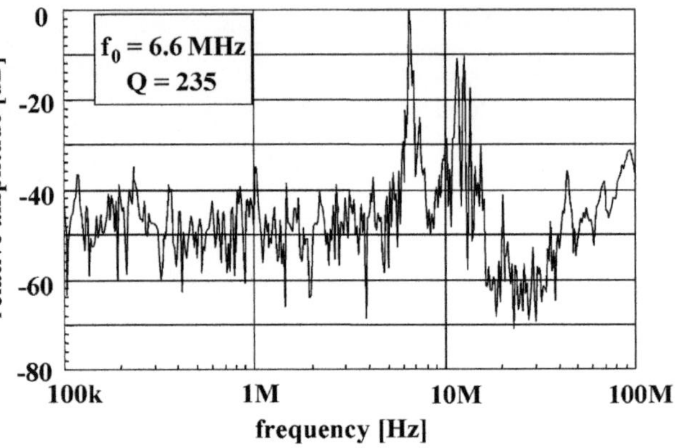

Fig. 13.36. Frequency spectrum of the developed cantilever (length: 10 μm, thickness: 0.2 μm)

contact mode AFM operation makes resonant frequency of the cantilever increase to 4 times higher than that in the noncontact mode.

Using this cantilever, we readout a standard line and space pattern with 1 μm and 100 nm in width. The experiment was done by the rotation-type AFM recording with new laser beam deflection system. In a case of the 1-μm line and space, 2-MHz periodic signal was obtained at a linear velocity of 2 m s^{-1}, when the rotation speed was about 1,200 rpm (Fig. 13.37a). Besides, about 15-MHz resonant fre-

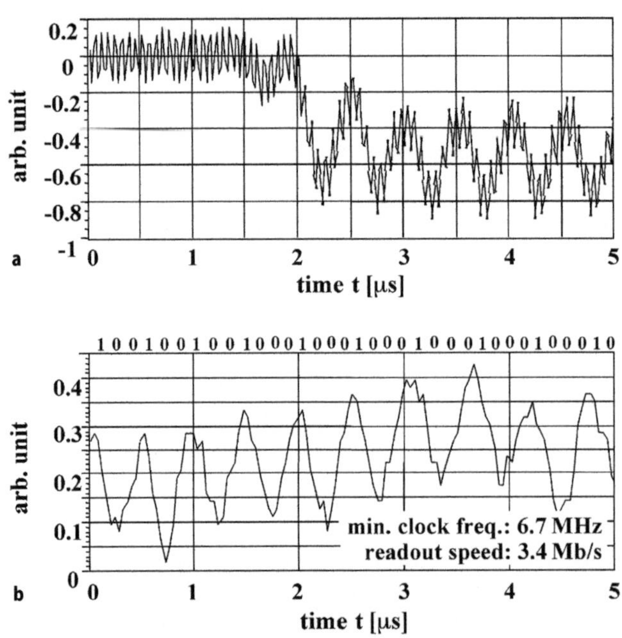

Fig. 13.37a, b. Readout signals of AFM-based storage using nano-indentation; **a** at a velocity of 2 m s^{-1} (revolution of 1,200 rpm) with 1-μm pit length pattern and **b** at 0.2 m s^{-1} (120 rpm) with 100 nm

quency signal is superimposed on the readout signal. This is due to the cross-section of the periodic pattern that has a sharp etched structure which produces a high-frequency component. This experiment was done by using a 3-MHz resonant frequency cantilever. A resonant frequency of the cantilever changes from 3 MHz to 15 MHz in contact mode because the edge-fixed condition changes from one edge fix to both edges fix. On the other hand, in the 100-nm line and space pattern on a sample, only the readout signal of the pattern with 2 MHz was obtained without the resonant frequency of 15 MHz (Fig. 13.37b). The experiment was performed at a rotation speed of 120 rpm, which corresponds to 0.2 m s^{-1}. This is because the pattern signal does not include such a high frequency component due to loose structure of the cross-section. These signals mean a potential for high-speed readout at higher than 5 Mb s^{-1} when using EFM (eight fourteen modulation). The results show that the readout of AFM-based storage has a potential to achieve a high-speed readout with higher than 10 Mb s^{-1} when using high resonant cantilever of over 5 MHz.

13.6.3 Summary

We proposed force modulation (FM)-AFM pit recording and demonstrated the possibility of achieving ultrahigh-density recording with a high-speed readout. We reached such a conclusion as our proposed FM-AFM pit recording is expected to be an attractive candidate for future ultrahigh-density recording.

We also achieved the following:

(1) A minimum pit size of around 10 nm in diameter was formed by cold plastic deformation of polycarbonate disc surface.
(2) A rotation-type FM-AFM pit recording system was prototyped.
(3) Using the prototyping, an ultrahigh recording density of 1.5 Tb in^{-2} when using 2–7 code modulation.
(4) A high-resonant-frequency AFM cantilever with a frequency of about 6.6 MHz is developed.
(5) Using the prototyping, a readout speed of faster than 5 Mb s^{-1} is demonstrated using the 100-nm line and space pattern.

However, there are still many challenges to be solved for data storage. We need to develop technologies in the future such as nano-tracking, higher speed readout, nano-tribology, etc.

13.7 Perspective and Summary

As mentioned above, we consider various SPM-based storages using atomic manipulation and surface modification. AFM- and SNOM-based storages have the potential to achieve an ultrahigh recording density of larger than 1 Tb in^{-2} and 170 Gb in^{-2}, respectively. Also, they have the potential to readout the bit at a high speed of over 5 Mb s^{-1}. Therefore, we predict that their positions becomes close to the future data storage, and we should focus efforts on the development of the future data storage on the AFM- and SNOM-based storages in an area density of higher than 100 Gb in^{-2}. However, there are still some problems such as a systematic engineering, a recording material, reliability, etc. We need to work hard to improve these techniques.

References

1. Albrecht TR, Dovek MM, Kirk MD, Lang CA, Quate CF, Smith DPE (1989) Appl Phys Lett 55:1727
2. Mamin HJ, Guethner PH, Rugar D (1990) Phys Rev Lett 65:2418
3. Hosoki S, Hosaka S, Hasegawa T (1992) Appl Surf Sci 60/61:643
4. Hosaka S, Koyanagi H, Kikukawa A (1993) Jpn J Appl Phys 32B:L464
5. Ohkubo T, Kishigami J, Yanagisawa K, Kaneko R (1991) IEEE Trans Magn 27:5286
6. Hosaka S, Koyanagi H, Kikukawa A, Miyamoto M, Imura R, Ushiyama J (1995) J Vac Sci Technol B 13:1307
7. Betzig E, Trautman JK, Wolfe R, Gyorgy EM, Finn PL, Kryder MH, Chang C-H (1992) Appl Phys Lett 61:142
8. Hosaka S, Hosoki S, Hasegawa T, Koyanagi H, Shintani T, Miyamoto M (1995) J Vac Sci Technol B13:2813
9. Binnig G, Rohrer H, Gerber C, Weibel E (1983) Phys Rev Lett 50:120
10. Fuchs H, Laschinski R (1990) Scanning 12:126
11. van Loenen EJ, Dijkkamp D, Hoeven AJ, Lenssinck JM, Dieleman J (1989) Appl Phys Lett 55:1312
12. Eigler DM, Schweizer EK (1990) Nature 344:524
13. Yau S-T, Saltz D, Nayfeh MH (1990) Appl Phys Lett 57:2913
14. Utsugi Y (1990) Nature 347:747
15. Binnig G, Quate CF, Gerber C (1986) Phys Rev Lett 56:930
16. Meyer G, Amer NM (1988) Appl Phys Lett 53:1045
17. Hosaka S, Hosoki S, Takata K (1993) Japanese Patent No. P2533728 (submitted: 10 July 1987); Hosaka S, Koyanagi H, Kikukawa A (1993) Jpn J Appl Phys 32:L464
18. Duclos SJ, Hadden RC, Glarum SH, Hebard AF, Lyons KB (1991) Solid State Commun 80:481
19. Tsong Tien T (1991) Phys Rev B 44:13703
20. Miskovsky NM, Tsong Tien T (1992) Phys Rev B 46:2640
21. Hosaka S, Kikukawa A, Honda Y, Koyanagi H, Tanaka S (1992) Jpn J Appl Phys 31:L904
22. Jung TA, Moser A, Hug HJ, Brodbeck D, Hofer R, Hidber HR, Schwartz UD (1992) Ultramicroscopy 42–44:1446
23. Hamada E, Kaneko R (1992) J Phys D Appl Phys 25:A53
24. Hosaka S, Koyanagi H, Kikukawa A, Miyamoto M, Nakamura K, Etoh K (1997) J Vac Sci Technol B15:788
25. Konno C, Saji M, Sagawa N, Umetani Y (1986) Proc Fall Joint Comput Conf, 2–8 Nov
26. Simmons JG (1963) J Appl Phys 34:1793
27. Miyamoto M, Ushiyama J, Hosaka S, Imura R (1994) Proc Mag Opt Rec Int Symp (MORIS '94). Tokyo, 1994, pp 141–144
28. Miyamoto M, Shintani T, Hosaka S, Imura R (1996) Jpn J Appl Phys 35:L584
29. Born M, Wolf E (1980) Principle of optics, 6th edn. Pergamon Press, chaps 1 and 5
30. Kawata S (1992) Jpn J Opt 21:766
31. Hosaka S (1996) IEEE Trans Mag 32:1873
32. Hosaka S, Kikukawa A, Koyanagi H, Shintani T, Miyamoto M, Nakamura K, Etoh K (1997) Nanotechnology 8:A58
33. Hosaka S, Itoh K, Shintani T, Issiki F (1999) O plus E 21:279 (in Japanese)
34. Hosaka S, Etoh K, Kikukawa A, Koyanagi H, Itoh K (1998) Abst MNE '98. Leuven, pp 19–20
35. Kikukawa A et al (1998) Abst 45th AVS Int Symp Baltimore, p 102

14 Super Density Optical Data Storage by Near-Field Optics

Jun Tominaga

14.1 Introduction

Optical near-field application has long been desired to realize subterabyte optical data storage free from optical diffraction limits [1, 2]. So far, two main streams of research have been carried out and demonstrated.

One of the techniques is the use of a metal-coated fiber probe by a near-field scanning optical microscope (NSOM) and the other is the use of a solid immersion lens (SIL) high-speed-flying above a recording medium with the aid of hard disk technologies. The resolution of the first technique has the potential to increase data recording density to subterabyte levels by sharpening the metal-coated fiber probe. The head is very fragile, however, and it becomes almost unfeasible, especially rotating a medium like a CD or a digital versatile disk (DVD) at high speed. This limits the data transfer rate and recording area in microscopic systems. The fabrication error of the apertures is very large, so this technique remains to be made practical for super-density optical data storage. Improving the fragility of fiber probes, instead, planar aperture heads have been studied [3, 4]. The use of a SIL is not restricted by data transfer rate nor recording area, but it produces another difficulty on increasing the numerical aperture (NA) more than 2.5 in the visible wavelength region. That is, the recording density will reach a limit in the future. The need to overcome such difficulties left in current near-field optical data storage is very urgent.

14.2 The Principle of Super-RENS

A reconsideration of near-field recording points out the intrinsic problems hidden in the systems. First, conventional NSOMs were developed with objectives for optical data storage. Secondly, scanning probe microscopes (SPMs) use piezo actuators to precisely control the distance between the probe and the medium. NSOM is not exceptional. Therefore, the systems never operate at high-speed recording as does a CD or a DVD. The great problem with SPMs is space (usually air or liquid) between the probe and the medium. Why not replaced the air or liquid with a solid? This was our simple proposal to overcome these difficulties. A structure retrieving optical near-field signals through a solid film is now called a "super-resolution near-field structure (super-RENS)" that generates a near-field in

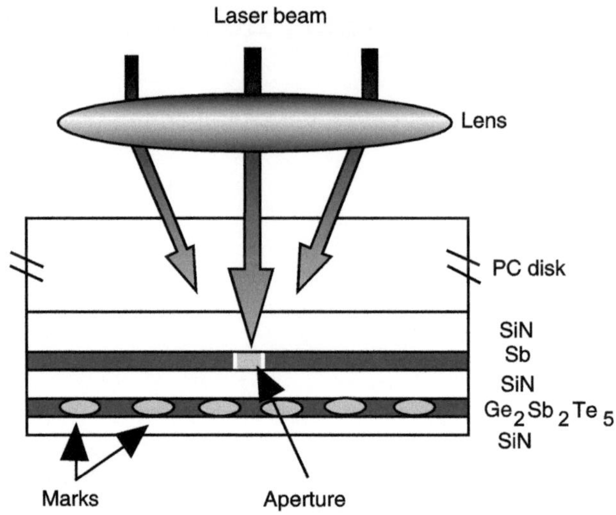

Fig. 14.1. The super-RENS disk structure. The laser beam is irradiated through the substrate. The focused beam has a spot size determined by the diffraction limit of the optics; however, a tiny optical aperture is generated due to the Gaussian profile of the laser intensity. The aperture can produce optical near-field, which interacts with the recorded marks in the phase change layer through a solid film with a 20-nm to 60-nm thickness [7]

the optical disk itself by using an antimony (Sb) layer [5, 6]. In the super-RENS, an optical near-field generated in an Sb layer interacts with the recording layer (phase change or magneto-optical) through the solid layer. The structure is depicted in Fig. 14.1.

A near-field aperture is generated in the Sb layer due to the Gaussian intensity profile on a focused laser spot far from the disk surface. The most important advantage of the super-RENS is that the intermediate solid layer (SiN) prevents the near-field aperture from crashes with the recording layer. This is because the intermediate layer is a solid film, which is deposited over a wide area by vacuum techniques. With this structure, the disk is rotated at a speed as fast as a DVD (≈ 6 m s^{-1}). Data is recorded or reproduced over a 120-mm-disk surface at a constant space:film thickness. This unique technique has gradually attracted much research in optical data storage, since we introduced the proposal and principle [5, 6]. The key to the super-RENS depends on the reversible phase transition of an Sb film sandwiched between SiN dielectric layers. An Sb film has an interesting optical characteristic between the crystalline and the molten states. The change under a static condition has not been observable, however, because the multilayer is very fragile in heat accumulation. The heat accumulation can be avoided by rotating the substrate because the heat produced in the laser spot is quickly dissipated by spot movement. In the Sb layer, only a small region is melted and aggregated by different laser intensities in the focused spot; the tiny aperture exists stable with the aid of elastic deformation of dielectric layers. As shown in Fig. 14.2, the Sb film shows a high optical nonlinearity, especially in high-speed disk rotation; this change generates a tiny semi-transparent aperture in the multilayer.

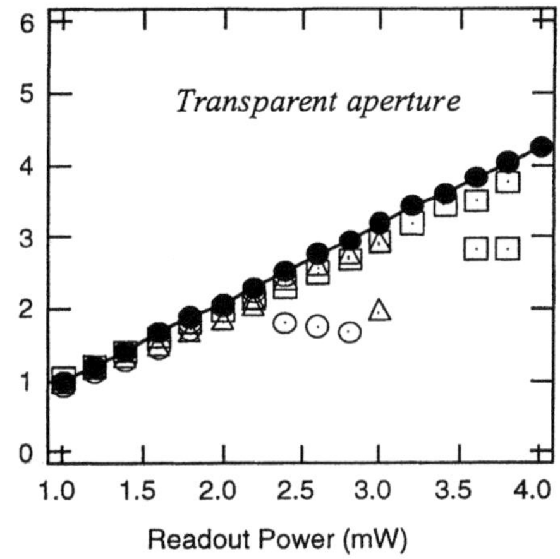

Fig. 14.2. Relationship between reflection and the readout power of the laser beam for the disk with the Sb film under the different disk rotational speeds 2.0, 4.0, 8.0, and 14.0 m s^{-1}. Disk structure PC/ SiN(170 nm)/Sb(15 nm)/ SiN (20 nm). By reducing the speed, the nonlinearity becomes clearer. The reflection from the disk gradually drops against the readout power increase

14.3 The Characteristics of Super-RENS Disks Using Sb Layers

Marks are usually recorded and read out on the super-RENS disks using a commercialized optical disk drive tester DDU-1000 (Pulstec Industrial Co. Ltd.) with a laser wavelength of 635 nm and a lens NA of 0.6. The diffraction limit of our system is at around 500 nm, and the minimum resolution of recorded marks in the phase change layer $Ge_2Sb_2Te_5$ becomes half the diffraction limit: 250–300 nm. The system configuration is depicted in Fig. 14.3. The laser beam is focused by the objective lens controlling the focus on the disk surface with a voice coil. The disk can be rotated at a constant linear velocity from 0.6 to 16.0 m s^{-1}. Reflected or scattered light is captured by the lens and directed to a photodetector. The detector is connected to a spectrum analyzer. In optical recording, we usually use the decibel (dB) as the signal intensity unit to background noise. dB is the power ratio of two levels and described as $10 \log_{10}(P_1/P_2)$. Here, P_1 and P_2 are signal intensities. The level is observed in the band range of 30 kHz.

In the super-RENS disk experiments, marks were first recorded through the Sb layer to the phase change layer by intensity (≈5.5 mW) higher than readout power. The recorded marks were then read out by the same disk drive tester, as increasing readout power. Due to the optical nonlinear characteristics of the Sb layer, the refractive index gradually changes at a center portion in the laser spot with increasing readout laser power, because the laser intensity has a Gaussian profile. Therefore, the index change is little with low laser intensity, while the change becomes large with high power intensity. The results observed at a linear velocity of 6.0 m s^{-1} are shown in Fig. 14.4.

Curve A is obtained at 1.0 mW readout, while curve B is obtained at 3.5 mW. No signals were retrieved less than 300 nm at 1.0 mW because of the resolution limit of the optics. Due to near-field interaction between recorded marks and a tiny aperture generated in the Sb layer, however, smaller marks were detected down to 60 nm far beyond the diffraction limit. A signal actually obtained from 100-nm-

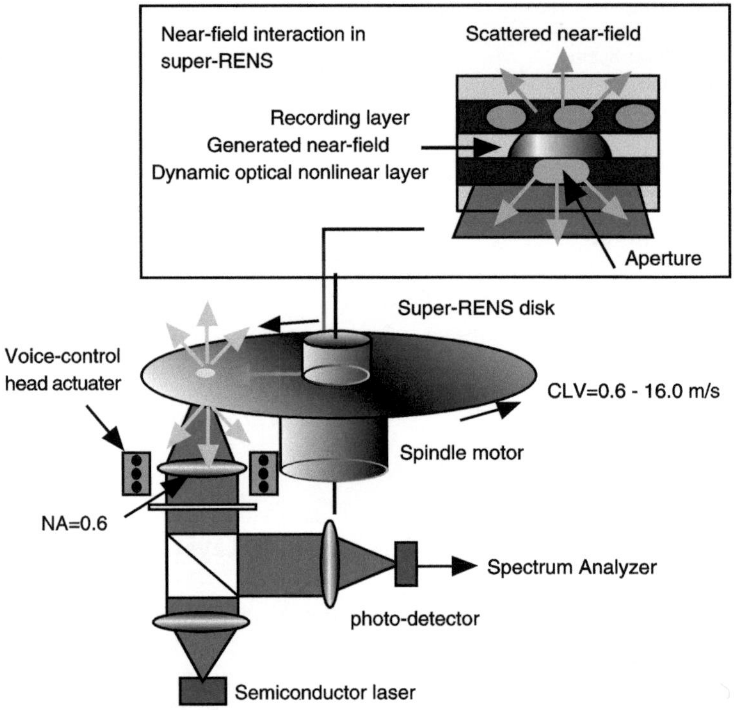

Fig. 14.3. System setup for dynamic optical nonlinearity and recording/readout with super-RENS disks. Wavelength = 635 nm, power modulation = 3–20 MHz, recording power = 1.0–15.0 mW, readout power = 1.0–4.0 mW

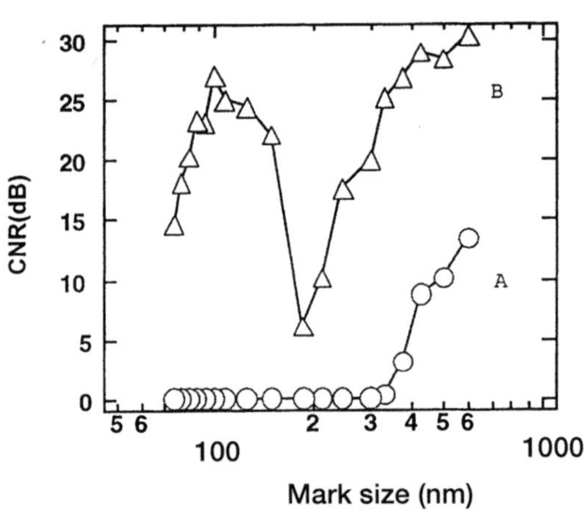

Fig. 14.4A, B. The reproduced CNR from a super-RENS disk using an Sb layer. **A** far-field detection (1.0 mW power), and **B** near-field detection (3.5-mW power). In this experiment, a 635-nm laser source and a 0.6-NA lens were applied for recording and readout. In the far-field detection, the resolution limit was clearly shown in the half of the diffraction limit; however, near-field detection showed much higher resolution until 60 nm far beyond the diffraction limit [7]

Fig. 14.5. Readout signals of 100-nm marks from the super-RENS disk. **a** Readout power: P_r = 1.0 mW at 6.0 m s^{-1} (far-field detection). **b** P_r = 3.5 mW (near-field detection) [7]

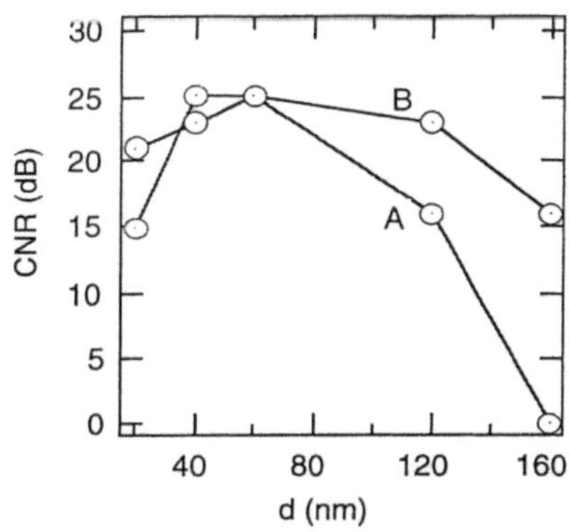

Fig. 14.6. Relationship between the intermediate SiN layer thickness and CNRs of Super-RENS disks. A: 100 nm marks and B: 300 nm marks. The signal intensity of the curve A strongly depends on the intermediate dielectric layer thickness, while the intensity a little decreases against the thickness. The resolution limit of the optics is about 250–300 nm

sized mark trains is shown in Fig. 14.5. Figure 14.5a is the spectrum by the readout with 1.0 mW, and 14.5b the spectrum with 3.5 mW.

It is clear that the refractive index change generates a strong near-field interaction between the aperture produced in the Sb layer and the phase change recording layer through the thin SiN layer. That is, the super-RENS can work just like a SNOM, but with a scanning speed one million times faster than the SNOM. Figure 14.6 shows the relationship between the signal intensity (carrier-to-noise ratio, CNR) and the intermediate solid layer thickness. The intensity of recorded marks in the resolution limit (half the diffraction limit) is little changed against the thickness, while marks beyond the resolution limit strongly depend on the thickness.

Optical near-field intensity is known to reduce exponentially with distance, so the characteristics of the super-RENS disks agree well with near-field properties.

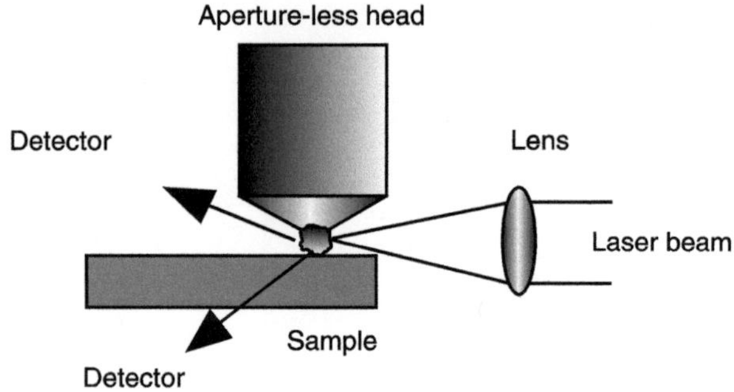

Fig. 7. Depicted configuration of aperture-less NSOM

14.4 The Proposal of Light-Scattering-Mode Super-RENS

Many NSOM studies have been done using aperture-less tips as well as aperture probes [8, 9]. In an aperture-less system, a laser beam is externally introduced on the surface of the aperture-less tip, as shown in Fig. 14.7.

The tip is usually metal-coated with a silver film to generate a surface plasmon by laser irradiation. The surface plasmon interacts with an electrical field on a sample surface, and then is scattered as far-field signals. In order to introduce the aperture-less tip into the super-RENS, one may think that a silver layer is the best material. However, it does not show little localized optical change without total reflection technique. Instead, we tried to use silver particles generated by the chemical reaction of silver oxide (AgOx). AgOx layers were first applied to optical recording as a recording material for CD-R using "nano-explosion" [10, 11]. In the study, a chemical reaction was used to produce pits on the surface of the polycarbonate disk by the heat and the high pressure, when the film was decomposed. Unfortunately, the chemical reaction is usually irreversible in a free space. We found later that the reaction becomes reversible in the same sandwich structure as that of the Sb-super-RENS, and is named "light-scattering-mode" super-RENS.

14.5 The Characteristics of Light-Scattering-Mode Super-RENS Disks

The light-scattering-mode super-RENS disk has almost the same structure as that of the Sb-super-RENS disk (now "transmission-mode" super-RENS disk), except for using a silver oxide layer instead of an Sb layer.

Figure 14.8 shows the refractive index change of AgOx films deposited by reactive r.f. magnetron sputtering, as increasing O_2 gas ratio to the total gas flow: $Ar + O_2$. Within a small O_2 gas ratio (0–0.3), the real part (n) of the index rapidly increased, and the imaginary part (k) oppositely decreased. Both indices n and k became mostly constant at a gas ratio between 0.3 and 0.6. On further increasing the ratio, k gradually increased at about 1.0, while n changed little.

Raman spectra of the specimen without O_2 showed few peaks over the wide wavenumber range. With increasing the ratio, one clear peak and another sub-peak were identified at 964 and 1,609 cm^{-1}. Schmidt et al. estimated the refractive in-

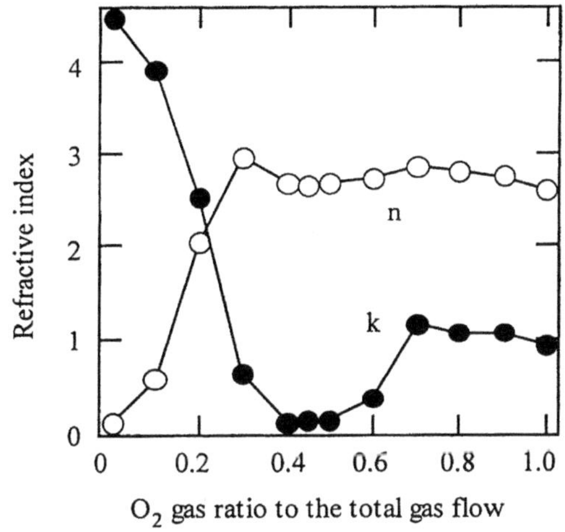

Fig. 8. AgO_x refractive index change by the oxygen gas ratio. n: real index and k: imaginary. All films were deposited by r.f. magnetron sputtering with the power of 200 W at the pressure of 0.5 Pa [13]. In Optical Storage and Optical Information Processing, Han-Ping D. Shieh, Tom D. Milster, Eds, Proceeding of SPIE 4081 (2000)

dex of an Ag_2O film by the envelope method, obtaining indices of $n = 2.5 \pm 0.1$ and $k = 0.11 \pm 0.02$ [12]. The results in this work showed $n \approx 2.7$ and $k \approx 0.08$ in a ratio between 0.3 and 0.5. In comparison with the above results, the peak at 964 cm^{-1} was contributed to by Ag_2O and the other peak at 1,609 cm^{-1} may be contributed to by AgO. Figure 14.9 shows the dynamic optical nonlinearity observed by rotating the disk sample at different speeds.

The relationship between the reflection output of the detector and readout laser power is usually linear in the absence of optical nonlinearity. The relationship was mostly linear at 8.0 m s^{-1}; however, nonlinearity was clearly observed at lower disk rotation speeds. Note that the nonlinearity is the opposite of that for Sb layers as shown in Fig. 14.2. Consequently, AgOx films produce a reflective spot by the focused laser beam. It was very interesting to apply AgOx layers, instead of Sb layers, as the mask of super-RENS disks. We produced super-RENS disks with

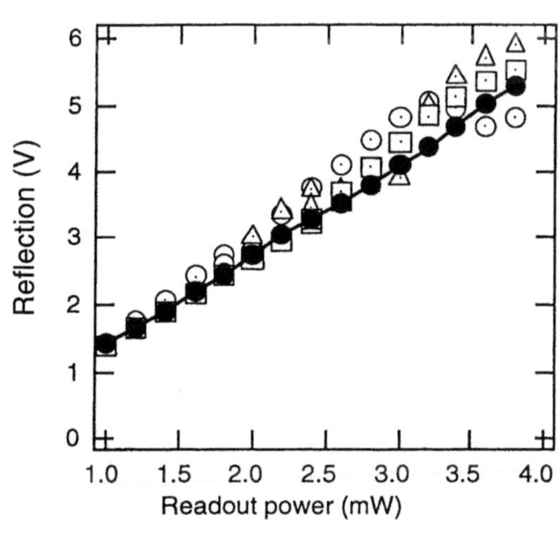

Fig. 9. Dynamic optical nonlinearity of a Ag_2O film under different disk rotation speeds. Circles: 0.6 m/s, triangles: 2.0 m/s, squares: 4.0 m/s and filled circles: 8.0 m/s. The reflection was the output voltage of the detector [13]. In Optical Storage and Optical Information Processing, Han-Ping D. Shieh, Tom D. Milster, Eds, Proceeding of SPIE 4081 (2000)

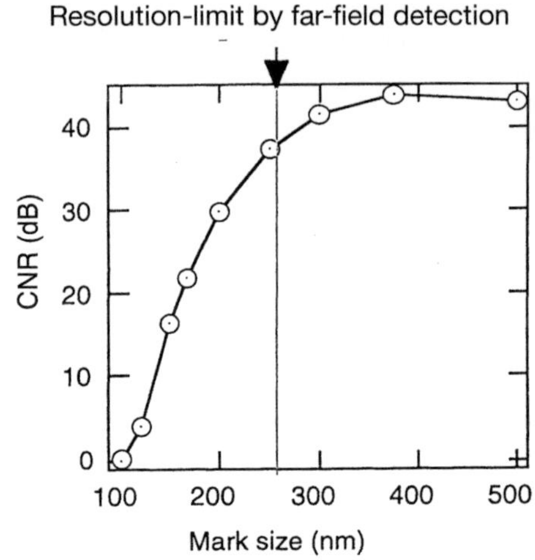

Fig. 10. The resolution limit of the light-scattering-mode super-RENS disk. Recording and readout powers were fixed at 6.5 mW and 2.5 mW. The disk rotation speed was 6.0 m s^{-1}. The disk structure was composed of ZnS-SiO_2 (170 nm)/Ag_2O (15 nm)/ ZnS-SiO_2 (20 nm)/GeSbTe- (20 nm)/ZnS-SiO_2(20 nm)

several AgOx layers. The thickness of one AgOx layer was fixed at 15 nm. Other layer thickesses were mostly the same as reported elsewhere [5, 6]. The resolution of the light-scattering-mode super-RENS disk with an O_2 gas ratio of 0.5 is shown in Fig. 14.10.

We found that small marks of 100 nm were detected using the AgOx film. As the gas ratio decreased, the threshold power for generating the scattering aperture increased. Threshold power became nearly constant at a ratio >0.3. On further reducing the gas ratio, the threshold power rapidly increased to >4.0 mW, as shown in Fig. 14.11.

This result suggests that this chemical reaction occurs in Ag_2O or a further O_2-rich composition, while the light-scattering center is not generated in a silver-rich state. We also found that the reaction is reversible and very stable in cyclability

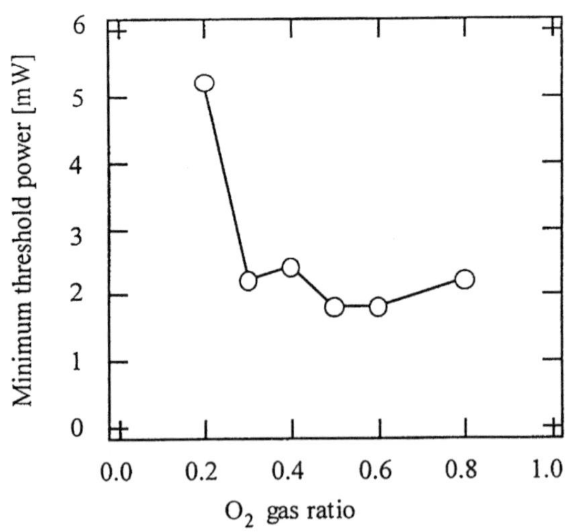

Fig. 11. The relationship between the minimum laser power to generate the near-field light scattering center and the oxygen gas ratio of AgO_x formation. Each disk structure was fixed at that shown in Fig. 10

readout. The super-RENS disk thus has a great advantage over the super-RENS disk using the Sb layer because the threshold power to generate near-field scattering is 1.0 mW lower than that of the Sb-type disk. This greatly reduces heat damage to the recording layer.

14.6 Summary

We proposed principles and described the basic characteristics of super-RENS disks. Super-RENS now has two different types: transmission and light-scattering modes. Disks of both modes could resolve marks to 100 nm far beyond the diffraction limit, with the aid of an optical near-field. At this moment, the CNRs observed from both super-RENS disks are not enough to be applied soon to actual optical recording systems because more than 40 dB is needed to reduce errors generated from laser, media, and heat. Further research is therefore necessary to improve the CNRs. However, it is valuable that the small marks less than 100 nm could be recorded and detected at m s^{-1} speeds beyond the diffraction limit using this technique. Once we can improve the low CNRs from the tiny marks, in principle, the storage density will increase to a few 100 s GB in a CD-size disk. We expect that this technique will become part of the main stream of optical data storage with optical near-field.

References

1. Beztig E, Trautman J (1992) Science 257:189
2. Milster T, Jo J, Hirota K, Shimura K, Zhang Y (1999) Jpn J Appl Phys 38:1793
3. Goto K (1998) Jpn J Appl Phys 37:2274
4. Lee MB, Kourogi M, Yatsui T, Tsutsui K, Atoda N, Ohtsu M (1999) Appl Opt 38:3566
5. TominagaJ, Nakano T, Atoda N (1998) Appl Phys Lett 73:2078
6. Tominaga J, Fuji H, Sato A, Nakano T, Fukaya T, Atoda N (1998) Jpn J Appl Phys 37:L1323
7. Tominaga J, Fuji H, Sato A, Nakano T, Atoda N (2000) Jpn J Appl Phys 39:957
8. Inouye Y, Kawata S (1994) Opt Lett 19:159
9. Fischer UC, Pohl DW (1989) Phys Rev Lett 62:458
10. Tominaga J, Haratani S, Uchiyama K, Takayama S (1992) Jpn J Appl Phys 31:2757
11. Haratani S, Tominaga J, Dohi H, Takayama S (1994) J Appl Phys 76:1297
12. Schmidt AA, Offermann J, Anton R (1996) Thin Solid Films 281–282:105
13. Tominaga J, Büchel D, Nakano T, Fuji H, Fukaya T, Atoda N (2000) SPIE 4081:86

15 Capacitance Storage Using a Ferroelectric Medium and a Scanning Capacitance Microscope (SCM)

Ryoichi Yamamoto

15.1 Introduction

15.1.1 "Near-Field" Data Storage Technology

Recently, a lot of research has been carried out concerning future data storage technologies, which are mainly centered on magnetic recording, optical recording, and scanning probe microscope (SPM) data storage.

Data storage technologies can be categorized as "near-field" technology and "far-field" technology depending on the distance from a probe tip to a medium. In the former the probe tip is very close to the medium. Magnetic recording and SPM data storage are typical of this category, and near-field optics are also included. The latter is provided with a focused radiation beam as a probe. Conventional optical recording and electron beam data storage are typical of this category.

The author considers the "near-field" technology to be more important, because fundamental limits on recording density do not exist. In the "near-field" technology, basically, recording bit size can be reduced to size that is about equal to the distance between the probe tip and the medium.

In the "far-field" technology, on the other hand, there is a limit on recording density essentially, which is a wave diffraction limit. Spot size of a focused beam cannot be reduced to less than about the wavelength. In order to overcome the limit, various research has been carried out. These include, for example, frequency-domain multiple recording using a multi-layered photochromic media [1], super-resolution reproduction using a small aperture [2], etc. However, they would only increase the recording density by about less than ten times.

15.1.2 Scanning Capacitance Microscope

Since the invention of the scanning tunnel microscope (STM) [3], many studies of data storage have been carried out using various types of scanning probe microscopes, such as, for example, the STM [4], an atomic force microscope (AFM) [5], a magnetic force microscope [6], a scanning near-field optical microscope [7], etc. One of them is a study using a scanning capacitance microscope (SCM), in which

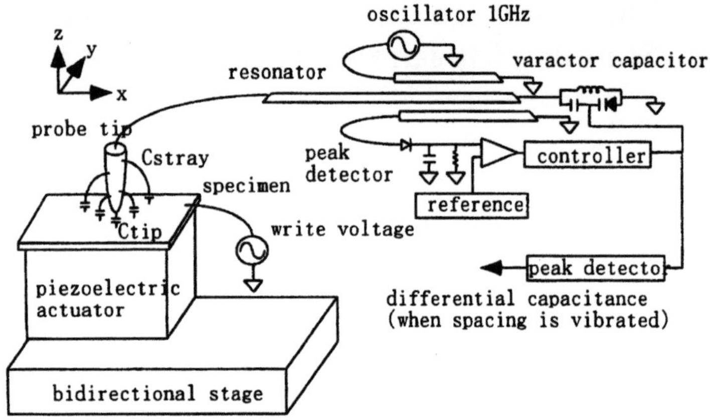

Fig. 15.1. Schematic diagram of the capacitance detection circuit

the probe is scanned over a surface of an object while measuring the capacitance between the probe tip and the object.

The first SCM was developed by RCA researchers [8], and several SCMs have since been developed by many researchers [9–18]. In the SCMs, a so-called RCA-type capacitance sensor is often used, which had been at first developed for videodisk systems such as CED and VHD videodisk systems [19]. The sensor has very high sensitivity which is typically about $\Delta C = 0.1$ fF on stray capacitance 0.1 pF. Currently zeptofarad (10^{-21} F) sensitivity has even been achieved [16].

A schematic circuit of the sensor is shown in Fig. 15.1 [20]. The heart of this detection method is an LC resonator circuit, the L of which is composed of a strip line and a fly lead involving kinds of probe tips at the end, while C is composed of C_{tip} (capacitance due to the gap between the tip and the medium in which an equivalent gap of the signal is contained) and C_{stray} (somewhat large stray capacitance). A UHF (typically 1 GHz) oscillator is coupled to the detection circuit through this resonator. The small changes in C_{tip} shift the center frequency of the resonator, thus changing the output voltage.

The RCA-type capacitance sensor has two advantages for data storage technologies. The first one is wide bandwidth. Usually, frequency of the UHF oscillator in Fig. 15.1 is about 1 GHz. In this case, a capacitance signal of the order of 10 MHz can be detected. Actually in the VHD videodisk system, the highest frequency of FM signal was about 8 MHz [21]. The second one is high spatial resolution. In the late 1970s, the VHD videodisk system had been already provided with recording density the same as current optical memories [22], in which the capacitance sensor was able to reproduce 0.25-μm pit size. Currently the spatial resolution is of the order of 10–50 nm [17, 18].

15.1.3 Capacitance Storage

The first capacitance storage was the abovementioned CED videodisk. In this videodisk bumps were formed at a surface of the disk as data and caused change of capacitance between the probe tip of the capacitance sensor and the disk.

But the CED or VHD video disk systems were read-only systems. As a data storage system, some additional functions would be required. The "charge storage" method has been studied in order to have the ability to record, erase, and rewrite data. Figure 15.2 shows the principle of the charge storage and data reproduction method using an SCM [23], which is the same as what is used in a solid-state nonvolatile memory called erasable-programmable ROM (EPROM).

Its medium typically comprises a silicon substrate (e.g., p-Si), 1st insulator layer (typically silicon oxide), 2nd insulator layer (typically silicon nitride), and a gate electrode. In order to store charge, negative voltage (e.g., for p-Si) is applied to the gate electrode. Some holes (e.g., for p-Si) are injected into the 2nd insulator through the 1st insulator, and these are stored at traps in the 2nd insulator. By erasing the stored charge, the reverse process may be done. In the case of a system using the SCM, a scanning probe tip of the SCM is substituted for the gate electrode.

To reproduce the data caused by the stored charge using an SCM, capacitance between the probe tip (i.e., gate electrode) and the silicon substrate is measured. In the figure, when charge is stored in the 2nd insulator, a depletion region is built up in the silicon and the total capacitance is reduced. On the other hand, when there is not the trapped charge in the 2nd insulator, the depletion capacitance does not exist.

Several media were studied for the charge storage method, for example, silicon nitride/silicon oxide/silicon (NOS)-layered structure [24–26], only silicon oxide on silicon [27], ONOS [28], and metal clusters in SiO_2 [29]. But it should be noted that the charge stored in these media cannot be retained for a long time in comparison with, for example, magnetic recording generally.

In the mid-1990s, another method of capacitance storage was proposed [20, 30]. In the method a ferroelectrics/silicon (F/S) structure medium is used instead of the

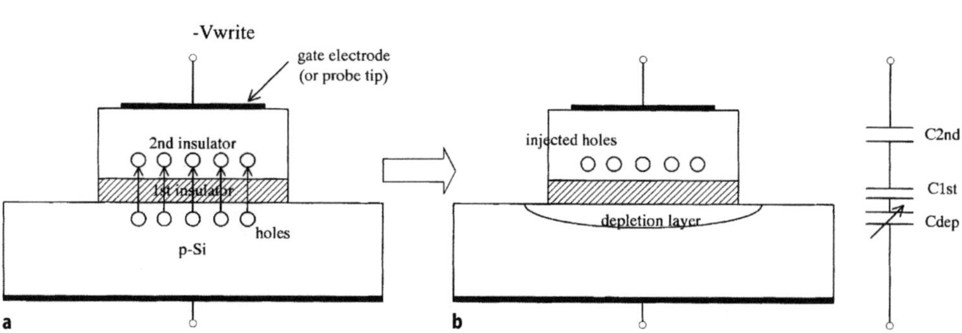

Fig. 15.2a,b. Principle of the charge storage method: **a** shows charge injection into 2nd insulator and **b** shows that depletion capacitance is built up in the p-Si and the total capacitance is reduced

charge storage media and data is stored by a spontaneous ferroelectric polarization. The capacitance storage using this F/S medium and an SCM, which is especially in a non-contact mode in order to avoid wear of the probe tip and the medium, have the following features;

(1) Higher recording density
(2) Wide bandwidth
(3) Pre-pit tracking
(4) Wear free
(5) Long-term data retention
(6) Direct overwriting.

It keeps merits using SCMs ((1)–(4)), and further, improves some poor properties using charge storage ((5), and (6)). In this chapter, the author describes the capacitance storage using the F/S medium and the non-contact-mode SCM in detail.

15.2 A Non-Contact-Mode SCM

First of all, the author describes a non-contact-mode SCM developed by Sanada et al. [20], and a typical capacitance image obtained is also shown.

15.2.1 Capacitance Detection Method

In the non-contact-mode SCM, the RCA-type sensor was used as a capacitance detector. The principle of capacitance detection was described in the previous section. In Fig. 15.3 [20], for capacitance detection, a varactor capacitor is used to adjust the resonator center frequency so that the oscillator always drives the same slope of the resonant curve. This method allows signal detection utilizing the

Fig. 15.3. Signal response of detector output

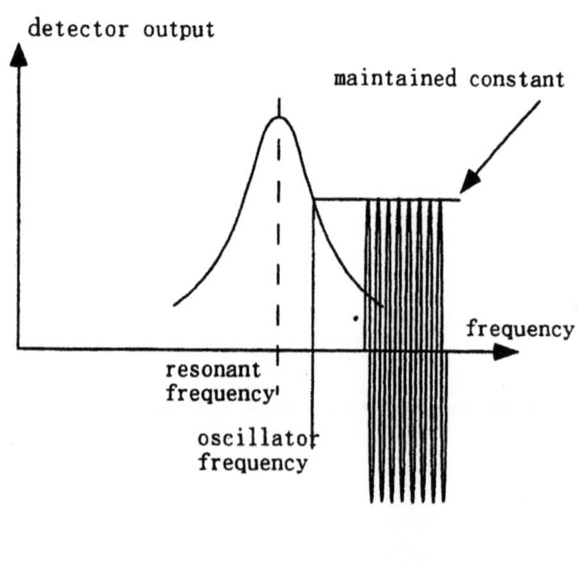

drive signal of the varactor and ensures the detection of capacitance even if the probe is located as far as several mm from the specimen. In the conventional videodisk system, this stray capacitance was approximately 0.1 pF and the signal was 0.1 fF.

In the system, Q was 25, center frequency was 1 GHz, and the –3 dB point of the resonator was chosen for signal detection. Noise level at the detector was estimated to be 6×10^{-19} F $\mathrm{Hz}^{-1/2}$ around 1 kHz and much lower at higher frequencies (1×10^{-21} F $\mathrm{Hz}^{-1/2}$ at 1 MHz). This noise was found to be of oscillator or of resonator fluctuation because of stray capacitance.

15.2.2 Spacing Control Method Using Differential Capacitance

In the non-contact-mode SCM, in order to control probe tip-medium spacing differential capacitance was used [20]. A piezoelectric stack actuator for vertical fine motion was vibrated 3 $\mathrm{nm_{pp}}$ in the vertical direction at 1 kHz for measuring differential capacitance. The peak detector output of the drive signal of the varactor indicated differential capacitance (Fig. 15.1).

The feedback system was designed as follows. Differential capacitance was maintained constant at frequencies below the bandwidth of 70 Hz. Since the feedback stabilizes with a steady-state error inversely proportional to the closed-loop gain, an integrator was utilized in the electrical feedback circuitry to minimize the steady-state error. Gain at low frequencies was set to approximately 400.

Experimental setup is shown in Fig. 15.1. A piezostack for z motion was placed on a bidirectional stage. Over the stack was a mechanically removable specimen plate, which was connected to a voltage source for poling. A specimen of the medium was cut into a 10 mm × 10 mm square and was fixed electrically and mechanically to this plate. The mechanical resonant frequency of this stage including the specimen was about 3 kHz.

Sanada et al. [20] have shown that typically it is possible to control the spacing in a height below 30 nm, where the remaining total spacing error on feedback was less than 10 $\mathrm{nm_{pp}}$.

15.2.3 Non-Contact Mapping of Surface Topography

Sanada et al. [20] evaluated the system performance as a surface profilometer utilizing capacitance. Capacitance and differential capacitance were simultaneously used to construct a topographical image.

In general, there were two ways of constructing surface images using SCM: first, spacing was kept constant using capacitance or differential capacitance feedback and the vertical motion of the actuator or $\mathrm{d}C/\mathrm{d}V$ was displayed as images [9–11]; second, the capacitance signal was displayed as images while the probe tip was in contact with the specimen as a contact mode AFM [8, 24, 25].

In the first case, the bandwidth of the signal is limited by that of the servo system and/or $\mathrm{d}C/\mathrm{d}V$ detection, while in the second case, the tip and the medium get a lot of wear.

Fig. 15.4. Capacitance image for a topography reference, where two-dimensional arrays of 5 μm × 5 μm square pits of depth 0.1 μm are located 10 μm apart

In order to avoid frequency limitations and to maintain non-contact operation, the capacitance signal for image was used while activating differential capacitance feedback.

Figure 15.4 [20] shows a typical image obtained by the abovementioned non-contact-mode SCM with an electrochemically etched tungsten probe 12 μm in tip radius.

The surface topography reference for STM was used as the specimen. The surface of this specimen was coated with platinum and had square pits of 5 μm × 5 μm (depth 0.1 μm), each of which was separated by a 10-μm pitch in both x and y directions. This figure is the raster-scanned view of capacitance, while the differential capacitance was kept constant at about 50 nm.

15.3 Data Storage Method Using the F/S Medium

In this section, the author describes a data storage method using the F/S medium. As the author mentioned above, it is analogous to the charge storage.

Figure 15.5 [30] shows a schematic illustration of the basic concept of a data storage method using the F/S medium. Data are recorded as direction of polarization of ferroelectric film. As recording data, voltage is applied between the probe tip and the semiconductor substrate. The applied voltage is poling the spontaneous polarization P_s of the ferroelectric film along the field, and then data are memorized in the medium as polarization. In order to erase the data, a reverse voltage is applied. For data reproduction, capacitance between the medium and the probe tip is detected the same as the charge storage described previously.

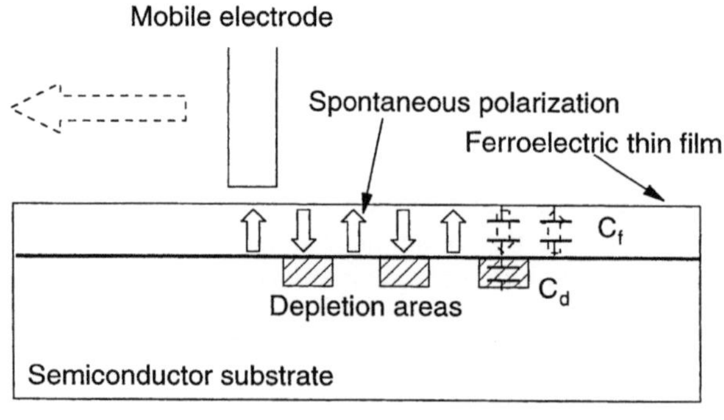

Fig. 15.5. Schematic illustration of basic concept of data storage method using the F/S medium

For example, when a p-type semiconductor is used and the polarization is directed downward, a depletion region is built up and the capacitance is reduced like in a metal-ferroelectric-semiconductor (MFS) diode [31]. On the other hand, as the polarization is directed upward, an accumulated layer is created and the depletion capacitance disappears. Thus, data stored as the ferroelectric polarizations can be reproduced by capacitance measurement.

15.4 Static Characteristics of the F/S Medium

Yamamoto et al. [30] used polymer ferroelectrics in the F/S media and studied its static characteristics. The author describes materials of the F/S medium and its static characteristics, which are ferroelectric properties and *C-V* characteristics. The *C-V* characteristics demonstrate the concept of a data storage method using the F/S medium. Both a fixed electrode on the medium and a non-contact probe electrode of the non-contact-mode SCM were used to carry out all experiments described in this section.

15.4.1 Materials and Ferroelectric Properties of the F/S Medium

Until now, as F/S medium there have been only combinations of a copolymer of vinylidenefluoride-trifluoroethylene (VDF-TrFE) and Si [30]. A VDF (58 mol%)-TrFE is known to adopt an all-*trans* structure which is a ferroelectric phase in melt-crystallized or solution-cast films [32], even though on silicon.

F/S media described in this section comprises a VDF-TrFE film as ferroelectrics which is cast with 0.1- to 1.0-μm thick film by the spin-coat technique [33], and a p-type silicon (p-Si) wafer as the semiconductor substrate, which had (100) orientation and 0.35- to 5-Ωcm resistivity.

The specimens consisting of only VDF-TrFE and p-Si were used for non-contact evaluations of the static characteristics and evaluations of dynamic recording characteristics as described later. In order to measure ferroelectric properties and static *C-V* characteristics, additional aluminum top electrodes (0.5 mm–5 mm in diameter) were prepared by vacuum deposition.

Figure 15.6 [30] shows a typical ferroelectric property of VDF-TrFE. The measurement was carried out by a modified Sawyer-Tower circuit, which is shown in Fig. 15.7a, under 20-mHz triangular electric field at room temperature.

Figure 15.6a shows a *D-E* curve of a VDF-TrFE film with 1.0-μm thickness. The hysteresis loop shows good squareness, and the values of remanent polarization (P_r) and coercive field (E_c) were 50 mC m^{-2} and 50 MV m^{-1}, respectively, which values were close to those of the previous report [32]. Figure 15.6b shows a *D-E* hysteresis curve of that with submicron thickness (0.22 μm) film. Even for the 0.22-μm-thick films a clear hysteresis loop was observed.

In Fig. 15.6c, the *D-E* hysteresis curve of the F/S medium with VDF-TrFE film of 1.0-μm thickness and 5-Ωcm-resistivity p-Si also shows. It should be noted that the curve shows an asymmetrical shape, which is different from Fig. 15.6a. The reason for this will be described at the end of this subsection.

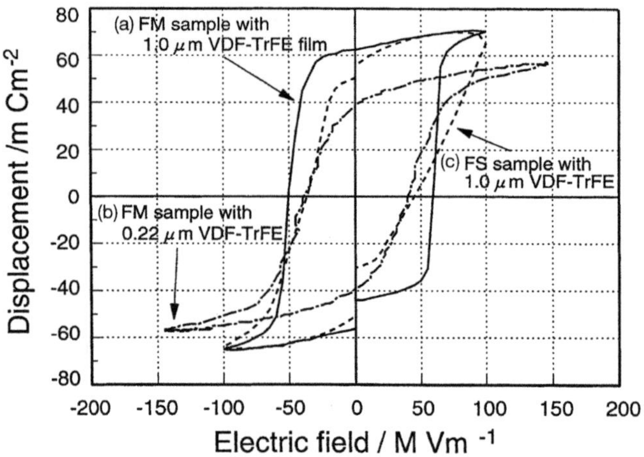

Fig. 6a–c. *D-E* hysteresis characteristics. **a** and **b** show the results of VDF-TrFE film 1.0-μm and 0.22-μm thick and are measured by applying slow-scanned voltage (20 mHz) and fast-scanned voltage (50 Hz), respectively; **c** shows the result of the 1.0-μm-thick VDF-TrFE film under the slow-scanned voltage (20 mHz)

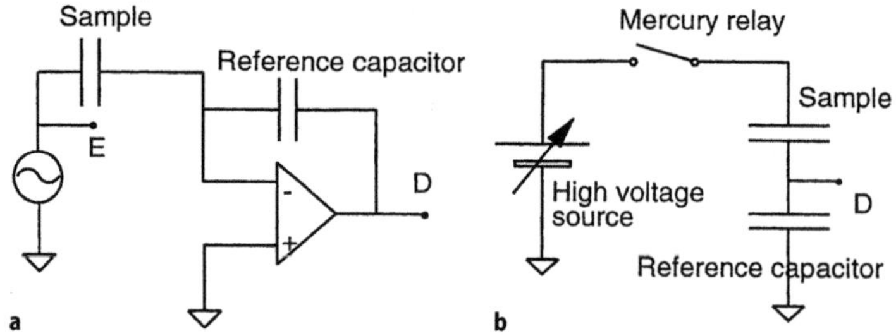

Fig. 15.7a, b. Schematic diagrams of measurement circuits. A modified Sawyer-Tower circuit is shown by **a**. The polarization switching characteristics were measured by circuit **b**

Figure 15.8 [30] shows typical switching characteristics of an F/S medium with a 1.0-μm-thick VDF-TrFE film and additional intermediate oxide layer with a 50-nm thickness between the VDF-TrFE film and the Si wafer. The measurement was carried out in which a circuit shown in Fig. 15.7b was used.

In this figure, electric displacement (*D*) was plotted against the time (*t*). Figures 15.8a and 15.8b show results when +150 V and −150 V step-function voltages were respectively applied to a top electrode 5 mm in diameter. It should be noted that *D* increased linearly and slowly when positive voltage was applied (see Fig. 15.8a).

It was considered that the asymmetric nature of the F/S medium shown in *D-E* characteristics and switching characteristics would occur for the same origin, which would be an effect of minority carriers, and would be not an intrinsic property of the VDF-TrFE films. It was suggested that, for example, the polarization switching time increased because of increasing the generation ratio of the minority car-

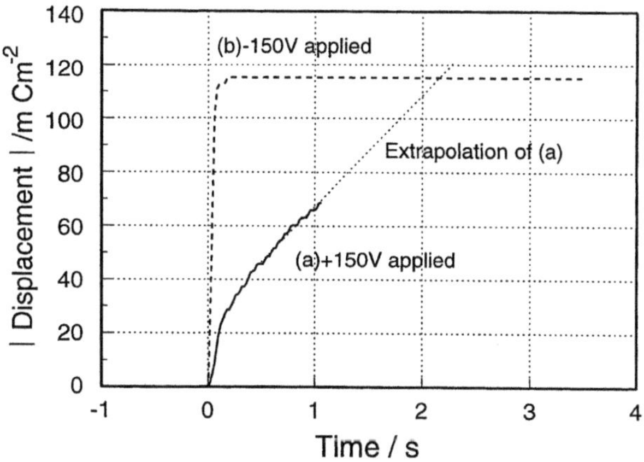

Fig. 15.8a, b. Typical polarization switching behavior of F/S medium with 1.0-μm VDF-TrFE film in thickness, intermediate oxide layer and 5-mm-diameter top electrode. **a** and **b** show results obtained by applying +150 V and −150 V, respectively

riers by irradiation [30]. There was also a report that the minority carriers affected ferroelectric characteristics [34].

15.4.2 Shift of C-V Characteristics Using Fixed Electrodes

Next, the author shows typical results of the *C-V* characteristics of the F/S medium. Figure 15.9 [30] shows typical *C-V* curves of the F/S specimen with VDF-TrFE film 1.0 μm in thickness. Figure 15.9a shows a *C-V* curve of the as-prepared specimen. It is close to the ideal high frequency *C-V* curve of a metal-insulator-semiconductor (MIS) capacitor. Figures 15.9b and 15.9c show *C-V* curves of the specimens poled by +100 V and −100 V, respectively. The poling could be regarded as data recording. The positive voltage directed the polarization downward; then a positive charge due to polarization appeared at the semiconductor surface and caused the *C-V* curve to shift toward the negative direction of the *x* axis (see Fig. 15.9b). In the same way, when the negative voltage was applied, the polariza-

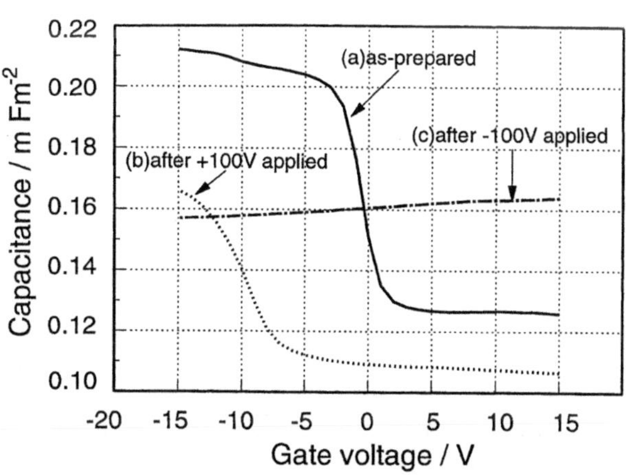

Fig. 15.9a–c. High-frequency *C-V* characteristics of the F/S specimen with 1.0-μm VDF-TrFE film in thickness; **a** shows the result of as-prepared specimen; **b** and **c** are the results of the specimen poled by +100 V and −100 V, respectively

tion was ordered upward. The *C-V* curve was then shifted toward the positive direction (see Fig. 15.9c).

The capacitance measurement at zero gate voltage could be regarded as data reproduction. The lower capacitance shows that a depletion region associated with the downward polarization occurs and the higher capacitance shows that holes are accumulated by the upward polarization. The amount of the difference in "remanent" capacitance (ΔC) corresponded to a reproduction signal amplitude. In the current experiments it was found to be 50 μF m^{-2}.

15.4.3 Non-Contact Demonstration

In a non-contact manner, polings of the ferroelectrics and capacitance variation due to the polarization observed in a *C-Z* plot are demonstrated [30]. The polings were carried out while the spacing was maintained at 100 nm using a feedback loop. The effect of attractive electrostatic force between the tip and the medium at high voltage application on the surface of the medium was avoided by the feedback loop. Since a typical *C-V* curve shift caused by change in polarization was observed at voltage application of ± 50 V_{pp} when the tip and the medium were in contact, a voltage application 4 times larger (± 200 V) was applied for polarization at 100-nm spacing when $\varepsilon = 10$ and thickness = 0.3 μm.

Figure 15.10 [30] shows capacitance versus spacing after poling at +200 V (ferroelectrics are poled upward) and –200 V (poled downward). In the figure, plots at minus spacing indicated that the tip and the medium were in contact and that the tip was pressing the medium. Figure 15.10b shows the subtraction of the two plots ΔC versus spacing. The capacitance variation, which is considered as the signal, was gradually increased as spacing was decreased. This capacitance variation could be observed at the maximum spacing of 100 nm using the method and the apparatus dimensions described above.

In Fig. 15.10, difference in capacitance ΔC between upward-poled medium and downward-poled medium indicates that polings and readings can be made while the tip is not in contact with the medium. It is important that the signal can be observed at certain values of spacing.

The intensity of the signal should be further discussed from the viewpoint of signal-to-noise ratio. According to the parallel-plate model, capacitance difference ΔC is estimated to be

$$\Delta C = \varepsilon_0 \frac{S}{z + d/\varepsilon_f} - \varepsilon_0 \frac{S}{z + d/\varepsilon_f + w_{\max}/\varepsilon_S} \tag{15.1}$$

where z is the spacing between the tip and the specimen, d is the thickness of the ferroelectric layer, $w_{\max}$ is the depth of the depletion layer in the silicon, and ε_f and ε_S are dielectric constants of the ferroelectric layer and silicon, respectively.

Assuming the area of the parallel plate to be $S = 1$ μm^2, this calculated curve can be fitted to the experimental curve (Fig. 15.10b). The actual surface of the tip concerned seems to be smaller than the tip radius, where only the apex of the tip must be having an effect. This is desirable for realizing high-density memories utilizing comparably large tips for probing.

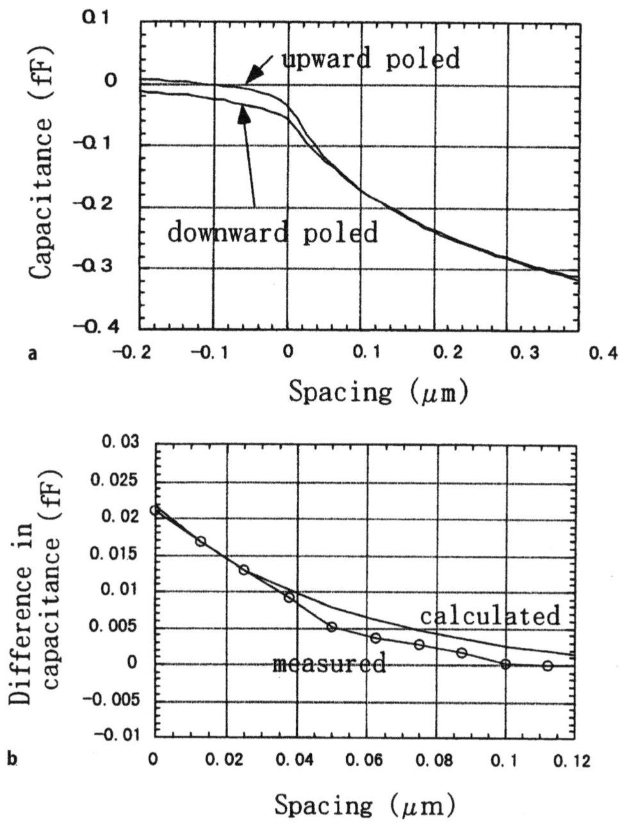

Fig. 15.10a, b. Capacitance versus spacing for upward and downward poling **a** and difference in capacitance ΔC between upward poling and downward poling versus spacing **b**

15.5 Dynamic Characteristics of the F/S Medium

In this section some dynamic characteristics and the recorded bits shape observed by xerographic method are shown. Experiments shown in this section were carried out by using a "contact-mode" SCM. In the contact-mode SCM the contact force was controlled by a rubber spring which is used in the conventional VHD system [21] and other parts were the same as the non-contact-mode SCM described earlier.

15.5.1 Dynamic Recording Characteristics

In the evaluations of dynamic recording characteristics two different type probes were used. One of them was an electrochemically etched tungsten probe with tip radii of curvature 2, 5, and 12 μm. Another was a diamond probe with a thin-film carbon electrode about 0.3 μm in thickness and about 0.9 μm in width, which was used in commercial VHD videodisk systems.

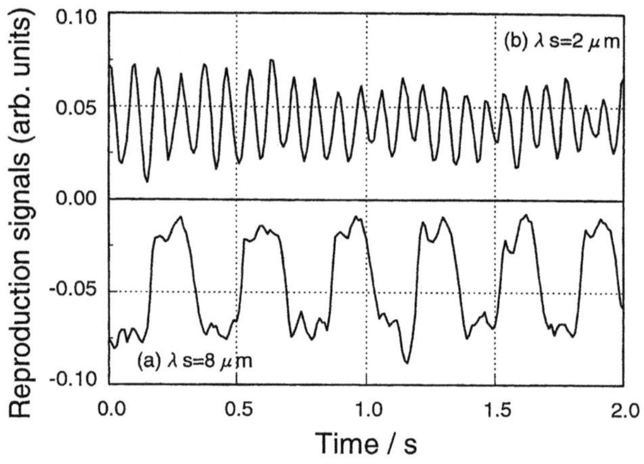

Fig. 15.11. Typical real-time reproduction signals with 8 μm and 2 μm in spatial wavelength. In the measurement, F/S specimens with VDF-TrFE film 0.3 μm in thickness, a diamond probe and ±50 V alternating voltage as recording signal were used

Figure 15.11 [30] shows the typical real-time reproduction signals obtained by the contact-mode SCM. Recording was done by applying alternating voltage pulses with ±50 V as the signals. The specimen has a 0.3-μm-thick VDF-TrFE film and 0.35-Ωcm p-Si. Measurements were carried out, in which the diamond probe was used and relative velocity between the probe and the specimen was 20 μm s^{-1}. The lower and upper curves in Fig. 15.11 show the reproduction signal with 8 μm and 2 μm in spatial wavelength (λ_s).

Evaluations at various frequencies were also carried out. The reproduction signals were analyzed by fast Fourier transformation (FFT). Figure 15.12 [30] shows the signal amplitude obtained by FFT analysis against the λ_s.

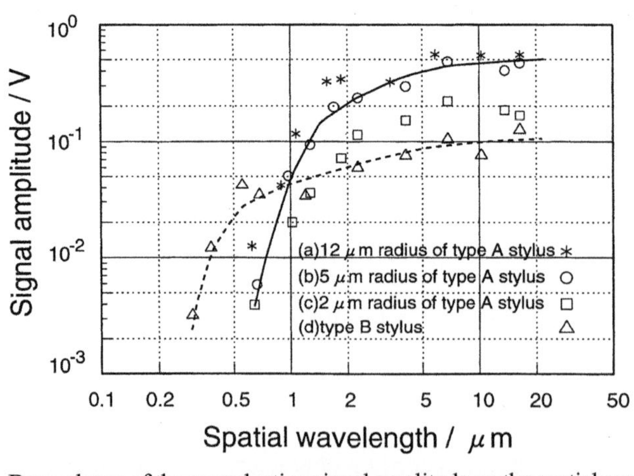

Fig. 15.12a–d. Dependence of the reproduction signal amplitude on the spatial wavelength. An F/S specimen with a VDF-TrFE film 0.3 μm in thickness was used. **a**, **b**, and **c** show the results of using the tungsten probes with 12-, 5-, or 2-μm tip radius of curvature, which are denoted by *, ○, and ◇, respectively, and a *solid line* also shows the result of using the probe with the 5-μm tip radius; **d** shows the result of using the diamond probe denoted by Δ and a *dashed line*

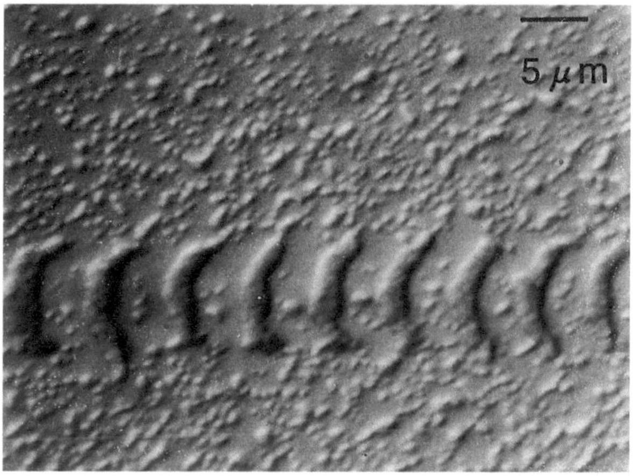

Fig. 15.13. Typical photo of developed recording bits. Data recording was carried out by applying ±100 V alternating voltage and using tungsten probes with 12-μm tip radius on the F/S specimen with 0.5-μm VDF-TrFE film

From Fig. 15.12, the minimum recording wavelength was suggested to be about 0.4-μm wavelength. This will be considered to be the value that would be determined by the tip size of the probe tip. It will be discussed again later.

15.5.2 Bit Shape Observation

In order to observe the recorded bit shape, Hatano et al. tried to develop latent images of the bit using xerographic liquid toner [35], which is provided with polymer beads 0.1 μm in radius carrying positive charges.

Figure 15.13 [30] shows a typical photo of a developed bit image, in which the bits were recorded on the specimen with 0.5 μm VDF-TrFE film, to which was applied ±100 V alternating voltage by a tungsten probe with 12-μm tip radius of curvature. In this figure, the bit shapes with spatial wavelength of about 5.0–6.5 μm and width of about 9 μm are clearly shown. The shape of the bits are significantly like a crescent. By this method, much smaller bits with about 1.0-μm wavelength could be observed.

The crescent shape of the bits is analogous to the shape of "magnetic field modulation" recording bits in magneto-optical (MO) recording [36]. It also indicates that the F/S medium was provided with the ability of "direct overwriting".

15.6 Numerical Simulation Studies of F/S Medium

Yamamoto et al. [39] have studied capacitance modulation of the F/S medium using numerical simulation compared with capacitance modulation of the charge storage. The study showed that characteristics of the F/S medium were different from those of the charge storage. In this section the author shows their numerical simulation studies.

15.6.1 Calculation Method

Electrostatic capacitance (C) is defined as follows;

$$C = \frac{dQ_e}{dV_e} \tag{15.2}$$

where Q_e and V_e are the amount of induced charge in the electrode and its potential, respectively.

In a semiconductor, dynamic behavior of carriers is described by three groups of basic equations: [37] Poisson's equation, current-density equations, and continuity equations:

$$\mathrm{div}(\mathrm{grad}\psi) = -\frac{\rho}{\varepsilon} \tag{15.3}$$

$$J_{\mathrm{n}} = q\mu_{\mathrm{n}} n E + q D_{\mathrm{n}} \nabla n \tag{15.4}$$

$$J_{\mathrm{p}} = q\mu_{\mathrm{p}} p E - q D_{\mathrm{p}} \nabla p \tag{15.5}$$

$$\nabla \cdot J_{\mathrm{n}} - q\frac{\partial n}{\partial t} = -qGR_{\mathrm{n}} \tag{15.6}$$

$$\nabla \cdot J_{\mathrm{p}} + q\frac{\partial p}{\partial t} = qGR_{\mathrm{p}} \tag{15.7}$$

where ψ, ρ, and ε are potential, charge density which includes ionized acceptors or donors, thermally generated carriers, and fixed charges which are due to remanent polarization in a ferroelectric, and permittivity, n and p are electron and hole densities, q is elemental charge, and J, μ, and D values are currents, mobilities, and diffusion constants, in which suffixes n and p correspond to electrons and holes, GRs are generation/recombination rates of each type of carrier, respectively.

From Eqs. (15.4)–(15.7) and $E = -\nabla\psi$, we obtain

$$\frac{\partial n}{\partial t} + \mathrm{div}(\mu_{\mathrm{n}} \cdot n \cdot \mathrm{grad}\psi - D_{\mathrm{n}} \cdot \mathrm{grad}) = GR_{\mathrm{n}} \tag{15.8}$$

$$\frac{\partial p}{\partial t} - \mathrm{div}(\mu_{\mathrm{p}} \cdot n \cdot \mathrm{grad}\psi + D_{\mathrm{p}} \cdot \mathrm{grad}) = GR_{\mathrm{p}} \tag{15.9}$$

Generally, for calculating transient behavior of electron/hole density a "quasi-equilibrium" assumption is used [38]. In this assumption the electron/hole densities given by

$$n = n_{\mathrm{i}} \exp \frac{q(\psi - \phi_{\mathrm{n}})}{kT} \tag{15.10}$$

$$p = n_{\mathrm{i}} \exp \frac{q(\phi_{\mathrm{p}} - \psi)}{kT} \tag{15.11}$$

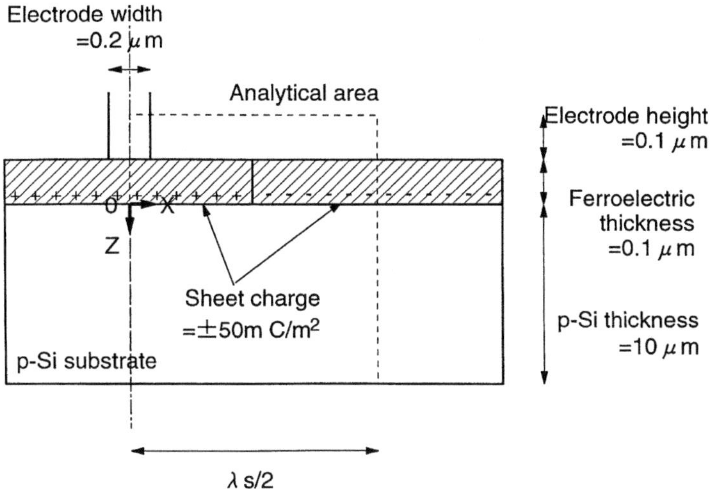

Fig. 15.14. Schematic cross-section of a calculation model. Area shown by *dashed lines* indicates the calculated area

where ϕ_n and ϕ_p are quasi-Fermi levels for each type of carrier; k, n_i, and T are the Boltzmann constant, intrinsic carrier density, and absolute temperature, respectively.

In order to simplify the calculation, a "high-frequency" approximation was used [37], in which the minority carriers do not respond to change of the potential with frequency higher than several hundred Hz. Since all measurements were carried out at about 1 GHz in the previous experiments, the high-frequency approximation should be valid. The capacitance was calculated in that time-dependent terms are neglected in Eqs. (15.3) and (15.8)–(15.11).

Figure 15.14 [39] shows a calculation model. The model is two-dimensional and the area shown by dashed lines in the figure indicates the calculated area. The origin of the z axis is defined as the interface between the ferroelectric and p-Si, and that of the x axis is the leftmost side of the calculated area. It is assumed that the fixed charges due to the remanent polarization of the ferroelectric film appear only at the interface with the semiconductor.

Important geometry and parameters used in these calculations are the probe tip width of 0.2 μm, remanent polarization of 50 mC m^{-2}, thickness of the ferroelectric of 0.1 μm, λ of 0.3–2.0 μm, probe tip-medium distance (d) of 0–1.0 μm, and acceptor concentration (N_a) of 10^{15} and 10^{16} cm^{-3}.

As boundary conditions, Dirichlet's condition along the z axis and Neumann's condition along the x axis are adopted. The calculations are performed by the finite element method typically with a 30 × 40 grid inside the area shown by dashed lines.

Commercially available software was used to solve Eqs. (15.3) and (15.8)–(15.11) numerically.

Procedure of capacitance calculation is as follows: first, the amounts of induced charge in the electrode ($Q_e(0)$) and accumulated minority carriers at the interface

($Q_m(0)$) are calculated under the condition that the electrode potential is zero, next the electrode potential is changed to a finite value V_1, and the dependence of $Q_m(V_1)$ on the quasi-Fermi level is calculated. In the high-frequency approximation, the quasi-Fermi level is adjusted so as to make $Q_m(V_1)$ equal to $Q_m(0)$. $Q_e(V_1)$ is calculated in which the adjusted quasi-Fermi level (ϕ_a) is used. Capacitance (C) is given by

$$C = \frac{Q_e(V_1) - Q_e(0)}{V_1} \tag{15.12}$$

Capacitance modulation (ΔC) is obtained as

$$\Delta C = C_- - C_+ \tag{15.13}$$

where C_- and C_+ are the capacitances when the probe is just above the center of an upward poled domain and a downward poled domain, respectively.

15.6.2 Capacitance Modulation of the F/S Medium

Figure 15.15 [39] shows a typical result of the dependence of ΔC on spatial wavelength (λ). From the figure, the two natures of ΔC of the F/S medium should be noted.

At first, ΔC would not disappear in the F/S medium as the spatial wavelength is reduced to less than "punch-through" wavelength (λ_p) at which the depletion areas connect with each other.

In the charge storage, it had been considered that ΔC would disappear when the spatial wavelength is shorter than λ_p [25]. However the numerical simulation showed that ΔC remained for an F/S medium [39].

This reason is considered to be as follows. In the F/S medium the recorded bits consist of alternate domains with opposite polarization in the ferroelectric, and thus areas of majority carrier depletion and accumulation always appear alternately at the surface of the semiconductor. As shown in Fig. 15.16 [39], even if the depletion areas connect with each other, the connected region exists only in the inner part

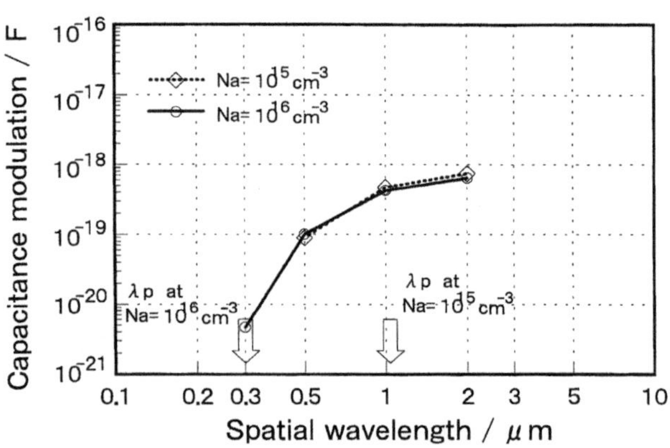

Fig. 15.15. λ-dependence of ΔC for $d = 0.03$ μm, where the *dashed line* and the *solid line* correspond to the results for $N_a = 10^{15}$ cm^{-3} and $N_a = 10^{16}$ cm^{-3}, respectively, and *arrows* indicate "punch-through" wavelength (λ_p) for each N_a

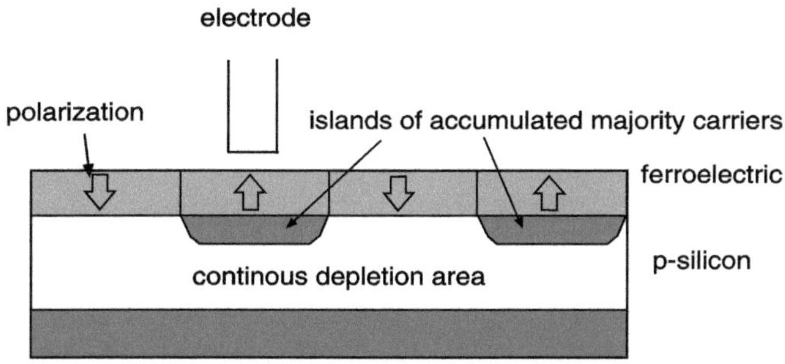

Fig. 15.16. Schematic cross-section of an F/S medium when depletion areas connect in the semiconductor and form a single depletion area. Regions of majority carrier accumulation remain and have finite thickness

of the semiconductor, and at the surface islands of accumulation regions with finite thickness always remain. Therefore, ΔC has a finite value.

On the other hand, the charge storages mostly store only electrons or holes [23, 25]. In such a case, there are no accumulated island areas, and so the capacitance modulation disappears in the "punch-through" state. This suggests potential of the F/S medium for higher-density recording compared with the charge storages.

Second, in Fig. 15.16, the dependence of ΔC on λ looks like that of magnetic recording. It is well known that for magnetic recording the dependence of output signal on recording wavelength is affected by the reproducing head gap. When the wavelength is equal to the head gap, the output signal vanishes [40]. In the Fig. 15.16, the ΔC steeply reduces as λ reduces to 0.2–0.3 µm when the probe tip thickness is 0.2 µm.

The experimental results (see Fig. 15.12) are also similar to the calculated results. In the experiments, a probe with 0.3-µm-thick tip is used and ΔC vanishes around $\lambda = 0.3$–0.4 µm. Thus, it suggests that the minimum recording wavelength shown in Fig. 15.12 would be mainly determined by tip size of the probe.

15.7 Summary and Possibilities of Capacitance Storage

15.7.1 Summary

In this chapter, the author described a capacitance storage using a ferroelectric/silicon medium and a non-contact-mode SCM.

A non-contact-mode SCM was developed in order to avoid wear of the probe tip and the medium. In the non-contact-mode SCM, probe tip-medium spacing had been actively controlled in which differential capacitance between the probe tip and the medium had been used as an active feedback signal. In this system, the probe tip-medium spacing was actually controlled to be as low as 30 nm, which was less than that of a magnetic hard disk. In addition, a topographical image was obtained in which capacitance and differential capacitance are simultaneously used.

It was experimentally demonstrated that a concept of the data storage using the F/S medium, in which a spin-on VDF-TrFE copolymer thin film and p-type silicon wafers were used as the ferroelectrics and the semiconductor, respectively. While the probe tip is not in contact with the medium and at the spacing below 100 nm in which the non-contact-mode SCM is used, it is shown that the ferroelectrics were poled and capacitance modulation caused by the direction of the polarization was detected. It had been dynamically demonstrated that bits with 0.3- to 0.4-μm spatial wavelength were recorded and reproduced in which a contact-mode SCM is used. This corresponds to >1 Gbit cm^{-2} in area recording density.

According to the numerical simulation, the F/S medium has a potential for higher-density recording compared with the charge storage method, because the change of capacitance in the F/S medium does not disappear even if the depletion regions are joined to each other.

Advantages of the capacitance storage using the ferroelectric/silicon medium and the non-contact-mode SCM are summarized as follows:

(1) Higher recording density
(2) Wide bandwidth
(3) Pre-pit tracking
(4) Long-term data retention
(5) Direct overwriting
(6) Wear free.

In Table 15.1, qualitative comparison to other data storage systems in the near-field technology is also shown.

15.7.2 Possibilities of the Capacitance Storage

The capacitance storage using the SCM and the F/S medium described in this chapter has potential for higher recording density and higher data transferring rate.

It is possible to achieve recording density of over 1 Tbit cm^{-2}. In general polarization domain size of ferroelectric materials is smaller than magnetic domain size.

Table 15.1. Qualitative comparison to otherdata storage system in the near-field technology

	Higher recording density	Wide band width	Pre-pit tracking	Long-term data retention	Direct over writing	Wear free
Non-contact-mode SCM and the F/S medium	better	good	good	good	good	good
SCM and charge storage	good	good	good	poor	poor	good
AFM data storage	excellent	poor	poor	good	poor	poor
Current magnetic recordings	good	good	impossible	excellent	good	good

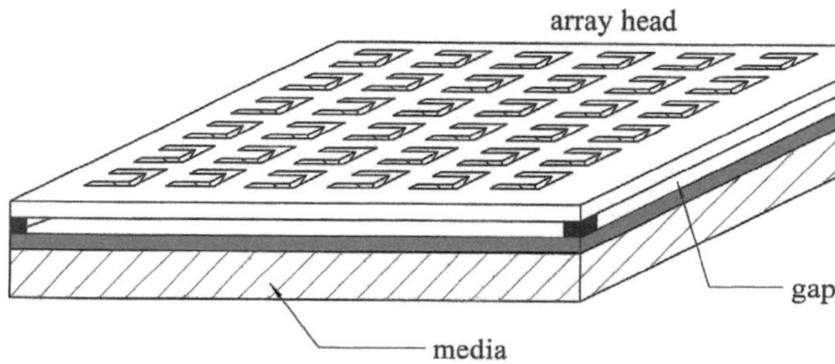

Fig. 15.17. Schematic illustration of a fixed card memory in future

Cho et al. demonstrated writing 6-nm domain dots in $LiTaO_3$ [41, 42]. It correspond to >1 Tbit cm^{-2}. Even though resolution of the present SCM is >10 nm, it would be possible to improve this by reducing the size of the probe tip in SCM.

The SCM had a reproducing signal bandwidth of about 10 MHz. Writing speed would also be of the same order of magnitude. There have been some reports that the switching time of polarization in Perovskite ferroelectrics is about 10^{-9} s [43].

Furthermore the SCM would be provided with an array probe head, which would be required for all near-field data storages in the future. As probe tip-medium distance is reduced, it would be difficult to increase relative velocity between a probe and a medium. Thus in order to improve data transferring rate, probe number should be increased necessarily. Actually an array probe head with a thousand cantilevers [44] is being developed for an AFM data storage. Consequently, it is considered that practical systems using such an array probe head would change to a fixed card (shown in Fig. 15.17 schematically) like solid-state memories.

Finally, it would be considered that a capacitance storage using a ferroelectrics/silicon medium and a non-contact-mode SCM is a candidate for one of the future data storage technologies.

Acknowledgments. The author would like to thank Dr S. Umemura and Mr K. Sanada of Fuji Photo Film Co., Ltd. for co-working this capacitance storage and suggestions for this paper and Mr K. Kawajiri of FUJIFILM Microdevices Co., Ltd. for useful discussions and his help in operating the simulation software.

References

1. Ando E (1985) Proc Int Symp Electron Devices, pp 47
2. Fukumoto A, Kubota S (1992) Jpn J Phys Lett 31:529
3. Binnig G, Rohrer H, Gerber CH, Weibel E (1982) Phys Rev Lett 49:57
4. Eigler DM, Schweizer EK (1990) Nature 344:524
5. Mamin HJ, Ruger D (1992) Appl Phys Lett 61:1003
6. Moreland J, Rice P (1990) Appl Phys Lett 57:310
7. Betzig E, Trautman JK (1992) Science 257:189
8. Matey JR, Blanc J (1985) J Appl Phys 57:1437
9. Bugg CD, King PJ (1988) J Phys E 21:147
10. Williams CC, Hough WP, Rishton SA (1989) Appl Phys Lett 55:203
11. Williams CC, Slinkman J, Hough WP, Wickramsinghe HK (1990) J Vac Sci Technol A 8:895
12. Dreyer M, Wiesendanger R (1995) Appl Phys A 61:357

13. Tomiye H, Kawamia H, Izawa M, Yoshimura M, Yao T (1995) Jpn J Appl Phys 34:3376
14. Nakagiri N, Yamamoto T, Sugimura H, Suzuki Y (1996) J Vac Sci Technol B 14:887
15. Goto K, Hane K (1997) Rev Sci Instrum 68:120
16. Tran T, Oliver DR, Thomson DJ, Bridges GE (2001) Rev Sci Instrum 72:2618
17. De Wolf P, Brazel E, Erickson A (2001) Mat Sci Sem Proc 4:71
18. Ciampolini L, Ciappa M, Malberti P, Fitchtner W (2002) Solid-State Elec 46:445
19. Palmer RC, Denlinger EJ, Kawamoto H (1982) RCA Rev 43:194
20. Sanada K, Yamamoto R, Umemura S (1994) Jpn J Appl Phys 33:6383
21. Tsuji S (1981) J Inst Television Eng Jpn 35:1061
22. IEC #845 (1988)
23. Chang JJ (1976) Proc IEEE 64:1039
24. Iwamura S, Nishida Y, Hashimoto K (1981) IEEE Trans Electron Devices ED-28:854
25. Barrett RC, Quate CF (1991) J Appl Phys 70:2725
26. Born A, Wiesendanger R (1999) Appl Phys A 68:131
27. Tomiye H, Yao T, Kawami H (1997) Appl Surf Sci 117/118:166
28. Fujiwara I, Kojima S, Seto J (1996) Jpn J Appl Phys 35:2764
29. Schaadt DM, Yu ET, Sanker S, Berkowitz AE (1999) Appl Phys Lett 74:472
30. Yamamoto R, Sanada K, Umemura S (1994) Jpn J Appl Phys 33:5829
31. Yamauchi N, Kato K, Wada T (1984) Jpn J Appl Phys 23:L671
32. Furukawa T (1989) Phase Transitions 18:143
33. Furukawa T, Matsuzaki H, Shiina M, Tajitsu Y (1985) Jpn J Appl Phys 24:L661
34. Wurfel P, Batra IP (1973) Phys Rev B 8:5126
35. Hatano J, Suda F, Futama H (1973) Jpn J Appl Phys 12:1644
36. Greidanus FJAM, Jacobs BAJ, Spruit JHM (1989) IEEE Trans Magn MAG-25:3524
37. Sze SM (1981) Physics of semiconductor devices, 2nd edn. Wiley, New York
38. Grove AS (1967) Physics and technology of semiconductor devices. Wiley, New York
39. Yamamoto R, Sanada K, Umemura S (1996) Jpn J Appl Phys 35:5284
40. Iwasaki S, Nakamura Y, Muraoka H (1981) IEEE Trans Magn MAG-17:2535
41. Cho Y, Matsuura K, Kazuta S, Kitamura K (2001) Tech Report of IEICE. MR2001–8, p 49
42. Hiranaga Y, Fujimoto K, Wagatsuma Y, Cho Y, Onoe A, Terabe K, Kitamura K (2002) Tech Report of IEICE. MR2002–6, p 13
43. Larsen PK, Kampschoer GLM, Ulenaers MJE, Spierings GACM, Cuppens R (1991) Appl Phys Lett 59:611
44. Despont M, Brugger J, Drechsler U, Durig U, Haberle W, Lutwyche M, Rothuizen H, Widmer R, Binnig GK, Vettiger P (1999) Proc IEEE MEMS ‘99, pp 564

16 Room-Temperature Single-Electron Devices formed by AFM Nano-Oxidation Process

Kazuhiko Matsumoto

16.1 Introduction

So far, various types of single-electron memories have been studied. Due to the difficulty of their artificial fabrication process, almost all of them used the self-organized nanostructure. Some examples are the polycrystalline silicon film [1], the poly-crystal silicon dot [2], and the squeezed delta-doped GaAlAs/GaAs layer [3] for the multi-tunnel junctions and the memory node. They showed clear memory effects such as digitized threshold voltage shifts [1] and hysteresis loops [3]. However, due to the fact that the nanostructures are spontaneously formed, these devices have the problem of poor reproducibility of their characteristics. In this paper we describe the operation of the room-temperature single-electron memory fabricated using the atomic force microscopy (AFM) nano-oxidation process [4–6].

16.2 AFM Nano-Oxidation Process

Figure 16.1 shows the principle of the atomic force microscopy (AFM) nano-oxidation process. The process uses the AFM cantilever as an ultra-small cathode, and DC bias was applied between the cantilever and the titanium (Ti) metal.

The surface of the Ti metal was anodized to form the oxidized titanium line (TiO_x) which works as a tunnel junction for the single-electron devices [4–6]. The chemical reactions at the AFM tip-water interface and Ti-water interface are described as follows:

$$4H_2O + 4e^- \rightarrow 4OH^- + 2H_2 \tag{16.a}$$

$$Ti + 4h^+ + 2OH^- \rightarrow TiO_2 + 2H^+ \tag{16.b}$$

The 2.5-nm-thick Ti thin film was on the SiO_2/Si or atomically flat α-Al_2O_3 substrate. In order to get the uniform and smooth TiO_x line, the latter substrate is preferable.

During this DC-mode AFM nano-oxidation process, however, the positive hydrogen ions (H^+) formed the space charge at the front end of the TiO_x-Ti interface, which prevents the further oxidation. This causes difficulty for the deeper oxidation.

In order to solve the problem, the ± 1.5–2 V pulse bias was applied instead of the DC bias between the conductive AFM cantilever and the Ti metal [7]. The nega-

Fig. 16.1. Schematic diagram of pulse-mode AFM nano-oxidation process on atomically flat α- Al_2O_3 substrate

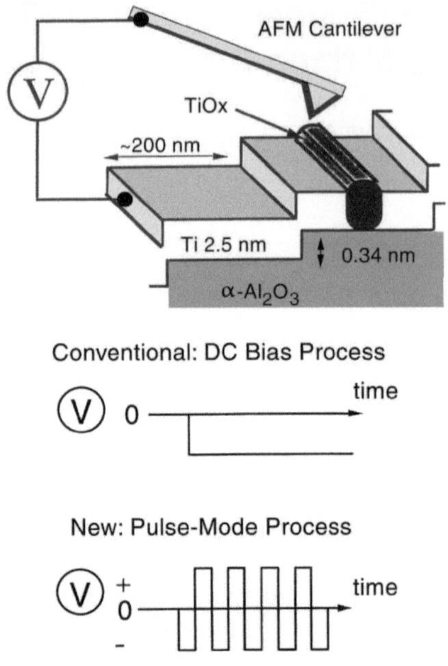

tive pulse anodized the surface of the metal and the positive pulse neutralized the positive space charge from the front end of TiO_x. Therefore, by using the pulse bias for the oxidation, the deeper oxidation becomes more possible than with the DC bias at the same applied bias. In another words, a narrower and deeper oxidized Ti line could be formed using the lower applied pulse bias. Figure 16.2 shows the AFM image of the TiO_x line folded six times on the atomically flat Ti/α-Al_2O_3 substrate fabricated by the present new AFM nano-oxidation process. The line width

Fig. 16.2. AFM image of TiO_x line folded six times on atomically flat Ti/α-Al_2O_3 substrate made by pulse-mode AFM nano-oxidation process. Linewidth is 20 nm, and line height 1.2 nm

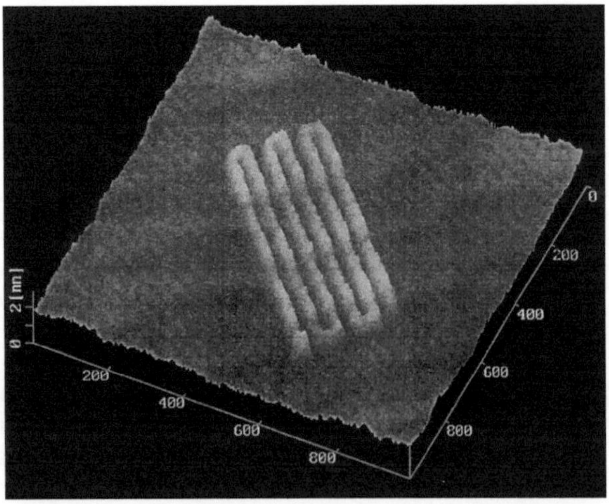

is 20 nm, and line height 1.2 nm. The smooth and uniform TiO_x line was obtained reproducibly. By using this new process, the single-electron transistor and the single-electron memory were fabricated.

16.3 Single-Electron Transistor

Figure 16.3a shows the schematic structure and Fig. 16.3b shows the AFM image of the fabricated one-island side-gate single-electron transistor (SET) on the atomically flat α-Al_2O_3 substrate. By using the pulse-mode AFM nano-oxidation process, two tunnel junctions for SET were formed by TiO_x lines. The size of the tunnel junction is 19 nm (width) × 26 nm (length) × 2 nm (thickness) and the island size is 8 nm × 26 nm. The width of the gate insulator for the side-gate SET was increased up to 964 nm to completely prevent the gate leakage current at room temperature.

For the suppression of the co-tunneling leakage current, the number of the islands of the SET could be increased. Figure 16.4 shows the drain current-drain

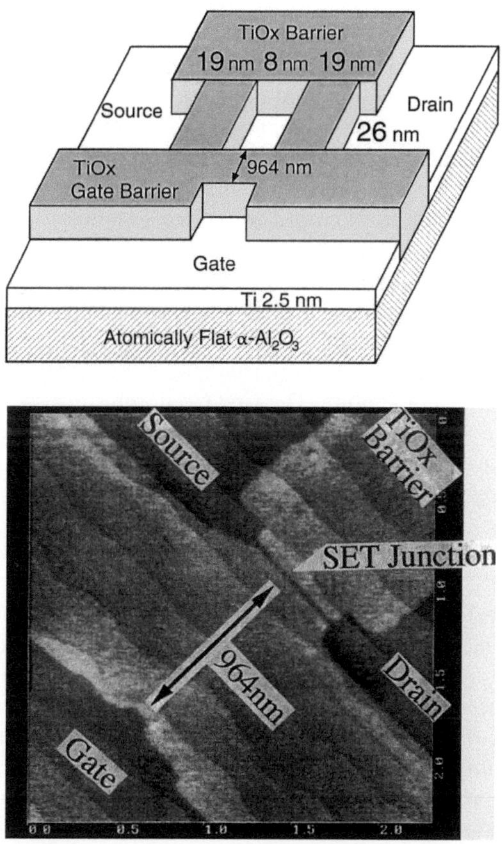

Fig. 16.3. **a** Schematic structure of the fabricated one-island side-gate SET on the atomically flat substrate. The two tunnel junction size is 19 nm (width) × 26 nm (length) × 2 nm (thickness) and the island size is 8 nm × 26 nm. **b** AFM image of the fabricated one-island side-gate SET on the atomically flat substrate

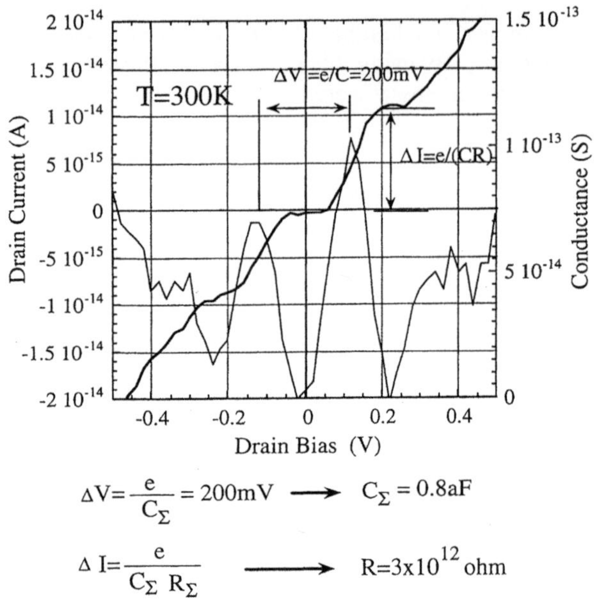

Fig. 16.4. Dependence of drain current of SET on drain bias at room temperature. Gate electrode is kept open. Clear Coulomb gap and Coulomb staircase are observed

voltage characteristics of the five-islands side-gate SET at room temperature with the gate terminal kept open.

In Fig. 16.4, the drain current is shown as a fat line and its differential (i.e., the conductance) is shown as a fine line. The clear Coulomb gap is seen at around zero drain voltage. There is no leakage current seen in the Coulomb gap. This may be attributed to the multi-tunneling junctions which can drastically decrease the co-tunneling current. The conductance shows zero Siemence at zero drain voltage, meaning the existence of a Coulomb gap, and oscillates with periods of 200 mV. From the value of the 200-mV Coulomb gap, the total capacitance of the SET is calculated to be $C_\Sigma = 0.8$ aF.

The Coulomb staircases with almost the same periods are also obtained even at room temperature. The first Coulomb staircases at approximately +180 mV and approximately −180 mV are clearly seen. From the height of the first Coulomb staircase ΔI, the total tunneling junction resistance can be calculated from the following relationship:

$$\Delta I = e / (C_\Sigma R_\Sigma) \tag{16.c}$$

and is $R_\Sigma = 1.8 \times 10^{13}\ \Omega$. Due to the large number of tunneling junctions (i.e., six junctions), the total tunneling junction resistance becomes high.

Figure 16.5a shows the drain current-gate bias characteristics of the SET at room temperature using the drain bias as a parameter.

The drain current of the SET oscillates with a period of 2.1 V at room temperature in the gate bias range of 0–10 V at the drain bias (V_D) of 0.25 V and 0.3 V. Five peaks are clearly seen for the each drain bias condition. The current modulation rate is from 20% to 30%. Even at the different drain bias, the drain current

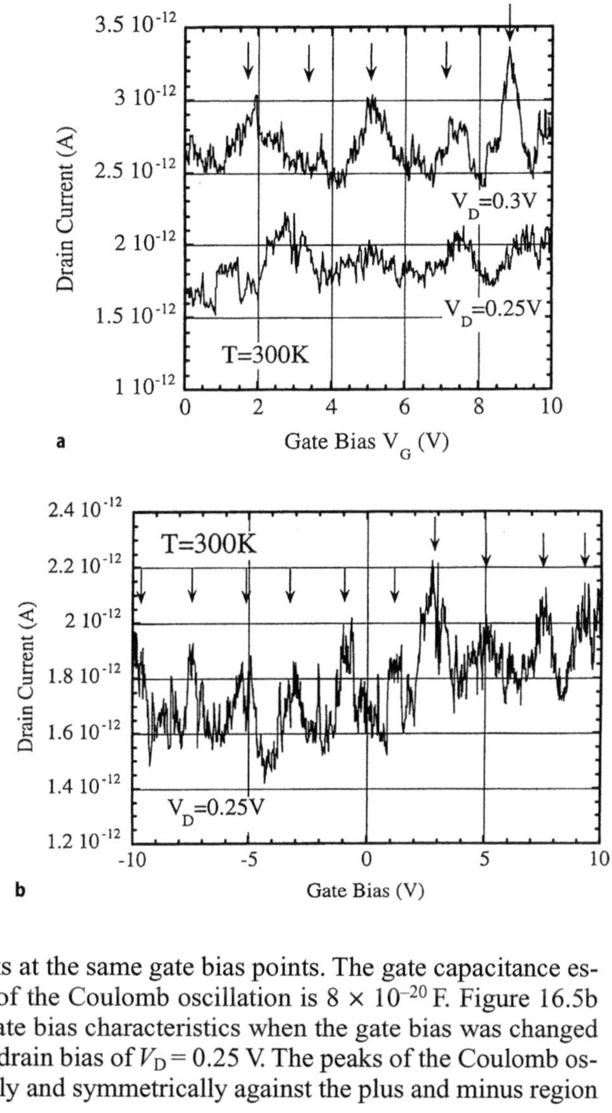

Fig. 16.5. **a** Dependence of Coulomb oscillation on drain bias at room temperature. $V_D = 0.25$ V and 0.3 V. **b** Symmetrical characteristics of Coulomb oscillation against gate bias at room temperature

shows the oscillation peaks at the same gate bias points. The gate capacitance estimated from the periods of the Coulomb oscillation is 8×10^{-20} F. Figure 16.5b shows the drain current-gate bias characteristics when the gate bias was changed from -10 V to 10 V at the drain bias of $V_D = 0.25$ V. The peaks of the Coulomb oscillation appear periodically and symmetrically against the plus and minus region in the gate bias.

Thus, the fabricated SET shows the clear gate action even at room temperature.

16.4 Single-Electron Memory

Figure 16.6 shows the equivalent circuit of the single-electron memory which consists of the multi-tunnel junction and the normal memory capacitance.

The memory node of the single-electron memory where few numbers of electrons are stored is connected by the single-electron transistor(SET) through the normal capacitance C_g. By using the equivalent circuit of the single-electron memory

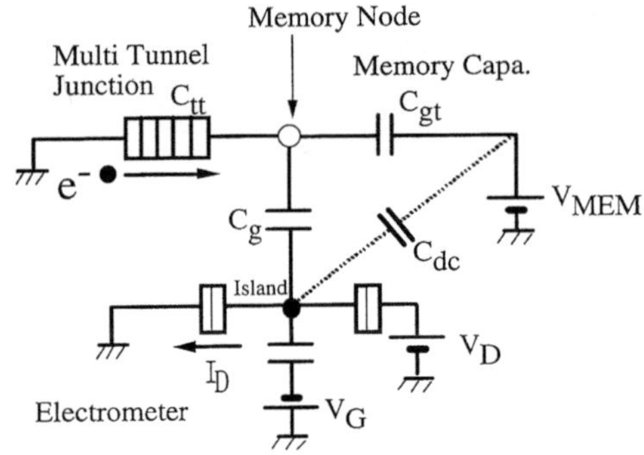

Fig. 16.6. Equivalent circuit of single-electron memory

shown in Fig. 16.6, the principle of the operation of the single-electron memory is explained as follows. By increasing the memory bias (V_{MEM}), the potential of the memory node V_t also increases in proportion to the ratio of the memory capacitance C_{gt} and the multi-tunnel junction capacitance C_{tt}, that is:

$$V_t = V_{MEM} C_{gt} / (C_{tt} + C_{gt}) \tag{16.1}$$

When the potential of the memory node V_t becomes larger than the Coulomb gap bias of the multi-tunnel junction V_0, one electron can pass through the multi-tunnel junction to reach the memory node, which lowers the potential of the memory node as $e/(C_{tt}+C_{gt})$, and the second electron cannot pass through the multi-tunnel junction. By increasing the memory bias further, the potential of the memory node V_t again reaches to the Coulomb gap bias (V_0), then the second electron can pass through the multi-tunnel junction to reach the memory node. Thus the one-by-one electron transfer becomes possible through the multi-tunnel junction to the memory node. When the n electrons are stored at the memory node, the potential of the memory node which is lowered by n electrons is described as follows:

$$V_t = (V_{MEM} C_{gt} - ne) / (C_{tt} + C_{gt}) \tag{16.2}$$

where e is the electron charge. When the memory bias, V_{MEM} comes back to the lower bias, the potential of the memory node V_t goes into the negative region because of the stored n electrons. When V_t becomes smaller than the negative Coulomb gap bias of the multi-tunnel junction $-V_0$, the electron is ejected (one by one) to the ground. Therefore, the potential of the memory node V_t shows a hysteresis loop. The relationship between the memory node potential V_t and the memory bias V_{MEM} is shown in Fig. 16.7.

Thus, few numbers of electrons are memorized at the memory node for a few hundred seconds and this device works as a single-electron memory. The change of the potential at the memory node by the injection and the ejection of electrons through the multi-tunnel junction is detected by the change of the drain current of the SET, which works as an electrometer. The change of the drain current of the electrometer is considered to be linearly proportional to the potential change of the memory node.

Figure 16.8 shows the schematic structure of the fabricated single-electron memory which sits on the atomically flat α-Al_2O_3 substrate. Figure 16.9 is the

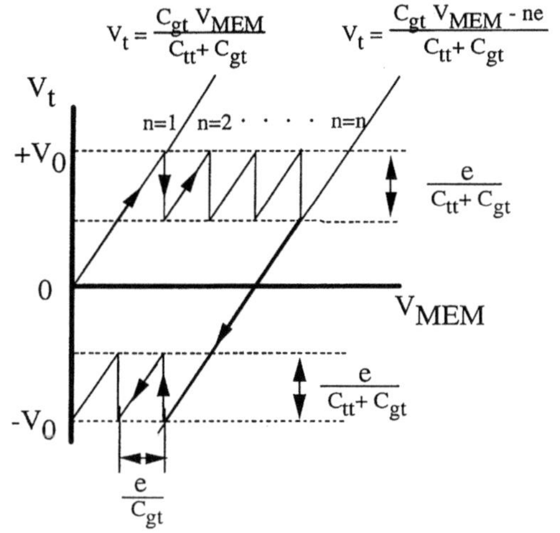

Fig. 16.7. Dependence of memory node potential (V_t) on memory bias (V_{MEM})

AFM image of the multi-tunnel junction area and two tunnel SET junctions of the fabricated single-electron memory.

The device consists of the multi-tunnel junction and normal capacitance for the memory and two tunnel junctions for the single-electron transistor. There are four electrodes (i.e., the memory bias electrode for the memory, the source, drain, and gate electrodes) for the single-electron transistor. The multi-tunnel junction consists of five or seven tunnel junctions which are approximately 15–25 nm in width, 15–25 nm in length, 2 nm in thickness, and the spacing between them is 10–15 nm. The normal memory capacitance is 341 nm in width and approximately 15–25 nm in length and 2 nm in thickness. The area between the multi-tunnel junction and the memory capacitance is designated as a memory node where few numbers of electrons are stored. The distance between the memory node and the island of the single-electron transistor is 486 nm which determines the sensitivity of the SET.

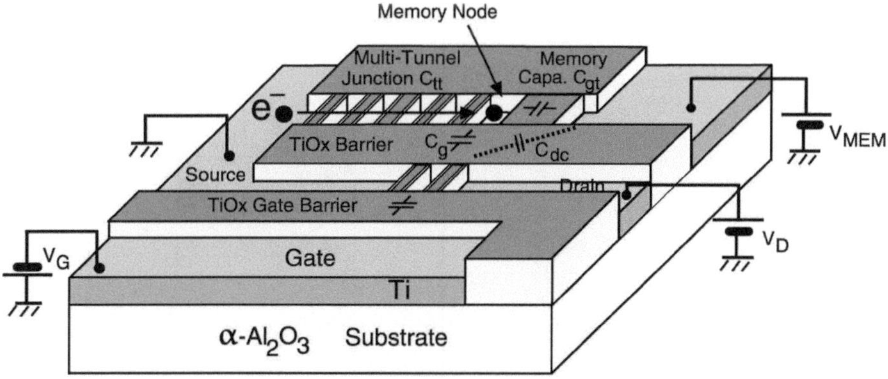

Fig. 16.8. Schematic structure of single-electron memory on atomically flat α-Al_2O_3 substrate made by pulse-mode AFM nano-oxidation process

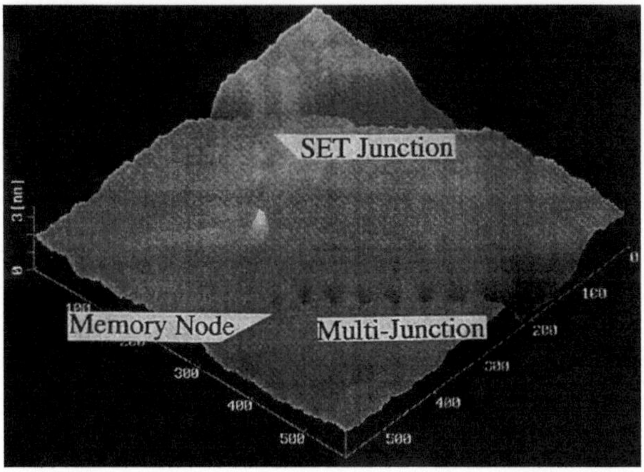

Fig. 16.9. Plain AFM image of the fabricated single-electron memory on atomically flat substrate. Multi-tunnel junctions for memory and SET junctions for electrometer are seen

The size of the two tunnel junctions for SET is approximately 15–25 nm in width, 15–25 nm in length, 2 nm in thickness, and the spacing between them is approximately 10–15 nm (i.e., the island size is approximately 10–15 nm × 15–25 nm).

For the measurement of the single-electron memory, the condition of the electrometer was set as the drain bias of $V_D = 3$ V, and the gate bias of $V_G = 1.5$ V to detect the potential change of the memory node. Figure 16.10 shows the drain current of the electrometer, that is, the single-electron transistor versus the memory bias characteristics at room temperature.

When the memory bias V_{MEM} of the single-electron memory increases from 0 V to 10 V, the drain current of the electrometer increases at the beginning and begins to oscillate. The drain current oscillation is attributed to the two effects: one is owing to the injection of the individual electrons through the multi-tunnel junction to the memory node which lowers the potential of the memory node, and the other is owing to the direct coupling of the memory bias to the island of the single-electron transistor which causes the Coulomb oscillation. When the memory bias

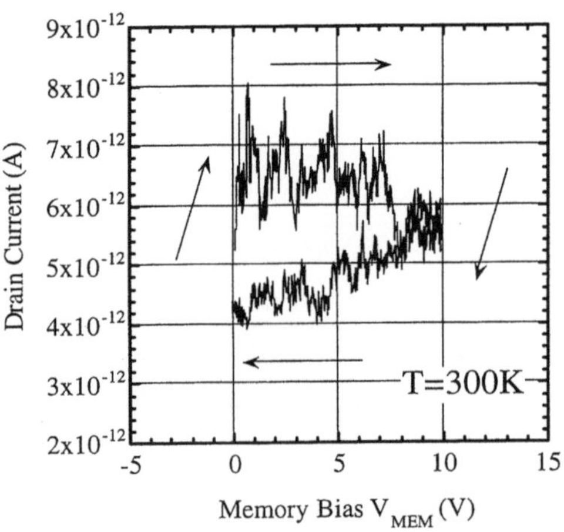

Fig. 16.10. Hysteresis loop of single-electron memory at room temperature. Gate bias and drain bias for SET were set at $V_G = 1.5$ V and $V_D = 3$ V, respectively. Memory bias was applied from $V_{MEM} = 0$ V to 10 V and back to 0 V

comes back from 10 V to 0 V, the drain current follows the different trace owing to the stored electrons at the memory node and shows a clear hysteresis loop even at room temperature. The measured hysteresis loop is quite similar to that shown in Fig. 16.7. Owing to the large noise level in Fig. 16.10, however, it is difficult to distinguish the oscillation peaks owing to the electron injection to the memory node from noise peaks.

The time transient of the drain current at room temperature when the memory bias was suddenly cut off from 10 V to 0 V showed the step-like transient trace indicating the one-by-one electron ejection from the memory node to the ground. The retention time is 600 s.

The number of electrons stored in the memory node when the memory bias of 10 V was applied could be calculated from Eq. (16.2) and

$$n - \left\{C_{gt} V_{MEM} - \left(C_{tt} + C_{gt}\right) V_t\right\} / e , \tag{16.3}$$

where $V_{MEM} \gg V_t$ when $V_{MEM} = 10$ V. Therefore, n is roughly calculated as

$$n = C_{gt} V_{MEM} / e . \tag{16.4}$$

By using the three-dimensional simulator, the three-dimensional Poisson's equation was solved for the single-electron memory structure and the memory capacitance is calculated to be $C_{gt} = 4 \times 10^{-19}$ F. By using this value, the number of electrons stored in the memory node is calculated to be about $n = 25$.

By using the improved pulse-mode AFM nano-oxidation process, the single-electron memory was fabricated on the atomically flat α-Al_2O_3 substrate. The single-electron memory stored about 25 electrons at the memory node with the applied memory bias of 10 V, and showed a hysteresis loop even at room temperature. The retention time of the single-electron memory is 600 s.

16.5 Summary

By using the improved pulse-mode AFM nano-oxidation process, the single-electron transistor and single-electron memory were fabricated on the atomically flat α-Al_2O_3 substrate. The single-electron transistor shows the Coulomb oscillation with periods of 2.1 V at room temperature. The single-electron memory shows a hysteresis loop even at room temperature.

References

1. Yano K, Ishii T, Hashimoto T, Kobayashi T, Murai F, Seki K (1993) Technical digest of international electron device meeting, p 541
2. Tiwari S, Rana F, Chan K, Hanafi H, Chan W, Buchanan D (1995) Technical digest of international electron device meeting, p 521
3. Nakazato K, Thornton TJ, White J, Ahmed H (1993) Electronics Lett 29:384
4. Matsumoto K, Takahashi S, Ishii M, Hoshi M, Kurokawa A, Ichimura S, Ando A (1995) Jpn J Appl Phys 34:1387
5. Matsumoto K, Ishii M, Segawa K, Oka Y, Vartanian BJ, Harris JS (1996) Appl Phys Lett 68:34
6. Matsumoto K (1997) Proc IEEE 85:612
7. Dagata JA, Inoue T, Itoh J, Matsumoto K, Yokoyama H (1998) J Appl Phys 84:6891

Subject Index

MIX
Papier aus verantwortungsvollen Quellen
Paper from responsible sources
FSC® C105338

If you have any concerns about our products,
you can contact us on
ProductSafety@springernature.com

In case Publisher is established outside the EU,
the EU authorized representative is:
Springer Nature Customer Service Center GmbH
Europaplatz 3, 69115 Heidelberg, Germany

Printed by Libri Plureos GmbH
in Hamburg, Germany